W9-ANN-684

Matroid Theory

Matroid Theory

JAMES G. OXLEY
Department of Mathematics
Louisiana State University

OXFORD NEW YORK TOKYO
OXFORD UNIVERSITY PRESS
1992

Oxford University Press, Walton Street, Oxford OX2 6DP
Oxford New York Toronto
Delhi Bombay Calcutta Madras Karachi
Petaling Jaya Singapore Hong Kong Tokyo
Nairobi Dar es Salaam Cape Town
Melbourne Auckland
and associated companies in
Berlin Ibadan

Oxford is a trade mark of Oxford University Press

Published in the United States
by Oxford University Press, New York

A catalogue record for this book is available from the British Library

Library of Congress Cataloging in Publication Data
Oxley, J. G.
Matroid theory / James G. Oxley.
Includes bibliographical references and index.
1. Matroids. I. title.
QA166.6.095 1992
511'.6 dc20
ISBN 0–19–853563–5

Typeset by the author using AMSTeX
Printed in Great Britain by
Bookcraft (Bath) Ltd
Midsomer Norton, Avon

Preface

I first heard mention of matroids in 1973 from a friend at the University of Tasmania who had just read Wilson's (1973) enticing survey paper. Since then, several people have taught me about the many fascinating properties of these structures: Don Row in Tasmania; Dominic Welsh and Aubrey Ingleton in Oxford; and Tom Brylawski in Chapel Hill. While I learnt much from all of my teachers, it was my doctoral supervisor, Dominic Welsh, who was the strongest single influence on this book. His own book, *Matroid theory* (1976), appeared during my second year at Oxford and it has been my constant companion ever since. When I contemplated writing this book, the first question I had to answer was how should it differ from Welsh's book.

This book attempts to blend Welsh's very graph-theoretic approach to matroids with the geometric approach of Rota's school that I learnt from Brylawski. Unfortunately, I cannot emulate Welsh's feat of providing, in a single volume, a complete survey of the current state of knowledge in matroid theory; the subject has grown too much. Therefore I have had to be selective. While the basic topics virtually select themselves, the more advanced topics covered here reflect my own research interests.

Chapters 1–6 are intended to provide a basic overview of matroid theory. By omitting certain sections and skimming through others, one can cover this material in a one-semester graduate course. The later chapters are intended for reference and for those who have the opportunity to teach a full year's course on matroids. These chapters treat more specialized topics or present difficult proofs delayed from earlier chapters. One of the exciting developments in matroid theory in the last fifteen years has been the discovery of relatively short proofs of all of the major theorems except Seymour's decomposition theorem for regular matroids. One of the main tasks of the second half of the book is to present these proofs in reasonably full detail. I have never enjoyed reading proofs in which numerous intermediate steps are left to the reader, so I have tried to avoid writing such proofs.

Exercises have been included at the end of each section except in the last chapter. The first few exercises in each set tend to be straightforward; the last few are often quite difficult. I have used an asterisk (*) to denote harder problems and a degree symbol (°) to denote unsolved problems. Some of the harder problems present theorems that have not been included in the text and these may have quite intricate proofs. In such instances, I have tried to provide an appropriate reference. In many places, proofs have been omitted from the text. Except where otherwise indicated, these proofs are relatively straightforward and they provide a good additional source of exercises.

In order to distinguish the relative importance of the many results in the book, I have adopted the convention of reserving the term 'theorem' for the main results; those results of lesser importance are called 'propositions'.

The last chapter of the book gathers together numerous unsolved problems, some of which have appeared earlier in the book. It seemed desirable to collect all these together in one place. The beginning of that chapter also contains a brief review of the various books on matroid theory that have appeared in the last few years. The references appear after the last chapter, each reference being accompanied by a list of the sections in which it is cited. Following the references, there is an appendix which summarizes the properties of certain interesting matroids that have appeared in earlier sections. The book concludes with an index of the most frequently used notation and with an index of terms. One convention that will apply throughout is that, except where otherwise stated, M will denote an arbitrary matroid.

One of my colleagues commented that it is somewhat intimidating to begin this book with a chapter entitled 'Basic definitions and examples' which is 61 pages long. With this in mind, the following remarks on Chapters 1–6 are made to guide the reader who is seeking a shorter introduction to matroid theory than these chapters provide. For completeness, the proofs of most of the results in the early chapters have been included. Many of these can be omitted on a first reading or by someone who is mainly interested in getting a feel for the subject.

The main ideas in Chapter 1 are that matroids can be axiomatized in many equivalent ways and that the two fundamental classes of examples of matroids arise from graphs and matrices. The proofs of equivalence of the various axiom systems could certainly be omitted as indeed could most of Sections 1.6–1.8. The first of these three sections deals with a class of matroids that arises in the consideration of scheduling problems. The reader omitting this section should then omit Section 2.4 and the last part of Section 3.2. All of Section 1.7 could be omitted beyond the first page. If this is done, then 3.3.2–3.3.4 and 3.3.7, along with the material on lattices in Chapter 6, should also be left out. Section 1.8 indicates why matroids are of basic importance in combinatorial optimization but stands alone from what follows it.

Chapter 2 is much shorter than Chapter 1. The reader seeking to shorten it further could omit from 2.1.12 to the end of Section 2.1, from 2.2.11 to the end of Section 2.2, and all of Section 2.4.

To shorten Chapter 3, the reader should omit those parts noted above that depend on Section 1.7. In Chapter 4, Section 4.3 contains more specialized results than the other two sections of that chapter and it too could be omitted. For a similar reason, the last two sections of Chapter 5 could be omitted.

Chapter 6, which deals with matroid representations over fields, is the longest in the book but the reader should concentrate on Section 6.1 and Sections 6.3–6.6. Even within these sections, material can be omitted. In Section 6.1, everything after 6.1.3 except 6.1.10 could be left out. If one restricts attention to representations over prime fields or $\mathbb{R}$, then one can omit 6.3.7–6.3.11 in Section 6.3. In Section 6.4, the main result is Theorem 6.4.7 and, after that, the rest of the section consists of examples illustrating the use of that result. Sections 6.5 and 6.6 survey the major theorems and unsolved problems in matroid representability. As such, they deserve the attention of even the casual reader. They are not difficult going since the proofs of the main results from these sections are postponed until later in the book.

Many people have assisted in the preparation of this book. I have lectured from most of it at Louisiana State University, and the students in those classes have helped to eliminate many mistakes. Several people have also read substantial parts of the manuscript and have offered many valuable suggestions. In particular, I thank Safwan Akkari, Jon Lee, Manoel Lemos, James Reid, and Geoff Whittle for their help in this regard. I am also very grateful to Mike Newman who first taught me about how to write mathematics. Nell Castleberry spent many painstaking hours preparing a first typescript of this book and then the final version in LaTeX. I thank her for the care and dedication she brought to this task.

During the time that this book was being written my work was partially supported by a grant from the Louisiana Education Quality Support Fund through the Board of Regents.

Finally, I thank Dominic Welsh who was instrumental in getting me to start this book and who provided both advice and encouragement along the way. This book is dedicated to him.

Contents

Preliminaries

Matroid theory draws heavily on both graph theory and linear algebra for its motivation, its basic examples, and its notation. The ideal background for a student using this book would include undergraduate courses in linear algebra, graph theory, and abstract algebra. However, I have used the book with students having no previous exposure to graphs and very little abstract algebra; most of the ideas needed from these areas, particularly in the earlier chapters, are relatively straightforward. The rest of this section introduces some notation and outlines some basic concepts. After reading this, the reader who is familiar with vector spaces should have enough background to handle at least the first half of the book.

Except where otherwise indicated all sets considered in this book will be *finite*. If E is a set, then its collection of subsets and its cardinality will be denoted by 2^E and $|E|$, respectively. Frequently throughout the book, when given a set $\mathcal{A}$ of subsets of E, we shall be interested in the *maximal* or *minimal members* of $\mathcal{A}$. The former are those members of $\mathcal{A}$ that are not properly contained in any member of $\mathcal{A}$; the latter are the members of $\mathcal{A}$ that do not properly contain any member of $\mathcal{A}$. The sets of positive integers, integers, rational numbers, real numbers, and complex numbers will be denoted by $\mathbb{Z}^+$, $\mathbb{Z}$, $\mathbb{Q}$, $\mathbb{R}$, and $\mathbb{C}$, respectively. If X and Y are sets, then $X - Y$ will denote the set $\{x \in X : x \notin Y\}$. For sets $X_1, X_2, \ldots, X_n$, the notation $X_1 \dot{\cup} X_2 \dot{\cup} \cdots \dot{\cup} X_n$ will refer to the set $X_1 \cup X_2 \cup \cdots \cup X_n$ and will also imply that $X_1, X_2, \ldots, X_n$ are pairwise disjoint. Frequently we shall want to add a single element to a set X or to remove a single element from X. In such cases, we shall often abbreviate $X \cup \{e\}$ and $X - \{e\}$ to $X \cup e$ and $X - e$, respectively.

The concept of a multiset is useful when working with collections of objects in which repetitions can occur. Formally, if S is a set, a *multiset* chosen from S is a function m from S into the set of non-negative integers. For example, $\{a, a, b, b, b, d\}$ denotes the multiset with $S = \{a, b, c, d, e\}$ and $m(a) = 2$, $m(b) = 3$, $m(c) = 0$, $m(d) = 1$, $m(e) = 0$.

A permutation of a finite set will often be written using cycle notation. So, for example, if $S = \{1, 2, \ldots, 9\}$, then the permutation $(1, 5, 2, 9)(7, 8)$ maps 1 to 5, 5 to 2, 2 to 9, 9 to 1, 7 to 8, 8 to 7, 3 to 3, 4 to 4, and 6 to 6.

If f and g are functions that are defined for all sufficiently large positive integers n, then $g(n) = O(f(n))$ means that $g(n)/f(n)$ remains bounded as $n \to \infty$.

Fields will appear quite often in the book, even in the earlier chapters. We now note some basic facts about these structures. Proofs of these facts can be found in most graduate algebra texts. If F is a finite field, then F has exactly p^k elements for some prime number p and some positive integer k. Indeed, for all such p and k, there is a unique field $GF(p^k)$ having exactly p^k elements. This field is called the *Galois field of order* p^k. When $k = 1$, $GF(p^k)$ coincides with $\mathbb{Z}_p$, the ring of integers modulo p. Thus $GF(p)$ has as its elements $0, 1, \ldots, p-1$, and addition and multiplication of these elements is performed modulo p. When $k > 1$, $GF(p^k)$ can be constructed as follows. Let $h(\omega)$ be a polynomial of degree k with coefficients in $GF(p)$ and suppose that this polynomial is irreducible, that is, $h(\omega)$ is not the product of two lower-degree polynomials over $GF(p)$. Such irreducible polynomials are known to exist. Consider the set S of all polynomials in ω that have degree at most $k-1$ and have coefficients in $GF(p)$. There are exactly p choices for each of the k coefficients of a member of S. Hence $|S| = p^k$. Moreover, under addition and multiplication, both of which are performed modulo $h(\omega)$, S forms a field, namely $GF(p^k)$.

The finite fields that will appear most often in this book are $GF(2)$, $GF(3)$, $GF(5)$, and $GF(4)$. The first three of these are $\mathbb{Z}_2$, $\mathbb{Z}_3$, and $\mathbb{Z}_5$, respectively. The elements of $\mathbb{Z}_3$ will usually be written as 0, 1, and -1 rather than 0, 1, and 2. We shall now use the above construction to obtain $GF(4)$, which the reader will easily see is not isomorphic to $\mathbb{Z}_4$. Let $h(\omega) = \omega^2 + \omega + 1$. It is straightforward to show that this polynomial is irreducible over $GF(2)$, and that the addition and multiplication tables for $GF(4)$ are as follows:

$+$	0	1	ω	$\omega+1$
0	0	1	ω	$\omega+1$
1	1	0	$\omega+1$	ω
ω	ω	$\omega+1$	0	1
$\omega+1$	$\omega+1$	ω	1	0

$\times$	0	1	ω	$\omega+1$
0	0	0	0	0
1	0	1	ω	$\omega+1$
ω	0	ω	$\omega+1$	1
$\omega+1$	0	$\omega+1$	1	ω

Let F be a field and consider the sequence $1, 1+1, 1+1+1, \ldots$ in F. If these elements are all distinct, then F is said to have *characteristic* 0; otherwise the *characteristic* of F is the smallest positive integer p for which the sum of p ones is zero. In the latter case, p is clearly a prime; the subset $\{0, 1, \ldots, p-1\}$ of F is a subfield of F isomorphic to $\mathbb{Z}_p$; and $p\alpha = 0$ for all α in F.

The intersection of all subfields of a field F is itself a field called the *prime subfield* of F. If F has characteristic 0, then its prime subfield is

$\mathbb{Q}$, the field of rational numbers. If F has characteristic p where $p \neq 0$, then its prime subfield is $\mathbb{Z}_p$.

Throughout the book, an arbitrary finite field will be denoted by $GF(q)$, it being implicit that q is a power of a prime. The multiplicative group of non-zero elements of $GF(q)$ will be denoted by $GF(q)^*$. It is well known that $GF(q)^*$ is a cyclic group.

If F is a field and r is a natural number, then $V(r, F)$ will denote the r-dimensional vector space over F. This vector space will also be written as $V(r, q)$ when $F = GF(q)$. A *non-singular linear transformation* of $V(r, F)$ is a permutation η of $V(r, F)$ such that $\eta(c\underline{x}+d\underline{y}) = c\eta(\underline{x})+d\eta(\underline{y})$ for all $\underline{x}$ and $\underline{y}$ in $V(r, F)$ and all c and d in F.

The matrix notation used here is relatively standard. The matrices I_r, J_r, and 0_r are, respectively, the $r \times r$ identity matrix, the $r \times r$ matrix of all ones, and the $r \times r$ matrix of all zeros. The transpose of a matrix A will be denoted by A^T. Some use will be made here of determinants and a quick summary of their properties can be found in Section 2.2. If $\{\underline{v}_1, \underline{v}_2, \ldots, \underline{v}_n\}$ is a set X of vectors in a vector space Vover a field F, then $\langle \underline{v}_1, \underline{v}_2, \ldots, \underline{v}_n \rangle$ denotes the subspace of V *spanned* by X, that is, the set of all linear combinations of $\underline{v}_1, \underline{v}_2, \ldots, \underline{v}_n$. The *dimension* of V is denoted by $\dim V$. If W is also a vector space over F, then $V+W$ denotes the vector space over F that is spanned by $\{\underline{v}+\underline{w} : \underline{v} \in V, \underline{w} \in W\}$.

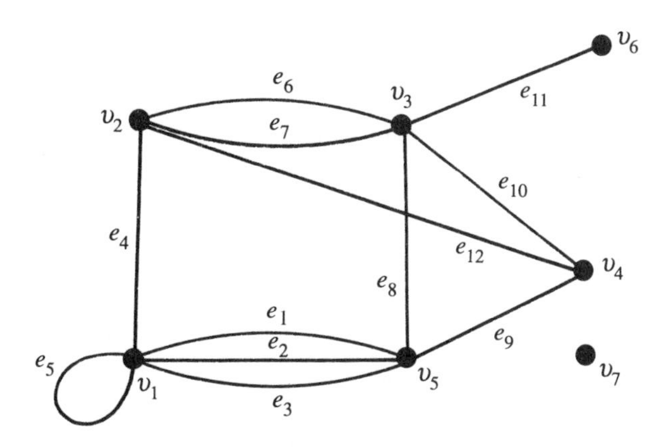

Fig. 0.1. An example of a graph.

Figure 0.1 is a pictorial representation of a particular graph. The vertex set and edge set of this graph are $\{v_1, v_2, \ldots, v_7\}$ and $\{e_1, e_2, \ldots, e_{12}\}$, respectively. An edge such as e_5, which joins a vertex to itself, is called a *loop*. Edges such as e_1, e_2, and e_3, which join the same pair of distinct vertices, are called *parallel edges*. The vertex v_7, which does not meet any edges, is an *isolated vertex*. The *ends* of the edge e_9 are v_4 and v_5.

Formally, a *graph* G consists of a non-empty set $V(G)$ of *vertices* and a multiset $E(G)$ of *edges* each of which consists of an unordered pair of (possibly identical) vertices. If $e \in E(G)$ and $e = \{u, v\}$ where u and v are in $V(G)$, then we say that u and v are *neighbours* or are *adjacent*, and e is *incident* with u and v. For example, in Figure 0.1, e_8 is the unique edge incident with both v_3 and v_5, so we could denote it unambiguously by $v_3 v_5$, or equivalently, by $v_5 v_3$. We call $V(G)$ and $E(G)$ the *vertex set* and *edge set*, respectively, of the graph G. A graph is *simple* if it has no loops and no parallel edges.

Sometimes a graph is specified by its *vertex–edge incidence matrix*. For the graph in Figure 0.1, this matrix is the following:

$$\begin{array}{c} \\ v_1 \\ v_2 \\ v_3 \\ v_4 \\ v_5 \\ v_6 \\ v_7 \end{array} \begin{array}{c} \begin{array}{cccccccccccc} e_1 & e_2 & e_3 & e_4 & e_5 & e_6 & e_7 & e_8 & e_9 & e_{10} & e_{11} & e_{12} \end{array} \\ \left[\begin{array}{cccccccccccc} 1 & 1 & 1 & 1 & 2 & 0 & 0 & 0 & 0 & 0 & 0 & 0 \\ 0 & 0 & 0 & 1 & 0 & 1 & 1 & 0 & 0 & 0 & 0 & 1 \\ 0 & 0 & 0 & 0 & 0 & 1 & 1 & 1 & 0 & 1 & 1 & 0 \\ 0 & 0 & 0 & 0 & 0 & 0 & 0 & 0 & 1 & 1 & 0 & 1 \\ 1 & 1 & 1 & 0 & 0 & 0 & 0 & 1 & 1 & 0 & 0 & 0 \\ 0 & 0 & 0 & 0 & 0 & 0 & 0 & 0 & 0 & 0 & 1 & 0 \\ 0 & 0 & 0 & 0 & 0 & 0 & 0 & 0 & 0 & 0 & 0 & 0 \end{array}\right] \end{array}.$$

In general, the rows of such a matrix $[a_{ij}]$ are indexed by the vertices of the graph, the columns are indexed by the edges, and a_{ij} is the number of times that the jth edge, e_j, is incident with the ith vertex, v_i. Hence a_{ij} is 0 or 1 unless e_j is a loop incident with v_i, in which case, $a_{ij} = 2$. The *degree*, $d(v_i)$, of the vertex v_i is the sum of the entries in the ith row of the incidence matrix. Equivalently, $d(v_i)$ is the number of edges incident with v_i, each loop counting as two edges.

A graph H is a *subgraph* of a graph G if $V(H)$ and $E(H)$ are subsets of $V(G)$ and $E(G)$, respectively. If V' is a non-empty subset of $V(G)$, then $G[V']$ denotes the subgraph of G whose vertex set is V' and whose edge set consists of those edges of G that have both endpoints in V'. We say that $G[V']$ is the subgraph of G *induced* by V'. Similarly, if E' is a non-empty subset of $E(G)$, then $G[E']$, the subgraph of G *induced* by E', has E' as its edge set and the set of endpoints of edges in E' as its vertex set.

If G_1 and G_2 are graphs, their *union* $G_1 \cup G_2$ is the graph with vertex set $V(G_1) \cup V(G_2)$ and edge set $E(G_1) \cup E(G_2)$. If $V(G_1)$ and $V(G_2)$ are disjoint, then so are $E(G_1)$ and $E(G_2)$, and G_1 and G_2 are called *disjoint graphs.*

Graphs G and H are *isomorphic*, written $G \cong H$, if there are bijections $\psi : V(G) \to V(H)$ and $\theta : E(G) \to E(H)$ such that a vertex v of G is incident with an edge e of G if and only if $\psi(v)$ is incident with $\theta(e)$. Two graphs that play an important role in matroid theory, and indeed

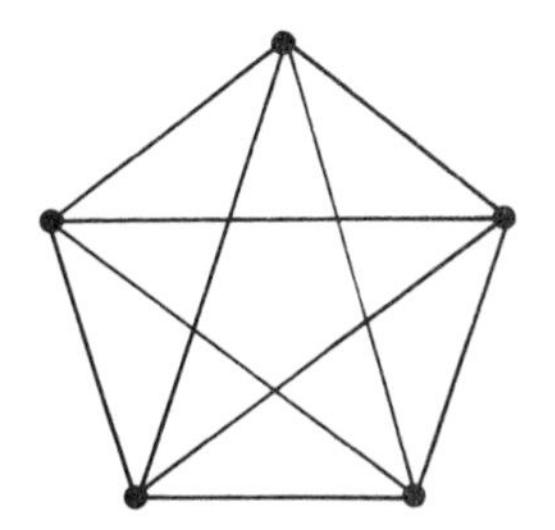
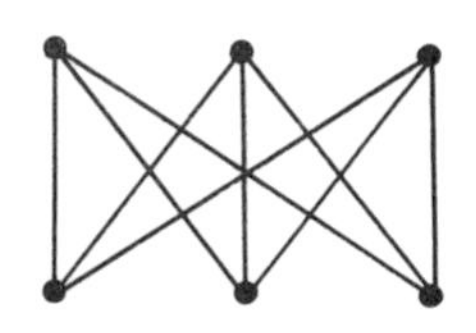

Fig. 0.2. K_5 and $K_{3,3}$.

throughout graph theory, are K_5 and $K_{3,3}$. These graphs are shown in Figure 0.2. In general, if n is a positive integer, there is, up to isomorphism, a unique graph on n vertices in which each pair of distinct vertices is joined by a single edge. This graph, K_n, is called the *complete graph* on n vertices. $K_{3,3}$ is a special type of *bipartite graph*, the latter being a graph whose vertex set can be partitioned into two *classes* X and Y such that each edge has one endpoint in X and the other in Y.

In Chapter 2, we shall consider the role played by K_5 and $K_{3,3}$ when one considers embedding graphs in the Euclidean plane. In this discussion, the topology of the plane is important. We shall not go into this in detail here although we shall require some basic facts. A *Jordan curve* is a continuous curve in the plane which does not intersect itself. A *closed Jordan curve* is a Jordan curve whose endpoints coincide. If all the points of such a closed curve are deleted from the plane, the remaining points are partitioned into two open sets. The *Jordan Curve Theorem* asserts that a line joining a point in one of these open sets to a point in the other open set must intersect the closed Jordan curve.

A *walk* W in a graph G is a sequence $v_0e_1v_1e_2 \cdots v_{k-1}e_kv_k$ such that $v_0, v_1, \ldots, v_k$ are vertices, $e_1, e_2, \ldots, e_k$ are edges, and each vertex or edge in the sequence, except v_k, is incident with its successor in the sequence. Now suppose that the vertices $v_0, v_1, \ldots, v_k$ are distinct. Then $e_1, e_2, \ldots, e_k$ are also distinct and W is a *path*. The *end-vertices* of this path are v_0 and v_k, and the path is said to be a (v_0, v_k)-*path* or to *join* v_0 and v_k. The vertices $v_1, v_2, \ldots, v_{k-1}$ are the *internal vertices* of the path. The *length* of a path is the number of edges that it contains. In a simple graph, a path is uniquely determined by its vertex set and so will usually be specified by listing the appropriate sequence of vertices.

A graph is *connected* if each pair of distinct vertices is joined by a path. A graph that is not connected is *disconnected*. In any graph G, the maximal connected subgraphs are called (*connected*) *components*. Evidently the vertex sets of the components of G partition $V(G)$. The number of these components will be denoted by $\omega(G)$.

If P is a (u, v)-path in a graph G and e is an edge of G that joins u to v but is not in P, then the subgraph of G whose vertex set is $V(P)$ and whose edge set in $E(P) \cup e$ is called a *cycle*. A connected graph having no cycles is a *tree*, while a union of trees is a *forest*. Clearly a graph is a forest if and only if it has no cycles. A *spanning tree* of a connected graph G is a subgraph T of G such that T is a tree and $V(T) = V(G)$. Trees have many attractive properties. In particular, for all trees T,

$$|E(T)| = |V(T)| - 1.$$

Hence if T is a spanning tree of a graph G, then

$$|E(T)| = |V(G)| - 1.$$

A directed graph is a graph in which every edge is directed from one endpoint to the other as, for example, in Figure 0.3. More formally, a *directed graph* or *digraph* D consists of a non-empty set $V(D)$ of *vertices* and a multiset $A(D)$ of *arcs*, each of which consists of an ordered pair of (possibly identical) vertices. If $a \in A(D)$ and $a = (u, v)$, then a *joins* u *to* v and we say that u is the *tail* of a and v is its *head*. If $u = v$, then a is a *loop* of D. In general, the terminology for directed graphs mimics that for graphs. For instance, a *directed path* is a sequence $(v_0, a_1, v_1, a_2, \ldots, v_{k-1}, a_k, v_k)$ where $v_0, v_1, \ldots, v_k$ are vertices, $a_1, a_2, \ldots, a_k$ are arcs, and, for all i in $\{1, 2, \ldots, k\}$, a_i joins v_{i-1} to v_i. A digraph D' is a *subdigraph* of D if $V(D') \subseteq V(D)$ and $A(D') \subseteq A(D)$.

A digraph that is obtained from a graph G by specifying an order on the endpoints of each edge is called an *orientation* of G. On the other hand, if D is a digraph and G is the graph that is obtained from D by replacing each arc by an edge with the same endpoints, then G is the *underlying graph* of D.

Numerous other graph-theoretic concepts that appear in the book will be defined as needed. For any remaining undefined terminology involving graphs, we refer the reader to Bondy and Murty's (1976) book.

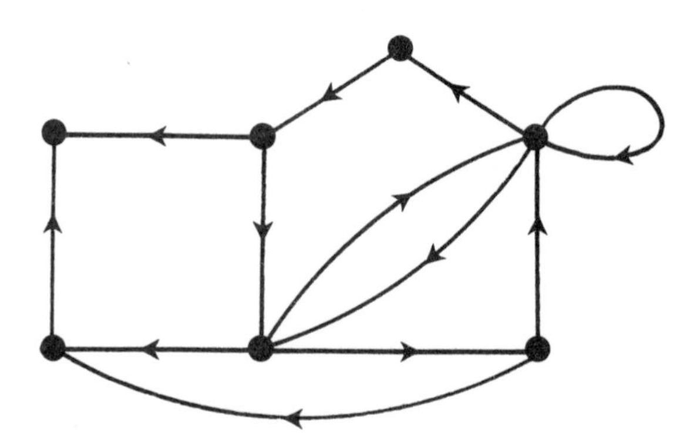

Fig. 0.3. An example of a digraph.

1
Basic definitions and examples

The study of matroids is an analysis of an abstract theory of dependence. Indeed, Whitney's (1935) founding paper in the subject was entitled 'On the abstract properties of linear dependence'. In defining a matroid, Whitney tried to capture the fundamental properties of dependence that are common to graphs and matrices. He was not alone in such attempts to extract the essence of dependence. The second edition of van der Waerden's *Moderne Algebra* (1937) distinguished three fundamental properties that are common to algebraic and linear dependence. Other pioneering work in matroid theory was done by Birkhoff (1935) and MacLane (1936,1938). We make no attempt here to survey these important early contributions but instead refer the interested reader to Kung's book (1986), which not only reprints the papers of Whitney, Birkhoff, and Mac Lane cited above, but also gives a comprehensive survey of the matroid theory literature that appeared before 1945.

A characteristic of matroids is that they can be defined in many different but equivalent ways. One important task of this opening chapter is to introduce these different axiom systems and to prove their equivalence. The other main purpose of the chapter is to present various fundamental examples of matroids and to illustrate the basic concepts in the context of these examples.

1.1 Independent sets and circuits

The two fundamental classes of matroids that appear in Whitney's paper (1935) arise from matrices and from graphs. In this section, we introduce two equivalent axiom systems for matroids and indicate how each is naturally associated with one of these classes.

A *matroid* M is an ordered pair $(E, \mathcal{I})$ consisting of a finite set E and a collection $\mathcal{I}$ of subsets of E satisfying the following three conditions:

(I1) $\emptyset \in \mathcal{I}$.

(I2) *If* $I \in \mathcal{I}$ *and* $I' \subseteq I$, *then* $I' \in \mathcal{I}$.

(I3) *If* I_1 *and* I_2 *are in* $\mathcal{I}$ *and* $|I_1| < |I_2|$, *then there is an element* e *of* $I_2 - I_1$ *such that* $I_1 \cup e \in \mathcal{I}$.

Condition (I3) is called the *independence augmentation axiom.*

If M is the matroid $(E, \mathcal{I})$, then M is called a matroid *on* E. The members of $\mathcal{I}$ are the *independent sets* of M, and E is the *ground set* of M. We shall often write $\mathcal{I}(M)$ for $\mathcal{I}$ and $E(M)$ for E, particularly when several matroids are being considered. A subset of E that is not in $\mathcal{I}$ is called *dependent.*

The name 'matroid' was coined by Whitney (1935) because a class of fundamental examples of such objects arises from matrices in the following way.

1.1.1 **Proposition.** *Let* E *be the set of column labels of an* $m \times n$ *matrix* A *over a field* F, *and let* $\mathcal{I}$ *be the set of subsets* X *of* E *for which the multiset of columns labelled by* X *is linearly independent in the vector space* $V(m, F)$. *Then* $(E, \mathcal{I})$ *is a matroid.*

Proof. Evidently $\mathcal{I}$ satisfies (I1) and (I2). To prove that (I3) holds, let I_1 and I_2 be linearly independent subsets of E such that $|I_1| < |I_2|$. Let W be the subspace of $V(m, F)$ spanned by $I_1 \cup I_2$. Then $\dim W$, the dimension of W, is at least $|I_2|$. Now suppose that $I_1 \cup e$ is linearly dependent for all e in $I_2 - I_1$. Then W is contained in the span of I_1. Thus $|I_2| \leq \dim\ W \leq |I_1| < |I_2|$; a contradiction. We conclude that $I_2 - I_1$ contains an element e such that $I_1 \cup e \in \mathcal{I}$, that is, (I3) holds. □

The matroid obtained as above from the matrix A will be denoted by $M[A]$. This matroid is called the *vector matroid* of A. As a particular example of such a matroid, consider the following.

1.1.2 **Example.** Let A be the matrix

$$\begin{array}{c} \begin{matrix} 1 & 2 & 3 & 4 & 5 \end{matrix} \\ \begin{bmatrix} 1 & 0 & 0 & 1 & 1 \\ 0 & 1 & 0 & 0 & 1 \end{bmatrix} \end{array}$$

over the field $\mathbb{R}$ of real numbers. Then $E = \{1, 2, 3, 4, 5\}$ and $\mathcal{I} = \{\emptyset, \{1\}, \{2\}, \{4\}, \{5\}, \{1, 2\}, \{1, 5\}, \{2, 4\}, \{2, 5\}, \{4, 5\}\}$. Thus the set of dependent sets of this matroid is

$$\{\{3\}, \{1, 3\}, \{1, 4\}, \{2, 3\}, \{3, 4\}, \{3, 5\}\} \cup \{X \subseteq E : |X| \geq 3\}.$$

The set of *minimal dependent sets*, that is, dependent sets all of whose proper subsets are independent, is $\{\{3\},\{1,4\},\{1,2,5\},\{2,4,5\}\}$. □

A minimal dependent set in an arbitrary matroid M will be called a *circuit* of M and we shall denote the set of circuits of M by $\mathcal{C}$ or $\mathcal{C}(M)$. A circuit of M having n elements will also be called an *n-circuit.* Evidently, as in the last example, once $\mathcal{I}(M)$ has been specified, $\mathcal{C}(M)$ can be determined. Similarly, $\mathcal{I}(M)$ can be determined from $\mathcal{C}(M)$: the members of $\mathcal{I}(M)$ are those subsets of $E(M)$ that contain no member of $\mathcal{C}(M)$. Thus a matroid is uniquely determined by its set $\mathcal{C}$ of circuits.

We now examine some properties of $\mathcal{C}$ with a view to characterizing those subsets of 2^E that can occur as the set of circuits of a matroid on E. It is easy to see that

(C1) $\emptyset \notin \mathcal{C}$; and

(C2) *if C_1 and C_2 are members of $\mathcal{C}$ and $C_1 \subseteq C_2$, then $C_1 = C_2$.*

Furthermore:

1.1.3 Lemma. *The set $\mathcal{C}$ of circuits of a matroid has the following property:*

(C3) *If C_1 and C_2 are distinct members of $\mathcal{C}$ and $e \in C_1 \cap C_2$, then there is a member C_3 of $\mathcal{C}$ such that $C_3 \subseteq (C_1 \cup C_2) - e$.*

Proof. Assume that $(C_1 \cup C_2) - e$ does not contain a circuit. Then $(C_1 \cup C_2) - e \in \mathcal{I}$. By (C2), $C_2 - C_1$ is non-empty, so we can choose an element f from this set. As C_2 is a minimal dependent set, $C_2 - f \in \mathcal{I}$. Now choose a subset I of $C_1 \cup C_2$ which is maximal with the properties that it contains $C_2 - f$ and is independent. Evidently $f \notin I$. Moreover, as C_1 is a circuit, some element g of C_1 is not in I. As $f \in C_2 - C_1$, the elements f and g are distinct. Hence

$$|I| \leq |(C_1 \cup C_2) - \{f,g\}| = |C_1 \cup C_2| - 2 < |(C_1 \cup C_2) - e|.$$

Now apply (I3), taking I_1 and I_2 to be I and $(C_1 \cup C_2) - e$, respectively. The resulting independent set contradicts the maximality of I. □

Condition (C3) is called the *circuit elimination axiom* or, sometimes, the *weak circuit elimination axiom* to contrast it with another circuit elimination axiom which will be introduced in Section 1.4 (see also Exercise 11). We show next that (C1)–(C3) characterize those collections of sets that can occur as the set of circuits of a matroid. In this and many other matroid arguments, drawing a Venn diagram makes the argument easier to follow.

1.1.4 Theorem. *Let E be a set and $\mathcal{C}$ be a collection of subsets of E satisfying (C1)–(C3). Let $\mathcal{I}$ be the collection of subsets of E that contain no member of $\mathcal{C}$. Then $(E, \mathcal{I})$ is a matroid having $\mathcal{C}$ as its collection of circuits.*

Proof. We shall first show that $\mathcal{I}$ satisfies (I1)–(I3). By (C1), $\emptyset$ does not contain a member of $\mathcal{C}$, so $\emptyset \in \mathcal{I}$ and (I1) holds. If I contains no member of $\mathcal{C}$ and I' is contained in I, then I' contains no member of $\mathcal{C}$. Thus (I2) holds.

To prove (I3), suppose that I_1 and I_2 are members of $\mathcal{I}$ and $|I_1| < |I_2|$. Assume that (I3) fails for the pair (I_1, I_2). Now $\mathcal{I}$ has a member that is a subset of $I_1 \cup I_2$ and has more elements than I_1. Choose such a subset I_3 for which $|I_1 - I_3|$ is minimal. As (I3) fails, $I_1 - I_3$ is non-empty, so we can choose an element e from this set. Now, for each element f of $I_3 - I_1$, let T_f be $(I_3 \cup e) - f$. Then $T_f \subseteq I_1 \cup I_2$ and $|I_1 - T_f| < |I_1 - I_3|$. Therefore $T_f \notin \mathcal{I}$, so T_f contains a member C_f of $\mathcal{C}$. Evidently $f \notin C_f$. Moreover, $e \in C_f$, otherwise $C_f \subseteq I_3$ contradicting the fact that $I_3 \in \mathcal{I}$.

Let g be an element of $I_3 - I_1$. If $C_g \cap (I_3 - I_1) = \emptyset$, then $C_g \subseteq ((I_1 \cap I_3) \cup e) - g \subseteq I_1$; a contradiction. Therefore there is an element h in $C_g \cap (I_3 - I_1)$. Now $e \in C_g \cap C_h$, so (C3) implies that there is a member C of $\mathcal{C}$ such that $C \subseteq (C_g \cup C_h) - e$. But, both C_g and C_h are subsets of $I_3 \cup e$ and hence $C \subseteq I_3$; a contradiction. We conclude that (I3) holds. Thus (E,$\mathcal{I}$) is a matroid M.

To prove that $\mathcal{C}$ is the set $\mathcal{C}(M)$ of circuits of M, we note that the following statements are equivalent:

(i) C is a circuit of M;

(ii) $C \notin \mathcal{I}(M)$ and $C - x \in \mathcal{I}(M)$ for all x in C;

(iii) C has a member C' of $\mathcal{C}$ as a subset, but C' is not a proper subset of C;

(iv) $C \in \mathcal{C}$. □

On combining Theorem 1.1.4 with Lemma 1.1.3 and the remarks preceding it, we obtain the following result:

1.1.5 Corollary. *Let $\mathcal{C}$ be a set of subsets of a set E. Then $\mathcal{C}$ is the collection of circuits of a matroid on E if and only if $\mathcal{C}$ satisfies the following conditions:*

(C1) $\emptyset \notin \mathcal{C}$.

(C2) *If C_1 and C_2 are members of $\mathcal{C}$ and $C_1 \subseteq C_2$, then $C_1 = C_2$.*

(C3) *If C_1 and C_2 are distinct members of $\mathcal{C}$ and $e \in C_1 \cap C_2$, then there is a member C_3 of $\mathcal{C}$ such that $C_3 \subseteq (C_1 \cup C_2) - e$.* □

The next proposition contains an elementary but frequently used property of matroids.

1.1.6 **Proposition.** *Suppose that I is an independent set in a matroid M and e is an element of M such that $I \cup e$ is dependent. Then M has a unique circuit contained in $I \cup e$ and this circuit contains e.*

Proof. Evidently $I \cup e$ contains a circuit C. Moreover, every circuit contained in $I \cup e$ must contain e. If C' is such a circuit and C' is distinct from C, then, by (C3), $(C \cup C') - e$ contains a circuit. Since $(C \cup C') - e \subseteq I$, this is a contradiction. Hence C is unique. □

We noted earlier that the class of vector matroids is one of two fundamental classes of matroids that were considered in Whitney's (1935) seminal paper. The second such class consists of matroids derived from graphs in a way that will be described in a moment. One of the more obvious indications of the pervasive influence of these two classes of examples throughout matroid theory is in the terminology of the subject, which borrows heavily from both linear algebra and graph theory.

Recall that a cycle of a graph is a closed walk all of whose vertices have degree two.

1.1.7 **Proposition.** *Let E be the set of edges of a graph G and $\mathcal{C}$ be the set of edge sets of cycles of G. Then $\mathcal{C}$ is the set of circuits of a matroid on E.*

Proof. Clearly $\mathcal{C}$ satisfies (C1) and (C2). To prove that it satisfies (C3), let C_1 and C_2 be the edge sets of two distinct cycles of G that have e as a common edge. Let u and v be the endpoints of e. We now construct a cycle of G whose edge set is contained in $(C_1 \cup C_2) - e$. For $i = 1, 2$, let P_i be the path from u to v in G whose edge set is $C_i - e$. Beginning at u, traverse P_1 towards v letting w be the first vertex at which the next edge of P_1 is not in P_2. Continue traversing P_1 from w towards v until the first time a vertex x is reached that is distinct from w but is also in P_2. Since P_1 and P_2 both end at v, such a vertex must exist. Now adjoin the section of P_1 from w to x to the section of P_2 from x to w. The result is a cycle (see Figure 1.1), the edge set of which is contained in $(C_1 \cup C_2) - e$. Hence $\mathcal{C}$ satisfies (C3). □

The matroid derived above from the graph G is called the *cycle matroid* or *polygon matroid* of G. It is denoted by $M(G)$. Clearly a set X of edges is independent in $M(G)$ if and only if X does not contain the edge set of a cycle or, equivalently, $G[X]$, the subgraph induced by X, is a forest. Since the ground set of $M(G)$ is the edge set of G, we shall often refer to certain subgraphs of G, such as cycles, when we mean just their edge sets. This commonplace practice should not cause confusion.

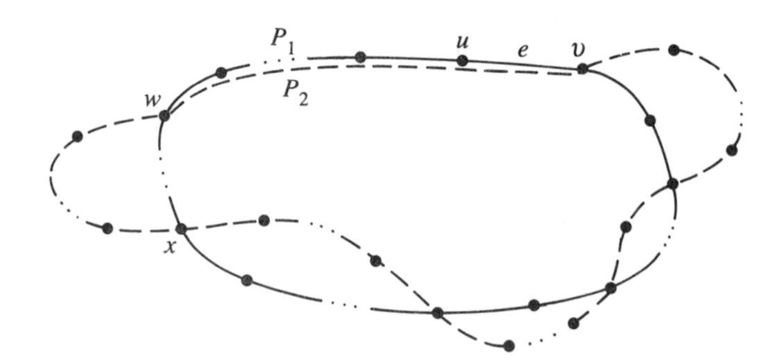

Fig. 1.1. The cycles in a graph satisfy the circuit elimination axiom.

1.1.8 Example. Let G be the graph shown in Figure 1.2 and let $M = M(G)$. Then $E(M) = \{e_1, e_2, e_3, e_4, e_5\}$ and $\mathcal{C}(M) = \{\{e_3\}, \{e_1, e_4\}, \{e_1, e_2, e_5\}, \{e_2, e_4, e_5\}\}$. Comparing M with the matroid $M[A]$ in Example 1.1.2, we see that, under the bijection ψ from $\{1, 2, 3, 4, 5\}$ to $\{e_1, e_2, e_3, e_4, e_5\}$ defined by $\psi(i) = e_i$, a set X is a circuit in $M[A]$ if and only if $\psi(X)$ is a circuit in M. Equivalently, a set Y is independent in $M[A]$ if and only if $\psi(Y)$ is independent in M. Thus the matroids $M[A]$ and M have the same structure or are isomorphic. Formally, two matroids M_1 and M_2 are *isomorphic*, written $M_1 \cong M_2$, if there is a bijection ψ from $E(M_1)$ to $E(M_2)$ such that, for all $X \subseteq E(M_1)$, $\psi(X)$ is independent in M_2 if and only if X is independent in M_1. □

A matroid that is isomorphic to the cycle matroid of a graph is called *graphic*. So, for instance, the matroid $M[A]$ in Example 1.1.2 is graphic. If M is isomorphic to the vector matroid of a matrix D over a field F, then M is said to be *representable over* F or F*-representable*; D is called a *representation* for M *over* F or an F*-representation* for M. A matroid is *representable* or, for some authors, *matric* if it is representable over some field. Thus the matroid $M(G)$ in Example 1.1.8 is representable, the matrix A being an $\mathbb{R}$-representation for it. Indeed, as we shall see in Chapter 5, every graphic matroid is representable over every field.

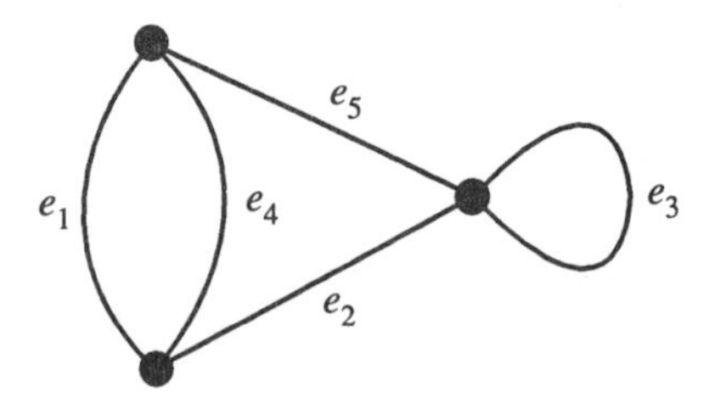

Fig. 1.2. This graph's cycle matroid is isomorphic to $M[A]$ in 1.1.2.

When looking for examples of matroids with certain properties, one tends not to look for collections of subsets of a set satisfying (I1)–(I3). Instead, one looks for matroids corresponding to certain graphs or matrices. In such examples, the graph or matrix provides a relatively compact way of presenting the information that is needed to determine the matroid. Not every matroid is graphic or representable, but, as we shall see later, several other classes of matroids have similarly compact presentations. Rarely does one specify a matroid by explicitly listing every independent set or every circuit.

All matroids on three or fewer elements are graphic. In Table 1.1 (p. 14), we list these matroids by presenting, for each, a corresponding graph. The latter need not be unique. The reader is urged to check the completeness of this table. Doubtless, the reader is now convinced that there are exactly 2^n non-isomorphic matroids on an n-element set! In fact, however, the four values of n shown in the table are the only four values for which there are exactly 2^n non-isomorphic matroids. Later in this chapter, we shall present a technique that helps one to show that there are exactly 17 non-isomorphic matroids on a 4-element set. Indeed, if $f(n)$ denotes the number of non-isomorphic matroids on an n-element set, then $f(n)$ is much closer to 2^{2^n} than it is to 2^n. For the reader interested in a more precise statement here, we note that, on combining bounds of Piff (1973) and Knuth (1974), one gets:

$$\begin{aligned} n - (3/2)\log_2 n + O(\log_2\log_2 n) &\leq \log_2\log_2 f(n) \\ &\leq n - \log_2 n + O(\log_2\log_2 n). \end{aligned}$$

In Example 1.1.8, the loop e_3 and the pair $\{e_1, e_4\}$ of parallel edges give rise to circuits in $M(G)$ of sizes one and two, respectively. Borrowing this graph-theoretic terminology, we call an element e a *loop* of an arbitrary matroid M if $\{e\}$ is a circuit of M. Moreover, if f and g are elements of M such that $\{f, g\}$ is a circuit, then f and g are said to be *parallel* in M. A *parallel class* of M is a maximal subset X of $E(M)$ such that any two distinct members of X are parallel and no member of X is a loop. A parallel class is *trivial* if it contains just one element. If M has no loops and no non-trivial parallel classes, it is called a *simple matroid* or a *combinatorial geometry*.

We have already seen that a matroid M on a fixed ground set E can be specified by a list of its independent sets or by a list of its circuits. Evidently M is also uniquely determined by its collection of maximal independent sets and, in the next section, we shall use this collection to give a third characterization of matroids. In subsequent sections, we shall look at yet other ways to characterize matroids. It is one of the great beauties of the subject of matroid theory that there are so many equivalent descriptions of matroids. These many different approaches, motivated as

Table 1.1. Matroids with three or fewer elements

Number n of elements	A corresponding graph	Number of non-isomorphic n-element matroids
0		1
1		2
1		
2		4
2		
2		
2		
3		8
3		
3		
3		
3	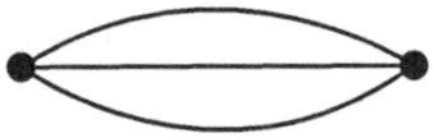	
3		
3	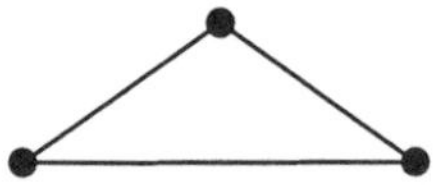	
3		

they are by numerous different examples of matroids, enable one to bring to the subject intuition developed in other areas of mathematics including graph theory, linear algebra, and geometry.

Like matroids, topological spaces have several equivalent descriptions. For example, they can be defined in terms of their open sets or their closed sets or via a (topological) closure operator. Just as each of these approaches is indispensable in topology, each of the many descriptions of matroids that we shall introduce provides an invaluable tool for the study of these objects.

Exercises

1. Prove that $(E, \mathcal{I})$ is a matroid if and only if $\mathcal{I}$ satisfies (I2) and the following two conditions:

(I1)′ $\mathcal{I} \neq \emptyset$.

(I3)′ *If* I_1 *and* I_2 *are in* $\mathcal{I}$ *and* $|I_2| = |I_1| + 1$, *then there is an element* e *of* $I_2 - I_1$ *such that* $I_1 \cup e \in \mathcal{I}$.

2. Let A be the matrix

$$\begin{array}{cccccc} 1 & 2 & 3 & 4 & 5 & 6 \end{array}$$
$$\begin{bmatrix} 1 & 0 & 0 & 1 & 1 & 0 \\ 0 & 1 & 0 & 1 & 0 & 1 \\ 0 & 0 & 1 & 0 & 1 & 1 \end{bmatrix}.$$

For q in $\{2, 3\}$, let $M_q[A]$ be the vector matroid of A when A is viewed over $GF(q)$, the field of q elements. Show that:

(i) The sets of circuits of $M_2[A]$ and $M_3[A]$ are different.

(ii) $M_2[A]$ is graphic but $M_3[A]$ is not.

(iii) $M_2[A]$ is representable over $GF(3)$, but $M_3[A]$ is not representable over $GF(2)$.

3. Prove that $(E, \mathcal{I})$ is a matroid if and only if $\mathcal{I}$ satisfies (I1), (I2), and the following condition:

(I3)″ *If* $X \subseteq E$ *and* I_1 *and* I_2 *are maximal members of* $\{I : I \in \mathcal{I}$ *and* $I \subseteq X\}$, *then* $|I_1| = |I_2|$.

4. Let A be a matrix over a field F. In terms of A, specify precisely when an element of $M[A]$ is a loop, and when two elements of $M[A]$ are parallel.

5. Let C_1 and C_2 be circuits of a matroid M such that $C_1 \cup C_2 = E(M)$ and $C_1 - C_2 = \{e\}$. Prove that if C_3 is a circuit of M, then either $C_3 = C_1$ or $C_3 \supseteq C_2 - C_1$.

6. For each of the matroids M in Table 1.1 (p. 14), find the following:

   (i) all graphs G for which $M(G) \cong M$; and

   (ii) a matrix A so that $M[A] \cong M$ when A is viewed over any field.

7. Let M_1 and M_2 be matroids on disjoint sets E_1 and E_2. Let $E = E_1 \cup E_2$ and $\mathcal{I} = \{I_1 \cup I_2 : I_1 \in \mathcal{I}(M_1),\ I_2 \in \mathcal{I}(M_2)\}$. Prove that $(E, \mathcal{I})$ is a matroid. This matroid, the *direct sum* $M_1 \oplus M_2$ of M_1 and M_2, will be looked at more closely in Chapter 4.

8. Let $\mathcal{D}$ be a set of subsets of a set E. Characterize when $\mathcal{D}$ is the set of dependent sets of a matroid on E.

9. Let M_1 and M_2 be matroids on a set E. Give an example to show that $(E, \mathcal{I}(M_1) \cap \mathcal{I}(M_2))$ need not be a matroid.

10. Let G be the graph obtained from K_5 by deleting two non-adjacent edges. Find representations for $M(G)$ over $GF(2)$ and $GF(3)$.

11.* Prove that the circuits of a matroid M satisfy the following strong elimination axiom: if C_1 and C_2 are circuits such that $e \in C_1 \cap C_2$ and $f \in C_1 - C_2$, then M has a circuit C_3 such that $f \in C_3 \subseteq (C_1 \cup C_2) - e$. (Hint: Let C_1 and C_2 be a pair of circuits for which the proposition fails so that, among such pairs, $|C_1 \cup C_2|$ is minimal. Apply (C3) several times.)

1.2 Bases

A list of the maximal independent sets in a matroid M is clearly a more efficient way to specify M than a list of all the independent sets. We call a maximal independent set in M a *basis* or a *base* of M. In this section, we study the bases of a matroid, showing that these sets have much in common with the bases in a vector space. For instance:

1.2.1 Lemma. *If B_1 and B_2 are bases of a matroid M, then $|B_1| = |B_2|$.*

Proof. Suppose that $|B_1| < |B_2|$. Then, as B_1 and B_2 are both independent in M, (I3) implies that there is an element e of $B_2 - B_1$ such that $B_1 \cup e \in \mathcal{I}$. This contradicts the maximality of B_1. Hence $|B_1| \geq |B_2|$ and, similarly, $|B_2| \geq |B_1|$. □

If M is a matroid and $\mathcal{B}$ is its collection of bases, then, by (I1),

(B1) *$\mathcal{B}$ is non-empty.*

The next result contains a far less obvious property of $\mathcal{B}$.

1.2.2 Lemma. *$\mathcal{B}$ satisfies the following condition:*

(B2) *If B_1 and B_2 are members of $\mathcal{B}$ and $x \in B_1 - B_2$, then there is an element y of $B_2 - B_1$ such that $(B_1 - x) \cup y \in \mathcal{B}$.*

Proof. Both $B_1 - x$ and B_2 are independent sets. Moreover, $|B_1 - x| < |B_2|$ since, by the preceding lemma, $|B_1| = |B_2|$. Therefore, by (I3), there is an element y of $B_2 - (B_1 - x)$ such that $(B_1 - x) \cup y \in \mathcal{I}$. Evidently $y \in B_2 - B_1$. Furthermore, as $(B_1 - x) \cup y$ is independent, it is contained in a maximal independent set B_1'. By Lemma 1.2.1 again, $|B_1'| = |B_1|$. Moreover, $|B_1| = |(B_1 - x) \cup y|$. Hence $(B_1 - x) \cup y = B_1'$, that is, $(B_1 - x) \cup y$ is a basis of M. Thus $\mathcal{B}$ satisfies (B2). □

Condition (B2) is one of several *basis exchange axioms* obeyed by matroids. We shall see further examples of such conditions in Chapter 2 and in the exercises. We now prove that (B1) and (B2) characterize the bases of a matroid.

1.2.3 Theorem. *Let E be a set and $\mathcal{B}$ be a collection of subsets of E satisfying* (B1) *and* (B2). *Let $\mathcal{I}$ be the collection of subsets of E that are contained in some member of $\mathcal{B}$. Then $(E, \mathcal{I})$ is a matroid having $\mathcal{B}$ as its collection of bases.*

Proof. Since $\mathcal{B}$ satisfies (B1), $\mathcal{I}$ satisfies (I1). Moreover, if $I \in \mathcal{I}$, then $I \subseteq B$ for some set B in $\mathcal{B}$. Thus if $I' \subseteq I$, then $I' \subseteq B$, so $I' \in \mathcal{I}$. Hence $\mathcal{I}$ satisfies (I2). The proof that $\mathcal{I}$ satisfies (I3) uses the following.

1.2.4 Lemma. *The members of $\mathcal{B}$ are equicardinal.*

Proof. Suppose that B_1 and B_2 are distinct members of $\mathcal{B}$ for which $|B_1| > |B_2|$ so that, among all such pairs, $|B_1 - B_2|$ is minimal. Clearly $B_1 - B_2 \neq \emptyset$. Thus, choosing x in $B_1 - B_2$, we can find an element y of $B_2 - B_1$ so that $(B_1 - x) \cup y \in \mathcal{B}$. Evidently $|(B_1 - x) \cup y| = |B_1| > |B_2|$ and $|((B_1 - x) \cup y) - B_2| < |B_1 - B_2|$. Thus the choice of B_1 and B_2 is contradicted and the lemma is proved. □

Returning to the proof of Theorem 1.2.3, suppose that (I3) fails for $\mathcal{I}$. Then $\mathcal{I}$ has members I_1 and I_2 with $|I_1| < |I_2|$ such that, for all e in $I_2 - I_1$, the set $I_1 \cup e \notin \mathcal{I}$. By definition, $\mathcal{B}$ contains members B_1 and B_2 such that $I_1 \subseteq B_1$ and $I_2 \subseteq B_2$. Assume that such a set B_2 is chosen so that $|B_2 - (I_2 \cup B_1)|$ is minimal. By the choice of I_1 and I_2,

$$I_2 - B_1 = I_2 - I_1. \tag{1}$$

Now suppose that $B_2 - (I_2 \cup B_1)$ is non-empty. Then we can choose an element x from this set. By (B2), there is an element y of $B_1 - B_2$ such that $(B_2 - x) \cup y \in \mathcal{B}$. But then $|((B_2 - x) \cup y) - (I_2 \cup B_1)| < |B_2 - (I_2 \cup B_1)|$ and the choice of B_2 is contradicted. Hence $B_2 - (I_2 \cup B_1)$ is empty and so $B_2 - B_1 = I_2 - B_1$. Thus, by (1),

$$B_2 - B_1 = I_2 - I_1. \tag{2}$$

Next we show that $B_1 - (I_1 \cup B_2)$ is empty. If not, then there is an element x in this set and an element y in $B_2 - B_1$ so that $(B_1 - x) \cup y \in \mathcal{B}$. Now $I_1 \cup y \subseteq (B_1 - x) \cup y$ so $I_1 \cup y \in \mathcal{I}$. Since $y \in B_2 - B_1$, it follows by (2) that $y \in I_2 - I_1$ and so we have a contradiction to our assumption that (I3) fails. We conclude that $B_1 - (I_1 \cup B_2)$ is empty. Hence $B_1 - B_2 = I_1 - B_2$. Since the last set is contained in $I_1 - I_2$, it follows that

$$B_1 - B_2 \subseteq I_1 - I_2. \tag{3}$$

By Lemma 1.2.4, $|B_1| = |B_2|$, so $|B_1 - B_2| = |B_2 - B_1|$. Therefore, by (2) and (3), $|I_1 - I_2| \geq |I_2 - I_1|$, so $|I_1| \geq |I_2|$. This contradiction completes the proof that $(E, \mathcal{I})$ is a matroid. Since $\mathcal{B}$ is clearly the set of bases of this matroid, the theorem is proved. □

On combining Theorem 1.2.3 with Lemma 1.2.2 and the remarks preceding it, we get the following.

1.2.5 Corollary. *Let $\mathcal{B}$ be a set of subsets of a set E. Then $\mathcal{B}$ is the collection of bases of a matroid on E if and only if it satisfies the following conditions:*

(B1) *$\mathcal{B}$ is non-empty.*

(B2) *If B_1 and B_2 are members of $\mathcal{B}$ and $x \in B_1 - B_2$, then there is an element y of $B_2 - B_1$ such that $(B_1 - x) \cup y \in \mathcal{B}$.* □

It was noted in Proposition 1.1.6 that if I is an independent set in a matroid M and $I \cup e$ is dependent in M, then M has a unique circuit contained in $I \cup e$, and this circuit contains e. The following result is an immediate consequence of this.

1.2.6 Corollary. *Let B be a basis of a matroid M. If $e \in E(M) - B$, then $B \cup e$ contains a unique circuit, $C(e, B)$. Moreover, $e \in C(e, B)$.* □

We call $C(e, B)$ the *fundamental circuit of e with respect to B* and we leave the reader to show that every circuit in a matroid is the fundamental circuit of some element with respect to some basis (Exercise 4).

1.2.7 **Example.** Let m and n be non-negative integers such that $m \leq n$. Let E be an n-element set and $\mathcal{B}$ be the collection of m-element subsets of E. Then it is easy to check that $\mathcal{B}$ is the set of bases of a matroid on E. We denote this matroid by $U_{m,n}$ and call it the *uniform matroid* of rank m on an n-element set. Clearly

$$\mathcal{I}(U_{m,n}) = \{X \subseteq E : |X| \leq m\}$$

and

$$\mathcal{C}(U_{m,n}) = \begin{cases} \emptyset, & \text{if } m = n, \\ \{X \subseteq E : |X| = m+1\}, & \text{if } m < n. \end{cases}$$

If m is 0, every element of $U_{m,n}$ is a loop, while if m is 1, $U_{m,n}$ consists of a single parallel class of elements. For $m \geq 2$, $U_{m,n}$ is simple. The matroids of the form $U_{n,n}$ are precisely the matroids having no dependent sets. We call such matroids *free*. One of these matroids, $U_{0,0}$, is the unique matroid on the empty set. It is called the *empty matroid*. □

In a matroid M, we have already specified the relationship between $\mathcal{B}(M)$ and $\mathcal{I}(M)$, and between $\mathcal{I}(M)$ and $\mathcal{C}(M)$. To relate $\mathcal{B}(M)$ and $\mathcal{C}(M)$ directly, we note that $\mathcal{B}(M)$ is the collection of maximal subsets of $E(M)$ that contain no member of $\mathcal{C}(M)$, while $\mathcal{C}(M)$ is the collection of minimal sets that are contained in no member of $\mathcal{B}(M)$.

The fact that all bases of a matroid are equicardinal enables one to define a matroid generalization of the dimension function in vector spaces. In the next section, we shall examine this function in detail. Before doing this, however, let us distinguish the bases in graphic and representable matroids. Let G be a graph. We observed earlier that a subset X of $E(G)$ is independent in $M(G)$ exactly when $G[X]$, the subgraph of G induced by X, is a forest. Thus X is a basis of $M(G)$ precisely when $G[X]$ is a forest and, for all $e \notin X$, $G[X \cup e]$ contains a cycle. It follows that, when G is connected, X is a basis of $M(G)$ if and only if $G[X]$ is a spanning tree of G. In general, X is a basis of $M(G)$ if and only if, for each component H of G having at least one non-loop edge, $H[X \cap E(H)]$ is a spanning tree of H.

As an example, consider the graphs G_1 and G_2 shown in Figure 1.3. Each has edge set $\{1,2,3,4,5\}$. Moreover, each of $M(G_1)$ and $M(G_2)$ has $\{2,3,4,5\}$ as its unique basis. Thus $M(G_1) = M(G_2)$, although clearly $G_1 \not\cong G_2$. In Chapter 5, we shall solve the problem of determining precisely when two non-isomorphic graphs have isomorphic cycle matroids. Before leaving this problem for the moment, we make the following elementary but useful observation.

1.2.8 **Proposition.** *Let M be a graphic matroid. Then $M \cong M(G)$ for some connected graph G.*

Proof. As M is graphic, $M \cong M(H)$ for some graph H. If H is connected, the result is proved. If not, suppose $H_1, H_2, \ldots, H_n$ are the connected components of H. For each i in $\{1, 2, ..., n\}$, choose a vertex v_i in H_i. Form a new graph G by identifying $v_1, v_2, \ldots, v_n$ as a single vertex. Clearly $E(H) = E(G)$ and G is connected. Moreover, if $X \subseteq E(H)$, then X is the set of edges of a cycle in H if and only if X is the set of edges of a cycle in G. Thus $M(G) \cong M(H)$ and the proposition is proved. □

We remark that, in the preceding proof, the key point about the way in which $H_1, H_2, \ldots, H_n$ were joined is that no new cycles were formed. Indeed, provided this condition is satisfied when components are stuck together, the cycle matroid of the resulting graph is isomorphic to $M(H)$.

We now turn attention to representable matroids. Let A be an $m \times n$ matrix over the field F. The columns of A span a subspace W of $V(m, F)$ of dimension r, say. If B is a subset of the column labels of A, then B is a basis of $M[A]$ if and only if $|B| = r$ and the columns labelled by B form a basis for W. For example, let A be the matrix

$$\begin{array}{cccc} 1 & 2 & 3 & 4 \end{array}$$
$$\begin{bmatrix} 1 & 0 & 1 & 1 \\ 0 & 1 & 1 & -1 \end{bmatrix}$$

over $GF(3)$, the field of three elements. In this case, dim W is 2 and the bases of $M[A]$ are all of the 2-element subsets of $\{1, 2, 3, 4\}$. Thus $M[A] \cong U_{2,4}$. Hence $U_{2,4}$ is *ternary*, that is, it is representable over $GF(3)$.

Ternary matroids are one of the most commonly studied classes of representable matroids. Two other important such classes are binary and regular matroids. A *binary matroid* is one that is representable over $GF(2)$. A *regular matroid* is one that can be represented by a *totally unimodular matrix*, the latter being a matrix over $\mathbb{R}$ for which every square submatrix has determinant in $\{0, 1, -1\}$. Some authors, for example, White (1987a), refer to regular matroids as *unimodular matroids*.

We leave it to the reader to check that $U_{2,4}$ is *not* binary. Indeed, as we shall see in Chapter 6, $U_{2,4}$ is the unique obstruction to a matroid

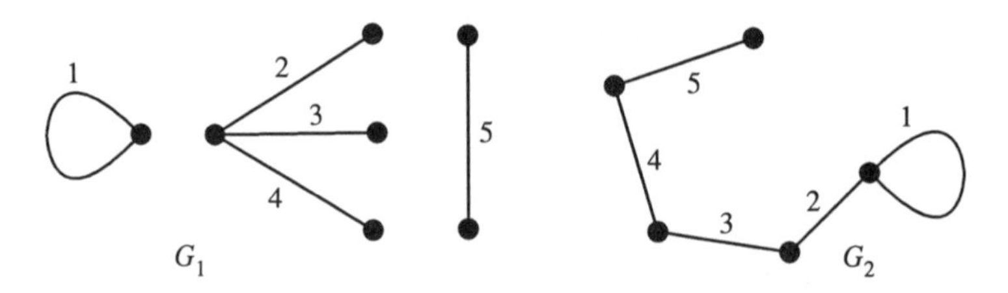

Fig. 1.3. Two non-isomorphic graphs with equal cycle matroids.

being binary where, of course, we shall need to be precise about what we mean here by the term 'obstruction'. In Chapters 5 and 6, we shall prove that every graphic matroid is regular, and that a matroid is regular if and only if it is representable over every field. It will follow from these results that $U_{2,4}$ is neither graphic nor regular, although both of these are easy to show directly (Exercise 5).

Exercises

1. Prove that $\mathcal{B}$ is the collection of bases of a matroid on E if and only if $\mathcal{B}$ satisfies (B1) and the following two conditions:

(B2)′ *If $B_1, B_2 \in \mathcal{B}$ and $e \in B_1$, then there is an element f of B_2 such that $(B_1 - e) \cup f \in \mathcal{B}$.*

(B3) *If $B_1, B_2 \in \mathcal{B}$ and $B_1 \subseteq B_2$, then $B_1 = B_2$.*

2. Suppose that $X \subseteq Y \subseteq E$. Prove that if E is the ground set of a matroid M that has bases B_1 and B_2 with $X \subseteq B_1$ and $B_2 \subseteq Y$, then M has a basis B_3 such that $X \subseteq B_3 \subseteq Y$.

3. Prove that the following statements are equivalent for an element e of a matroid M:

(a) e is a loop.

(b) e is in no bases.

(c) e is in no independent sets.

4. Prove that if C is a circuit of a matroid M and $e \in C$, then M has a basis B such that $C = C(e, B)$.

5. **(i)** Prove that $U_{2,4}$ is representable over a field F if and only if $|F| \geq 3$.

(ii) For $n \geq 2$, generalize (i) by finding necessary and sufficient conditions for $U_{2,n}$ to be F-representable.

(iii) Prove that $U_{2,4}$ is neither graphic nor regular.

6. Suppose B is a basis of a matroid M, $f \in E(M)$, and $e \in E(M) - B$. Prove that $(B \cup e) - f$ is a basis of M if and only if $f \in C(e, B)$.

7. Let M be a matroid and X and Y be subsets of $E(M)$, each of which contains a basis. Prove that if $|X| > |Y|$, then, for some element x of $X - Y$, there is a basis contained in $X - x$.

8. (Bixby 1981) Show that a collection $\mathcal{B}$ of subsets of a set E satisfying (B1), (B3), and the following condition need not be the set of bases of a matroid on E:

(B2)$^-$ *If $B_1 \in \mathcal{B}$ and $e \in E - B_1$, then there is an element f such that $(B_1 - f) \cup e \in \mathcal{B}$.*

9. Find the values of m and n for which $U_{m,n}$ is graphic.

10. **(i)** Give an example of two matroids M_1 and M_2 on a set E such that a subset X of E is a circuit of M_1 if and only if it is a basis of M_2.

(ii) Characterize all pairs (M_1, M_2) of matroids for which the condition in (i) holds.

11.* (Brualdi 1969) Prove that if B_1 and B_2 are bases of a matroid M and $e \in B_1 - B_2$, then there is an element f of $B_2 - B_1$ such that both $(B_1 - e) \cup f$ and $(B_2 - f) \cup e$ are bases of M.

12.* Let M be a matroid and $\mathcal{D}$ be the set of minimal subsets of $E(M)$ that meet every circuit of M. Prove that $\mathcal{D}$ is the set of bases of a matroid on $E(M)$.

1.3 Rank

Two very useful notions in linear algebra are those of dimension of a vector space and span of a set of vectors. In this section and the next, we introduce the matroid generalizations of these ideas.

We begin by defining a fundamental and very natural matroid construction. Let M be the matroid $(E, \mathcal{I})$ and suppose that $X \subseteq E$. Let $\mathcal{I}|X$ be $\{I \subseteq X : I \in \mathcal{I}\}$. Then it is easy to see that the pair $(X, \mathcal{I}|X)$ is a matroid. We call this matroid the *restriction of M to X* or the *deletion of $E - X$ from M*. It is denoted by $M|X$ or $M\backslash(E - X)$. One easily checks that

$$\mathcal{C}(M|X) = \{C \subseteq X : C \in \mathcal{C}(M)\}.$$

As $M|X$ is a matroid, Lemma 1.2.1 implies that all its bases are equicardinal. We define the *rank* $r(X)$ of X to be the size of a basis B of $M|X$ and call such a set B a *basis of X*. Evidently r maps 2^E into the set of non-negative integers. This function, the *rank function* of M, will often be written as r_M. In addition, we shall usually write $r(M)$ for $r(E(M))$. It is clear that r has the following properties:

(R1) *If $X \subseteq E$, then $0 \leq r(X) \leq |X|$.*

(R2) *If* $X \subseteq Y \subseteq E$, *then* $r(X) \leq r(Y)$.

Furthermore:

1.3.1 Lemma. *The rank function* r *of a matroid* M *on a set* E *satisfies the following condition:*

(R3) *If* X *and* Y *are subsets of* E, *then*

$$r(X \cup Y) + r(X \cap Y) \leq r(X) + r(Y).$$

Before proving this lemma, we remark that (R3) is reminiscent of the identity

$$\dim(V + W) + \dim(V \cap W) = \dim V + \dim W$$

that holds for vector spaces V and W, where $V + W$ consists of all vectors of the form $v+w$ with v in V and w in W. The inequality in (R3) is referred to as the *semimodular* or *submodular inequality.* It is straightforward to find examples where this inequality is strict (Exercise 2).

Proof of Lemma 1.3.1. Let $B_{X \cap Y}$ be a basis for $X \cap Y$. Then $B_{X \cap Y}$ is an independent set in $M|(X \cup Y)$. It is therefore contained in a basis $B_{X \cup Y}$ of this matroid. Now $B_{X \cup Y} \cap X$ and $B_{X \cup Y} \cap Y$ are independent in $M|X$ and $M|Y$, respectively. Therefore

$$|B_{X \cup Y} \cap X| \leq r(X) \quad \text{and} \quad |B_{X \cup Y} \cap Y| \leq r(Y).$$

So

$$\begin{aligned} r(X) + r(Y) &\geq |B_{X \cup Y} \cap X| + |B_{X \cup Y} \cap Y| \\ &= |(B_{X \cup Y} \cap X) \cup (B_{X \cup Y} \cap Y)| \\ &\quad + |(B_{X \cup Y} \cap X) \cap (B_{X \cup Y} \cap Y)| \\ &= |B_{X \cup Y} \cap (X \cup Y)| + |B_{X \cup Y} \cap X \cap Y|. \end{aligned}$$

But $B_{X \cup Y} \cap (X \cup Y) = B_{X \cup Y}$ and $B_{X \cup Y} \cap X \cap Y = B_{X \cap Y}$. Therefore

$$\begin{aligned} r(X) + r(Y) &\geq |B_{X \cup Y}| + |B_{X \cap Y}| \\ &= r(X \cup Y) + r(X \cap Y), \quad \text{as required.} \end{aligned}$$ □

Following the pattern of the earlier sections, we now establish that conditions (R1)–(R3) characterize the rank function of a matroid.

1.3.2 Theorem. *Let* E *be a set and* r *be a function that maps* 2^E *into the set of non-negative integers and satisfies* (R1)–(R3). *Let* $\mathcal{I}$ *be the collection of subsets* X *of* E *for which* $r(X) = |X|$. *Then* $(E, \mathcal{I})$ *is a matroid having rank function* r.

The proof of this theorem will use the following.

1.3.3 Lemma. *Let E be a set and r be a function on 2^E satisfying* (R2) *and* (R3). *If X and Y are subsets of E such that, for all y in $Y-X$, $r(X \cup y) = r(X)$, then $r(X \cup Y) = r(X)$.*

Proof. Let $Y - X = \{y_1, y_2, ..., y_k\}$. We shall argue by induction on k. If $k = 1$, the result is immediate. Assume it is true for $k = n$ and let $k = n+1$. Then, by the induction assumption and (R3),

$$\begin{aligned} r(X) + r(X) &= r(X \cup \{y_1, y_2, ..., y_n\}) + r(X \cup y_{n+1}) \\ &\geq r((X \cup \{y_1, y_2, ..., y_n\}) \cup (X \cup y_{n+1})) \\ &\quad + r((X \cup \{y_1, y_2, ..., y_n\}) \cap (X \cup y_{n+1})) \\ &= r(X \cup \{y_1, y_2, ..., y_{n+1}\}) + r(X) \\ &\geq r(X) + r(X), \end{aligned}$$

where the last step follows by (R2). Since the first and last lines of the above are equal, equality must hold throughout and, therefore,

$$r(X \cup \{y_1, y_2, ..., y_{n+1}\}) = r(X).$$

Thus, by induction, the lemma holds. □

Proof of Theorem 1.3.2. By (R1), $0 \leq r(\emptyset) \leq |\emptyset| = 0$, so $r(\emptyset) = |\emptyset|$ and $\emptyset \in \mathcal{I}$. Hence $\mathcal{I}$ satisfies (I1). Now suppose that $I \in \mathcal{I}$ and $I' \subseteq I$. Then $r(I) = |I|$. By (R3),

$$r(I' \cup (I - I')) + r(I' \cap (I - I')) \leq r(I') + r(I - I'),$$

that is,

$$r(I) + r(\emptyset) \leq r(I') + r(I - I'). \tag{1}$$

But $r(I) = |I|$ and $r(\emptyset) = 0$. Moreover, by (R2), $r(I') \leq |I'|$ and $r(I - I') \leq |I - I'|$. Therefore, by (1),

$$|I| \leq r(I') + r(I - I') \leq |I'| + |I - I'| = |I|.$$

Since the first and last quantities here are equal, equality must hold throughout. Therefore $r(I') = |I'|$, that is, $I' \in \mathcal{I}$.

To prove that $\mathcal{I}$ satisfies (I3), we assume the contrary, that is, let I_1 and I_2 be members of $\mathcal{I}$ with $|I_1| < |I_2|$ and suppose that, for all e in $I_2 - I_1$, $I_1 \cup e \notin \mathcal{I}$. Then, for all such e, $r(I_1 \cup e) \neq |I_1 \cup e|$. Hence, by (R1), (R2), and the fact that $I_1 \in \mathcal{I}$, we get that, for all such e,

$$|I_1| + 1 > r(I_1 \cup e) \geq r(I_1) = |I_1|,$$

and so,

$$r(I_1 \cup e) = |I_1|.$$

Now applying Lemma 1.3.3 with $X = I_1$ and $Y = I_2$, we immediately get that $r(I_1) = r(I_1 \cup I_2)$. But $|I_1| = r(I_1)$ and $r(I_1 \cup I_2) \geq r(I_2) = |I_2|$, so $|I_1| \geq |I_2|$; a contradiction. We conclude that $\mathcal{I}$ satisfies (I3) and therefore that $(E, \mathcal{I})$ is a matroid.

To complete the proof of the theorem, we need to show that r is the rank function r_M of M. Suppose that $X \subseteq E$. If $X \in \mathcal{I}$, then $r(X) = |X|$ and, as X is a basis of $M|X$, $r_M(X) = |X|$. Thus, if $X \in \mathcal{I}$, then $r(X) = r_M(X)$. Now suppose that $X \notin \mathcal{I}$ and let B be a basis for $M|X$. Then $r_M(X) = |B|$. Moreover, $B \cup x \notin \mathcal{I}$ for all x in $X - B$. Hence $|B| = r(B) \leq r(B \cup x) < |B \cup x|$, so $r(B \cup x) = r(B)$. By Lemma 1.3.3, it follows that $r(B \cup X) = r(B)$, that is, $r(X) = r(B) = |B|$. Thus if $X \notin \mathcal{I}$, then $r(X) = r_M(X)$. We conclude that $r = r_M$. □

The next corollary is obtained by combining the last theorem with Lemma 1.3.1 and the remarks preceding it.

1.3.4 Corollary. *Let E be a set. A function $r : 2^E \to \mathbb{Z}^+ \cup \{0\}$ is the rank function of a matroid on E if and only if r satisfies the following conditions:*

(R1) *If $X \subseteq E$, then $0 \leq r(X) \leq |X|$.*

(R2) *If $X \subseteq Y \subseteq E$, then $r(X) \leq r(Y)$.*

(R3) *If X and Y are subsets of E, then*

$$r(X \cup Y) + r(X \cap Y) \leq r(X) + r(Y).$$ □

Independent sets, bases, and circuits are easily characterized in terms of the rank function. We leave it to the reader to prove the following.

1.3.5 Proposition. *Let M be a matroid with rank function r and suppose that $X \subseteq E(M)$. Then*

(i) *X is independent if and only if $|X| = r(X)$;*

(ii) *X is a basis if and only if $|X| = r(X) = r(M)$; and*

(iii) *X is a circuit if and only if X is non-empty and, for all x in X, $r(X - x) = |X| - 1 = r(X)$.* □

We now specify the rank function explicitly for the classes of uniform, representable, and graphic matroids. In each case, we shall suppose that $X \subseteq E(M)$. If $M = U_{m,n}$, then clearly

$$r(X) = \begin{cases} |X|, & \text{if } |X| < m, \\ m\,, & \text{if } |X| \geq m. \end{cases}$$

If $M = M[A]$ where A is an $m \times n$ matrix over the field F, then, clearly, $r(X)$ is the rank of A_X, the $m \times |X|$ submatrix of A consisting of those columns of A that are labelled by members of X. Equivalently, $r(X)$ equals the dimension of the subspace of $V(m, F)$ that is spanned by the columns of A_X.

Now let $M = M(G)$ where G is a graph. First we determine $r(M)$. Assume initially that G is connected. Then a basis of G is the set of edges of a spanning tree in G. It is well known and can easily be proved by induction that, for a tree T,

$$|V(T)| = |E(T)| + 1.$$

Hence if G is connected, then

1.3.6 $$r(M) = |V(G)| - 1.$$

Extending this, it is clear that, if G has $\omega(G)$ connected components, then

1.3.7 $$r(M) = |V(G)| - \omega(G).$$

It follows that if $X \subseteq E(G)$, then

1.3.8 $$r(X) = |V(G[X])| - \omega(G[X]).$$

1.3.9 Example. Let $M = M(G)$ where G is the graph in Figure 1.4(a). Then, as G is connected, $r(M) = |V(G)| - 1 = 4$. If $X = \{4,5,6,7,8\}$, then a basis for it is $\{4,5,6\}$, so $r(\{4,5,6,7,8\}) = 3$. Equivalently, we can see from Figure 1.4(b) that $|V(G[X])| = 4$ and $\omega(G[X]) = 1$, so, by 1.3.8, $r(X) = 3$. If $Y = \{1,3,7\}$, then $|V(G[Y])| = 4$ and $\omega(G[Y]) = 2$, so $r(Y) = 2$. Finally, if $Z = \{1,2\}$, then $r(Z) = 0$ because $\emptyset$ is a basis for $M|Z$. □

We close this section by introducing another important class of matroids. It is straightforward to show (Exercise 4) that a matroid M is uniform if and only if it has no circuits of size less than $r(M) + 1$. Thus the class of uniform matroids is contained in the class of paving matroids, where a matroid M is *paving* if it has no circuits of size less than $r(M)$. It is not difficult to find examples of paving matroids that are not uniform,

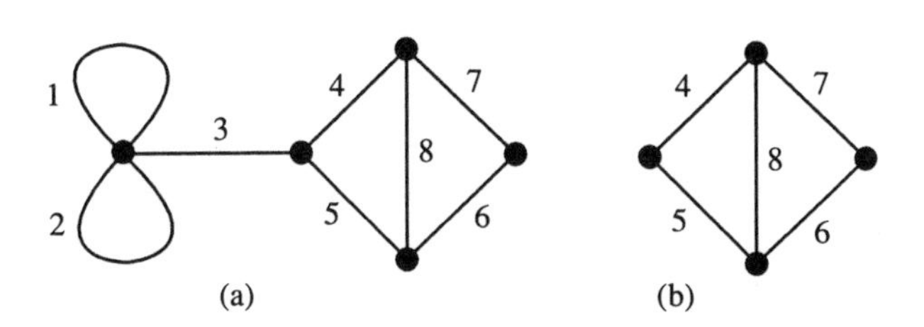

Fig. 1.4. (a) A graph G. (b) An edge-induced subgraph $G[X]$ of G.

for instance $M(K_4)$ is such a matroid. The following result characterizes paving matroids in terms of their collections of circuits. The proof is left to the reader.

1.3.10 Proposition. *Let $\mathcal{D}$ be a collection of non-empty subsets of a set E. Then $\mathcal{D}$ is the set of circuits of a rank-r paving matroid on E if and only if, for some positive integer k, there is a partition $(\mathcal{D}', \mathcal{D}'')$ of $\mathcal{D}$ such that*

(i) *every member of $\mathcal{D}'$ has k elements, and if two distinct members D_1 and D_2 of $\mathcal{D}'$ have $k-1$ common elements, then every k-element subset of $D_1 \cup D_2$ is in $\mathcal{D}'$; and*

(ii) *$\mathcal{D}''$ consists of all of the $(k+1)$-element subsets of E that contain no member of $\mathcal{D}'$.* □

Exercises

1. In terms of the rank function, characterize when an element is a loop of a matroid and when two elements are parallel.

2. **(i)** Find a matroid in which equality does not always hold in (R3).

(ii) Find all matroids for which equality does always hold in (R3).

3. Let M be a matroid on the set E and k be a non-negative integer not exceeding $r(M)$. Define $r_{(k)} : 2^E \to \mathbb{Z}^+ \cup \{0\}$ by $r_{(k)}(X) = \min\{k, r(X)\}$.

(i) Prove that $r_{(k)}$ is the rank function of a matroid on E of rank k. This matroid is obtained from M by a sequence of truncations. The operation of *truncation* will be discussed in Section 7.3.

(ii) Find the collection of independent sets of the matroid in (i).

4. Prove that a matroid M is uniform if and only if it has no circuits of size less than $r(M)+1$.

5. (i) Characterize paving matroids in terms of their collections of independent sets and in terms of their collections of bases.

   (ii) Characterize uniform matroids in terms of their collections of circuits.

   (iii) Modify the statement of 1.3.10 so that it characterizes the set of circuits of a non-uniform rank-r paving matroid.

6. Let M be a matroid on the set E and k be an integer with $r(M) \leq k \leq |E|$. Let $\mathcal{S}_k$ denote the set of k-element subsets of E having rank $r(M)$. Prove that $\mathcal{S}_k$ is the set of bases of a matroid on E. This matroid is called the *elongation of M to rank k* (Welsh 1976) or, when $k = r(M) + 1$, the *Higgs lift* of M (Brylawski 1986a).

7.* (Edmonds and Rota 1966) Let f be a function from 2^E into $\mathbb{Z}^+ \cup \{0\}$ such that

   (a) $f(\emptyset) = 0$;

   (b) if $X \subseteq Y \subseteq E$, then $f(X) \leq f(Y)$; and

   (c) if $X, Y \subseteq E$, then $f(X \cup Y) + f(X \cap Y) \leq f(X) + f(Y)$.

   Let $\mathcal{I}(f) = \{I \subseteq E : f(X) \geq |X| \text{ for all } X \subseteq I\}$. Prove that $(E, \mathcal{I}(f))$ is a matroid, and find its rank function. Matroids of this type will be looked at more closely in Chapter 12.

1.4 Closure

In a vector space V, the vector $\underline{v}$ is in the span of $\{\underline{v}_1, \underline{v}_2, \ldots, \underline{v}_m\}$ if the subspaces spanned by $\{\underline{v}_1, \underline{v}_2, \ldots, \underline{v}_m\}$ and $\{\underline{v}_1, \underline{v}_2, \ldots, \underline{v}_m, \underline{v}\}$ have the same dimension. Now let M be an arbitrary matroid having ground set E and rank function r. Let cl be the function from 2^E into 2^E defined, for all $X \subseteq E$, by

1.4.1 $$\text{cl}(X) = \{x \in E : r(X \cup x) = r(X)\}.$$

This function is called the *closure operator* of M. Thus, in Example 1.3.9, $\text{cl}(\emptyset) = \{1, 2\}$, $\text{cl}(\{1, 3, 5\}) = \{1, 2, 3, 5\}$, and $\text{cl}(\{4, 5, 6\}) = \{1, 2, 4, 5, 6, 7, 8\}$.

One of the main results of this section will characterize matroids in terms of their closure operators. The following is an important preliminary to that result.

1.4.2 Lemma. *The closure operator of a matroid on the set E has the following properties:*

(CL1) *If* $X \subseteq E$, *then* $X \subseteq \mathrm{cl}(X)$.

(CL2) *If* $X \subseteq Y \subseteq E$, *then* $\mathrm{cl}(X) \subseteq \mathrm{cl}(Y)$.

(CL3) *If* $X \subseteq E$, *then* $\mathrm{cl}(\mathrm{cl}(X)) = \mathrm{cl}(X)$.

(CL4) *If* $X \subseteq E$, $x \in E$, *and* $y \in \mathrm{cl}(X \cup x) - \mathrm{cl}(X)$, *then* $x \in \mathrm{cl}(X \cup y)$.

Proof. The fact that the closure operator satisfies (CL1) is immediate from the definition. Now, to show that (CL2) holds, suppose that $X \subseteq Y$ and $x \in \mathrm{cl}(X) - X$. Then $r(X \cup x) = r(X)$. Thus, if B_X is a basis of X, then B_X is a basis of $X \cup x$. Therefore $Y \cup x$ has a basis $B_{Y \cup x}$ that contains B_X but does not contain x. Since $B_{Y \cup x}$ must also be a basis of Y, it follows that $r(Y \cup x) = |B_{Y \cup x}| = r(Y)$ and therefore, $x \in \mathrm{cl}(Y)$. We conclude that $\mathrm{cl}(X) \subseteq \mathrm{cl}(Y)$, that is, (CL2) holds.

To prove (CL3), we first observe that, by (CL1), $\mathrm{cl}(X) \subseteq \mathrm{cl}(\mathrm{cl}(X))$. Now, to establish the reverse inclusion, choose x in $\mathrm{cl}(\mathrm{cl}(X))$. Then $r(\mathrm{cl}(X) \cup x) = r(\mathrm{cl}(X))$. But, for all y in $\mathrm{cl}(X) - X$, $r(X \cup y) = r(X)$. Thus, by Lemma 1.3.3, $r(X) = r(X \cup (\mathrm{cl}(X) - X)) = r(\mathrm{cl}(X))$. Hence $r(\mathrm{cl}(X) \cup x) = r(X)$. But, by (R2), $r(\mathrm{cl}(X) \cup x) \geq r(X \cup x) \geq r(X)$. Therefore equality holds throughout the last statement and so $x \in \mathrm{cl}(X)$. Thus $\mathrm{cl}(\mathrm{cl}(X)) \subseteq \mathrm{cl}(X)$ and (CL3) follows.

The proof of (CL4) will use the following straightforward result.

1.4.3 Lemma. *If* $X \subseteq E$ *and* $x \in E$, *then* $r(X) \leq r(X \cup x) \leq r(X) + 1$.

Proof. Let B_X be a basis of X. Then either B_X or $B_X \cup x$ is a basis of $X \cup x$ and so $r(X \cup x)$ is $r(X)$ or $r(X) + 1$. □

Now suppose $y \in \mathrm{cl}(X \cup x) - \mathrm{cl}(X)$. Then $r(X \cup x \cup y) = r(X \cup x)$ and $r(X \cup y) \neq r(X)$. From the last inequality and Lemma 1.4.3, we deduce that $r(X \cup y) = r(X) + 1$. Thus

$$r(X) + 1 = r(X \cup y) \leq r(X \cup y \cup x) = r(X \cup x) \leq r(X) + 1.$$

Hence $r(X \cup y \cup x) = r(X \cup y)$, so $x \in \mathrm{cl}(X \cup y)$ as required. □

Conditions (CL1)–(CL4) characterize those functions that can be closure operators of matroids. Sometimes (CL4) is referred to as the *Mac Lane–Steinitz exchange property.*

1.4.4 Theorem. *Let E be a set and* cl *be a function from 2^E into 2^E satisfying* (CL1)–(CL4). *Let*

$$\mathcal{I} = \{X \subseteq E : x \notin \mathrm{cl}(X - x) \ \text{ for all } \ x \ \text{ in } \ X\}.$$

Then $(E, \mathcal{I})$ is a matroid having closure operator cl.

Proof. Evidently $\emptyset \in \mathcal{I}$, so $\mathcal{I}$ satisfies (I1). Now suppose $I \in \mathcal{I}$ and $I' \subseteq I$. If $x \in I'$, then $x \in I$ so $x \notin \mathrm{cl}(I-x)$. By (CL2), $\mathrm{cl}(I-x)$ contains $\mathrm{cl}(I'-x)$, so $x \notin \mathrm{cl}(I'-x)$ and therefore $I' \in \mathcal{I}$. Thus $\mathcal{I}$ satisfies (I2).

The rest of the proof will make frequent use of the following result.

1.4.5 Lemma. *Suppose $X \subseteq E$ and $x \in E$. If X is in $\mathcal{I}$, but $X \cup x$ is not, then $x \in \mathrm{cl}(X)$.*

Proof. As $X \cup x \notin \mathcal{I}$, there is an element y of $X \cup x$ such that $y \in \mathrm{cl}((X \cup x)-y)$. If $y = x$, then the lemma holds. If $y \neq x$, then $(X \cup x)-y = (X-y) \cup x$ and $y \in \mathrm{cl}((X-y) \cup x) - \mathrm{cl}(X-y)$. Therefore, by (CL4), $x \in \mathrm{cl}((X-y) \cup y) = \mathrm{cl}(X)$. □

We now prove that $\mathcal{I}$ satisfies (I3). Suppose that I_1 and I_2 are members of $\mathcal{I}$ such that $|I_1| < |I_2|$ and (I3) fails for the pair (I_1, I_2). Assume, moreover, that among all such pairs, $|I_1 \cap I_2|$ is maximal. Choose y in $I_2 - I_1$ and consider $I_2 - y$. Assume that $I_1 \subseteq \mathrm{cl}(I_2 - y)$. Then, by (CL2) and (CL3), $\mathrm{cl}(I_1) \subseteq \mathrm{cl}(I_2 - y)$. As $y \notin \mathrm{cl}(I_2 - y)$, $y \notin \mathrm{cl}(I_1)$. Therefore, by Lemma 1.4.5, $I_1 \cup y \in \mathcal{I}$ and so (I3) holds for the pair (I_1, I_2); a contradiction. We conclude that $I_1 \not\subseteq \mathrm{cl}(I_2 - y)$ and so some element t of I_1 is not in $\mathrm{cl}(I_2 - y)$. Evidently $t \in I_1 - I_2$. Moreover, by Lemma 1.4.5 again, $(I_2 - y) \cup t \in \mathcal{I}$. Since $|I_1 \cap ((I_2 - y) \cup t)| > |I_1 \cap I_2|$, (I3) holds for the pair $(I_1, (I_2 - y) \cup t)$, that is, for some x in $((I_2 - y) \cup t) - I_1$, the set $I_1 \cup x \in \mathcal{I}$. But $x \in I_2 - I_1$, so (I3) holds for (I_1, I_2). This contradiction completes the proof that $(E, \mathcal{I})$ is a matroid M.

We must now check that cl and the closure operator cl_M of M coincide. Suppose first that $x \in \mathrm{cl}_M(X) - X$. Then $r_M(X \cup x) = r_M(X)$. Let B be a basis of X. Then $B \in \mathcal{I}$ and $B \cup x \notin \mathcal{I}$, so, by Lemma 1.4.5, $x \in \mathrm{cl}(B)$. But, by (CL2), $\mathrm{cl}(B) \subseteq \mathrm{cl}(X)$. Therefore $x \in \mathrm{cl}(X)$ and so $\mathrm{cl}_M(X) \subseteq \mathrm{cl}(X)$. To prove the reverse inequality, suppose that x is in $\mathrm{cl}(X) - X$. Let B be a basis for X. Then $B \cup y \notin \mathcal{I}$ for all y in $X - B$, so, by Lemma 1.4.5, $X \subseteq \mathrm{cl}(B)$. Hence $\mathrm{cl}(X) \subseteq \mathrm{cl}(B)$. Since $x \in \mathrm{cl}(X)$, it follows that $x \in \mathrm{cl}(B)$. Thus $B \cup x \notin \mathcal{I}$, so B is a basis for $X \cup x$ and $r_M(X \cup x) = |B| = r_M(X)$. Hence $x \in \mathrm{cl}_M(X)$ and we conclude that $\mathrm{cl}(X) \subseteq \mathrm{cl}_M(X)$ and therefore that $\mathrm{cl}(X) = \mathrm{cl}_M(X)$. □

1.4.6 Corollary. *Let E be a set. A function $\mathrm{cl} : 2^E \to 2^E$ is the closure operator of a matroid on E if and only if it satisfies the following conditions:*

(CL1) *If $X \subseteq E$, then $X \subseteq \mathrm{cl}(X)$.*

(CL2) *If $X \subseteq Y \subseteq E$, then $\mathrm{cl}(X) \subseteq \mathrm{cl}(Y)$.*

(CL3) *If $X \subseteq E$, then $\mathrm{cl}(\mathrm{cl}(X)) = \mathrm{cl}(X)$.*

(CL4) *If* $X \subseteq E$, $x \in E$, *and* $y \in \text{cl}(X \cup x) - \text{cl}(X)$, *then* $x \in \text{cl}(X \cup y)$. □

If M is a matroid and $X \subseteq E(M)$, we call $\text{cl}(X)$ the *closure* or *span* of X in M, and we sometimes write this as $\text{cl}_M(X)$. Some authors denote $\text{cl}(X)$ by $\overline{X}$. If $X = \text{cl}(X)$, then X is called a *flat* or a *closed set* of M. A *hyperplane* of M is a flat of rank $r(M) - 1$. A subset X of $E(M)$ is a *spanning set* of M if $\text{cl}(X) = E(M)$.

1.4.7 Example. Let $M = M(K_5)$ where the edges of K_5 are as labelled in Figure 1.5. Then M has a unique flat of rank 0, namely $\emptyset$; M has ten rank-1 flats, the ten edges of K_5. The rank-2 flats of M are of two types: edge sets of triangles, of which there are ten; and pairs of non-adjacent edges, of which there are fifteen. The rank-3 flats or hyperplanes are also of two types: edge sets of K_4-subgraphs, of which there are five; and edge sets of subgraphs isomorphic to the disjoint union of a K_2 and a K_3, of which there are ten. Finally, the only rank-4 flat is $E(K_5)$. □

The next proposition shows that the bases and hyperplanes are special spanning and non-spanning sets, respectively. To prove this, we shall use the following lemma whose straightforward proof is left as an exercise.

1.4.8 Lemma. *Suppose that* M *is a matroid and* $X \subseteq E(M)$. *If* $x \in \text{cl}(X)$, *then* $\text{cl}(X \cup x) = \text{cl}(X)$. □

1.4.9 Proposition. *Let* M *be a matroid and* X *be a subset of* $E(M)$. *Then*

(i) X *is a spanning set if and only if* $r(X) = r(M)$;

(ii) X *is a basis if and only if it is a minimal spanning set; and*

(iii) X *is a hyperplane if and only if it is a maximal non-spanning set.*

Proof. (i) Suppose that X is a spanning set of M. Then $\text{cl}(X) = E$. Let B be a basis of X. Then, for all x in $X - B$, the set $B \cup x \notin \mathcal{I}(M)$.

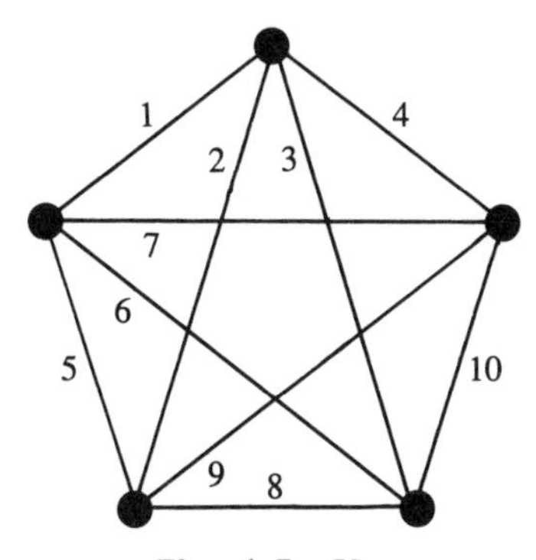

Fig. 1.5. K_5.

Thus, by Lemma 1.4.5, $x \in \mathrm{cl}(B)$. Hence $X \subseteq \mathrm{cl}(B)$ and so $E = \mathrm{cl}(X) \subseteq \mathrm{cl}(B) \subseteq E$. Therefore, for all y in $E - B$, $y \in \mathrm{cl}(B)$, so $B \cup y \notin \mathcal{I}$. Hence B is a basis of M. Thus $r(X) = |B| = r(M)$. The converse is similar and is left as an exercise.

(ii) Suppose that X is a minimal spanning set of M. Then, by (i), $r(X) = r(M)$. Moreover, for all x in X, the set $X - x$ is not spanning and so, by Lemma 1.4.8, $x \notin \mathrm{cl}(X - x)$. Thus $X \in \mathcal{I}(M)$. Since $r(X) = r(M)$, it follows that X is a basis of M.

The proofs of (iii) and the converse of (ii) are left as exercises. □

From this result, we deduce that if G is a connected graph, a subset X of $E(G)$ is spanning in $M(G)$ if and only if X contains the set of edges of a spanning tree of G.

We have already related the closure operator to independent sets, bases, spanning sets, and hyperplanes. Next we relate it to circuits.

1.4.10 Proposition. *Let M be a matroid and X be a subset of $E(M)$. Then*

(i) *X is a circuit if and only if X is a minimal set with the property that, for all x in X, $x \in \mathrm{cl}(X - x)$.*

(ii) $\mathrm{cl}(X) = X \cup \{x : M$ *has a circuit* C *such that* $x \in C \subseteq X \cup x\}$.

Proof. Part (i) just restates the fact that a circuit is a minimal dependent set. To prove (ii), suppose first that $x \in \mathrm{cl}(X) - X$. Then $r(X \cup x) = r(X)$. Hence if B is a basis of X, then $B \cup x$ is dependent. Thus, by Corollary 1.2.6, there is a circuit C such that $x \in C \subseteq B \cup x$. Hence $x \in C \subseteq X \cup x$. On the other hand, if such a circuit exists, then, by (i) and (CL2), $x \in \mathrm{cl}(C - x) \subseteq \mathrm{cl}(X)$, so (ii) holds. □

We now use 1.4.10(ii) to obtain a strengthening of (C3) and thereby a different characterization of matroids in terms of circuits.

1.4.11 Proposition. *Let $\mathcal{C}$ be the set of circuits of a matroid M. Then $\mathcal{C}$ satisfies the following condition:*

(C3)′ *If C_1 and C_2 are members of $\mathcal{C}$, $e \in C_1 \cap C_2$, and $f \in C_1 - C_2$, then there is a member C_3 of $\mathcal{C}$ such that $f \in C_3 \subseteq (C_1 \cup C_2) - e$.*

Proof. As $e \in \mathrm{cl}(C_2 - e)$ and $C_2 - e \subseteq (C_1 \cup C_2) - \{e, f\}$, the element e is in $\mathrm{cl}((C_1 \cup C_2) - \{e, f\})$. Thus, by Lemma 1.4.8, $\mathrm{cl}((C_1 \cup C_2) - \{e, f\}) = \mathrm{cl}((C_1 \cup C_2) - f)$. But $f \in \mathrm{cl}(C_1 - f) \subseteq \mathrm{cl}((C_1 \cup C_2) - f)$, so $f \in \mathrm{cl}((C_1 \cup C_2) - \{e, f\})$. Thus, by Proposition 1.4.10(ii), M has a circuit C such that $f \in C \subseteq (C_1 \cup C_2) - e$, that is, $\mathcal{C}$ satisfies (C3)′. □

Condition (C3)$'$ is known as the *strong circuit elimination axiom.* An alternative proof of it is sketched in Exercise 11 of Section 1.1. On combining Proposition 1.4.11 with Corollary 1.1.5, we deduce the following.

1.4.12 Corollary. *Let $\mathcal{C}$ be a collection of subsets of a set E. Then $\mathcal{C}$ is the set of circuits of a matroid on E if and only if $\mathcal{C}$ satisfies* (C1), (C2), *and* (C3)$'$. □

How can we recognize flats in general and hyperplanes in particular in graphic, representable, and uniform matroids? In $U_{m,n}$, the flats consist of all sets with fewer than m elements together with $E(U_{m,n})$ itself. The hyperplanes are thus the sets with exactly $m-1$ elements. If $M = M[A]$ where A is an $m \times n$ matrix over the field F, then X is a flat of M if and only if there is a subspace W of $V(m,F)$ such that X is the set of labels for the multiset consisting of those columns of A that are members of W. Moreover, although the subspace W can be chosen to have dimension $r(X)$, it is clear that, in general, the intersection of a subspace of $V(m,F)$ with the multiset of columns of A need not span the subspace.

If $M = M(G)$ for some graph G, then, by Proposition 1.4.10(ii), a subset X of $E(G)$ is a flat in M if and only if G has no cycles C such that X contains all but exactly one edge of C. The latter holds if and only if, for some positive integer n, there is a partition $\{V_1, V_2, \ldots, V_n\}$ of $V(G)$ such that X is the union of the edge sets of $G[V_1], G[V_2], \ldots, G[V_n]$. We now characterize the hyperplanes of $M(G)$ in terms of their complements in $E(G)$. As we shall see in Chapter 2, hyperplane complements play an important role in matroid theory.

1.4.13 Proposition. *Let G be a graph. Then H is a hyperplane in $M(G)$ if and only if $E(G) - H$ is a minimal set of edges whose removal from G increases the number of connected components.*

Proof. Let $E(G) - H$ be a minimal set of edges whose removal from G increases the number of connected components. Then an edge e of G that is not in H joins two distinct components G_1 and G_2 of $G[H]$. Moreover, every other edge of $E(G) - H$ must also join G_1 and G_2. Hence $H \cup e$ spans $M(G)$. Thus H is a maximal non-spanning set of $M(G)$, so, by Proposition 1.4.9(iii), H is a hyperplane. The proof of the converse is similar and is left to the reader. □

The final result of this section gives an alternative characterization of the rank function r of a matroid in terms of conditions that are more local than (R1)–(R3). We have already noted in Corollary 1.3.4 and Lemmas 1.3.3 and 1.4.3 that r obeys the three conditions below. We leave the proof of the converse to the reader (Exercise 9).

1.4.14 Theorem. *Let E be a set. A function r from 2^E into the set of non-negative integers is the rank function of a matroid on E if and only if it satisfies the following conditions:*

(R1)′ $r(\emptyset) = 0$.

(R2)′ *If $X \subseteq E$ and $x \in E$, then $r(X) \le r(X \cup x) \le r(X) + 1$.*

(R3)′ *If $X \subseteq E$ and x and y are elements of E such that $r(X \cup x) = r(X \cup y) = r(X)$, then $r(X \cup x \cup y) = r(X)$.* □

Exercises

1. Let M be a matroid and r and cl be its rank function and its closure operator. Prove the following:

   (i) If $X \subseteq \mathrm{cl}(Y)$ and $\mathrm{cl}(Y) \subseteq \mathrm{cl}(X)$, then $\mathrm{cl}(X) = \mathrm{cl}(Y)$.

   (ii) If $Y \subseteq \mathrm{cl}(X)$, then $\mathrm{cl}(X \cup Y) = \mathrm{cl}(X)$.

   (iii) The intersection of all of the flats containing X equals $\mathrm{cl}(X)$.

   (iv) $r(X \cup Y) = r(X \cup \mathrm{cl}(Y)) = r(\mathrm{cl}(X) \cup \mathrm{cl}(Y)) = r(\mathrm{cl}(X \cup Y))$.

   (v) If $X \subseteq Y$ and $r(X) = r(Y)$, then $\mathrm{cl}(X) = \mathrm{cl}(Y)$.

2. Show that a subset X of a matroid is a basis if and only if X is both independent and spanning.

3. **(i)** Find all matroids having a unique basis.

   (ii) Find all matroids having a unique spanning set.

4. Find all simple connected graphs G such that, whenever a simple connected graph G_1 has G as a subgraph, $E(G)$ is a flat of $M(G_1)$.

5. The *nullity* of a set X in a matroid M is $|X| - r_M(X)$. Find two sets of axioms that characterize the nullity function of a matroid, one corresponding to Corollary 1.3.4 and the other to Theorem 1.4.14.

6. Prove that statements (a)–(g) below are equivalent for an element e of a matroid M:

   (a) e is in every basis.

   (b) e is in no circuits.

   (c) If $X \subseteq E(M)$ and $e \in \mathrm{cl}(X)$, then $e \in X$.

   (d) $r(E(M) - e) = r(E(M)) - 1$.

   (e) $E(M) - e$ is a flat.

(f) $E(M) - e$ is a hyperplane.

(g) If I is an independent set, then so is $I \cup e$.

7. Let X and Y be flats of a matroid M such that $Y \subseteq X$ and $r(Y) = r(X) - 1$. Prove that M has a hyperplane H such that $Y = H \cap X$.

8. Let X be a subset of a matroid M. Prove that X is a hyperplane if and only if $E(M) - X$ is a minimal set intersecting every basis.

9. Prove Theorem 1.4.14.

10. Let $\mathcal{S}$ be a collection of subsets of a set E. Prove that $\mathcal{S}$ is the set of spanning sets of a matroid on E if and only if the following conditions hold:

(S1) *$\mathcal{S}$ is non-empty.*

(S2) *If $S_1 \in \mathcal{S}$ and $S_2 \supseteq S_1$, then $S_2 \in \mathcal{S}$.*

(S3) *If $S_1, S_2 \in \mathcal{S}$ and $|S_1| > |S_2|$, then there is an element e in $S_1 - S_2$ such that $S_1 - e \in \mathcal{S}$.*

11. Let $\mathcal{F}$ be a collection of subsets of a set E. Prove that $\mathcal{F}$ is the set of flats of a matroid on E if and only if the following conditions hold:

(F1) *$E \in \mathcal{F}$.*

(F2) *If $F_1, F_2 \in \mathcal{F}$, then $F_1 \cap F_2 \in \mathcal{F}$.*

(F3) *If $F \in \mathcal{F}$ and $\{F_1, F_2, \ldots, F_k\}$ is the set of minimal members of $\mathcal{F}$ that properly contain F, then the sets $F_1 - F$, $F_2 - F, \ldots$, $F_k - F$ partition $E - F$.*

12. (Asche 1966) Let $C_1, C_2, \ldots, C_n$ be distinct circuits of a matroid M such that, for all j in $\{1, 2, \ldots, n\}$, $C_j \subseteq \bigcup_{i \neq j} C_i$. Prove that if $D \subseteq E(M)$ and $|D| < n$, then M has a circuit C such that $C \subseteq (\bigcup_{i=1}^{n} C_i) - D$.

13. (Seymour 1980a) Let $C_1, C_2, \ldots, C_k$ be pairwise disjoint circuits of a matroid M and suppose that, for all i in $\{1, 2, \ldots, k\}$, $x_i \in C_i$. Assume also that M has at least one more circuit.

(i) Prove that M has a circuit C such that $C \cap \{x_1, x_2, \ldots, x_k\} = \emptyset$.

(ii)* Prove that M has a circuit D that is not in $\{C_1, C_2, \ldots, C_k\}$ such that, for all i in $\{1, 2, \ldots, k\}$, either $x_i \in D$, or $D \cap C_i = \emptyset$.

14.* (Brylawski 1973, Greene 1973, Woodall 1974) Let B and B' be bases of a matroid M and $(B_1, B_2, \ldots, B_k)$ be a partition of B. Prove that there is a partition $(B'_1, B'_2, \ldots, B'_k)$ of B' such that $(B - B_i) \cup B'_i$ is a basis for all i in $\{1, 2, ..., k\}$.

1.5 Geometric representations of matroids of small rank

One attractive feature of graphic matroids is that one can determine many properties of such matroids from the pictures of the graphs. In this section we show that all matroids of small rank have a geometric representation that is similarly useful.

We begin our discussion by introducing another class of matroids. A multiset $\{\underline{v}_1, \underline{v}_2, \ldots, \underline{v}_k\}$, the members of which are in $V(m, F)$, is *affinely dependent* if $k \geq 1$ and there are elements $a_1, a_2, \ldots, a_k$ of F, not all zero, such that $\sum_{i=1}^{k} a_i \underline{v}_i = \underline{0}$ and $\sum_{i=1}^{k} a_i = 0$. Equivalently, $\{\underline{v}_1, \underline{v}_2, \ldots, \underline{v}_k\}$ is affinely dependent if the multiset $\{(1, \underline{v}_1), (1, \underline{v}_2), \ldots, (1, \underline{v}_k)\}$ is linearly dependent in $V(m + 1, F)$, where $(1, \underline{v}_i)$ is the $(m + 1)$-tuple of elements of F whose first entry is 1 and whose remaining entries are the entries of $\underline{v}_i$. A multiset of elements from $V(m, F)$ is *affinely independent* if it is not affinely dependent. Clearly an affinely independent multiset must be a set.

1.5.1 Proposition. *Suppose that E is a set that labels a multiset of elements from $V(m, F)$. Let $\mathcal{I}$ be the collection of subsets X of E such that X labels an affinely independent subset of $V(m, F)$. Then $(E, \mathcal{I})$ is a matroid.*

Proof. Suppose that E labels the multiset $\{\underline{v}_1, \underline{v}_2, \ldots, \underline{v}_n\}$. Then, from the second definition of affine dependence, we deduce that $(E, \mathcal{I}) = M[A]$ where A is the $(m + 1) \times n$ matrix over F, the ith column of which is $(1, \underline{v}_i)^T$. Alternatively, one can prove this result directly by using the first definition of affine dependence, and we leave this to the reader (Exercise 2). □

The matroid $(E, \mathcal{I})$ in the last proposition is called the *affine matroid* on E, and if M is isomorphic to such a matroid, we say that M is affine *over* F.

1.5.2 Example. Let E be the subset $\{(0, 0), (1, 0), (2, 0), (0, 1), (0, 2), (1, 1)\}$ of $V(2, \mathbb{R})$ and consider the affine matroid M on E. The six elements of E can be represented as points in the Euclidean plane $\mathbb{R}^2$ as in Figure 1.6. It is not difficult to check that the dependent sets of M consist of all subsets of E with four or more elements together with all 3-element subsets of E such that the corresponding three points in Figure 1.6 are collinear. □

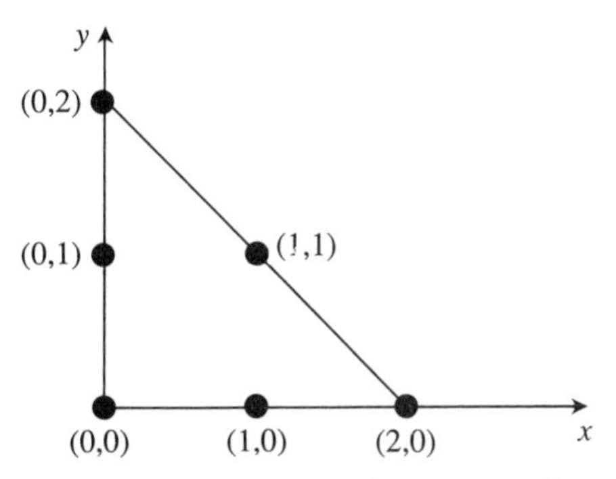

Fig. 1.6. A rank-3 affine matroid.

In general, if M is an affine matroid over $\mathbb{R}$ of rank $m+1$ where $m \leq 3$, then a subset X of $E(M)$ is dependent in M if, in the representation of X by points in $\mathbb{R}^m$, there are two identical points, or three collinear points, or four coplanar points, or five points in space. Hence the flats of M of ranks one, two, and three are represented geometrically by points, lines, and planes, respectively. A typical geometric representation of such an affine matroid is given in the next example.

1.5.3 Example. Consider the affine matroid M on the subset E of $V(3, \mathbb{R})$ where $E = \{(0,0,0), (1,0,0), (0,1,0), (0,0,1), (1,1,0), (0,1,1)\}$. M has the representation shown in Figure 1.7. From that diagram, we see that the only dependent subsets of E with fewer than five elements are the three planes $\{(0,0,0),\ (1,0,0),\ (0,1,0),\ (1,1,0)\}$, $\{(0,0,0),\ (0,1,0),\ (0,0,1),\ (0,1,1)\}$, and $\{(1,0,0),\ (1,1,0),\ (0,0,1),\ (0,1,1)\}$. □

We now have a geometric way to represent real affine matroids of rank at most four. Next we show how to extend the use of this type of diagram to represent arbitrary matroids of rank at most four.

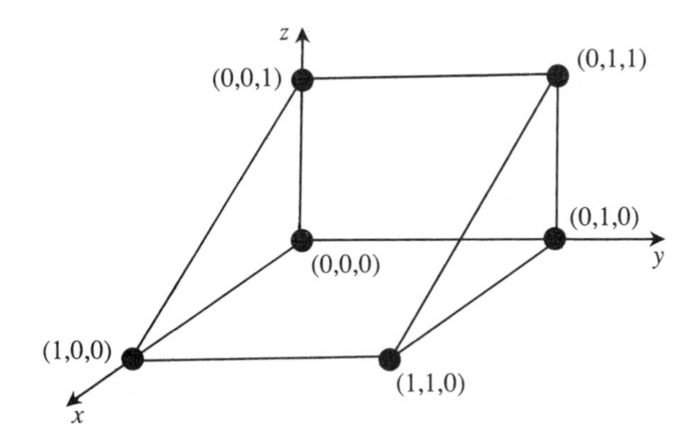

Fig. 1.7. A rank-4 affine matroid.

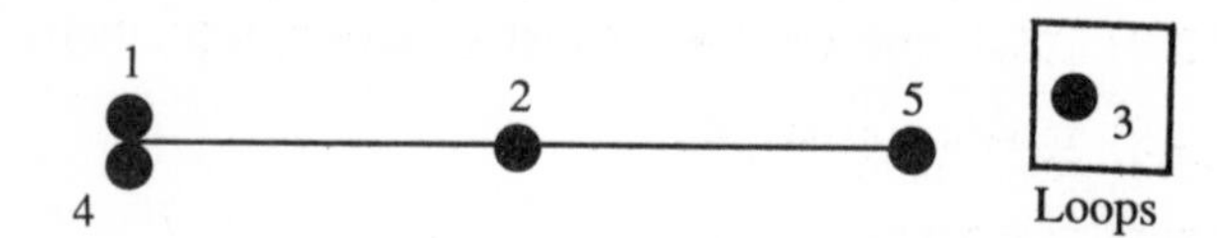

Fig. 1.8. A geometric representation of a rank-2 matroid.

1.5.4 Example. The matroid M in Example 1.1.2 can be represented by the diagram in Figure 1.8. In such a diagram, we represent a 2-element circuit by two touching points, and a 3-element circuit by a line through the corresponding points. Loops, which cannot occur in an affine matroid, are represented in an inset as shown. □

In general, such diagrams are governed by the following rules. All loops are marked in a single inset. Parallel elements are represented by touching points, or sometimes by a single point labelled by all the elements in the parallel class. Corresponding to each element that is not a loop and is not in a non-trivial parallel class, there is a distinct point in the diagram which touches no other points. If three elements form a circuit, the corresponding points are collinear. Likewise, if four elements form a circuit, the corresponding points are coplanar. In such a diagram, the lines need not be straight and the planes may be twisted. Moreover, sometimes, to simplify the diagram, certain lines and planes will be listed rather than drawn. At other times, certain lines with fewer than three points on them will be marked as part of the indication of a plane, or as construction lines. We call such a diagram a *geometric representation* for the matroid. The reader is warned that such a representation is not to be confused with the diagram of a graph. Where ambiguity could arise in what follows, we shall always indicate how a particular diagram is to be interpreted.

One needs to be careful not to assume that an arbitrary diagram involving points, lines, and planes is actually a geometric representation for some matroid.

1.5.5 Example. 'The Escher matroid' (Brylawski and Kelly 1980). Consider the diagram shown in Figure 1.9, the dependent lines being $\{1, 2, 3\}$

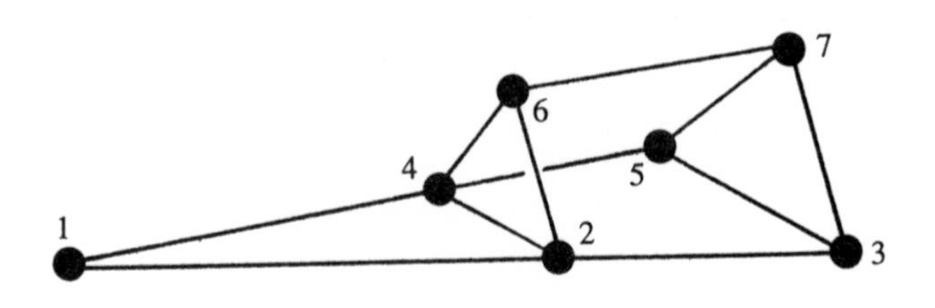

Fig. 1.9. The Escher matroid.

and $\{1,4,5\}$, and the dependent planes $\{1,2,3,4,5\}, \{1,2,3,6,7\}$, and $\{1,4,5,6,7\}$. With the rules that govern diagrams being as specified above, this diagram does not represent a matroid on $E = \{1,2,\ldots,7\}$. To see this, assume the contrary and let $X = \{1,2,3,6,7\}$ and $Y = \{1,4,5,6,7\}$. Then $r(X) = 3 = r(Y)$ and $r(X \cup Y) = 4$. Thus, by (R3), $r(\{1,6,7\}) = r(X \cap Y) \leq 2$. But 1, 6, and 7 are distinct non-collinear points, so $r(\{1,6,7\}) = 3$; a contradiction. If we make 1, 6, and 7 collinear as in Figure 1.10, the resulting diagram does represent a rank-4 matroid. We leave it to the reader to check this. □

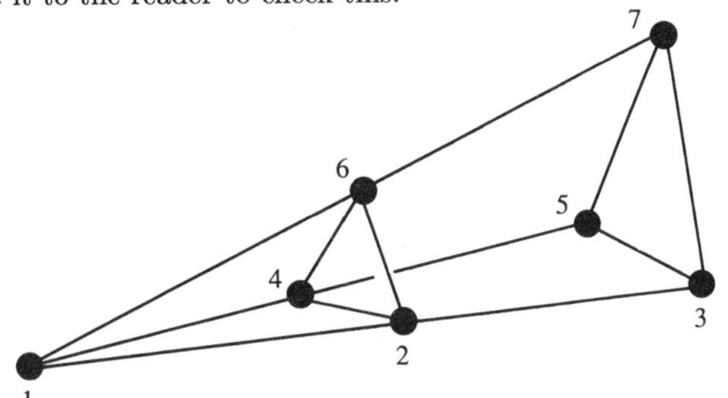

Fig. 1.10. The Escher matroid should have an extra line.

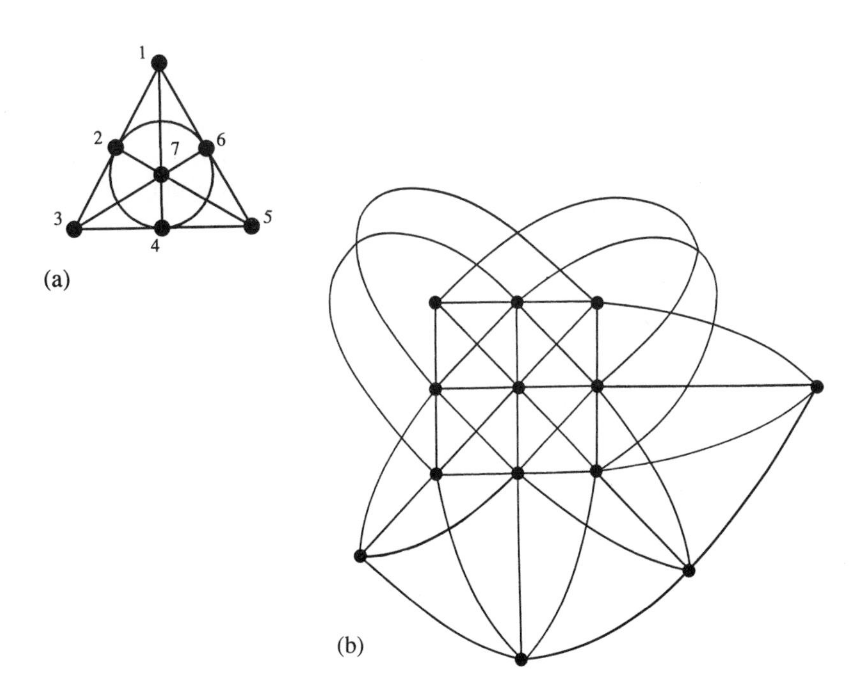

Fig. 1.11. (a) $PG(2,2)$ and (b) $PG(2,3)$.

1.5.6 **Example.** The pictures shown in Figure 1.11 are representations of the 7-point and 13-point projective planes, $PG(2,2)$ and $PG(2,3)$. Interpreting these pictures as diagrams subject to the above rules, it is not difficult to check that each represents a rank-3 matroid. The 7-point projective plane is called the *Fano plane*. The corresponding matroid, the *Fano matroid*, will be denoted by F_7 or $PG(2,2)$. This matroid is of fundamental importance and will occur frequently throughout this book. Indeed, all projective geometries play an important role in matroid theory; we shall discuss this in detail in Chapter 6.

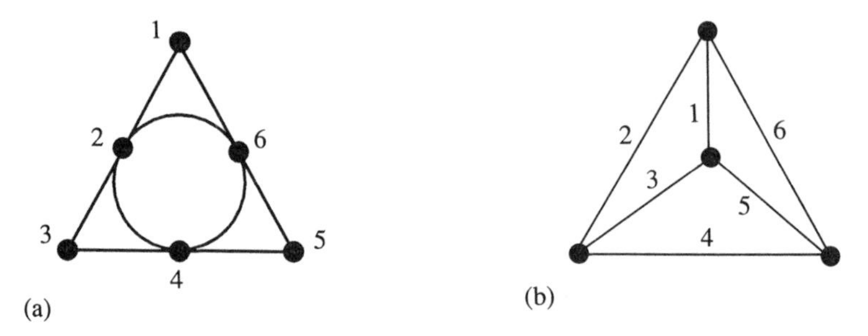

Fig. 1.12. (a) $F_7\backslash 7$. (b) A graph whose cycle matroid is $F_7\backslash 7$.

The diagram in Figure 1.12(a) represents the matroid that is obtained from F_7 by deleting the element 7. Notice that no line has been drawn through 3 and 6, or through 2 and 5, or through 1 and 4, even though $\{3,6\}$, $\{2,5\}$, and $\{1,4\}$ are rank-2 flats of the matroid. Such 2-point lines are usually omitted from these diagrams, as are 3-point planes.

It is not difficult to check that $F_7\backslash 7 \cong M(K_4)$, where the edges of the graph K_4 are labelled as in Figure 1.12(b). Similarly, $F_7\backslash 2$, for which a geometric representation is shown in Figure 1.13, is also isomorphic to $M(K_4)$. Indeed, as the reader can easily check, $F_7\backslash x \cong M(K_4)$ for all x in $E(F_7)$. This is one of the many attractive features of F_7. □

1.5.7 **Example.** Consider the *affine* matroid on the full vector space $V(3,2)$. We denote this matroid by $AG(3,2)$. It has eight elements, correspond-

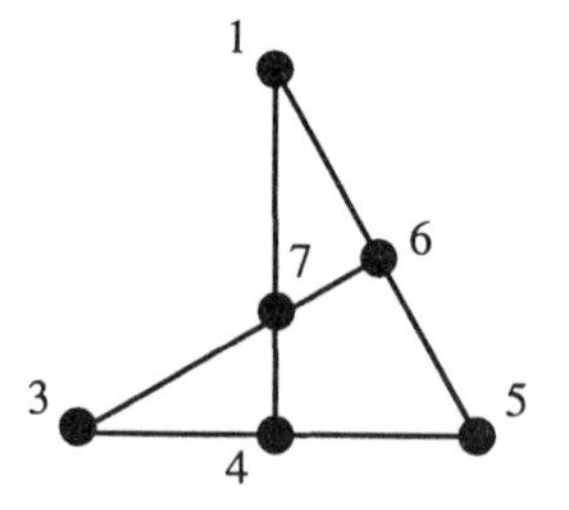

Fig. 1.13. $F_7\backslash 2$.

ing to the eight points in Figure 1.14(a). In addition, it has fourteen 4-point planes, not all of which are marked in Figure 1.14(a). These planes consist of the six faces of the cube, the six diagonal planes such as $\{(0,0,0),(1,0,0),(1,1,1),(0,1,1)\}$, and the two twisted planes, $\{(0,0,0),(1,0,1),(1,1,0),(0,1,1)\}$ and its complement. Note that if Figure 1.14(a) is viewed as an affine matroid over $\mathbb{R}$ instead of over $GF(2)$, then we get twelve rather than fourteen planes, the two twisted planes disappearing.

An alternative representation for $AG(3,2)$ can be obtained from the 11-point matroid shown in Figure 1.14(b). This 11-point matroid is obtained by 'sticking together' two copies of F_7 along a line. The restriction of this matroid to the set $\{1,2,\ldots,8\}$ is isomorphic to the binary affine cube in Figure 1.14(a). We leave it to the reader to check the details of this. To see the fourteen 4-point planes in the second representation for $AG(3,2)$, we first note that $\{1,2,3,4\}$ and $\{5,6,7,8\}$ are two of the fourteen. The other twelve break into three groups of four according to whether the corresponding planes in the original 11-point matroid contain a, b, or c. For example, the four planes containing a arise by taking the union of two lines containing a, one from each copy of F_7, and neither equal to $\{a,b,c\}$. □

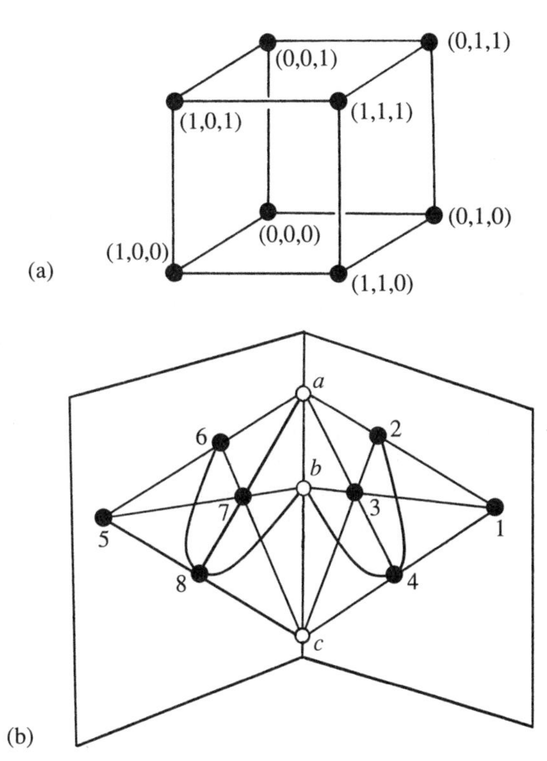

Fig. 1.14. Two geometric representations of $AG(3,2)$.

We have already seen in Example 1.5.5 that a diagram involving points, lines, and planes need not correspond to a matroid. Next we state necessary and sufficient conditions under which such a diagram is actually a geometric representation for a simple matroid of rank at most four in which the rank-1, rank-2, and rank-3 flats correspond to the points, lines, and planes in the diagram. These rules are stated just for simple matroids because we already know how to recognize loops and parallel elements in such a diagram. The rules, all of which are very natural from our current geometric perspective, include the following straightforward non-degeneracy conditions: the sets of points, lines, and planes are disjoint; there are no sets of touching points; every line contains at least two points; any two distinct points lie on a line; every plane contains at least three non-collinear points; and any three distinct non-collinear points lie on a plane. For a diagram having at most one plane, there is only one other condition:

1.5.8 *Any two distinct lines meet in at most one point.*

For a diagram having two or more planes, there are three rules in addition to the non-degeneracy conditions (Mason 1971):

1.5.9 *Any two distinct planes meeting in more than two points do so in a line.*

1.5.10 *Any two distinct lines meeting in a point do so in at most one point and lie on a common plane.*

1.5.11 *Any line not lying on a plane intersects it in at most one point.*

We leave the proofs of these results to the reader (Exercises 3 and 4). Using geometric representations, the reader should be able to check that there are 17 non-isomorphic matroids on a 4-set, and 35 non-isomorphic matroids on a 5-set.

1.5.12 **Example.** The diagram in Figure 1.15 obeys 1.5.8 and is therefore a geometric representation for a matroid N. Comparing Figure 1.15 with the geometric representation for F_7 in Figure 1.11(a), we see that $\{2,4,6\}$ is both a circuit and a hyperplane in F_7, whereas, in N, this set is a

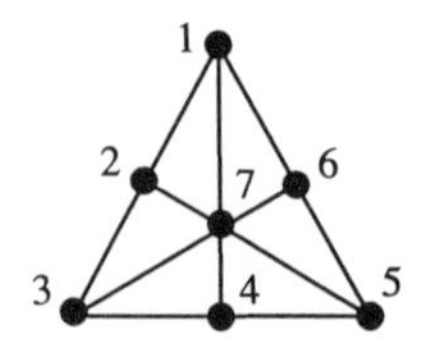

Fig. 1.15. F_7^-.

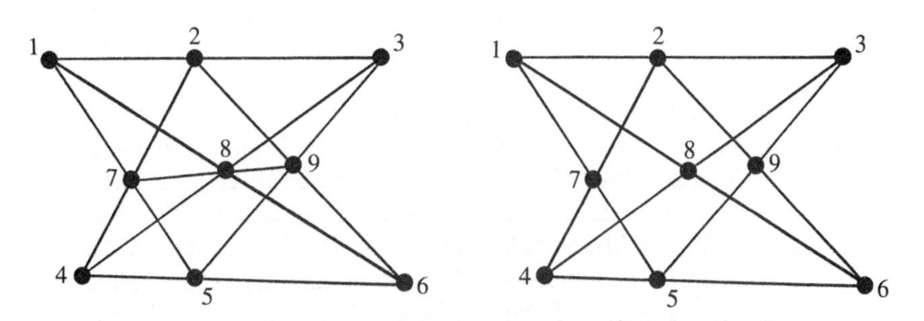

Fig. 1.16. The Pappus and non-Pappus matroids.

basis. We say that N has been obtained from F_7 by *relaxing* the circuit–hyperplane $\{2,4,6\}$. This operation can be performed on matroids in general. □

1.5.13 Proposition. *Let M be a matroid having a subset X that is both a circuit and a hyperplane. Let $\mathcal{B}' = \mathcal{B}(M) \cup \{X\}$. Then $\mathcal{B}'$ is the set of bases of a matroid M' on $E(M)$. Moreover,*

$$\mathcal{C}(M') = (\mathcal{C}(M) - \{X\}) \cup \{X \cup e : e \in E(M) - X\}.$$

Proof. We leave this as an exercise. □

The matroid M' in the last proposition is called a *relaxation* of M. Thus the matroid N in Figure 1.15 is a relaxation of F_7. We call N the *non-Fano matroid* and denote it by F_7^-.

1.5.14 Example. The diagrams in Figure 1.16 obey 1.5.8 and are therefore geometric representations for rank-3 matroids. We call these matroids the *Pappus* and *non-Pappus matroids*, respectively, because of their relationship to the well-known Pappus configuration in projective geometry. As we shall see in Chapter 6, the non-Pappus matroid is not representable over any field. A smallest matroid with this property can be obtained from $AG(3,2)$ by relaxing a circuit–hyperplane. The proof that this matroid is non-representable will also be delayed until Chapter 6. □

1.5.15 Example. Figure 1.17 contains another familiar object from projective geometry, the 3-dimensional Desargues configuration. One can check that the points, lines, and planes of this diagram obey 1.5.9–1.5.11 so that this diagram is indeed a geometric representation for a 10-element rank-4 matroid. Alternatively, one can show that this diagram is the geometric representation for $M(K_5)$ with the edges of K_5 being labelled as in Figure 1.5. □

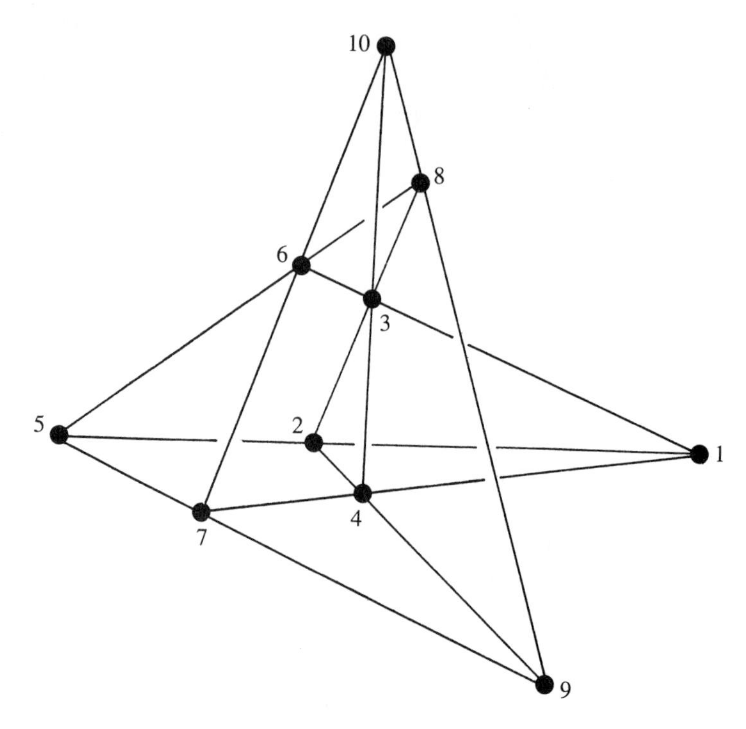

Fig. 1.17. $M(K_5)$.

In light of the geometric representations discussed in this section, it should not be surprising that the terms *point*, *line*, and *plane* are often used in an arbitrary matroid to refer to flats of ranks one, two, and three, respectively.

Exercises

1. Determine which of conditions 1.5.9–1.5.11 is violated by the diagram in Figure 1.9.

2. Give a direct proof of Proposition 1.5.1 by using the first definition of affine dependence.

3. Let D be a diagram involving points and lines in the plane and satisfying 1.5.8 as well as the non-degeneracy conditions stated before it. Prove that there is a simple matroid of rank at most three on the set of points of D whose rank-1 and rank-2 flats are the points and lines, respectively, of D.

4. Let D be a diagram involving points, lines, and planes and satisfying 1.5.9–1.5.11 as well as the non-degeneracy conditions. Prove that there is a simple matroid of rank at most four on the set of points of D that has as its rank-1, rank-2, and rank-3 flats the points, lines, and planes, respectively, of D.

5. Show that neither the Fano nor the non-Fano matroid is graphic.

6. Prove that an affine matroid over $GF(2)$ has no circuits with an odd number of elements.

7. Prove that every relaxation of F_7 is isomorphic to F_7^-.

8. Let M be a matroid and X be a circuit-hyperplane of M. Let M' be the matroid obtained from M by relaxing X. Find, in terms of M, the independent sets, the rank function, the hyperplanes, and the flats of M'.

9. Prove that the following statements are equivalent for a rank-r matroid M.

   (a) M is a relaxation of some matroid.

   (b) M has a basis B such that $C(e, B) = B \cup e$ for every e in $E(M) - B$ and neither B nor $E(M) - B$ is empty.

   (c) M has a non-empty basis B such that $B \neq E(M)$ and every $(r-1)$-element subset of B is a flat.

10. The matroids M_1 and M_2 for which geometric representations are shown in Figure 1.18 are graphic. For $i = 1, 2$, find a graph G_i for which $M(G_i) \cong M_i$.

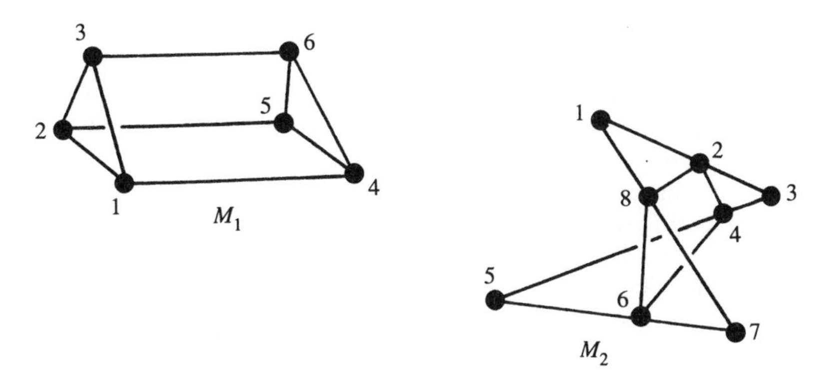

Fig. 1.18. Geometric representations of two graphic matroids.

11. Find geometric representations for the vector matroids of the following matrices where I_r is the $r \times r$ identity matrix, and each matrix is over $GF(3)$ except the first which is over $GF(2)$.

$$A_1 = \left[\begin{array}{c|ccccc} & 1 & 0 & 0 & 1 & 1 \\ I_4 & 1 & 1 & 0 & 0 & 1 \\ & 0 & 1 & 1 & 0 & 1 \\ & 0 & 0 & 1 & 1 & 1 \end{array}\right] \qquad A_2 = \left[\begin{array}{c|cccc} & 1 & 0 & 0 & 1 \\ I_4 & 1 & 1 & 0 & 0 \\ & 0 & 1 & 1 & 0 \\ & 0 & 0 & 1 & 1 \end{array}\right]$$

$$A_3 = \left[\begin{array}{c|cccc} & 1 & 0 & 0 & 1 \\ I_4 & 1 & 1 & 0 & 0 \\ & 0 & 1 & 1 & 0 \\ & 0 & 0 & 1 & -1 \end{array}\right] \qquad A_4 = \left[\begin{array}{c|cccc} & 0 & 1 & 1 & 1 \\ I_3 & 1 & 0 & 1 & -1 \\ & 1 & 1 & 0 & -1 \end{array}\right]$$

1.6 Transversal matroids

So far we have considered three important classes of matroids: graphic, representable, and uniform matroids. In this section, we consider a fourth such class, that of transversal matroids.

Let S be a finite set. A *family* of subsets of S is a finite sequence $(A_1, A_2, \ldots, A_m)$ such that, for all j in $\{1, 2, \ldots, m\}$, $A_j \subseteq S$. Note that the terms of this sequence, the *members* of the family, need not be distinct. If $J = \{1, 2, \ldots, m\}$, we shall frequently abbreviate $(A_1, A_2, \ldots, A_m)$ as $(A_j : j \in J)$. A *transversal* or *system of distinct representatives* of $(A_1, A_2, \ldots, A_m)$ is a subset $\{e_1, e_2, \ldots, e_m\}$ of S such that $e_j \in A_j$ for all j in J. More formally, T is a transversal of $(A_j : j \in J)$ if there is a bijection $\psi : J \to T$ such that $\psi(j) \in A_j$ for all j in J. If $X \subseteq S$, then X is a *partial transversal* of $(A_j : j \in J)$ if, for some subset K of J, X is a transversal of $(A_j : j \in K)$.

Another way to view partial transversals uses the idea of a matching in a bipartite graph. If $\mathcal{A}$ is the family $(A_1, A_2, \ldots, A_m)$ of subsets of a set S and $J = \{1, 2, \ldots, m\}$, then the *bipartite graph* $\Delta[\mathcal{A}]$ *associated with* $\mathcal{A}$ has vertex set $S \cup J$; its edge set is $\{xj : x \in S,\ j \in J, \text{ and } x \in A_j\}$. A *matching* in a graph is a set of edges in the graph no two of which have a common endpoint. It is not difficult to see that a subset X of S is a partial transversal of $\mathcal{A}$ if and only if there is a matching in $\Delta[\mathcal{A}]$ in which every edge has one endpoint in X. When such a matching exists, we say that X is matched *into* J. Moreover, if J_X consists of those vertices of J that meet an edge of the matching, we say that X is matched *onto* J_X.

1.6.1 Example. Let $S = \{x_1, x_2, \ldots, x_6\}$, $A_1 = \{x_1, x_2, x_6\}$, $A_2 = \{x_3, x_4, x_5, x_6\}$, $A_3 = \{x_2, x_3\}$, and $A_4 = \{x_2, x_4, x_6\}$. Then, for $\mathcal{A} = (A_1, A_2, A_3, A_4)$, the bipartite graph $\Delta[\mathcal{A}]$ is as shown in Figure 1.19.

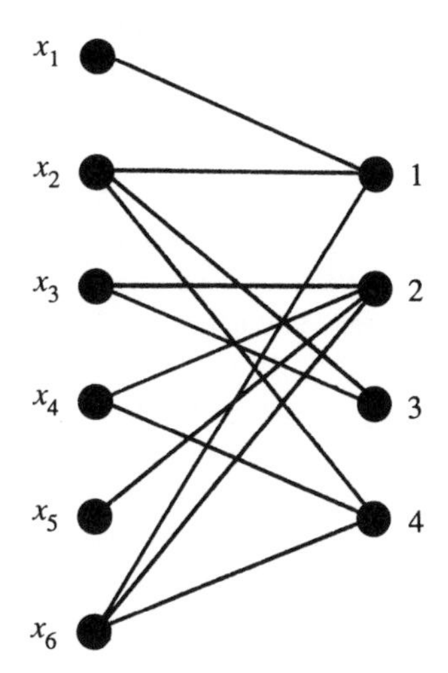

Fig. 1.19. The bipartite graph associated with a family of sets.

The set $\{x_1, x_2, x_3, x_4\}$ is a transversal of $\mathcal{A}$. To check this, one needs only check that $\{x_11, x_42, x_33, x_24\}$ is a matching in $\Delta[\mathcal{A}]$. Similarly, as $\{x_61, x_23, x_42\}$ is a matching in $\Delta[\mathcal{A}]$, $\{x_6, x_2, x_4\}$ is a partial transversal of $\mathcal{A}$. Evidently $\mathcal{A}$ has many other partial transversals. □

In solving scheduling or timetabling problems, one is led naturally to the consideration of partial transversals of families of sets. As we shall see in Section 1.8, the next result of Edmonds and Fulkerson (1965) provides a useful tool in the study of such problems.

1.6.2 Theorem. *Let $\mathcal{A}$ be a family $(A_1, A_2, \ldots, A_m)$ of subsets of a set S. Let $\mathcal{I}$ be the set of partial transversals of $\mathcal{A}$. Then $\mathcal{I}$ is the collection of independent sets of a matroid on S.*

Proof. The empty set is a transversal of the empty subfamily of $\mathcal{A}$, so (I1) holds. Moreover, if I is a partial transversal of $\mathcal{A}$ and $I' \subseteq I$, then I' is also a partial transversal of $\mathcal{A}$. Thus (I2) holds.

Now, to prove that $\mathcal{I}$ satisfies (I3), suppose that I_1 and I_2 are partial transversals of $\mathcal{A}$ such that $|I_1| < |I_2|$. Then, in $\Delta[\mathcal{A}]$, there are matchings W_1 and W_2 that match I_1 and I_2, respectively, into J. Colour the edges of $W_1 - W_2$, $W_2 - W_1$, and $W_1 \cap W_2$ red, blue, and purple, respectively, and let W be the subgraph of $\Delta[\mathcal{A}]$ induced by those edges that are red or blue. Since $|I_1| = |W_1|$ and $|I_2| = |W_2|$, there are more blue edges than red in W.

Because W_1 and W_2 are both matchings, every vertex in W has degree one or two. It is routine to check (see Exercise 2) that every connected component of W is a cycle or a path. Moreover, as W is bipartite, every cycle is even. As no two like-coloured edges meet at a vertex, there are equal numbers of red and blue edges in every cycle of W and in every even path. Since W has more blue edges than red, it must have as a component an odd path, P, whose first and last edges are blue. Let the vertices of P, in order, be $v_1, v_2, \ldots, v_{2k}$. Clearly one of v_1 and v_{2k} is in S and the other is in J. Assume, without loss of generality, that $v_1 \in S$. Then, since v_1 meets a blue but no red edge, $v_1 \in I_2 - I_1$. Moreover, $\{v_2, v_4, \ldots, v_{2k}\} \subseteq J$ and $\{v_3, v_5, \ldots, v_{2k-1}\} \subseteq I_1 \cap I_2$. Now interchange the colours on the red and blue edges of P leaving the rest of $\Delta[\mathcal{A}]$ unchanged. In the recoloured graph, there is now one more red edge than before. Indeed, every vertex in $I_1 \cup v_1$ is the endpoint of a red or a purple edge. Moreover, this set of red and purple edges forms a matching. We conclude that $I_1 \cup v_1$ is a partial transversal of $\mathcal{A}$. Hence (I3) holds and $\mathcal{I}$ is indeed the set of independent sets of a matroid. □

The matroid obtained above from the set of partial transversals of $\mathcal{A}$ will be denoted by $M[\mathcal{A}]$. Its ground set was denoted above by S to avoid the potential confusion of using E for a set of vertices in a graph. If M is an arbitrary matroid and $M \cong M[\mathcal{A}]$ for some family $\mathcal{A}$ of sets, then we shall call M a *transversal matroid* and $\mathcal{A}$ a *presentation* of M. It is not difficult to check that all uniform matroids are transversal (Exercise 1). Moreover, as we shall see in Chapter 12, every transversal matroid is representable over all sufficiently large fields. In the next example, we exhibit a graphic transversal matroid and a graphic non-transversal matroid. A characterization of precisely which graphic matroids are transversal will be presented in Chapter 13.

1.6.3 Example. Let G_1 and G_2 be the graphs shown in Figure 1.20. Let $A_1 = \{1,2,7\}$, $A_2 = \{3,4,7\}$, and $A_3 = \{5,6,7\}$. Then, for $\mathcal{A} = (A_1, A_2, A_3)$ and $S = \{1,2,\ldots,7\}$, it is routine to check that $M[\mathcal{A}] = M(G_1)$.

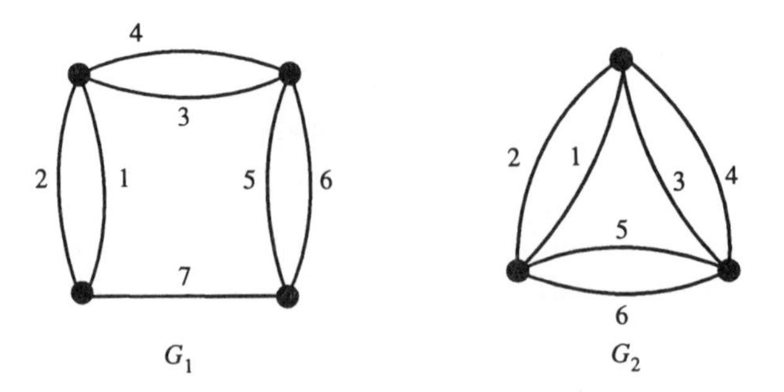

Fig. 1.20. $M(G_1)$ is transversal; $M(G_2)$ is not.

In contrast, $M(G_2)$ is not transversal. To see this, assume the contrary, that is, suppose that $M(G_2) = M[\mathcal{A}']$ for some family $\mathcal{A}'$ of subsets of $\{1, 2, \ldots, 6\}$. As $\{1\}$ and $\{2\}$ are independent but $\{1, 2\}$ is dependent, there is a unique member, say A'_1, of $\mathcal{A}'$ meeting $\{1, 2\}$. Moreover, A'_1 contains both 1 and 2. Similarly, $\mathcal{A}'$ has a unique member A'_2 meeting $\{3, 4\}$ and a unique member A'_3 meeting $\{5, 6\}$, and these members contain $\{3, 4\}$ and $\{5, 6\}$, respectively. As $\{1, 3\}$, $\{1, 5\}$, and $\{3, 5\}$ must be partial transversals of $\mathcal{A}'$, the sets A'_1, A'_2, and A'_3 are distinct. This implies that $\{1, 3, 5\}$ is a partial transversal of $\mathcal{A}'$; a contradiction. We conclude that $M(G_2)$ is indeed non-transversal. □

It will sometimes be convenient to view transversal matroids as coming directly from bipartite graphs. It follows from our earlier discussion that if Δ is a bipartite graph with vertex classes S and J, then the set of subsets X of S that are matched into J is precisely the set of independent sets of a transversal matroid. Indeed, this matroid is $M[\mathcal{A}]$ where $\mathcal{A}$ is *the family* $(A_j : j \in J)$ *of sets associated with* Δ, that is, $A_j = \{x \in S : xj \in E(\Delta)\}$ for all j in J.

Exercises

1. Show the following:

- **(i)** All uniform matroids are transversal.
- **(ii)** A transversal matroid need not be graphic.
- **(iii)** A paving matroid need not be transversal.

2. Prove the following for a graph G.

- **(i)** If every vertex of G has degree at most two, then G is a disjoint union of paths and cycles.
- **(ii)** If every vertex of G has degree at least two, then G has a cycle.
- **(iii)** If every vertex of G has degree exactly two, then G is a disjoint union of cycles.

3. Let $S = \{1, 2, \ldots, 6\}$ and $\mathcal{A} = (A_1, A_2, A_3)$ where $A_1 = \{1, 2, 3\}$, $A_2 = \{2, 3, 4\}$, and $A_3 = \{4, 5, 6\}$.

- **(i)** Find $\Delta[\mathcal{A}]$.
- **(ii)** Give a geometric representation for $M[\mathcal{A}]$.

4. Characterize the circuits of $M[\mathcal{A}]$ in terms of the bipartite graph $\Delta[\mathcal{A}]$.

5. Find a restriction of F_7 that is non-transversal and is minimal with this property.

6. Give a family $\mathcal{A}$ of subsets of $\{1, 2, \ldots, 7\}$ such that Figure 1.21 is a geometric representation for $M[\mathcal{A}]$.

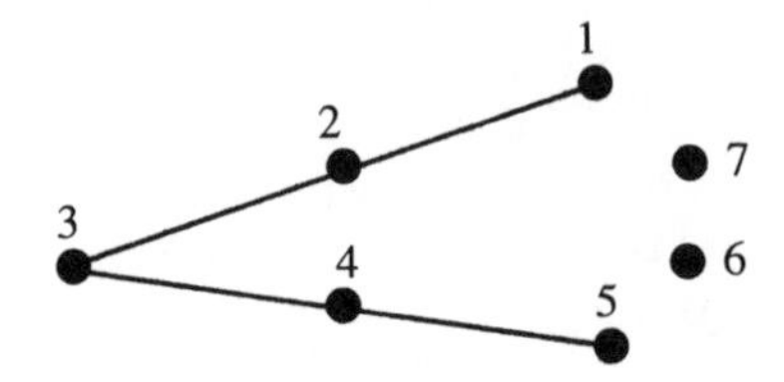

Fig. 1.21. A geometric representation of a transversal matroid.

7. Let n be an integer exceeding one. Give an example of two families $\mathcal{A}_1$ and $\mathcal{A}_2$ of distinct non-empty subsets of $\{1, 2, \ldots, n\}$ such that $M[\mathcal{A}_1] = M[\mathcal{A}_2]$, and $\mathcal{A}_2$ cannot be obtained by permuting the members of $\mathcal{A}_1$.

8. Let $\mathcal{A} = (A_j : j \in J)$ be a family of subsets of a set S. Prove that $S - A_j$ is a flat of $M[\mathcal{A}]$ for all j in J.

9. Let Δ be a bipartite graph with vertex classes S and J. Show that:

(i) A matching in Δ need not be contained in a maximum-sized matching.

(ii) If $\mathcal{A}$ is the family of sets associated with Δ, then every partial transversal of $\mathcal{A}$ is contained in a maximum-sized partial transversal of $\mathcal{A}$.

10. Show that both the matroids for which geometric representations are shown in Figure 1.22 are ternary, non-binary, non-graphic, and non-transversal.

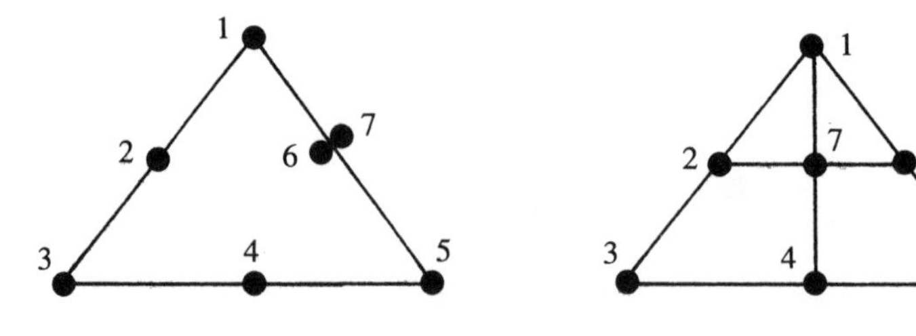

Fig. 1.22.

11. (Hoffman and Kuhn 1956) Let $\mathcal{A}$ be a family of subsets of a set S and suppose $X \subseteq S$. Prove that $\mathcal{A}$ has a transversal containing X if and only if $\mathcal{A}$ has some transversal and X is a partial transversal of $\mathcal{A}$.

12*. (Welsh 1969a) Define a presentation $(A_1, A_2, \ldots, A_m)$ of a transversal matroid to be *nested* if $A_1 \subseteq A_2 \subseteq \ldots \subseteq A_m$. Prove that, for a set S, there are exactly $2^{|S|}$ non-isomorphic transversal matroids on S having nested presentations.

1.7 The lattice of flats

The set of flats of a matroid, ordered by inclusion, has a special structure which will be characterized in this section. This approach to matroids has proved attractive to a number of authors (see, for example, Birkhoff 1935, Dilworth 1944, and Crapo and Rota 1970). It will not be our primary focus here, although we shall find it indispensable in certain contexts. After seeing the notion of the simple matroid associated with a given matroid, the reader unfamiliar with partially ordered sets may wish to skim the rest of this section, delaying detailed consideration of it until its techniques are first used in Chapter 6. This material has been included in this chapter both for reference and because of its historical importance in the development of matroid theory.

One feature that will emerge from our study of the order structure of flats is that this structure is unaffected by the presence of loops and parallel elements. We begin by looking at an elementary construction that removes loops and parallel elements from a matroid. This generalizes a graph-theoretic construction. The simple graph $\widetilde{G}$ associated with a given graph G is obtained from G by deleting all loops and, for each parallel class of edges, identifying all the edges of the class as a single edge. An example of this construction is given in Figure 1.23. Before giving a precise description of the corresponding matroid construction, we make the following elementary observations for an arbitrary matroid M. For the first of these, see Exercise 3 of Section 1.2.

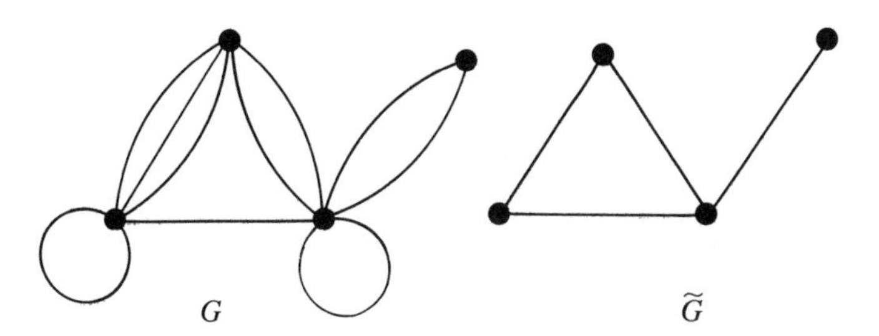

Fig. 1.23. A graph and its associated simple graph.

1.7.1 *The element e is a loop of M if and only if e is in no bases of M.*

1.7.2 *If e_1 is not a loop of M, then e_2 is parallel with e_1 if and only if, for every basis B containing e_1, the set $(B - e_1) \cup e_2$ is also a basis.*

Thus no basis intersects the set of loops of M. Now let X be a non-trivial parallel class of M and, for each x in X, let $\mathcal{B}_x$ denote the set of bases of M containing x. If x and y are distinct members of X, then $\mathcal{B}_x \cap \mathcal{B}_y = \emptyset$, but $\{B - x : B \in B_x\} = \{B - y : B \in \mathcal{B}_y\}$. Therefore, if we delete all the loops from M and then, for each non-trivial parallel class X, delete all but one distinguished element of X, the matroid we obtain is uniquely determined up to a renaming of the distinguished elements. This matroid will be denoted by $\widetilde{M}$ and will be called the *simple matroid associated with* M. Formally, the ground set of $\widetilde{M}$ is the set of all parallel classes of M, while a subset $\{X_1, X_2, \ldots, X_k\}$ of these parallel classes is in $\mathcal{I}(\widetilde{M})$ if and only if $r_M(X_1 \cup X_2 \cup \ldots \cup X_k) = k$. Evidently, for a graph G whose associated simple graph is $\widetilde{G}$, we have $\widetilde{M(G)} = M(\widetilde{G})$.

An equivalent way to view $\widetilde{M}$ is to take its ground set to be the set of rank-1 flats of M with a set $\{Y_1, Y_2, \ldots, Y_k\}$ of these flats being independent in $\widetilde{M}$ if and only if $r_M(Y_1 \cup Y_2 \cup \ldots \cup Y_k) = k$.

We now examine more closely the structure of the set of flats of a matroid. We shall need some more terminology. A *partially ordered set* or *poset* is a (possibly infinite) set P together with a binary relation, $\leq$, such that, for all x, y, and z in P, the following conditions hold.

(P1) $x \leq x$.

(P2) *If $x \leq y$ and $y \leq x$, then $x = y$.*

(P3) *If $x \leq y$ and $y \leq z$, then $x \leq z$.*

Posets abound in mathematics. As examples we may take the real numbers with the usual less-than-or-equal-to relation; the subsets of a set with the relation of set inclusion; and the set D_n of positive integral divisors of a positive integer n under the relation $x \leq y$ if x divides y.

In a partially ordered set P, if $x \leq y$ we shall sometimes write $y \geq x$. If $x \leq y$ but $x \neq y$, we write $x < y$ or $y > x$. If $x < y$ but there is no element z of P such that $x < z < y$, then we say that y *covers* x in P. We can represent P by a *Hasse diagram.* Such a diagram is a simple graph, the vertices of which correspond to elements of P. In this graph, if $x > y$, then the vertex corresponding to x is placed higher than that corresponding to y. Two vertices x and y are joined by an edge, a straight line segment, whenever x covers y. Figure 1.24 shows two Hasse diagrams, one corresponding to the set of subsets of $\{1, 2, 3, 4\}$ ordered by inclusion;

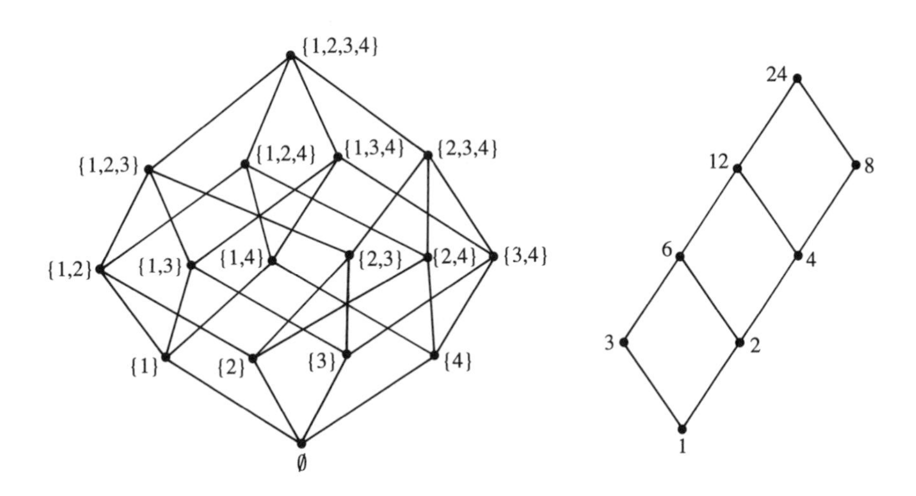

Fig. 1.24. Two Hasse diagrams.

and the other corresponding to D_{24}, the set of positive integral divisors of 24 under the divisibility order described earlier.

The Hasse diagram in Figure 1.25 represents both the partially ordered set D_{30} of positive integral divisors of 30, and the set of subsets of $\{1, 2, 3\}$ ordered by inclusion, these posets being isomorphic. In general, posets P_1 and P_2 are *isomorphic* if there is a bijection $\psi : P_1 \to P_2$ such that $x \leq y$ if and only if $\psi(x) \leq \psi(y)$.

The posets represented in Figures 1.24 and 1.25 are all examples of lattices. A *lattice* is a poset $\mathcal{L}$ such that, for every pair of elements, the least upper bound and greatest lower bound of the pair exists. Formally, if x and y are arbitrary elements of $\mathcal{L}$, then $\mathcal{L}$ contains elements $x \vee y$ and $x \wedge y$, the *join* and *meet* of x and y, such that

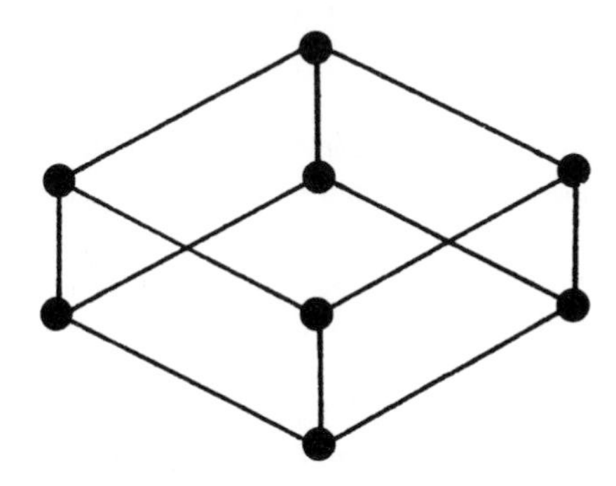

Fig. 1.25. D_{30} and the lattice of subsets of $\{1, 2, 3\}$.

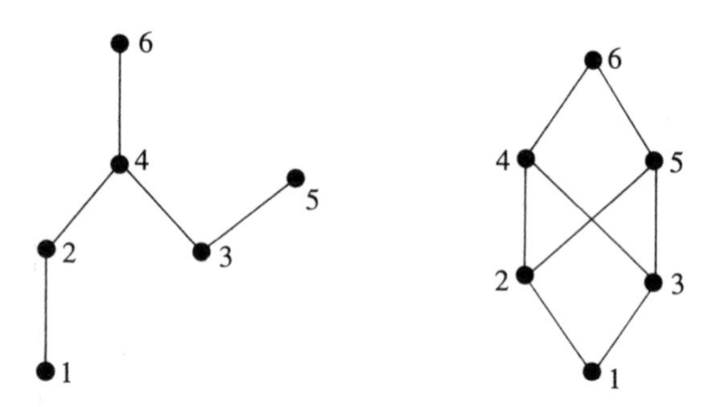

Fig. 1.26. Two posets that are not lattices.

(L1) $x \vee y \geq x$, $x \vee y \geq y$, *and if* $z \geq x$ *and* $z \geq y$, *then* $z \geq x \vee y$; *and*

(L2) $x \wedge y \leq x$, $x \wedge y \leq y$, *and if* $z \leq x$ *and* $z \leq y$, *then* $z \leq x \wedge y$.

It is routine to check that both of the operations $\vee$ and $\wedge$ are commutative and associative.

The Hasse diagrams shown in Figure 1.26 represent posets that are not lattices. In the first, $2 \wedge 3$ does not exist; in the second, $2 \vee 3$ does not exist.

If M is a matroid, then $\mathcal{L}(M)$ will denote the set of flats of M ordered by inclusion.

1.7.3 Lemma. *$\mathcal{L}(M)$ is a lattice and, for all flats X and Y of M,*

$$X \wedge Y = X \cap Y \quad \textit{and} \quad X \vee Y = \text{cl}(X \cup Y).$$

Proof. Evidently $\mathcal{L}(M)$ is a partially ordered set. Moreover, if X and Y are flats, so is $X \cap Y$. To see this, assume to the contrary that x is in $\text{cl}(X \cap Y) - (X \cap Y)$. Then, by Proposition 1.4.10(ii), there is a circuit C such that $x \in C \subseteq (X \cap Y) \cup x$. Then $x \in \text{cl}(X) \cap \text{cl}(Y) = X \cap Y$; a contradiction. We conclude that the meet of X and Y is $X \cap Y$. It is easy to see that the join of X and Y is $\text{cl}(X \cup Y)$. Although $X \cup Y$ need not be closed, $\text{cl}(X \cup Y)$ certainly must be. □

We shall show next that $\mathcal{L}(M)$ is in fact a rather special type of lattice. The Hasse diagrams shown in Figure 1.27 all represent lattices, but none is isomorphic to the lattice of flats of a matroid. To characterize matroid lattices, we shall require some more terminology. Let P be a finite partially ordered set. A *chain* in P *from* x_0 *to* x_n is a subset $\{x_0, x_1, \ldots, x_n\}$ of P such that $x_0 < x_1 < \ldots < x_n$. The *length* of such a chain is n, and the chain is *maximal* if x_i covers x_{i-1} for all i in $\{1, 2, \ldots, n\}$. If, for every pair $\{a, b\}$ of elements of P with $a < b$, all maximal chains from a to b have the same length, then P is said to satisfy the *Jordan–Dedekind*

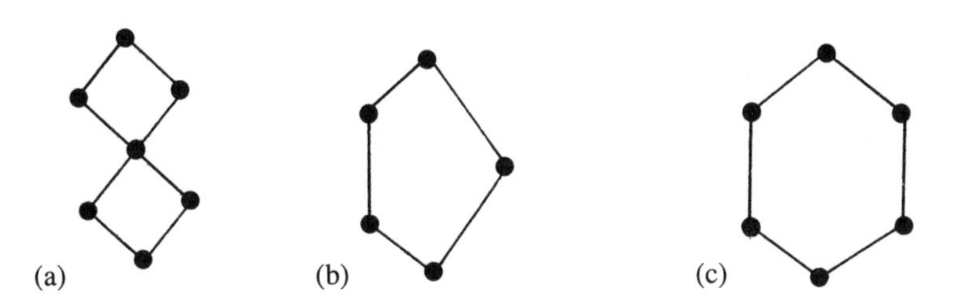

Fig. 1.27. Three posets that are lattices.

chain condition. Thus, of the three lattices represented in Figure 1.27, only (b) fails to satisfy this condition.

If the poset P has an element z such that $z \leq x$ for all x in P, then we call z the *zero* of P and denote it by 0. Similarly, if P has an element w such that $w \geq x$ for all x in P, then w is called the *one* of P. The first poset in Figure 1.26 has neither a zero nor a one; the second has both.

Now suppose that P is a partially ordered set having a zero. An element x is called an *atom* of P if x covers 0. The *height* $h(y)$ of an element y of P is the maximum length of a chain from 0 to y. Thus, in particular, the atoms of P are precisely the elements of height one.

It is not difficult to check that every finite lattice has a zero and a one. In particular, for a matroid M, the zero of $\mathcal{L}(M)$ is $\text{cl}(\emptyset)$, while the one is $E(M)$. A finite lattice $\mathcal{L}$ is called *semimodular* if it satisfies the Jordan–Dedekind chain condition and, for every pair x and y of elements of $\mathcal{L}$,

1.7.4 $$h(x) + h(y) \geq h(x \vee y) + h(x \wedge y).$$

A *geometric lattice* is a finite semimodular lattice in which every element is a join of atoms.

The following theorem, the main result of this section, motivates the lattice-theoretic approach to matroids.

1.7.5 Theorem. *A lattice $\mathcal{L}$ is geometric if and only if it is the lattice of flats of a matroid.*

Proof. Let M be a matroid. We begin by showing that $\mathcal{L}(M)$ is geometric. We shall use the following.

1.7.6 Lemma. *If X and Y are flats of M and $X \subseteq Y$, then every maximal chain of flats from X to Y has length $r(Y) - r(X)$.*

Proof. Suppose that Y covers X. The lemma will follow if we can show, in this case, that $r(Y) = r(X) + 1$. As $X \subsetneq Y$, we can choose an element

x from $Y - X$. Clearly $X \subsetneq \mathrm{cl}(X \cup x) \subseteq Y$. Therefore, since Y covers X, we must have that $\mathrm{cl}(X \cup x) = Y$. But, by (R2)$'$, $r(\mathrm{cl}(X \cup x)) = r(X \cup x) \leq r(X) + 1$. Hence $r(Y) = r(X) + 1$ as required. □

It is an immediate consequence of this lemma that $\mathcal{L}(M)$ satisfies the Jordan–Dedekind chain condition. Moreover, for all flats X, $h(X) = r(X)$. Thus, for flats X and Y,

$$\begin{aligned} h(X) + h(Y) = r(X) + r(Y) &\geq r(X \cup Y) + r(X \cap Y), \text{ by (R3)}; \\ &= r(\mathrm{cl}(X \cup Y)) + r(X \cap Y) \\ &= h(\mathrm{cl}(X \cup Y)) + h(X \cap Y) \\ &= h(X \vee Y) + h(X \wedge Y). \end{aligned}$$

Hence $\mathcal{L}(M)$ is a semimodular lattice. If X is a flat and $\{b_1, b_2, \ldots, b_k\}$ is a basis for X, then

$$X = \mathrm{cl}(\{b_1, b_2, \ldots, b_k\}) = \mathrm{cl}(\{b_1\}) \vee \mathrm{cl}(\{b_2\}) \vee \ldots \vee \mathrm{cl}(\{b_k\}).$$

Since $\mathrm{cl}(\{b_i\})$ is an atom, we conclude that every flat of M is a join of atoms, with $\mathrm{cl}(\emptyset)$ being the join of the empty set of atoms. Thus $\mathcal{L}(M)$ is indeed a geometric lattice.

Conversely, suppose that $\mathcal{L}$ is an arbitrary geometric lattice. If the zero and one of $\mathcal{L}$ coincide, then $\mathcal{L} \cong \mathcal{L}(U_{0,0})$. Now suppose that the zero and one of $\mathcal{L}$ are distinct. Then $\mathcal{L}$ has a non-empty set E of atoms. If $X \subseteq E$, define $r(X) = h(\bigvee_{x \in X} x)$. We shall show that r is the rank function of a matroid M on E whose lattice of flats is isomorphic to $\mathcal{L}$. First we note that $r(\emptyset) = 0$, that is, r satisfies (R1)$'$. Now suppose $X \subseteq E$ and $e \in E$. Then

$$r(X) = h(\bigvee_{x \in X} x) \leq h(\bigvee_{x \in X \cup e} x) = r(X \cup e).$$

Moreover,

$$r(X \cup e) \leq h(\bigvee_{x \in X} x) + h(e) - h((\bigvee_{x \in X} x) \wedge e) \leq r(X) + 1.$$

Thus r satisfies (R2)$'$. Finally, if $X \subseteq E$ and e, f are elements of E such that $r(X \cup e) = r(X \cup f) = r(X)$, then

$$\begin{aligned} r(X \cup e \cup f) &= h(\bigvee_{x \in X \cup e \cup f} x) \\ &= h((\bigvee_{x \in X \cup e} x) \vee (\bigvee_{x \in X \cup f} x)) \\ &\leq h(\bigvee_{x \in X \cup e} x) + h(\bigvee_{x \in X \cup f} x) \\ &\qquad - h((\bigvee_{x \in X \cup e} x) \wedge (\bigvee_{x \in X \cup f} x)) \\ &\leq r(X) + r(X) - h(\bigvee_{x \in X} x) = r(X). \end{aligned}$$

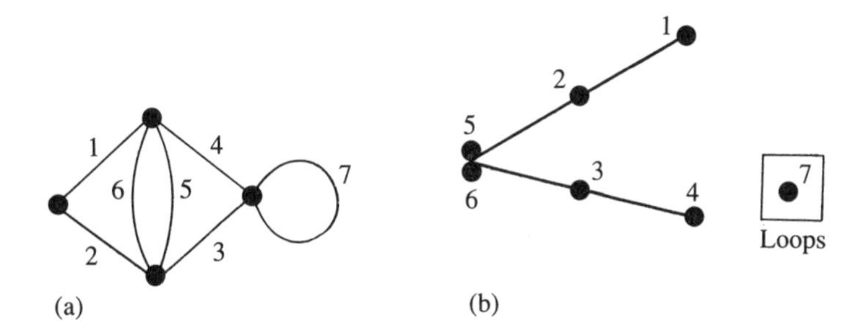

Fig. 1.28. (a) A graph G. (b) A geometric representation for $M(G)$.

Hence r satisfies (R3)$'$ and so, by Theorem 1.4.14, r is the rank function of a matroid M on E.

To complete the proof, we need to show that $\mathcal{L}(M)$ and $\mathcal{L}$ are isomorphic posets. To do this, let $\psi : \mathcal{L} \to \mathcal{L}(M)$ be defined by $\psi(X) = \{x \in E : x \leq X\}$. For ψ to be well-defined, we must check that $\{x \in E : x \leq X\}$ is a flat of M. If not, then there is an element y of E such that $y \not\leq X$ and $r(\{x \in E : x \leq X\} \cup y) = r(\{x \in E : x \leq X\})$. But then $h((\bigvee_{x \leq X} x) \vee y) = h(\bigvee_{x \leq X} x)$, and it follows by 1.7.4 that $y \leq \bigvee_{x \leq X} x = X$; a contradiction. Therefore ψ is well-defined. It clearly preserves order and is one-to-one. To show it is onto, suppose Y is a flat of $\mathcal{L}(M)$. We shall show that $Y = \psi(Y')$ where $Y' = \bigvee_{y \in Y} y$. Since $\psi(Y') = \{x \in E : x \leq Y'\}$, it is clear that $\psi(Y') \supseteq Y$. On the other hand, if $z \in \psi(Y')$, then $z \leq \bigvee_{y \in Y} y$. Therefore $h((\bigvee_{y \in Y} y) \vee z) = h(\bigvee_{y \in Y} y)$, that is, $r(Y \cup z) = r(Y)$, so $z \in \text{cl}(Y) = Y$. Hence $Y \supseteq \psi(Y')$. We conclude that ψ is onto. Hence, $\mathcal{L} \cong \mathcal{L}(M)$ and the theorem is proved. □

1.7.7 Example. Let $M = M(G)$ where G is the graph shown in Figure 1.28(a). A geometric representation for this matroid is shown in Figure 1.28(b), and $\mathcal{L}(M)$ is as shown in Figure 1.29. It is clear from that diagram that the Hasse diagram for $\mathcal{L}(M)$ can become quite cluttered even when M has a relatively small number of elements. □

A particularly important example of a geometric lattice is $\mathcal{P}_n$, the lattice of flats of the matroid $M(K_n)$. This lattice is often referred to as a *partition lattice* since it can also be derived as follows. Let V be the vertex set of K_n. If F is a flat of $M(K_n)$, we denote by π_F the partition of V in which i and j are in the same class if and only if the edge ij is in F. This determines a map from the set of flats of $M(K_n)$ into the set of partitions of the n-set V. Moreover, this map is easily shown to be a bijection. Indeed, the map is an isomorphism from $\mathcal{L}(M(K_n))$ to the set of partitions of V where, for partitions α and β, we define $\alpha \leq \beta$ if α is a refinement of β, that is, every class of α is contained in a class of β.

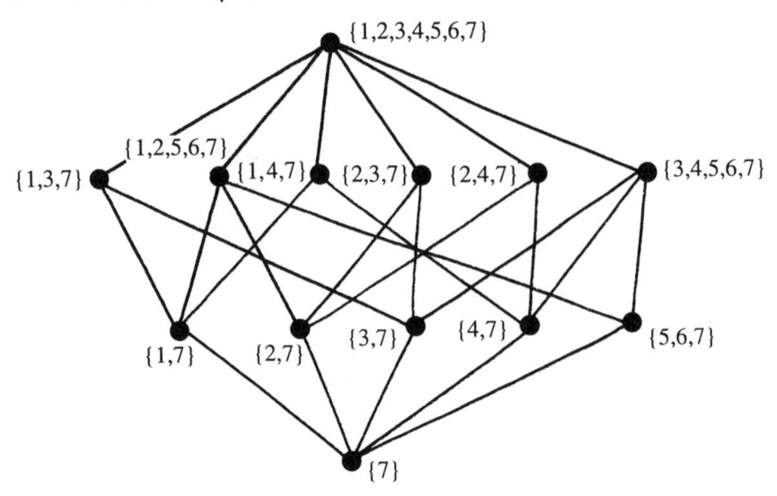

Fig. 1.29. $\mathcal{L}(M(G))$ where G is as shown in Figure 1.28(a).

The lattice shown in Figure 1.30(b) is both the lattice of flats of $M(K_4)$, with the edges labelled as in Figure 1.30(a), as well as the lattice $\mathcal{P}_4$ of partitions of the 4-set $\{a, b, c, d\}$. Every element of the lattice has been labelled by the corresponding flat and some representative elements have also been labelled by the corresponding partition.

The lattice of flats of a matroid M does not uniquely determine M for clearly $\mathcal{L}(\widetilde{M}) \cong \mathcal{L}(M)$. However, a consequence of the proof of Theorem 1.7.5 is that if M_1 and M_2 are simple matroids for which $\mathcal{L}(M_1) \cong \mathcal{L}(M_2)$, then $M_1 \cong M_2$.

Every element x in a geometric lattice $\mathcal{L}$ is a join of atoms. Indeed, if $h(x) = k$, there is a set of k atoms whose join is x. The next result asserts that the same result holds in the lattice obtained from $\mathcal{L}$ by reversing all order relations. We state it without explicit reference to the lattice.

1.7.8 Proposition. *Let X be a flat in a matroid M and suppose that $r(X) = r(M) - k$ where $k \geq 1$. Then M has a set $\{H_1, H_2, \ldots, H_k\}$ of hyperplanes such that $X = \cap_{i=1}^{k} H_i$.*

Proof. We argue by induction on k. The result is immediate for $k = 1$. Assume it true for $k = n$ and let $k = n + 1$. Evidently M has an element y that is not in X. As X is a flat, $\text{cl}(X \cup y)$ is a flat of M covering X. Thus $r(\text{cl}(X \cup y)) = r(M) - n$, so, by the induction assumption, M has hyperplanes $H_1, H_2, \ldots, H_n$ such that $\cap_{i=1}^{n} H_i = \text{cl}(X \cup y)$. Now either $E(M) - y$ is spanning in M, or not. In the latter case, let $H_{n+1} = E(M) - y$. In the former case, as X is non-spanning, M has a maximal non-spanning set that contains X and is contained in $E(M) - y$. In this case, let H_{n+1} be such a set. It follows that, in both cases, since H_{n+1}

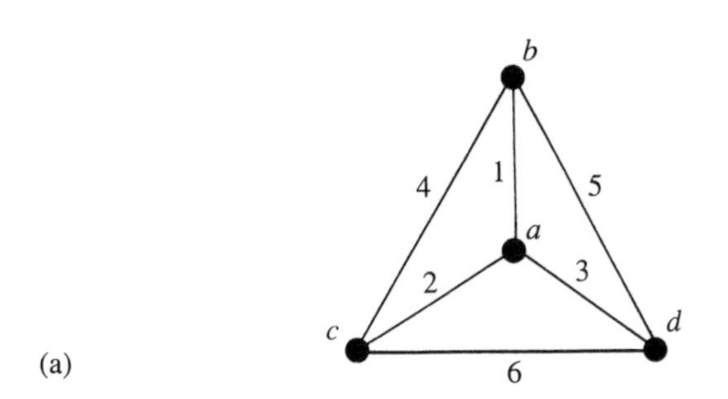

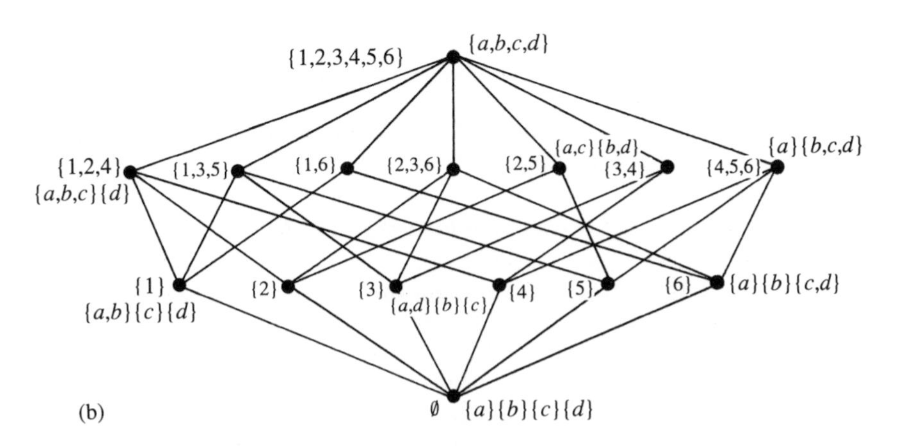

Fig. 1.30. (a)K_4. (b) $\mathcal{L}(M(K_4))$ is isomorphic to $\mathcal{P}_4$.

is a maximal non-spanning set, it is a hyperplane. Since $y \in \cap_{i=1}^{n} H_i$, but $y \notin H_{n+1}$, we have $\cap_{i=1}^{n+1} H_i \neq \cap_{i=1}^{n} H_i$. Thus $\mathrm{cl}(X \cup y) = \cap_{i=1}^{n} H_i \supsetneqq \cap_{i=1}^{n+1} H_i \supseteq X$. But $\mathrm{cl}(X \cup y)$ covers X, so $X = \cap_{i=1}^{n+1} H_i$. The proposition now follows by induction. □

While the intersection of two distinct hyperplanes in a rank-r matroid M has rank at most $r-2$, it may have rank as low as 0. For example, in $U_{k+1,2k}$, the hyperplanes are precisely the k-element subsets of the ground set and, for any j in $\{0, 1, 2, \ldots, k-2\}$, one can easily find two hyperplanes whose intersection has rank j. One may suspect that intersecting more hyperplanes will lower the rank. But, in $U_{4,n}$, one can easily find as many as $n-2$ distinct hyperplanes, the intersection of all of which has rank equal to $r(U_{4,n}) - 2$. We defer to the exercises consideration of some of the many other interesting properties of geometric lattices.

Exercises

1. Suppose that $A_1 = \{1,2,3\}$, $A_2 = \{3,4,5\}$, $A_3 = \{3,6,7\}$, and $A_4 = \{8\}$. Let $S = \{1,2,\ldots,10\}$ and $M[\mathcal{A}]$ be the transversal matroid on S associated with $\mathcal{A} = (A_1, A_2, A_3, A_4)$. Find the simple matroid associated with M, keeping the lowest-labelled element in every non-trivial parallel class.

2. **(i)** Draw the Hasse diagram for D_{54}, the set of positive integral divisors of 54 under the divisibility order.

(ii) Show that $D_{54} \cong D_{24}$ and generalize this to determine all n for which $D_n \cong D_{54}$.

3. Show that a matroid in which every two distinct hyperplanes are disjoint has rank at most two.

4. Find all positive integers n for which D_n is a geometric lattice.

5. Prove that a finite lattice $\mathcal{L}$ is semimodular if, for all x, y in $\mathcal{L}$, the following condition holds:

(†) *If both x and y cover $x \wedge y$, then $x \vee y$ covers both x and y.*

6. Draw Hasse diagrams for the lattices of flats of $U_{3,5}$ and the matroids M_1 and M_2 from Exercise 10 of Section 1.5.

7. A lattice $\mathcal{L}$ is *complemented* if, for every element x of $\mathcal{L}$, there is an element y, called a *complement* of x, such that $x \vee y$ and $x \wedge y$ are the one and zero, respectively, of $\mathcal{L}$. A lattice $\mathcal{L}$ is *relatively complemented* if every interval is complemented where an *interval* $[a,b]$ of $\mathcal{L}$ is the lattice induced on $\{z : a \leq z \leq b\}$ by the order on $\mathcal{L}$. Show that:

(i) Every geometric lattice is relatively complemented.

(ii) Every interval of a geometric lattice is a geometric lattice.

8. Let A be a matrix having $V(n+1,q)$ as its set of columns and let $\mathcal{L}$ be the lattice of flats of $M[A]$. Show that, for all elements x and y of $\mathcal{L}$, $h(x) + h(y) = h(x \vee y) + h(x \wedge y)$.

9. Find simple matroids M_1 and M_2 so that if $\mathcal{L}_1$ and $\mathcal{L}_2$ are the lattices in Figure 1.31, then $\mathcal{L}(M_1) \cong \mathcal{L}_1$ and $\mathcal{L}(M_2) \cong \mathcal{L}_2$.

10. Let $\mathcal{P}'$ be the poset obtained from the poset $\mathcal{P}$ by defining $x \leq y$ in $\mathcal{P}'$ if and only if $x \geq y$ in $\mathcal{P}$. Show that:

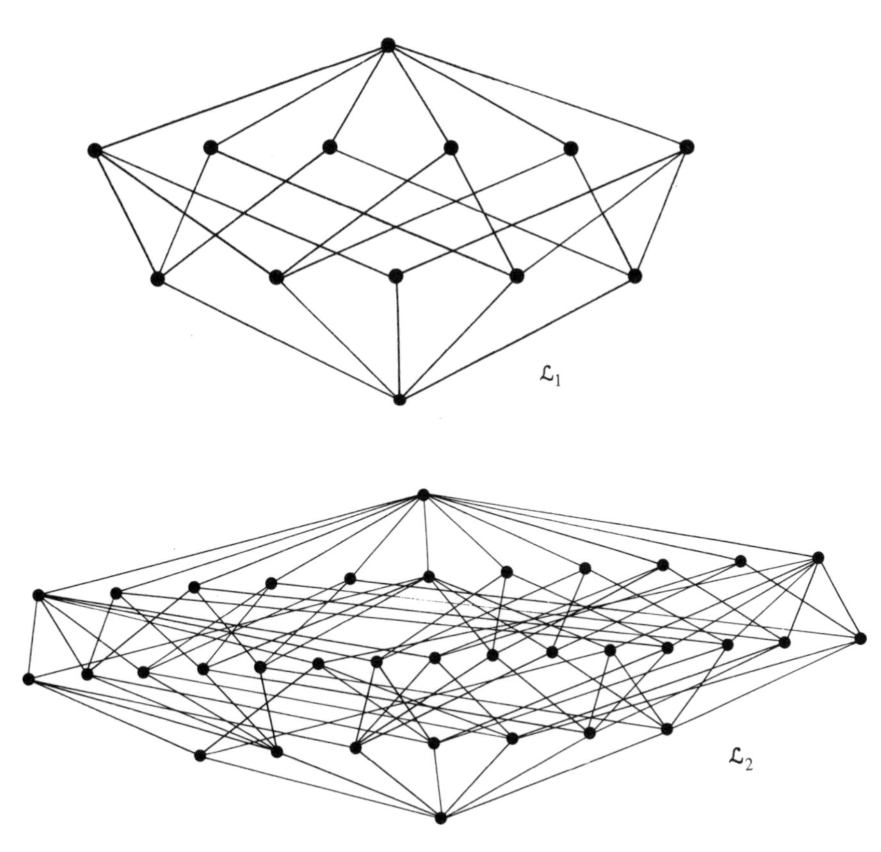

Fig. 1.31. Two geometric lattices.

(i) If $\mathcal{P}$ is a lattice, then so is $\mathcal{P}'$.

(ii) If $\mathcal{P}$ is a geometric lattice, then $\mathcal{P}'$ need not be geometric.

11. Prove that two lattices $\mathcal{L}_1$ and $\mathcal{L}_2$ are isomorphic if and only if there is a bijection $\psi : \mathcal{L}_1 \to \mathcal{L}_2$ such that, for all x, y in $\mathcal{L}_1$,

$$\psi(x \vee y) = \psi(x) \vee \psi(y) \quad \text{and} \quad \psi(x \wedge y) = \psi(x) \wedge \psi(y).$$

1.8 The greedy algorithm

In this section, we consider yet another characterization of matroids. The attractive feature of this characterization is that it indicates clearly why matroids arise naturally in a number of problems in combinatorial optimization.

We begin by discussing a well-known optimization problem for graphs. Let G be a connected graph and suppose that w is a function from $E(G)$ into $\mathbb{R}$. We call w a *weight function* on G and, for all $X \subseteq E(G)$, we define the *weight* $w(X)$ of X to be $\sum_{x \in X} w(x)$. For the given G and w, the problem here is to find a spanning tree of G of minimum weight. For instance, G could be K_n where the vertices of K_n correspond to towns to be linked by a railway network, and the weight on each edge is the cost of providing a direct link between the towns corresponding to the edge's endpoints. In this case, the minimum weight of a spanning tree in G corresponds to the minimum cost of providing a railway network that will link all n towns.

Kruskal's algorithm (1956) for solving this problem chooses edges one at a time. At each stage, the next edge e to be chosen is one of minimum weight such that

(i) e has not been previously chosen; and

(ii) e does not form a cycle with some set of previously chosen edges.

When no further edges can be chosen subject to these restrictions, the algorithm stops. At this stage, the set of chosen edges is the edge set of a spanning tree of minimum weight. We shall not prove the last statement directly. Instead, we shall deduce it later from the main result of this section.

1.8.1 **Example.** Let G be the graph in Figure 1.32 with edge weights as shown. It is not difficult to check that Kruskal's algorithm will produce one of the spanning trees T_1, T_2, and T_3 shown. Precisely which of these is obtained depends upon which two edges of weight three are chosen. □

The minimum-weight spanning tree problem is a special case of the following natural optimization problem (Bixby 1981). Let $\mathcal{I}$ be a collection of subsets of a set E and suppose $\mathcal{I}$ satisfies (I1) and (I2). Let w be a function from E into $\mathbb{R}$. As before, define the *weight* $w(X)$ of any non-empty subset X of E by

$$w(X) = \sum_{x \in X} w(x)$$

and let $w(\emptyset) = 0$. The *optimization problem* for the pair $(\mathcal{I}, w)$ is as follows:

1.8.2 *Find a maximal member B of $\mathcal{I}$ of maximum weight.*

We call the set B a *solution* to this optimization problem. If we replace the weight function w by its negative and solve the optimization problem

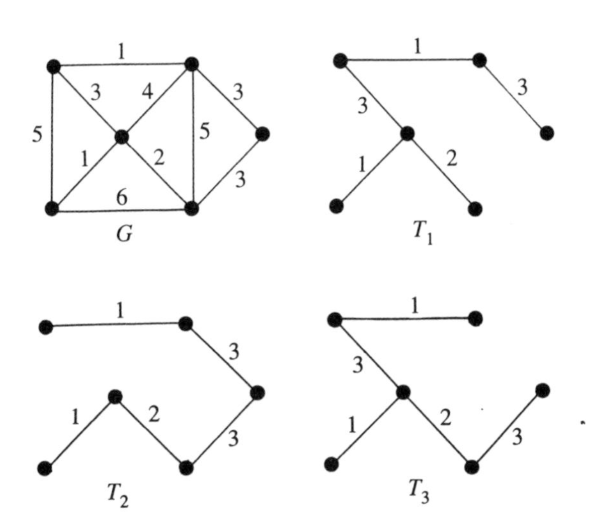

Fig. 1.32. G and its minimum-weight spanning trees.

for $(\mathcal{I}, -w)$, then we obtain a maximal member B' of $\mathcal{I}$ for which $w(B')$ is a minimum. Therefore, letting $\mathcal{I}$ be the collection of independent sets in $M(G)$ where G is a connected graph, we see that the minimum-weight spanning tree problem for graphs is a special case of 1.8.2.

The *greedy algorithm* for the pair $(\mathcal{I}, w)$ proceeds as follows:

(i) Set $X_0 = \emptyset$ and $j = 0$.

(ii) If $E - X_j$ contains an element e such that $X_j \cup e \in \mathcal{I}$, choose such an element e_{j+1} of maximum weight, let $X_{j+1} = X_j \cup e_{j+1}$, and go to (iii); otherwise, let $X_j = B_G$ and go to (iv).

(iii) Add 1 to j and go to (ii).

(iv) Stop.

1.8.3 Lemma. *If $(E, \mathcal{I})$ is a matroid M, then B_G is a solution to the optimization problem* 1.8.2.

Proof. If $r(M) = r$, then $B_G = \{e_1, e_2, \ldots, e_r\}$ and B_G is a basis of M. Let B be another basis of M, say $B = \{f_1, f_2, \ldots, f_r\}$ where $w(f_1) \geq w(f_2) \geq \ldots \geq w(f_r)$. Lemma 1.8.3 follows without difficulty from the next result which shows that not only is B_G a maximum-weight basis of M, but, for all j, the jth heaviest element of B_G has weight at least that of the jth heaviest element of B.

1.8.4 Lemma. *If $1 \leq j \leq r$, then $w(e_j) \geq w(f_j)$.*

Proof. Assume the contrary and let k be the least integer for which $w(e_k) < w(f_k)$. Take $I_1 = \{e_1, e_2, \ldots, e_{k-1}\}$ and $I_2 = \{f_1, f_2, \ldots, f_k\}$. Since $|I_1| < |I_2|$, (I3) implies that $I_1 \cup f_t \in \mathcal{I}$ for some f_t in $I_2 - I_1$. But $w(f_t) \geq w(f_k) > w(e_k)$. Hence the greedy algorithm would have chosen f_t in preference to e_k. This contradiction completes the proof of Lemma 1.8.4 and thereby that of Lemma 1.8.3. □

On combining Lemma 1.8.3 with the remarks following 1.8.2, we deduce that Kruskal's algorithm does indeed produce a spanning tree of minimum weight in a connected graph G. We remark here that if G has n vertices, a spanning tree of G has $n-1$ edges. Hence Kruskal's algorithm may be terminated once $n-1$ edges have been chosen.

We have seen that the greedy algorithm works for matroids. More strikingly, it fails for everything else.

1.8.5 Theorem. *Let $\mathcal{I}$ be a collection of subsets of a set E. Then $(E, \mathcal{I})$ is a matroid if and only if $\mathcal{I}$ satisfies the following conditions:*

(I1) $\emptyset \in \mathcal{I}$.

(I2) *If $I \in \mathcal{I}$ and $I' \subseteq I$, then $I' \in \mathcal{I}$.*

(G) *For all weight functions $w : E \to \mathbb{R}$, the greedy algorithm produces a maximal member of $\mathcal{I}$ of maximum weight.*

Proof. If $(E, \mathcal{I})$ is a matroid, then (I1) and (I2) certainly hold; by Lemma 1.8.3, so too does (G). Now suppose that $\mathcal{I}$ satisfies (I1), (I2), and (G). We want to show that $\mathcal{I}$ satisfies (I3). Assume the contrary, that is, suppose that I_1, I_2 are members of $\mathcal{I}$ with $|I_1| < |I_2|$ such that $I_1 \cup e \notin \mathcal{I}$ for all e in $I_2 - I_1$.

Now $|I_1 - I_2| < |I_2 - I_1|$ and $I_1 - I_2$ is non-empty, so we can choose a positive number ε such that

$$0 < (1+\varepsilon)|I_1 - I_2| < |I_2 - I_1|. \tag{1}$$

Define $w : E \to \mathbb{R}$ by

$$w(e) = \begin{cases} 2, & \text{if } e \in I_1 \cap I_2, \\ 1/|I_1 - I_2|, & \text{if } e \in I_1 - I_2, \\ (1+\varepsilon)/|I_2 - I_1|, & \text{if } e \in I_2 - I_1, \\ 0, & \text{otherwise.} \end{cases}$$

Then the greedy algorithm will first pick all the elements of $I_1 \cap I_2$ and then all the elements of $I_1 - I_2$. By assumption, it cannot then pick any element of $I_2 - I_1$. Thus the remaining elements of B_G will be in $E - (I_1 \cup I_2)$. Hence

$$\begin{aligned} w(B_G) &= 2|I_1 \cap I_2| + |I_1 - I_2|(1/|I_1 - I_2|) \\ &= 2|I_1 \cap I_2| + 1. \end{aligned} \tag{2}$$

But, by (I2), I_2 is contained in a maximal member I_2' of $\mathcal{I}$ and

$$\begin{aligned} w(I_2') \geq w(I_2) &= 2|I_1 \cap I_2| + |I_2 - I_1|[(1+\varepsilon)/|I_2 - I_1|] \\ &= 2|I_1 \cap I_2| + 1 + \varepsilon. \end{aligned} \tag{3}$$

From (2) and (3), we deduce that $w(I_2') > w(B_G)$, that is, the greedy algorithm fails for this weight function. This contradiction completes the proof of the theorem. □

Proofs of the last result have been published by a number of different authors. Curiously, however, the first of these seems to be due to Borůvka (1926) whose paper predates even Whitney's (1935) introduction of the term 'matroid'.

We have already seen an application of the greedy algorithm to graphic matroids. The next application, due to Gale (1968), is to transversal matroids. Suppose that a certain set of one-worker jobs has been arranged in order of importance and that we want to fill the jobs from a pool of workers each of whom is qualified to perform some subset of the jobs. We also assume that the jobs are to be done simultaneously so that no worker can be assigned to more than one job. In general, it will not be possible to fill all the jobs, so we seek a way of choosing the set of jobs to be filled that is optimal relative to the priority order on the jobs.

Before defining optimality in this context, we describe a reformulation of the problem in terms of matroids. Let S be the set of jobs and Y the set of workers. For each y in Y, let A_y be the set of jobs that worker y is qualified to perform. Let $\mathcal{A} = (A_y : y \in Y)$. Evidently the maximum number of jobs that can be done simultaneously is the size of a largest partial transversal of $\mathcal{A}$, or equivalently, the rank of $M[\mathcal{A}]$.

Now let $p : S \to \mathbb{R}$ be the function that corresponds to the priority order on the jobs, a lower p-value corresponding to a higher priority for the job. A *job assignment* is a basis of $M[\mathcal{A}]$. Let B be the job assignment $\{x_1, x_2, \ldots, x_r\}$ where $p(x_1) \leq p(x_2) \leq \ldots \leq p(x_r)$. Then B is called *optimal* if, for any other job assignment $\{z_1, z_2, \ldots, z_r\}$ where $p(z_1) \leq p(z_2) \leq \ldots \leq p(z_r)$, we have $p(x_i) \leq p(z_i)$ for all i in $\{1, 2, \ldots, r\}$. If we define $\mathcal{I}$ to be the collection of independent sets of $M[\mathcal{A}]$ and let $w = -p$, then, by Lemma 1.8.3, it follows that the greedy algorithm applied to the pair $(\mathcal{I}, w)$ will find an optimal job assignment.

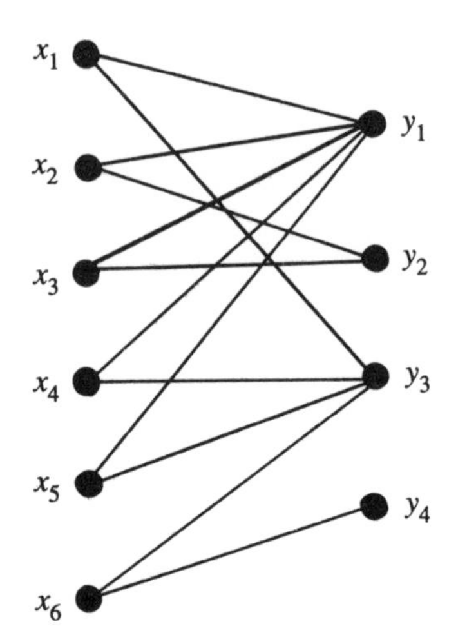

Fig. 1.33. Worker x_i can do job y_j if x_iy_j is an edge.

1.8.6 Example. In the bipartite graph in Figure 1.33, $\{y_1, y_2, y_3, y_4\}$ corresponds to the set of workers, $\{x_1, x_2, \ldots, x_6\}$ to the set of jobs, and two vertices are joined when the corresponding worker can perform the corresponding job. Assume that the priority order on the jobs is $p(x_1) < p(x_2) < \ldots < p(x_6)$. Then it is not difficult to check that the unique optimal job assignment in this case is $\{x_1, x_2, x_4, x_6\}$. □

We remark that, in this job assignment problem, the greedy algorithm finds an optimal basis of the corresponding transversal matroid without backtracking. However, we cannot pick the workers for the jobs in this optimal basis at the same time as the jobs are added to the basis. For example, suppose Figure 1.34 represents such a problem with workers y_1 and y_2 and jobs x_1 and x_2 such that $p(x_1) < p(x_2)$. Then the greedy algorithm picks x_1 first and x_2 second. But, if x_1 is paired with y_1, then x_2 cannot be paired. In general, once an optimal basis B has been found, the problem of assigning workers to the jobs in B is precisely the problem of finding a maximum-sized matching in the bipartite graph that is obtained from $\Delta[\mathcal{A}]$ by deleting the vertices in $S - B$. Algorithms for solving the latter problem can be found in numerous books (see, for example, Lawler 1976 or Bondy and Murty 1976).

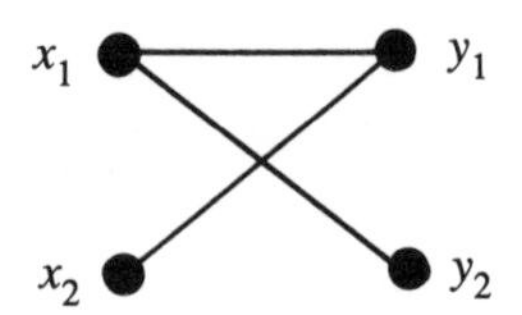

Fig. 1.34. The edge x_1y_1 is not in a maximum-sized matching.

Exercises

1. **(i)** Find a maximum-weight spanning tree of the graph in Figure 1.32. Is this the unique such tree?

   (ii) Find all maximum-weight spanning trees and all minimum-weight spanning trees of the graph in Figure 1.28(a), where the edge labels are interpreted as weights.

2. If the priority order on the jobs in Example 1.8.6 is changed to $p(x_6) < p(x_5) < \ldots < p(x_1)$, find an optimal job assignment.

3. Apply the greedy algorithm to the matroid in Figure 1.17 interpreting the element labels as weights. Find both a maximum-weight basis and a minimum-weight basis.

4. Let M be a matroid and $w : E(M) \rightarrow \mathbb{R}$ be a one-to-one function. Prove that M has a unique basis of maximum weight.

5. Let M be a matroid and $w : E(M) \rightarrow \mathbb{R}$. When the greedy algorithm is applied to the pair $(\mathcal{I}(M), w)$, each application of step (ii) involves a potential choice. Thus, in general, there are a number of different sets that the algorithm can produce as solutions to the optimization problem $(\mathcal{I}(M), w)$. Let $\mathcal{B}_G$ be the set of such sets and let $\mathcal{B}_{max}$ be the set of maximum-weight bases of M. Prove that $\mathcal{B}_G = \mathcal{B}_{max}$.

6. Modify the greedy algorithm to give an algorithm for finding a maximum-weight basis B of a matroid M subject to the constraint that B contains some fixed independent set I.

7. Let w be a weight function on a matroid M and $\mathcal{S}$ be the set of spanning sets of M. Prove that the set B_G produced by the following algorithm is a basis of M of minimum weight:

   (a) Set $Y_0 = E(M)$ and $j = 0$.

   (b) If Y_j contains an element e such that $Y_j - e \in \mathcal{S}$, then choose such an element e_{j+1} of maximum weight, let $Y_{j+1} = Y_j - e_{j+1}$, and go to (c); otherwise, let $Y_j = B_G$ and go to (d).

   (c) Add 1 to j and go to (b).

   (d) Stop.

8*. Let B be a maximum-weight basis of a matroid M and H be a hyperplane of M. Let X be the set of maximum-weight elements of $E(M) - H$. Prove that $X \cap B$ is non-empty.

2
Duality

One of the most attractive features of matroid theory is the existence of a theory of duality. This theory, which was introduced by Whitney (1935), extends both the notion of orthogonality in vector spaces and the concept of a planar dual of a plane graph. It dramatically increases the weapons at one's disposal in attacking any matroid problem. The theory of infinite matroids provides ample evidence of the vital role that duality plays in matroid theory. In the infinite context, there is no duality theory having the richness of the theory in the finite context. This seems to be one of the main reasons why attention is usually restricted to the finite case in most developments of matroid theory. For surveys of infinite matroids the reader is referred to Welsh (1976) and Oxley (1992).

2.1 The definition and basic properties

In this section we define the dual of a matroid, prove that it is also a matroid, and establish some fundamental links between matroids and their duals.

2.1.1 Theorem. *Let M be a matroid and $\mathcal{B}^*(M)$ be $\{E(M) - B : B \in \mathcal{B}(M)\}$. Then $\mathcal{B}^*(M)$ is the set of bases of a matroid on $E(M)$.*

The proof of this theorem will use the following result.

2.1.2 Lemma. *The set $\mathcal{B}$ of bases of a matroid M satisfies the following condition:*

(B2)* *If B_1 and B_2 are in $\mathcal{B}$ and $x \in B_2 - B_1$, then there is an element y of $B_1 - B_2$ such that $(B_1 - y) \cup x \in \mathcal{B}$.*

We remark here that there is a genuine difference between (B2)* and (B2) that cannot be achieved simply by relabelling.

Proof of Lemma 2.1.2. By Corollary 1.2.6, $B_1 \cup x$ contains a unique circuit, $C(x, B_1)$. As $C(x, B_1)$ is dependent and B_2 is independent, $C(x, B_1) - B_2$ is non-empty. Let y be an element of this set. Evidently $y \in B_1 - B_2$. Moreover, since $(B_1 - y) \cup x$ does not contain $C(x, B_1)$, the former is independent. Finally, since $(B_1 - y) \cup x$ has the same number of elements as B_1 and is independent, it is a basis. □

Proof of Theorem 2.1.1. As $\mathcal{B}(M)$ is non-empty, $\mathcal{B}^*(M)$ is non-empty. Hence $\mathcal{B}^*(M)$ satisfies (B1). Now suppose B_1^* and B_2^* are members of $\mathcal{B}^*(M)$ and $x \in B_1^* - B_2^*$. For $i = 1, 2$, let $B_i = E(M) - B_i^*$. Then $B_i \in \mathcal{B}(M)$ and $B_1^* - B_2^* = B_2 - B_1$. By (B2)*, as $x \in B_2 - B_1$, there is an element y of $B_1 - B_2$ such that $(B_1 - y) \cup x \in \mathcal{B}(M)$. It follows that $y \in B_2^* - B_1^*$ and that $E(M) - ((B_1 - y) \cup x) \in \mathcal{B}^*(M)$. But $E(M) - ((B_1 - y) \cup x) = ((E(M) - B_1) - x) \cup y = (B_1^* - x) \cup y$. We conclude that $\mathcal{B}^*(M)$ satisfies (B2). Hence $\mathcal{B}^*(M)$ is indeed the set of bases of a matroid on $E(M)$. □

The matroid in the last theorem, whose ground set is $E(M)$ and whose set of bases is $\mathcal{B}^*(M)$, is called the *dual* of M and is denoted by M^*. Thus $\mathcal{B}(M^*) = \mathcal{B}^*(M)$. Moreover, it is clear that

2.1.3 $$(M^*)^* = M.$$

2.1.4 **Example.** Consider $U_{m,n}$. Its bases are all of the m-element subsets of an n-element set E. Hence $\mathcal{B}^*(U_{m,n})$ consists of all the $(n - m)$-element subsets of E and so

$$U_{m,n}^* = U_{n-m,n}.$$ □

Corollary 1.2.5 gives one characterization of a matroid in terms of its collection of bases. A consequence of Theorem 2.1.1 is that an alternative such characterization is as follows.

2.1.5 **Corollary.** *Let $\mathcal{B}$ be a set of subsets of a set E. Then $\mathcal{B}$ is the collection of bases of a matroid on E if and only if $\mathcal{B}$ satisfies* (B1) *and* (B2)*.

Proof. If $\mathcal{B}$ is the collection of bases of a matroid, then, by Corollary 1.2.5 and Lemma 2.1.2, $\mathcal{B}$ satisfies (B1) and (B2)*. Now suppose that $\mathcal{B}$ satisfies (B1) and (B2)* and consider $\mathcal{B}' = \{E - B : B \in \mathcal{B}\}$. Clearly $\mathcal{B}'$ satisfies (B1) and (B2), so, by Corollary 1.2.5, $\mathcal{B}'$ is the set of bases of a matroid M on E. Thus $\mathcal{B}$ is the set of bases of a matroid on E, namely M^*, and the corollary is proved. □

There are a number of other alternative characterizations of matroids in terms of their bases. These are surveyed in Brylawski (1986b) and some of them are considered in the exercises.

The bases of M^* are called *cobases* of M. A similar convention applies to other distinguished subsets of $E(M^*)$. Hence, for example, the circuits, hyperplanes, independent sets, and spanning sets of M^* are called *cocircuits, cohyperplanes, coindependent sets*, and *cospanning sets* of M. The next result gives some elementary relationships between these sets.

2.1.6 Proposition. *Let M be a matroid on a set E and suppose $X \subseteq E$. Then*

(i) *X is independent if and only if $E - X$ is cospanning;*

(ii) *X is spanning if and only if $E - X$ is coindependent;*

(iii) *X is a hyperplane if and only if $E - X$ is a cocircuit; and*

(iv) *X is a circuit if and only if $E - X$ is a cohyperplane.*

Proof. Parts (i) and (ii) are left as exercises. For (iii), we have, by (ii) and Proposition 1.4.9(iii), that the following statements are equivalent:

(a) X is a hyperplane of M.

(b) X is a non-spanning set of M but $X \cup y$ is spanning for all $y \notin X$.

(c) $E - X$ is dependent in M^* but $(E - X) - y$ is independent in M^* for all y in $E - X$.

(d) $E - X$ is a cocircuit of M.

Part (iv) can be deduced by applying (iii) to M^*. □

A consequence of the last result is that if X is a circuit–hyperplane of a matroid M, then $E(M) - X$ is a circuit–hyperplane of M^*. Moreover, the operation of relaxation has the following attractive property.

2.1.7 Proposition. *If M' is obtained from M by relaxing the circuit–hyperplane X of M, then $(M')^*$ can be obtained from M^* by relaxing the circuit–hyperplane $E(M) - X$ of M^*.*

Proof. As $\mathcal{B}(M') = \mathcal{B}(M) \cup \{X\}$, we have $\mathcal{B}((M')^*) = \mathcal{B}(M^*) \cup \{E(M) - X\}$ and the result follows immediately. □

Proposition 2.1.7 is illustrated in Figure 2.1. The matroids in the second row are the duals of those in the first row. In both rows, each of the last four matroids is obtained from its predecessor by relaxing a circuit–hyperplane. Clearly each of $M(K_4)$, $\mathcal{W}^3$, Q_6, and P_6 is isomorphic to its dual, whereas $U_{3,6}$ is *equal* to its dual. We follow Bondy and Welsh (1971) in calling M *self-dual* if $M \cong M^*$, and *identically self-dual* if $M = M^*$.

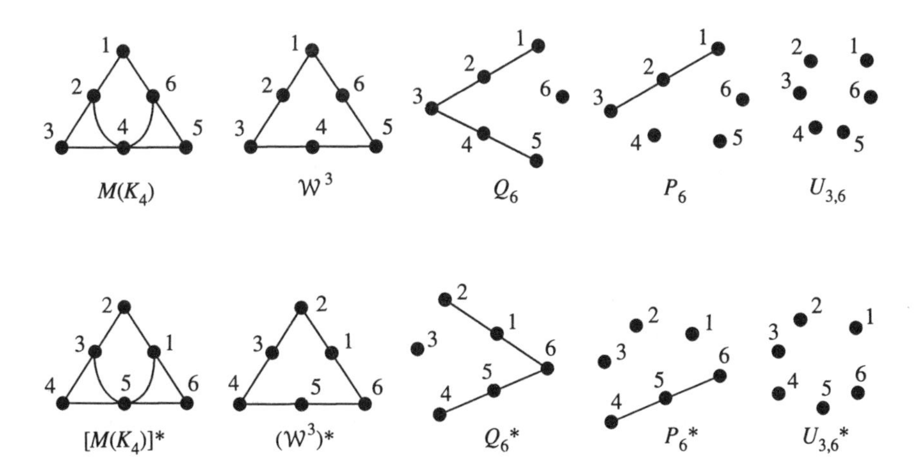

Fig. 2.1. Five matroids and their duals.

Note, however, that some authors reserve the term 'self-dual' for matroids M such that $M = M^*$.

The matroid M in Figure 2.2 is clearly simple. However, its dual M^* is not. It is mainly because the class of simple matroids is not closed under duality that matroid theorists do not, in general, confine their attention to simple matroids; duality is too important a part of matroid theory. This provides a contrast between matroid theory and graph theory, for, in the latter, many workers concentrate exclusively on simple graphs.

In Figure 2.2, the element 6, which is a loop of M^*, is a coloop of M. As 6 is in no basis of M^*, it is in every cobasis of M^*, that is, the element 6 is in every basis of M. We leave the reader to check that the last property actually characterizes coloops (Exercise 4).

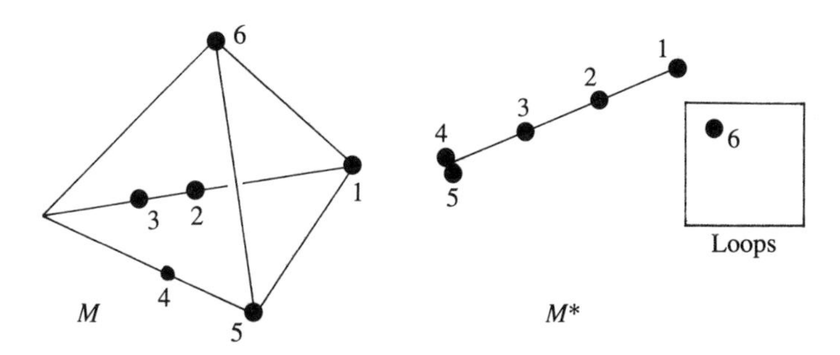

Fig. 2.2. A simple matroid with a non-simple dual.

In general, we attach an asterisk to a symbol to denote association with the dual. Thus, for example, r^* will denote the rank function of M^* while $\mathcal{C}^*$ denotes its set of circuits. Evidently

2.1.8 $$r(M) + r^*(M) = |E(M)|.$$

The geometric representations for matroids described in Section 1.5 apply only to matroids of rank at most four. However, such representations are also useful for matroids having rank more than four but corank at most four since, for these matroids, one can work with geometric representations for their duals.

The next result generalizes 2.1.8 to give a formula for r^*, the *corank function* of M. We caution the reader that some authors take the corank of a set X in a matroid M to be $r(M) - r(X)$ rather than $r^*(X)$ as is done here.

2.1.9 Proposition. *For all subsets X of the ground set E of a matroid M,*

$$r^*(X) = |X| - r(M) + r(E - X).$$

The proof of this will follow easily from the following augmentation result.

2.1.10 Lemma. *Let I and I^* be disjoint subsets of $E(M)$ such that I is independent and I^* is coindependent. Then M has a basis B and a cobasis B^* such that $I \subseteq B$, $I^* \subseteq B^*$, and B and B^* are disjoint.*

Proof. In $M|(E - I^*)$, the set I is independent. It is therefore contained in a basis B of this restriction. So $r(B) = r(E - I^*)$. But, by Proposition 2.1.6(ii), $E - I^*$ is spanning in M, hence $r(B) = r(M)$ and so B is a basis of M. Since $I \subseteq B$ and $I^* \subseteq E - B$, it follows that B and $E - B$ are the required basis and cobasis of M. □

Proof of Proposition 2.1.9. Let B^*_X be a basis for X in M^* and B_{E-X} be a basis for $E - X$ in M (see Figure 2.3). Then $r^*(X) = |B^*_X|$

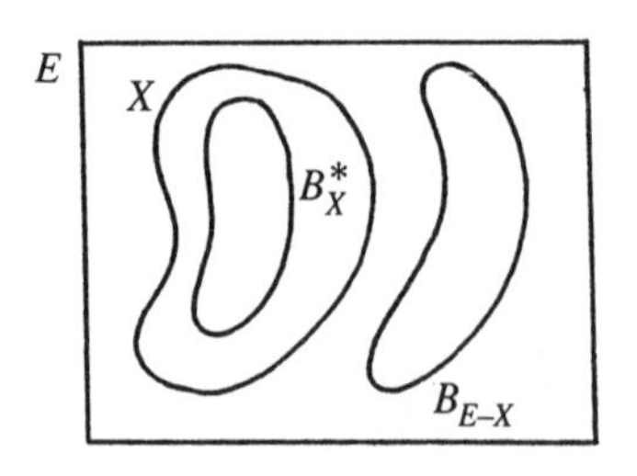

Fig. 2.3.

and $r(E-X) = |B_{E-X}|$. Now B_{E-X} and B^*_X are independent and coindependent, respectively, in M. Thus, by the last lemma, M has a basis B and a cobasis B^* such that B contains B_{E-X}, B^* contains B^*_X, and $B \cap B^* = \emptyset$. Because B_{E-X} is a basis for $M|(E-X)$, we have $B \cap (E-X) = B_{E-X}$. Similarly, $B^* \cap X = B^*_X$. Hence $B = B_{E-X} \dot\cup (B \cap X)$, so $|B \cap X| = |B| - |B_{E-X}|$. It follows that

$$\begin{aligned} |X| &= |X \cap B| + |X \cap B^*| \\ &= |B| - |B_{E-X}| + |B^*_X| \\ &= r(M) - r(E-X) + r^*(X). \end{aligned}$$

Rewriting this gives the required result. □

The following proposition expresses a fundamental link between circuits and cocircuits.

2.1.11 Proposition. *If C is a circuit and C^* is a cocircuit of the matroid M, then $|C \cap C^*| \neq 1$.*

Proof. Assume the contrary, letting $C \cap C^* = \{x\}$. Let $E(M) - C^* = H$. Then, by Proposition 2.1.6(iii), H is a hyperplane of M, so $\text{cl}(H) = H$. But $x \in C \subseteq H \cup x$, so, by 1.4.10(ii), $x \in \text{cl}(H) - H$; a contradiction. □

As we shall see later in this section, the last property of circuits is so basic that it can be used to characterize them.

A *clutter* is a collection of sets none of which is a proper subset of another. There are several clutters that are naturally associated with a matroid. These include the sets of bases, circuits, hyperplanes, cobases, cocircuits, and cohyperplanes. Indeed, several matroid axioms, such as (C2), are assertions that a particular collection of sets is a clutter. If $\mathcal{A}$ is a clutter of subsets of the set S, then clearly the set $\mathcal{A}'$ of complements of members of $\mathcal{A}$ is also a clutter on S. We call $\mathcal{A}'$ the *complementary clutter* of $\mathcal{A}$. The *blocker* $b(\mathcal{A})$ of $\mathcal{A}$ consists of those minimal subsets of S that have non-empty intersection with every member of $\mathcal{A}$. Evidently $b(\mathcal{A})$ is a clutter. Moreover, Edmonds and Fulkerson (1970) proved that clutters have the following attractive property.

2.1.12 Proposition. *If $\mathcal{A}$ is a clutter on the set S, then*

$$b(b(\mathcal{A})) = \mathcal{A}.$$

Proof. This is left to the reader (Exercise 16). □

Let M be a matroid and $\mathcal{B}(M)$, $\mathcal{C}(M)$, $\mathcal{H}(M)$, $\mathcal{B}^*(M)$, $\mathcal{C}^*(M)$, and $\mathcal{H}^*(M)$ be the clutters of bases, circuits, hyperplanes, cobases, cocircuits, and cohyperplanes of M. Then clearly

2.1.13 $\mathcal{B}(M)' = \mathcal{B}^*(M)$;

2.1.14 $\mathcal{H}(M)' = \mathcal{C}^*(M)$; and

2.1.15 $\mathcal{C}(M)' = \mathcal{H}^*(M)$.

Furthermore, the cocircuits of M are the minimal sets having non-empty intersection with every basis.

2.1.16 Proposition. *$\mathcal{C}^*(M) = b(\mathcal{B}(M))$ and $b(\mathcal{C}^*(M)) = \mathcal{B}(M)$.*

Proof. The following statements are equivalent:

(i) $C^* \in \mathcal{C}^*(M)$.

(ii) $E(M) - C^* \in \mathcal{H}(M)$.

(iii) $E(M) - C^*$ is a maximal non-spanning set in M.

(iv) $E(M) - C^*$ does not contain a basis of M but, for all x in C^*, $E(M) - (C^* - x)$ does contain a basis of M.

(v) C^* meets every basis of M but, for all x in C^*, $C^* - x$ avoids some basis of M.

(vi) $C^* \in b(\mathcal{B}(M))$.

We conclude that $\mathcal{C}^*(M) = b(\mathcal{B}(M))$, and the rest of the proposition follows easily from this using Proposition 2.1.12. □

If we replace M by M^* in the statement of the last proposition we obtain the statement:

$$\mathcal{C}^*(M^*) = b(\mathcal{B}(M^*)) \text{ and } b(\mathcal{C}^*(M^*)) = \mathcal{B}(M^*).$$

On rewriting this in terms of M, we get the following.

2.1.17 Corollary. *$\mathcal{C}(M) = b(\mathcal{B}^*(M))$ and $b(\mathcal{C}(M)) = \mathcal{B}^*(M)$.* □

This corollary is called the *dual* of Proposition 2.1.16. Expressed in words, the first part of the corollary asserts that the circuits of M are the minimal sets having non-empty intersection with every cobasis. This observation is the dual of the observation that preceded the statement of 2.1.16. In general, to form the dual of a statement or proposition, one follows the procedure exemplified above. Clearly a proposition is true if and only if the dual proposition is true. The dual of 2.1.11 is the same as the original except for renaming the sets C and C^*. Propositions such as this are called *self-dual.*

Since $\mathcal{H}(M)' = \mathcal{C}^*(M)$, we can use Corollary 1.1.5 to derive the following characterization of matroids in terms of their hyperplanes. The straightforward proof is left to the reader.

2.1.18 Proposition. *Let $\mathcal{H}$ be a set of subsets of a set E. Then $\mathcal{H}$ is the collection of hyperplanes of a matroid on E if and only if $\mathcal{H}$ satisfies the following conditions:*

(H1) $E \notin \mathcal{H}$.

(H2) *$\mathcal{H}$ is a clutter.*

(H3) *If H_1 and H_2 are distinct members of $\mathcal{H}$ and $e \in E - (H_1 \cup H_2)$, then there is a member H_3 of $\mathcal{H}$ such that $H_3 \supseteq (H_1 \cap H_2) \cup e$.* □

The next proposition extends 2.1.11 to give a characterization of the set of circuits of a matroid. The proof is based on the following lemma whose proof is left as an exercise.

2.1.19 Lemma. *Let $\mathcal{X}$ and $\mathcal{Y}$ be collections of subsets of a finite set E such that every member of $\mathcal{X}$ contains a member of $\mathcal{Y}$, and every member of $\mathcal{Y}$ contains a member of $\mathcal{X}$. Then the minimal members of $\mathcal{X}$ are precisely the minimal members of $\mathcal{Y}$.* □

2.1.20 Proposition. *Let M be a matroid having ground set E. Then D is a circuit of M if and only if D is a minimal non-empty subset of E such that $|D \cap C^*| \neq 1$ for every cocircuit C^* of M.*

Proof. We use the last lemma taking $\mathcal{X}$ equal to $\mathcal{C}(M)$ and $\mathcal{Y}$ equal to the set of non-empty subsets Y of E such that $|Y \cap C^*| \neq 1$ for all cocircuits C^*. By Proposition 2.1.11, if $X \in \mathcal{X}$, then $X \in \mathcal{Y}$. Now suppose Y is a minimal member of $\mathcal{Y}$. If Y is independent, then M has a basis B containing Y. Choose y in Y and consider $\text{cl}(B - y)$. This is a hyperplane of M. Its complement C_1^* is a cocircuit that meets Y in $\{y\}$; a contradiction. Hence Y is dependent. Thus Y contains a circuit C of M. If $Y \neq C$, then, by Proposition 2.1.11, we obtain a contradiction to the minimality of Y. We conclude that Y is a circuit of M, that is, $Y \in \mathcal{X}$. The proposition now follows immediately from Lemma 2.1.19. □

Recall from Section 1.3 that a paving matroid is a matroid M that has no circuits of size less than $r(M)$. Thus every matroid of rank less than two is paving. To close this section, we state a characterization, in terms of their hyperplanes, of the paving matroids of rank at least two (Hartmanis 1959). This result explains the reason for the name 'paving matroid', which was introduced by Welsh (1976) following a suggestion of G.-C. Rota. We have extended Welsh's definition slightly by dropping his requirement that such matroids have rank at least two. Let k and m be integers with $k > 1$ and $m > 0$. Suppose that $\mathcal{T}$ is a set $\{T_1, T_2, \ldots, T_k\}$ of subsets of a set E such that each member of $\mathcal{T}$ has at least m elements, and each m-element subset of E is contained in a unique member of $\mathcal{T}$. Such a set $\mathcal{T}$ is called an *m-partition* of E.

2.1.21 Proposition. *If $\mathcal{T}$ is an m-partition $\{T_1, T_2, \ldots, T_k\}$ of a set E, then $\mathcal{T}$ is the set of hyperplanes of a paving matroid of rank $m+1$ on E. Moreover, for $r \geq 2$, the set of hyperplanes of every rank-r paving matroid on E is an $(r-1)$-partition of E.*

Proof. It is straightforward to show that $\mathcal{T}$ satisfies (H1)–(H3). We leave the details of this, along with the rest of the proof, to the reader (Exercise 19). □

2.1.22 Example. Figure 2.4 is a geometric representation of an 8-element rank-4 matroid. To see that this diagram does indeed represent a matroid, one could check that it satisfies 1.5.9–1.5.11. Alternatively, let $\mathcal{T}_1 = \{\{1,2,3,4\}, \{1,4,5,6\}, \{2,3,5,6\}, \{1,4,7,8\}, \{2,3,7,8\}\}$ and $\mathcal{T} = \mathcal{T}_1 \cup \{T \subseteq \{1,2,\ldots,8\} : |T| = 3$ and T is not contained in any member of $\mathcal{T}_1\}$. Then one easily checks that $\mathcal{T}$ is a 3-partition of $\{1,2,\ldots,8\}$. Therefore, by the last proposition, $\mathcal{T}$ is the set of hyperplanes of a paving matroid on $\{1,2,\ldots,8\}$. Evidently Figure 2.4 is a geometric representation for this paving matroid. We call this matroid the *Vámos matroid* (Vámos 1968) and denote it by V_8. It occurs quite frequently throughout this book and has many interesting properties. For example, as we shall see in Chapter 6, it is not representable over any field.

A related matroid V_8^+ has the same ground set and is defined in the same way as V_8 but with $\{5,6,7,8\}$ added to $\mathcal{T}_1$. Again one can use Proposition 2.1.21 to check that V_8^+ is actually a matroid. In fact, V_8 is obtained from V_8^+ by relaxing $\{5,6,7,8\}$. It is straightforward to check that both V_8 and V_8^+ are self-dual with the latter, but not the former, being identically self-dual. □

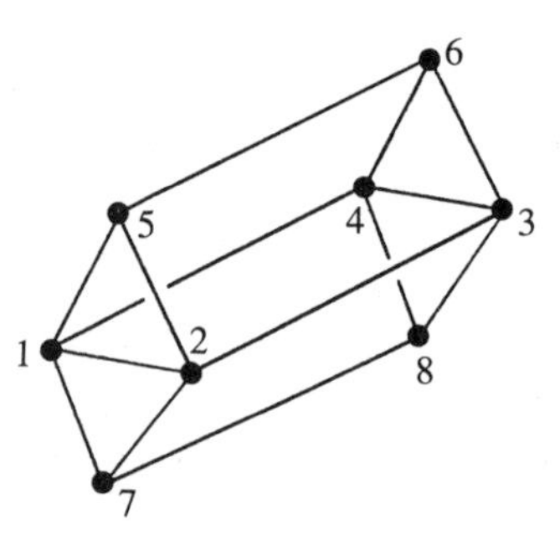

Fig. 2.4. The Vámos matroid, V_8.

Exercises

1. Find each of the following:

   (i) all self-dual uniform matroids;

   (ii) all identically self-dual uniform matroids;

   (iii) all self-dual graphic matroids on six or fewer elements;

   (iv) all identically self-dual graphic matroids on six or fewer elements;

   (v) an infinite family of simple graphic self-dual matroids.

2. Let M be a matroid. Show that M^* has two disjoint circuits if and only if M has two hyperplanes whose union is $E(M)$.

3. Let (X, Y) be a partition of the ground set of a matroid such that X is independent and Y is coindependent. Show that X is a basis and Y is a cobasis.

4. Show that an element e of a matroid M is a coloop of M if and only if e is in every basis of M. Now refer to Exercise 6 of Section 1.4 for a number of alternative characterizations of coloops.

5. For each of the following matroid properties, determine whether the property is necessary or sufficient (or both) for M to be identically self-dual:

   (i) Every basis of M is a cobasis of M.

   (ii) Every flat of M is a flat of M^*.

   (iii) The flats of M and M^* coincide.

   (iv) Every circuit of M is a cocircuit of M.

6. Let e and f be distinct elements of a matroid. Prove that every circuit containing e also contains f if and only if $\{e\}$ or $\{e, f\}$ is a cocircuit.

7. Let n be a positive integer. Show that the number of non-isomorphic matroids on an n-element set is at most twice the number of non-isomorphic self-dual matroids on a $(2n)$-element set.

8. Prove that a matroid is uniform if and only if every circuit meets every cocircuit.

9. Let M be a rank-r matroid.

(i) Show that if $r^*(M) \leq r$ and every hyperplane is a circuit, then every circuit is a hyperplane.

(ii) Use duality to show that if $r^*(M) \geq r$ and every circuit is a hyperplane, then every hyperplane is a circuit.

(iii) Give examples to show that (i) and (ii) can fail if the conditions on $r^*(M)$ are dropped.

10. Let B be a basis of a matroid M and B^* be $E(M) - B$. If $e \in B$, let $C^*(e, B^*)$ denote the *fundamental cocircuit of* e *with respect to the cobasis* B^* of M, that is, $C^*(e, B^*) = C_{M^*}(e, B^*)$.

(i) Show that $C^*(e, B^*)$ is the unique cocircuit that is disjoint from $B - e$.

(ii) If $f \in B^*$, prove that $f \in C^*(e, B^*)$ if and only if $e \in C(f, B)$.

11. (Bondy and Welsh 1971) Let $C_1^*, C_2^*, \ldots, C_r^*$ be cocircuits of a rank-r matroid M. Prove that the following are equivalent:

(a) M has a cobasis B^* such that $C_1^*, C_2^*, \ldots, C_r^*$ are all of the fundamental cocircuits with respect to B^*;

(b) for all j in $\{1, 2, \ldots, r\}$, C_j^* is not contained in $\bigcup_{i \neq j} C_i^*$.

12. Let M be a matroid on the set E and, for all subsets X of E, define $r_1(X) = |X| - r(M) + r(E - X)$.

(i) Prove directly that r_1 satisfies (R1)–(R3) and hence deduce that r_1 is the rank function of a matroid M_1 on E.

(ii) Show directly that B is a basis of M_1 if and only if $E - B$ is a basis of M.

13. A *cyclic flat* of a matroid M is a flat of M that is the union of a (possibly empty) set of circuits. Show that:

(i) X is a flat of M if and only if $E(M) - X$ is the union of a (possibly empty) set of cocircuits of M. Deduce that F is a cyclic flat of M if and only if $E(M) - F$ is a cyclic flat of M^*.

(ii) A flat F of M is cyclic if and only if $M|F$ has no coloops.

(iii) A matroid is uniquely determined by a list of its cyclic flats and their ranks.

14. (Brylawski 1986b) Let $\mathcal{B}$ be a collection of subsets of a set E and suppose $\mathcal{B}$ satisfies the following conditions:

(B1) $\mathcal{B} \neq \emptyset$.

(B3) *If $B_1, B_2 \in \mathcal{B}$ and $B_1 \subseteq B_2$, then $B_1 = B_2$.*

(B2)$^{*-}$ *If $B_1, B_2 \in \mathcal{B}$ and $e \in B_1$, then there is an element f of B_2 such that $(B_2 - f) \cup e \in \mathcal{B}$.*

(i) Give an example when $|E| = 5$ to show that $\mathcal{B}$ need not be the set of bases of a matroid on E.

(ii) Prove that the set of bases of a matroid on E satisfies (B2)$^{*-}$.

15. Find all graphs whose cycle matroids have a circuit–hyperplane.

16. (Dawson 1981) Let $\mathcal{A}$ and $\mathcal{D}$ be clutters on a set S. Prove that the following are equivalent:

(a) $\mathcal{D} = b(\mathcal{A})$.

(b) For all A in $\mathcal{A}$ and all D in $\mathcal{D}$, $A \cap D \neq \emptyset$. Moreover, if $d \in D$, then $\mathcal{A}$ has a member A_d such that $D \cap A_d = \{d\}$.

(c) For all $X \subseteq S$, X contains a member of $\mathcal{D}$ if and only if $S - X$ does not contain a member of $\mathcal{A}$.

(d) $\mathcal{A} = b(\mathcal{D})$.

(Hint: Show that (a) implies (b), that (b) implies (c), and that (c) implies (a). Then use the symmetry of (c).)

17. Show that $AG(3, 2)$ is identically self-dual.

18. Let G_1 and G_2 be the graphs shown in Figure 1.20. The duals of $M(G_1)$ and $M(G_2)$, which are denoted $M^*(G_1)$ and $M^*(G_2)$, respectively, are both graphic and are both transversal. Find the following:

(i) graphs H_1 and H_2 such that $M(H_1) \cong M^*(G_1)$ and $M(H_2) \cong M^*(G_2)$;

(ii) families of sets $\mathcal{A}_1$ and $\mathcal{A}_2$ such that $M[\mathcal{A}_i] \cong M^*(G_i)$ for $i = 1, 2$;

(iii) geometric representations for $M^*(G_1)$ and $M^*(G_2)$.

19. Prove Proposition 2.1.21.

20. Characterize paving matroids in terms of their collections of flats.

21. **(i)** If $(E, \mathcal{I})$ is a matroid M, $I \in \mathcal{I}$, and $e \in E - I$, prove that $I \cup e \in \mathcal{I}$ if and only if there is a cocircuit C^* of M such that $e \in C^*$ and $C^* \cap I$ is empty.

(ii)* (Dawson 1980) Let $\mathcal{B}$ be the set of bases and C^* be the set of cocircuits of a matroid M. Let $w : E \to \mathbb{R}$ be a weight function where $E = E(M)$. Prove that the set B_D produced by the following modified greedy algorithm is a maximum-weight basis of M.

(a) Set $Y_0 = \emptyset$ and $j = 0$.

(b) If $E - Y_j$ does not contain a member of $\mathcal{C}^*$, then let $Y_j = B_D$ and go to (d); otherwise, let C^*_{j+1} be a member of $\mathcal{C}$ contained in $E - Y_j$ and choose e_{j+1} to be an element of C^*_{j+1} of maximum weight. Let $Y_{j+1} = Y_j \cup e_{j+1}$ and go to (c).

(c) Add 1 to j and go to (b).

(d) Stop.

(iii) Consider the Pappus matroid and the three-dimensional Desargues matroid shown in Figures 1.16 and 1.17. Taking the element labels as weights, use the above algorithm to find maximum-weight bases in the *duals* of these two matroids.

2.2 Duals of representable matroids

In this section we prove that the dual of an F-representable matroid M is F-representable by constructing an explicit representation for M^* from a representation for M.

Let A be an $m \times n$ matrix over the field F and let $M[A]$ be the vector matroid of A. Then $M[A]$ has as its ground set the set E of column labels of A. In general, $M[A]$ does not uniquely determine A. Let $M = M[A]$. Then it is an elementary exercise to check that M remains unchanged if one performs any of the following operations on A.

2.2.1 *Interchange two rows.*

2.2.2 *Multiply a row by a non-zero member of F.*

2.2.3 *Replace a row by the sum of that row and another.*

2.2.4 *Delete a zero row (unless it is the only row).*

2.2.5 *Interchange two columns (the labels moving with the columns).*

2.2.6 *Multiply a non-zero column by a non-zero member of F.*

2.2.7 *Replace each entry of the matrix by its image under some automorphism of F.*

We call operations 2.2.1–2.2.3 *elementary row operations.* Operation 2.2.7 will be examined in detail in Section 6.3. It will not be considered further in this section.

Assume that the matrix A is non-zero. It is well known and not difficult to check that, by a sequence of operations of types 2.2.1–2.2.5, one can reduce A to the form $[I_r|D]$ where I_r is the $r \times r$ identity matrix and D is some $r \times (n-r)$ matrix over F. Evidently $r = r(M)$. The matrix $[I_r|D]$ is called a *standard representative matrix* for M. In the next theorem, we shall assume that the columns of $[I_r|D]$ are labelled, in order, $e_1, e_2, \ldots, e_n$ as in Figure 2.5(a), and that the matrix $[-D^T|I_{n-r}]$ has its columns labelled in the same order, as in Figure 2.5(b). We shall also assume that $n > r$. The proof of this theorem will use the familiar notion of rank of a matrix. Recall that this is the maximum size of a linearly independent set of rows of the matrix or, equivalently, the maximum size of a linearly independent set of columns.

$$\begin{array}{cc} \begin{array}{c} e_1e_2\cdots e_r \quad e_{r+1}e_{r+2}\cdots e_n \\ \left[\begin{array}{c|c} \quad I_r \quad & \quad D \quad \end{array}\right] \end{array} & \begin{array}{c} e_1e_2\cdots e_r \quad e_{r+1}e_{r+2}\cdots e_n \\ \left[\begin{array}{c|c} \quad -D^T \quad & \quad I_{n-r} \quad \end{array}\right] \end{array} \\ \text{(a)} & \text{(b)} \end{array}$$

Fig. 2.5. Representations for a matroid and its dual.

2.2.8 Theorem. *If M is the vector matroid of $[I_r|D]$, then M^* is the vector matroid of $[-D^T|I_{n-r}]$.*

Proof. Let B be a basis of M. We shall first show that $E - B$ is a basis of the vector matroid of $[-D^T|I_{n-r}]$. The only effect on $[-D^T|I_{n-r}]$ of rearranging rows and rearranging columns of $[I_r|D]$ is to rearrange columns and rearrange rows of $[-D^T|I_{n-r}]$. By such a rearrangement, we may assume that, for some t in $\{0, 1, \ldots, r\}$, $B = \{e_{r-t+1}, e_{r-t+2}, \ldots, e_r, e_{r+1}, \ldots, e_{2r-t}\}$. Thus $[I_r|D]$ can be partitioned into blocks as follows:

$$\begin{array}{cccc} e_1\cdots e_{r-t} & e_{r-t+1}\cdots e_r & e_{r+1}\cdots e_{2r-t} & e_{2r-t+1}\cdots e_n \end{array}$$
$$\left[\begin{array}{c||c|c||c} I_{r-t} & 0 & D_1 & D_2 \\ \hline 0 & I_t & D_3 & D_4 \end{array}\right].$$

The matrix

$$\left[\begin{array}{c|c} 0 & D_1 \\ \hline I_t & D_3 \end{array}\right]$$

has rank r because B is a basis. Thus D_1, and hence $-D_1^T$, has rank $r-t$. The partition of $[I_r|D]$ induces the following partition of $[-D^T|I_{n-r}]$:

$$\begin{array}{c} \quad e_1 \cdots e_{r-t} \quad e_{r-t+1} \cdots e_r \quad e_{r+1} \cdots e_{2r-t} \quad e_{2r-t+1} \cdots e_n \\ \left[\begin{array}{c||c|c||c} -D_1^T & -D_3^T & I_{r-t} & 0 \\ \hline -D_2^T & -D_4^T & 0 & I_{n-(2r-t)} \end{array}\right]. \end{array}$$

In this matrix, the submatrix corresponding to $E - B$ is

$$\left[\begin{array}{c|c} -D_1^T & 0 \\ \hline -D_2^T & I_{n-(2r-t)} \end{array}\right].$$

Its rank equals the sum of the ranks of $I_{n-(2r-t)}$ and $-D_1^T$, that is, $(n-(2r-t))+(r-t) = n-r$. Hence $E-B$ is indeed a basis of the vector matroid of $[-D^T|I_{n-r}]$. An easy extension of this argument shows that every basis of this matroid arises in this way. We conclude that $[-D^T|I_{n-r}]$ is indeed a representation for M^*. □

We remark here that the last theorem assumed that the rank r of the matrix A is neither 0 nor n. If $r = 0$, then $M \cong U_{0,n}$, so $M^* \cong U_{n,n} \cong M[I_n]$. If $r = n$, then $M \cong U_{n,n}$, so $M^* \cong U_{0,n} \cong M[0_n]$ where 0_n is the $n \times n$ zero matrix. On combining these observations with Theorem 2.2.8, we immediately obtain the following result.

2.2.9 Corollary. *If M is representable over the field F, then M^* is also representable over F.* □

In particular, the dual of a binary matroid is binary, and the dual of a ternary matroid is ternary.

2.2.10 Example. Consider the following matrix, A.

$$\left[\begin{array}{c|cccc} & 0 & 1 & 1 & 1 \\ I_3 & 1 & 0 & 1 & 1 \\ & 1 & 1 & 0 & 1 \end{array}\right]$$

It is not difficult to show that if A is viewed as a matrix over $GF(2)$, it represents F_7, whereas, viewed as a matrix over $GF(3)$, it represents F_7^-. By Theorem 2.2.8, F_7^* and $(F_7^-)^*$ are represented by the matrix

$$\left[\begin{array}{ccc|c} 0 & -1 & -1 & \\ -1 & 0 & -1 & \\ -1 & -1 & 0 & I_4 \\ -1 & -1 & -1 & \end{array}\right]$$

viewed over $GF(2)$ and $GF(3)$, respectively, where, in the former case, $-1 = 1$. When the last matrix is viewed over $GF(3)$, scalar multiplication of columns enables us to change the signs of all negative entries. Hence F_7^* and $(F_7^-)^*$ are represented by the matrix

$$\left[\begin{array}{ccc|c} 0 & 1 & 1 & \\ 1 & 0 & 1 & \\ 1 & 1 & 0 & I_4 \\ 1 & 1 & 1 & \end{array}\right]$$

viewed over $GF(2)$ and $GF(3)$, respectively. Before leaving this example, we remark that, by Proposition 2.1.7, as F_7^- is a relaxation of F_7, it follows that $(F_7^-)^*$ is a relaxation of F_7^*. □

We shall show next that the dual of a regular matroid is regular. Recall that M is regular if $M \cong M[A]$ for some totally unimodular matrix A, the latter being a real matrix for which every square submatrix has determinant in $\{0, 1, -1\}$. By the last corollary, if M is regular, then M^* is certainly $\mathbb{R}$-representable, so we need to check that M^* has a totally unimodular representation. We begin by recalling for reference, both here and throughout the rest of the book, the following basic facts about determinants of matrices. The proofs of these results can be found in most linear algebra texts.

Let A be the $n \times n$ matrix $[a_{ij}]$ over the field F. The *determinant* of A, denoted $\det A$, is defined to be

2.2.11 $$\sum_\sigma \operatorname{sgn} \sigma \, a_{1\sigma(1)} a_{2\sigma(2)} \cdots a_{n\sigma(n)}$$

where the summation is over all permutations σ of $\{1, 2, \ldots, n\}$ and sgn σ is $+1$ or -1 according to whether σ is a product of an even or an odd number of cycles of length two.

2.2.12 *For all i in $\{1, 2, \ldots, n\}$,*

$$\det A = \sum_{j=1}^{n} a_{ij}(-1)^{i+j} \det(A_{ij})$$

where A_{ij} is the matrix obtained from A by deleting row i and column j.

2.2.13 *For all j in $\{1, 2, \ldots, n\}$,*

$$\det A = \sum_{i=1}^{n} a_{ij}(-1)^{i+j} \det(A_{ij}).$$

2.2.14 $$\det(A^T) = \det A.$$

2.2.15 *If D is obtained from A by multiplying some row or column by a constant c, then* $\det D = c \det A$.

2.2.16 *If D is obtained from A by interchanging two rows or interchanging two columns, then* $\det D = -\det A$.

2.2.17 *If D is obtained from A by replacing* row i *by* row $i + c$ row j *for some constant c and some $j \neq i$, then* $\det D = \det A$.

On combining 2.2.15–2.2.17, it is not difficult to show that

2.2.18 $\det A = 0$ *if and only if the columns of A are linearly dependent.*

Finally,

2.2.19 $$\det(AD) = \det A \cdot \det D.$$

A familiar operation that forms part of the process of transforming a matrix into row-echelon form is pivoting. Given an $m \times n$ matrix X and a non-zero entry x_{st} of X, a *pivot* of X *on* x_{st} transforms the tth column of X into the sth unit vector. This is achieved by applying the following sequence of operations to X:

(i) For each i in $\{1, 2, \ldots, s-1, s+1, \ldots, m\}$, replace row i of X by row $i - (x_{it}/x_{st})$ row s.

(ii) Multiply row s by $1/x_{st}$.

The entry x_{st} of X is called the *pivot element*; its value in the transformed matrix is 1.

We shall need two more lemmas to show that the class of regular matroids is *closed under duality*, that is, the dual of every member of the class is also in the class.

2.2.20 Lemma. *Let X be a totally unimodular matrix. If Y is obtained from X by pivoting on the non-zero entry x_{st} of the latter, then Y is totally unimodular.*

Proof. Let X' and Y' be corresponding submatrices of X and Y, each having their rows and columns indexed by the sets J_R and J_C, respectively. We want to show that $\det Y' \in \{0, 1, -1\}$. If $s \in J_R$, then it follows easily from 2.2.15 and 2.2.17 that $|\det Y'| = |\det X'|$ and so $\det Y' \in \{0, 1, -1\}$. Thus we may assume that $s \notin J_R$. In that case, if $t \in J_C$, then Y' has a zero column, so $\det Y' = 0$. Hence we may also assume that $t \notin J_C$. Now let X'' and Y'' be the submatrices of X and Y whose rows and columns are indexed by $J_R \cup \{s\}$ and

$J_C \cup \{t\}$. As the only non-zero entry in the column of Y'' indexed by t is a 1, $|\det\ Y''| = |\det\ Y'|$. But, as above, $|\det Y''| = |\det X''|$. Hence again, $\det Y' \in \{0, 1, -1\}$ and the lemma is proved. □

2.2.21 Lemma. *A matroid M of non-zero rank r is regular if and only if M can be represented over $\mathbb{R}$ by some totally unimodular matrix of the form $[I_r|D]$.*

Proof. Let M be regular. Then $M \cong M[A]$ for some totally unimodular matrix A. But, by 2.2.16 and Lemma 2.2.20, A can be transformed, by a sequence of pivots and column interchanges, into a totally unimodular matrix in the form $[I_r|D]$. Hence $M \cong M[I_r|D]$. The converse is immediate from the definition of regularity. □

2.2.22 Proposition. *The dual of a regular matroid is regular.*

Proof. Let M be regular. If $r(M)$ or $r^*(M)$ is 0, then M^* is certainly regular, so we may assume that both $r(M)$ and $r^*(M)$ are positive. By Lemma 2.2.21, $M \cong M[I_r|D]$ for some totally unimodular matrix $[I_r|D]$. Thus, by Theorem 2.2.8, M^* is represented by $[-D^T|I_{n-r}]$ and, since the last matrix is certainly totally unimodular, the proposition follows. □

Some authors (see, for example, White 1986) call M^* the *orthogonal matroid* of M. This is because duality for representable matroids is actually an extension of the notion of orthogonality in vector spaces. Recall that if $\underline{v}$ and $\underline{w}$ are the vectors $(v_1, v_2, \ldots, v_n)$ and $(w_1, w_2, \ldots, w_n)$ of $V(n, F)$, then $\underline{v} \cdot \underline{w} = \sum_{i=1}^{n} v_i w_i$. These vectors are *orthogonal* if $\underline{v} \cdot \underline{w} = 0$. Moreover, $\underline{v}$ is *orthogonal to a subspace* W of $V(n, F)$ if v is orthogonal to every member of W. One easily checks that the set $W^{\perp}$ of vectors in $V(n, F)$ that are orthogonal to W forms a subspace; we call this the *orthogonal subspace* of W. Clearly $(W^{\perp})^{\perp} = W$. The link between matroid duality and vector spaces orthogonality, which is contained in the next result, is stated in terms of row spaces of matrices. In general, the *row space* $\mathcal{R}(A)$ of an $m \times n$ matrix A over a field F is the subspace of $V(n, F)$ that is spanned by the rows of A.

2.2.23 Proposition. *Let $[I_r|D]$ be an $r \times n$ matrix over a field F where $1 \leq r \leq n - 1$. Then the orthogonal subspace of $\mathcal{R}[I_r|D]$ is $\mathcal{R}[-D^T|I_{n-r}]$.*

Proof. Let $\underline{v}$ and $\underline{w}$ be rows g and h of $[I_r|D]$ and $[-D^T|I_{n-r}]$, respectively. Then, writing δ_{ij} for the Kronecker delta and letting $D = [d_{ij}]$, we have

$$\underline{v} = (\delta_{g1}, \delta_{g2}, \ldots, \delta_{gg}, \ldots, \delta_{gr}, d_{g1}, d_{g2}, \ldots d_{g(n-r)})$$

and

$$\underline{w} = (-d_{1h}, -d_{2h}, \ldots, -d_{rh}, \delta_{h1}, \delta_{h2}, \ldots, \delta_{hh}, \ldots, \delta_{h(n-r)}).$$

Thus $\underline{v} \cdot \underline{w} = -d_{gh} + d_{gh} = 0$, that is, $\underline{v}$ and $\underline{w}$ are orthogonal. Therefore the orthogonal subspace of $\mathcal{R}[I_r|D]$ contains $\mathcal{R}[-D^T|I_{n-r}]$.

Now suppose $\underline{u}$ is an arbitrary member $(u_1, u_2, \ldots, u_n)$ of the orthogonal subspace of $\mathcal{R}[I_r|D]$. Then, by taking the dot product of $\underline{u}$ with each row of $[I_r|D]$ we get that, for all i in $\{1, 2, \ldots, r\}$, $u_i = -\sum_{j=1}^{n-r} u_{r+j} d_{ij}$. Hence

$$\begin{aligned}\underline{u} &= u_{r+1}(-d_{11}, -d_{21}, \ldots, -d_{r1}, 1, 0, 0, \ldots, 0) \\ &\quad + u_{r+2}(-d_{12}, -d_{22}, \ldots, -d_{r2}, 0, 1, 0, \ldots, 0) + \ldots \\ &\quad + u_{r+(n-r)}(-d_{1(n-r)}, -d_{2(n-r)}, \ldots, -d_{r(n-r)}, 0, 0, 0, \ldots, 1).\end{aligned}$$

Since the $n-r$ vectors on the right-hand side are the rows of $[-D^T|I_{n-r}]$, we conclude that $\underline{u} \in \mathcal{R}[-D^T|I_{n-r}]$. □

The following is a straightforward consequence of the last proposition and we leave the proof to the reader.

2.2.24 Corollary. *Let W be an r-dimensional subspace of $V(n, F)$. Then $W^\perp$ has dimension $n - r$. Moreover, if F has characteristic zero, then $W^\perp \cap W = \{\underline{0}\}$.* □

By this corollary, if F has characteristic zero and $[I_r|D]$ is an $r \times n$ matrix over F where $1 \leq r \leq n-1$, then $\mathcal{R}[I_r|D]$ and $\mathcal{R}[-D^T|I_{n-r}]$ are complementary subspaces of $V(n, F)$. However, if the characteristic of F is not zero, these row spaces may have non-trivial intersection. For instance, if F is $GF(2)$ and $W = \mathcal{R}[I_r|I_r]$, then $W^\perp$ actually equals W.

We shall denote by $M^*[A]$ the dual of the vector matroid $M[A]$. Theorem 2.2.8 and the remarks following it give us a procedure for constructing a representation for $M^*[A]$ for any $m \times n$ matrix A of rank r. If r is 0 or n, then $M^*[A]$ is represented by I_n or 0_n, respectively. Otherwise, we reduce A to the form $[I_r|D]$. Then $M^*[A]$ is represented by $[-D^T|I_{n-r}]$. Notice that, when $0 < r < n$, $M^*[A]$ is also represented by $[D^T|I_{n-r}]$ since, as noted in 2.2.6, multiplying a column of a matrix by a non-zero scalar does not alter the associated vector matroid. In general, however, this alternative representation no longer has the property that its row space is the orthogonal subspace of $\mathcal{R}[I_r|D]$.

2.2.25 Example. Consider the vector matroid M of the following matrix

$$\begin{array}{c} \begin{matrix} 1 & 2 & 3 & 4 & \quad 5 & 6 & 7 & 8 \end{matrix} \\ \left[\begin{array}{c|cccc} & 0 & 1 & 1 & -1 \\ I_4 & 1 & 0 & 1 & 1 \\ & 1 & 1 & 0 & 1 \\ & -1 & 1 & 1 & 0 \end{array}\right] \end{array}$$

over $GF(3)$. Then M^* is the vector matroid of the matrix

$$\begin{array}{c} \begin{array}{cccc} 5 & 6 & 7 & 8 \end{array} \quad \begin{array}{cccc} 1 & 2 & 3 & 4 \end{array} \\ \left[\begin{array}{c|cccc} & 0 & 1 & 1 & -1 \\ I_4 & 1 & 0 & 1 & 1 \\ & 1 & 1 & 0 & 1 \\ & -1 & 1 & 1 & 0 \end{array}\right]. \end{array}$$

Hence M is self-dual. □

This example illustrates the fact that if $M = M[I_r|D]$ where D is a symmetric matrix, then M is self-dual. We shall return to this example several times in this book. We leave it now with the comment that, when the matrix representing M is viewed over $GF(2)$, it represents $AG(3,2)$, although M itself is certainly not isomorphic to $AG(3,2)$.

Exercises

1. For $i = 1,2$, let A_i be an $m_i \times n$ matrix over a field F. Suppose that every row of A_2 is a linear combination of rows of A_1. Prove that $M\left[\frac{A_1}{A_2}\right] = M[A_1]$.

2. Let A be an $m \times n$ matrix over a field F.

(i) For each of the row operations 2.2.1–2.2.3, specify an $m \times m$ matrix L such that multiplying A on the left by L has the same effect as the row operation.

(ii) For each of the column operations 2.2.4 and 2.2.5, specify an $n \times n$ matrix R such that multiplying A on the right by R has the same effect as the column operation.

3. Let A be a non-zero matrix over a field F and suppose $M[A]$ has no coloops. Show that no row of A contains exactly one non-zero entry.

4. Show that, in a binary matroid, a circuit and cocircuit cannot have an odd number of common elements.

5. Prove that $U_{n,n+2}$ is F-representable if and only if $|F| \geq n+1$.

6. Let M_1 and M_2 be matroids on a set E and cl_1 and cl_2 be their closure operators. Prove that $M_1 = M_2^*$ if and only if, for every partition $(\{e\}, X, Y)$ of E, the element e is in exactly one of $\mathrm{cl}_1(X)$ and $\mathrm{cl}_2(Y)$.

7. Find an $\mathbb{R}$-representation for M^* when M is the vector matroid of the following matrix over $\mathbb{R}$:

$$\begin{array}{c} \begin{array}{cccccc} 1 & 2 & 3 & & n{-}1 & n \end{array} \\ \begin{bmatrix} 1 & 0 & 0 & \dots & 0 & 1 \\ 1 & 1 & 0 & \dots & 0 & 0 \\ 0 & 1 & 1 & \dots & 0 & 0 \\ 0 & 0 & 1 & \dots & 0 & 0 \\ \vdots & \vdots & \vdots & \ddots & \vdots & \vdots \\ 0 & 0 & 0 & \dots & 1 & 0 \\ 0 & 0 & 0 & \dots & 1 & (-1)^n \end{bmatrix} \end{array}.$$

8. Let T_8 and R_8 be the vector matroids of the following matrices over $GF(3)$:

$$\begin{array}{cc} \begin{array}{cccc} 1 & 2 & 3 & 4 \end{array} & \begin{array}{cccc} 5 & 6 & 7 & 8 \end{array} \\ \left[\begin{array}{c} \\ I_4 \\ \\ \end{array}\right. & \left|\begin{array}{cccc} 0 & 1 & 1 & 1 \\ 1 & 0 & 1 & 1 \\ 1 & 1 & 0 & 1 \\ 1 & 1 & 1 & 0 \end{array}\right] \end{array}, \quad \begin{array}{cc} \begin{array}{cccc} 1 & 2 & 3 & 4 \end{array} & \begin{array}{cccc} 5 & 6 & 7 & 8 \end{array} \\ \left[\begin{array}{c} \\ I_4 \\ \\ \end{array}\right. & \left|\begin{array}{cccc} -1 & 1 & 1 & 1 \\ 1 & -1 & 1 & 1 \\ 1 & 1 & -1 & 1 \\ 1 & 1 & 1 & -1 \end{array}\right] \end{array}.$$

(i) Show that T_8 and R_8 are both self-dual.

(ii) Show that R_8 is identically self-dual but T_8 is not.

(iii) Give geometric representations for T_8 and R_8.

(iv) Show that if $M \in \{T_8, R_8\}$ and $X = E(M) - \{8\}$, then $(M|X)^* \cong F_7^-$.

(v) Consider the following matrices over $GF(3)$:

$$A_1 = \left[\begin{array}{c|cccc} & 0 & 1 & 1 & 1 \\ I_4 & 1 & 0 & 1 & 1 \\ & 1 & 1 & 0 & -1 \\ & 1 & 1 & 1 & 0 \end{array}\right], \quad A_2 = \left[\begin{array}{c|cccc} & 0 & 1 & 1 & -1 \\ I_4 & 1 & 0 & 1 & -1 \\ & 1 & 1 & 0 & -1 \\ & 1 & 1 & 1 & 1 \end{array}\right].$$

Show, by applying a sequence of the operations 2.2.1–2.2.6 to A_1 and A_2, that $M[A_1] \cong T_8$ and $M[A_2] \cong R_8$.

(vi) Show that R_8 can be obtained from $AG(3,2)$ by relaxing *two* disjoint circuit–hyperplanes.

9. Is the matroid M in Example 2.2.25 identically self-dual?

10. (i) Show that if $\{b_1, b_2, \ldots, b_r\}$ is a basis of an F-representable matroid M, then there is a matrix $[I_r|D]$ over F such that $M[I_r|D] = M$ and the columns of I_r are labelled, in order, by $b_1, b_2, \ldots, b_r$.

    (ii) Let A_1 and A_2 be matrices over a field F and suppose each has n columns. Assume that $\mathcal{R}(A_2)$ is orthogonal to $\mathcal{R}(A_1)$ and that the sum of the dimensions of these row spaces is n. Show that $M[A_1] \cong M^*[A_2]$.

11. Let M be a binary matroid and $B = \{b_1, b_2, \ldots, b_r\}$ be a basis of M where $r \geq 1$. Let $E(M) - B = \{e_1, e_2, \ldots, e_k\}$ and define the $r \times k$ matrix $[d_{ij}]$ by

$$d_{ij} = \begin{cases} 1, & \text{if } b_i \in C(e_j, B); \\ 0, & \text{otherwise.} \end{cases}$$

    (i) Prove that M equals the vector matroid of $[I_r|D]$ where this matrix is viewed over $GF(2)$, its columns being labelled, in order, by $b_1, b_2, \ldots, b_r$, $e_1, e_2, \ldots, e_k$.

    (ii) Find binary representations for $M(K_5)$ and $M(K_{3,3})$ and their duals. (Hint: Use (i) and the fact, to be proved in Chapter 5, that graphic matroids are binary.)

    (iii) Use (i) to show that none of $U_{2,4}$, F_7^-, V_8, and V_8^+ is binary.

2.3 Duals of graphic matroids

If G is a graph, we denote the dual of the cycle matroid of G by $M^*(G)$. This matroid is called the *bond matroid* of G or, sometimes, the *cocycle matroid* of G. An arbitrary matroid that is isomorphic to the bond matroid of some graph is called *cographic*. In this section we investigate cographic matroids, focusing particularly on the question of when such matroids are graphic.

If X is a set of edges in a graph G, we shall denote by $G \backslash X$ the subgraph of G obtained by deleting all the edges in X. If $G \backslash X$ has more connected components than G, we shall call X an *edge cut* of G. An edge e for which $\{e\}$ is an edge cut is called a *cut-edge*. A minimal edge cut will also be called a *bond* or a *cocycle* of G. In Proposition 1.4.13, we proved that the complement of a hyperplane in $M(G)$ is a bond of G. The next result follows immediately on combining that result with Proposition 2.1.6(iii).

2.3.1 Proposition. *The following statements are equivalent for a subset X of the set of edges of a graph G:*

(i) X *is a circuit of* $M^*(G)$.

(ii) X *is a cocircuit of* $M(G)$.

(iii) X *is a bond of* G. □

For an arbitrary graph G, the circuits of $M(G)$ are the edge sets of cycles of G, while the circuits of $M^*(G)$ are the edge sets of bonds. If v is a vertex of G and X is the set of edges meeting v, then X is an edge cut. If such an X is a minimal edge cut, we call it a *vertex bond* of G.

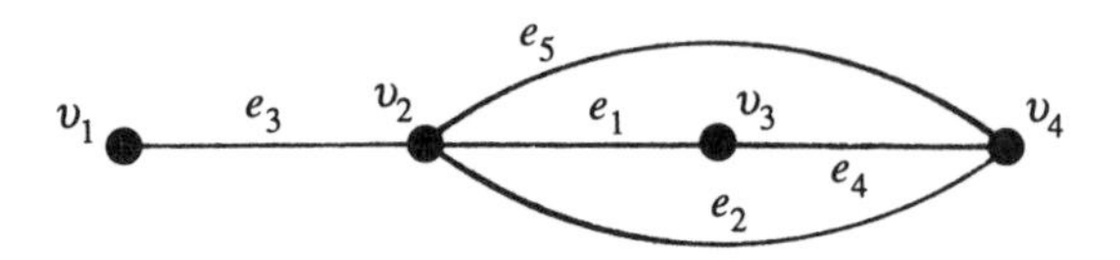

Fig. 2.6. The edges incident with a vertex need not be a bond.

2.3.2 **Example.** If H is the graph shown in Figure 2.6, then the vertex bonds are $\{e_3\}$, $\{e_1, e_4\}$, and $\{e_2, e_4, e_5\}$. Evidently, although $\{e_1, e_2, e_3, e_5\}$ is the set of edges meeting the vertex v_2, it is not a bond. Moreover, it is not difficult to check that $M(H) \cong M^*(G)$ where G is the graph in Example 1.1.8. □

The graphic matroid in the last example had the property that its dual is also graphic. We saw in the last section that the dual of a representable matroid is always representable. We now consider the general problem of characterizing precisely when the dual of a graphic matroid is graphic. The following result indicates that this need not always occur and hints at an answer to this problem.

2.3.3 **Proposition.** *Neither* $M^*(K_5)$ *nor* $M^*(K_{3,3})$ *is graphic.*

Proof. Let $M = M^*(K_5)$. We shall show that M is not graphic, leaving the similar argument for $M^*(K_{3,3})$ to the reader. Assume that $M \cong M(G)$ for some graph G. By Proposition 1.2.8, we may assume that G is connected. Now $M(K_5)$ has 10 elements and rank 4. Hence M has 10 elements and rank 6. Thus, by 1.3.6, G has 7 vertices and 10 edges. Hence the average vertex degree of G is $2|E(G)|/|V(G)|$, that is, $20/7$. Since this is less than three, G has a vertex v of degree at most two. Hence M^* has a circuit of size one or two. But this means that K_5 has a cycle of size one or two; a contradiction. We conclude that $M^*(K_5)$ is not graphic. □

The pairing of the graphs K_5 and $K_{3,3}$ in the last result is reminiscent of Kuratowski's well-known characterization of planar graphs. This link

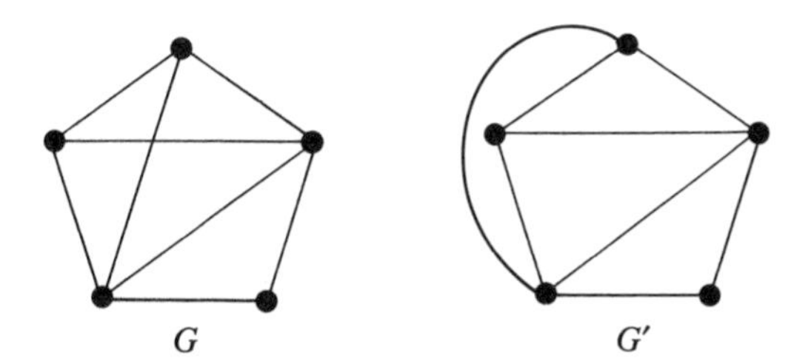

Fig. 2.7. A non-plane graph G and a planar embedding G' of it.

is no coincidence and its precise nature will be stated at the end of this section. To establish the details of this link, we shall need some more definitions. A *plane graph* is a graph drawn in the Euclidean plane so that the vertices are points of the plane, the edges are Jordan curves, and two distinct edges do not intersect except possibly at their endpoints. A graph G is *planar* if it is isomorphic to a plane graph G'. Under such circumstances, G' is called a *planar embedding* of G. The 5-vertex graph G shown in Figure 2.7 is not a plane graph. However, because it is isomorphic to the plane graph G', it is planar.

A plane graph G partitions the Euclidean plane into regions. Such regions, called *faces* of G, can be formally defined as follows: Suppose that P is the set of points of the plane that are not vertices of G and do not lie on edges of G. Two points x and y of P are in the same face of G if there is a Jordan curve joining x and y, all the points of which lie in P. A plane graph has exactly one unbounded face. We call this the *infinite face.*

Now recall that we are seeking graphs G whose bond matroids are graphic.

2.3.4 Theorem. *If G is planar, then $M^*(G)$ is graphic.*

To prove this result, we shall construct a graph whose cycle matroid is isomorphic to $M^*(G)$. We illustrate this construction by reference to the example in Figure 2.8. In that diagram, parts (a) and (c) show the plane graph G and its geometric dual G^*. In (b), G and G^* have been superimposed to show that G^* is formed by taking a vertex in each face of G and joining two such vertices when the corresponding faces share an edge. The edges of G^* are drawn so as to cross the corresponding edges of G. The graph G' in (d) is a different planar embedding of G^*.

Given an arbitrary plane graph G having at least one edge, the construction of G^*, the *geometric dual* of G, is formally described as follows: Choose a single point v_F in each face F of G. These points are to be the vertices of G^*. Suppose that the set of edges common to the boundaries

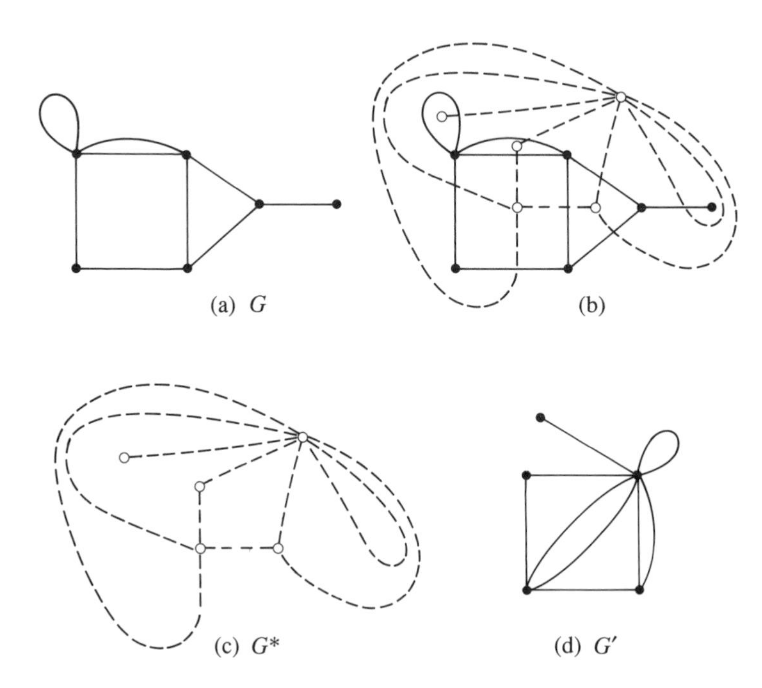

Fig. 2.8. Construction of the geometric dual of a plane graph.

of the faces F and F' is $\{e_1, e_2, \ldots, e_k\}$. Then we join v_F and $v_{F'}$ by k edges $e'_1, e'_2, \ldots, e'_k$, where e'_i crosses e_i but no other edge of G. The only points common to two distinct edges of G^* can be their endpoints. If the edge e of G lies on the boundary of a single face F, we add a loop e' at v_F that crosses e but no other edge of G or G^*.

To ensure that G^* is well-defined as a plane graph, we should specify when two plane graphs are equivalent. We shall not do this yet, preferring to delay a discussion of this until Section 8.2. It is not difficult to show that the graph G^* is connected. Moreover, provided G is connected,

2.3.5 $$(G^*)^* = G.$$

For an arbitrary planar graph G, a *geometric dual* of G is the geometric dual of some planar embedding of G. We observe that, because G may have several different planar embeddings, it may have several different geometric duals.

2.3.6 **Example.** In Figure 2.9, the plane graphs G_1 and G_2 are different planar embeddings of the same planar graph G. Thus G_1^* and G_2^* are geometric duals of G. Evidently $G_1^* \not\cong G_2^*$. □

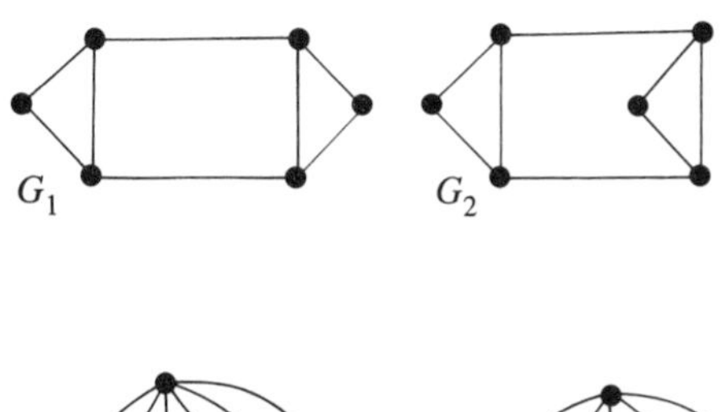

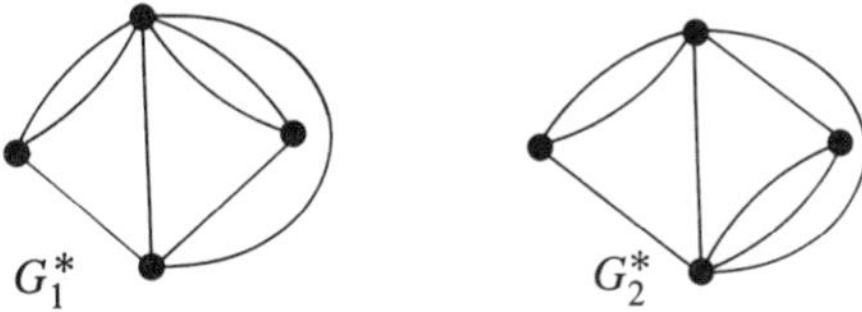

Fig. 2.9. Different embeddings giving different geometric duals.

Theorem 2.3.4 will follow immediately from the following result which shows that, although two geometric duals of a planar graph might not be isomorphic, they do have isomorphic cycle matroids.

2.3.7 Lemma. *If G^* is a geometric dual of the planar graph G, then*

$$M(G^*) \cong M^*(G).$$

Proof. Let G_0 be a planar embedding of G, and G^* be the geometric dual of G_0. The construction of G^* from G_0 determines a bijection α from $E(G_0)$ to $E(G^*)$. We shall show that, under the map α,

$$M(G_0) \cong M^*(G^*).$$

Since $G_0 \cong G$, the required result will then follow.

Let C be a circuit in $M(G_0)$. We want to show that $\alpha(C)$ is a bond in G^*. Now C forms a closed Jordan curve in the plane, and each edge in $\alpha(C)$ has one endpoint inside and the other endpoint outside this closed curve. Thus $\alpha(C)$ is an edge cut in G^*. The fact that $\alpha(C)$ is a minimal edge cut will follow from what we shall show next, namely, that if X is a minimal edge cut in G^*, then $\alpha^{-1}(X)$ contains a cycle of G_0. Evidently, on deleting the edges of X from G^*, we obtain a graph having two components, G_1^* and G_2^*. Moreover, every edge in X has one endpoint in G_1^* and the other in G_2^*. If $|X| = 1$, then it is clear from the construction of G^* that the single edge in $\alpha^{-1}(X)$ is a loop of G_0. Now suppose that $|X| > 1$ and let F be a face of G^* such that some edge, say x, of X is in the set F' of boundary edges of F. Then, as x is not a cut-edge of G^*, it is not difficult to check that x is not a cut-edge of $G^*[F']$. It is now an easy exercise in graph theory to show that F' contains a circuit C_x of $M(G^*)$ that contains x. Let the endpoints of x be u and v. Then, because X is a minimal edge cut of G^*, the vertices u and v must be in the same component of $G^*\backslash(X-x)$ but in different components of $G^*\backslash X$. Therefore, $|X \cap C_x| \geq 2$. Hence if a face of G^* meets an edge of X, it meets more than one such edge. But the faces of G^* correspond to vertices of G_0. Therefore, every vertex of G_0 meeting an edge of $\alpha^{-1}(X)$ meets at least two such edges. Now a graph in which every vertex has degree at least two contains a cycle (Exercise 2, Section 1.6). Therefore the induced graph $G_0[\alpha^{-1}(X)]$ contains a cycle of G_0. The lemma and Theorem 2.3.4 now follow without difficulty by using Lemma 2.1.19. □

The converse of Theorem 2.3.4 is also true. The proof relies on the following famous theorem of Kuratowski (1930), and we delay it until Chapter 5 where graphic matroids will be examined in more detail. A graph G' is a *subdivision* of a graph G if G' can be obtained from G by replacing non-loop edges of the latter by paths of non-zero lengths and replacing loop edges by cycles.

2.3.8 Kuratowski's Theorem. *A graph is planar if and only if it has no subgraph that is a subdivision of K_5 or $K_{3,3}$.* □

A proof of this theorem can be found in Bondy and Murty (1976, Section 9.5).

Exercises

1. (i) Show that the geometric dual of a plane graph is connected.

   (ii) Give an example of a plane graph G for which $(G^*)^* \neq G$.

2. Find geometric duals of the graphs obtained from K_5 and $K_{3,3}$ by deleting a single edge of each.

3. Find all identically self-dual graphic matroids.

4. For each positive integer n, find a planar graph G_n having at least n non-isomorphic geometric duals.

5. Complete the proof of Proposition 2.3.3 by showing that $M^*(K_{3,3})$ is not graphic.

6. Construct the geometric duals of the plane graphs in Figures 1.4(a) and 1.28(a).

7. Find all graphs for which every bond is a vertex bond.

8. If X is a set of edges in a graph G, specify the rank of X in $M^*(G)$ in terms of $G[X]$.

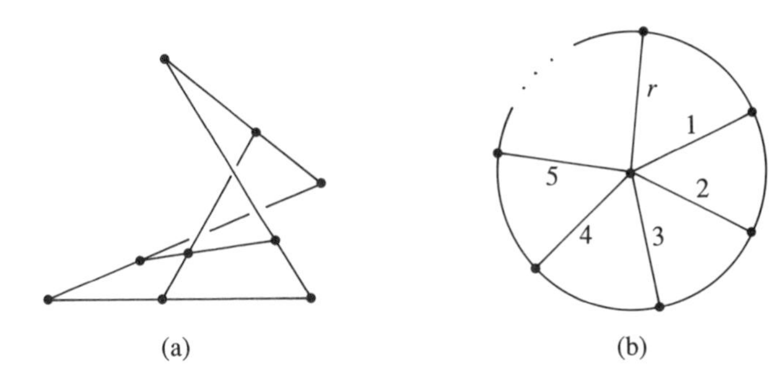

Fig. 2.10. (a) $M^*(K_{3,3})$. (b) $\mathcal{W}_r$.

9. Show that $M^*(K_{3,3})$ has the geometric representation shown in Figure 2.10(a).

10. Let r be an integer exceeding one and $\mathcal{W}_r$ be the graph in Figure 2.10(b). Show that:

(i) $\mathcal{W}_3 \cong K_4$.

(ii) $\mathcal{W}_r^* \cong \mathcal{W}_r$.

(iii) $M(\mathcal{W}_4)$ is isomorphic to a restriction of $M^*(K_{3,3})$.

2.4 Duals of transversal matroids

In this section we shall identify the class of duals of transversal matroids as being a class of matroids that arise from linkings in directed graphs. While this is an important result, it is both more difficult and more specialized than the earlier results in this chapter. Consequently, after taking note of the result, the reader may wish to delay reading its proof until Chapter 12 when some of the results of this section will be generalized.

We begin with a technical result for transversal matroids. Let $\mathcal{A}$ be a family $(A_j : j \in J)$ of sets and let $M = M[\mathcal{A}]$. Evidently $r(M) \leq |J|$. We shall show next that M has a presentation having exactly $r(M)$ members.

2.4.1 Lemma. *Let $\mathcal{A}$ be $(A_1, A_2, \ldots, A_m)$, a family of subsets of the set E, and suppose that T, a maximal partial transversal of $\mathcal{A}$, is a transversal of $\mathcal{A}' = (A_1, A_2, \ldots, A_t)$ where $t < m$. Then every maximal partial transversal of $\mathcal{A}$ is a transversal of $\mathcal{A}'$.*

Proof. Let $\Delta[\mathcal{A}]$ be the bipartite graph associated with $\mathcal{A}$. Let $J = \{1, 2, \ldots, m\}$ and $J' = \{1, 2, \ldots, t\}$. In $\Delta[\mathcal{A}]$, there is a matching of T

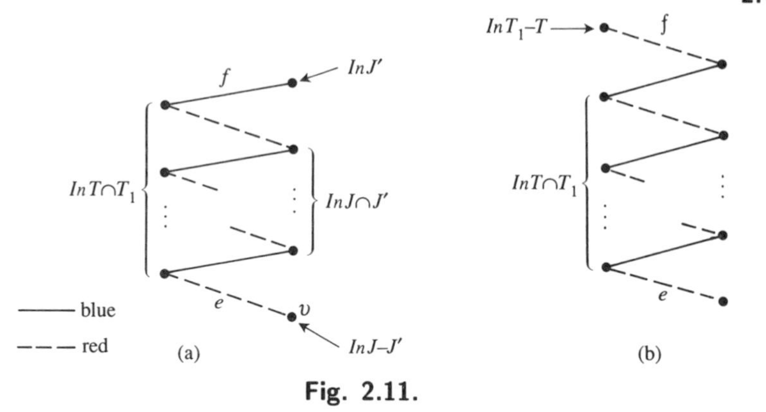

Fig. 2.11.

onto J'. Colour the edges of this matching blue. Now suppose that there is a maximal partial transversal T_1 of $\mathcal{A}$ that is not a transversal of $\mathcal{A}'$. Then there is a matching of T_1 into J. Choose this matching so that it meets as many vertices of J' as possible, and colour the edges of this matching red. We note that some edges may now be coloured both red and blue. As T and T_1 are both bases of $M[\mathcal{A}]$, $|T| = |T_1|$. Moreover, T_1 is not a transversal of $\mathcal{A}'$ and therefore, for some v in $J - J'$, there is a red edge e meeting v. Let P be a path of coloured edges beginning at v that has maximum length among such paths. The first edge in this path is e. It is red but not blue. Since both the red edges and the blue edges form matchings, the edges of P alternate red and blue with no edge being both red and blue. Let the last edge of P be f. Now either f is blue or f is red. Assume the former (see Figure 2.11(a)). Interchange the colours on the edges of P. Then, in the new colouring of $\Delta[\mathcal{A}]$, the red edges still match T_1 into J. But this new red matching meets more vertices of J' than the original red matching; a contradiction. We conclude that f is red. Thus P has more red edges than blue (see Figure 2.11(b)). Again we interchange the colours on the edges of P, this time looking at the blue edges of the recoloured graph. They still form a matching, but this matching has $|T|+1$ edges. This contradicts the fact that T is a maximal partial transversal of $\mathcal{A}$, thereby finishing the proof of the lemma. □

Our approach to identifying the duals of transversal matroids is that of Ingleton and Piff (1973). It uses the bipartite-graph view of transversal matroids. We begin, however, by introducing some new ideas. Suppose that G is a *directed* graph and X and Y are subsets of the vertex set V of G. We say that X is *linked to* Y if $|X| = |Y|$ and there are $|X|$ disjoint (directed) paths whose initial vertex is in X and whose final vertex is in Y. Observe here that such paths may consist of just a single vertex. If $Z \subseteq V$, then we say that X is *linked into* Z if X is linked to some subset of Z. In what follows, whenever we refer to a path in a directed graph, we

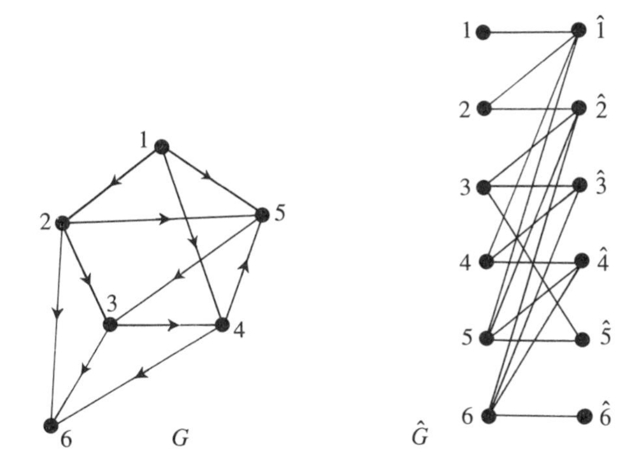

Fig. 2.12. A digraph G and the corresponding bipartite graph $\hat{G}$.

shall mean a *directed path.* Now let B_0 be a fixed subset of V, and denote by $L(G, B_0)$ those subsets of V that are linked into B_0. As we shall see, $L(G, B_0)$ is the set of independent sets of a matroid on V. The proof of this will be indirect and will rely heavily on the following construction of a bipartite graph $\hat{G}$ from G. We begin by taking a disjoint copy $\hat{V}$ of V. For each element v of V, we shall denote by $\hat{v}$ the corresponding element of $\hat{V}$. A similar convention will apply to subsets of V. The vertex set of $\hat{G}$ is $V \cup \hat{V}$. Its edge set is $\{v\hat{v} : v \in V\} \cup \{v\hat{u} : (u, v) \in E(G)\}$.

2.4.2 Example. The above construction is illustrated in Figure 2.12, which shows a directed graph G and its corresponding bipartite graph $\hat{G}$. □

The next result (Ingleton and Piff 1973) notes a fundamental connection between linkings in a directed graph G and matchings in the corresponding bipartite graph $\hat{G}$. This result is the key to the main theorem of the section, a characterization of the duals of transversal matroids.

2.4.3 Lemma. *Suppose that X and Y are subsets of V. Then X is linked to Y in G if and only if $V - X$ is matched to $\hat{V} - \hat{Y}$ in $\hat{G}$.*

Proof. Assume first that there is a set $\{P_x : x \in X\}$ of disjoint paths in G linking X to Y. Now define a map ψ from $\hat{V} - \hat{Y}$ onto $V - X$ as follows:

$$\psi(\hat{u}) = \begin{cases} v, & \text{if } u \text{ is on one of the paths } \{P_x : x \in X\} \\ & \text{and } v \text{ is its successor on this path;} \\ u, & \text{if } u \in V - Y \text{ but } u \text{ is not on any of} \\ & \text{the paths } \{P_x : x \in X\}. \end{cases}$$

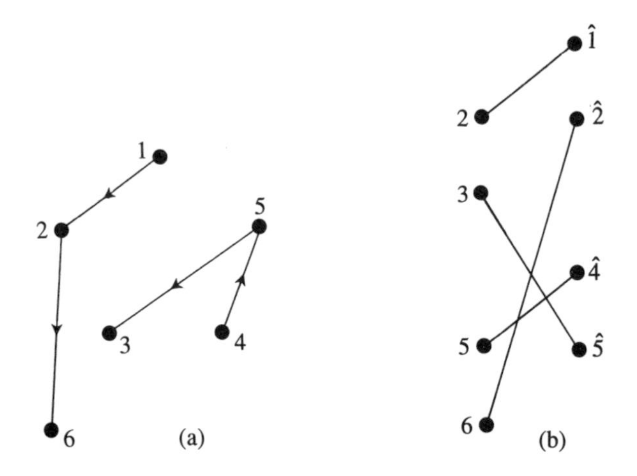

Fig. 2.13. Disjoint paths in G giving a matching in $\hat{G}$.

Since the paths $\{P_x : x \in X\}$ are disjoint, ψ is well-defined and one-to-one. Moreover, it follows easily from the definitions of $\hat{G}$ and ψ that $\{\psi(\hat{u})\hat{u} : \hat{u} \in \hat{V} - \hat{Y}\}$ is a subset of $E(\hat{G})$ and the edges of this subset match $V - X$ to $\hat{V} - \hat{Y}$.

Before proving the converse, we illustrate the above procedure on Example 2.4.2. Let $X = \{1, 4\}$ and $Y = \{6, 3\}$. Let the paths linking X to Y be as shown in Figure 2.13(a). Then $\psi(\hat{1}) = 2$, $\psi(\hat{2}) = 6$, $\psi(\hat{4}) = 5$, and $\psi(\hat{5}) = 3$, so the matching of $\hat{V} - \hat{Y}$ to $V - X$ is as shown in Figure 2.13(b).

Now, to prove the converse, suppose that $V - X$ is matched to $\hat{V} - \hat{Y}$ in $\hat{G}$. Colour the edges of this matching red. In addition, for all v in V, colour the edge $v\hat{v}$ blue. We observe that some edges may be coloured both red and blue.

We now define a set $\{P_x : x \in X\}$ of disjoint paths linking X to Y in G. This construction is illustrated with an example at the end of the proof. If a member x of X is also in Y, we let P_x consist of the single vertex x. Now suppose that $x \in X - Y$. Then x meets a blue edge but no red edge. Take a maximal path P'_x of coloured edges beginning at x.

We show next that P'_x must meet a vertex of $\hat{Y}$. Assume the contrary. The first edge in P'_x is blue and, as the red edges and the blue edges both form matchings, the edges of P'_x alternate in colour with none being coloured both red and blue. If the last edge of P'_x is blue, this edge is of the form $v\hat{v}$. Moreover, it follows, since P'_x does not meet a vertex of $\hat{Y}$, that $\hat{v} \notin \hat{Y}$. Hence there is a red edge meeting $\hat{v}$. Since the other endpoint of this edge cannot already be in P'_x, we may adjoin this red edge to P'_x thereby forming a longer path and contradicting the choice of P'_x. Thus we may assume that the last edge in P'_x is red. If this edge is $v\hat{u}$, then

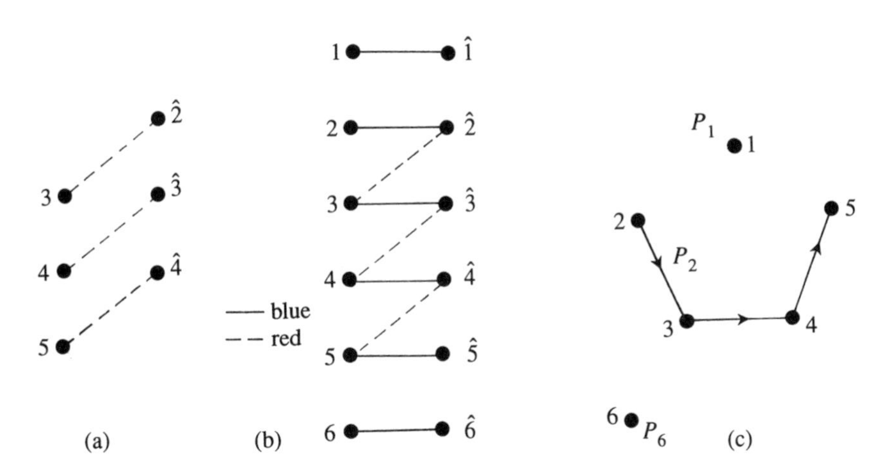

Fig. 2.14. Using a matching in $\hat{G}$ to produce disjoint paths in G.

$v\hat{u} \neq v\hat{v}$, and we may adjoin the blue edge $v\hat{v}$ to P'_x to again obtain a contradiction to the choice of P'_x. We conclude that P'_x does indeed meet a vertex in $\hat{Y}$.

Let $\hat{y}$ be the first vertex of $\hat{Y}$ met by P'_x. As $\hat{y} \notin \hat{V} - \hat{Y}$, no red edge meets $\hat{y}$. Hence P'_x ends at $\hat{y}$, the last edge being the blue edge $y\hat{y}$. Clearly P'_x is the unique maximal path of coloured edges in $\hat{G}$ beginning at x. Moreover, if $x = x_0$ and $\hat{y} = \hat{x}_n$, then P'_x is $x_0\hat{x}_0x_1\hat{x}_1\cdots x_n\hat{x}_n$. Let P_x be $x_0x_1x_2\cdots x_n$. Evidently P_x is a path in G.

This defines a set $\{P_x : x \in X\}$ of paths in G. If z_1 and z_2 are distinct members of $X-Y$, then it is easily checked that P'_{z_1} and P'_{z_2} are disjoint. Hence so are P_{z_1} and P_{z_2}. It follows that the paths $\{P_x : x \in X\}$ are disjoint. Since each links a vertex of X to a vertex of Y and $|X| = |Y|$, these paths link X to Y. □

To illustrate the construction of the set of paths $\{P_x : x \in X\}$ in the second part of the last proof, we again refer to Example 2.4.2. Take the matching in $\hat{G}$ shown in Figure 2.14(a). Then $V - X = \{3, 4, 5\}$ and $\hat{V} - \hat{Y} = \{\hat{2}, \hat{3}, \hat{4}\}$. Colour the edges in this matching red. In addition, colour blue all edges $v\hat{v}$ for v in V. This gives the edge-coloured graph in Figure 2.14(b). As $X = \{1, 2, 6\}$ and $Y = \{1, 5, 6\}$, both 1 and 6 are in $X \cap Y$, so P_1 and P_6 consist of the single vertices 1 and 6, respectively. On the other hand, P'_2 is $2\hat{2}3\hat{3}4\hat{4}5\hat{5}$, so P_2 is 2345. Figure 2.14(c) shows the three disjoint paths P_1, P_2, and P_6 in G.

We are now ready to prove the main result of this section (Ingleton and Piff 1973), a characterization of cotransversal matroids, the duals of transversal matroids.

2.4.4 Theorem. *Let G be a directed graph having vertex set V and let B_0 be a subset of V. Then $L(G, B_0)$ is the set of independent sets of a cotransversal matroid on V. Conversely, given any cotransversal matroid M on a set V and any basis B_0 of M, there is a directed graph G having vertex set V such that M has $L(G, B_0)$ as its set of independent sets.*

Proof. Given the directed graph G, construct the corresponding bipartite graph $\hat{G}$. By the last lemma, B is a maximal member of $L(G, B_0)$ if and only if $V - B$ is matched to $\hat{V} - \hat{B}_0$ in $\hat{G}$. Let $\hat{G}_1$ be the subgraph of $\hat{G}$ obtained by deleting the vertices in $\hat{B}_0$. Then $V - B$ is matched to $\hat{V} - \hat{B}_0$ in $\hat{G}_1$ if and only if B is a maximal member of $L(G, B_0)$. But the subsets of V that are matched to $\hat{V} - \hat{B}_0$ are precisely the bases of a transversal matroid N on V, and the above assertion is that the cobases of N are the maximal members of $L(G, B_0)$. Hence $L(G, B_0)$ is the set of independent sets of the cotransversal matroid N^* on V.

Now let N be an arbitrary transversal matroid on V and let $V - B_0$ be a basis of N. Then, by Lemma 2.4.1, $N = M[\mathcal{A}]$ for some family $\mathcal{A}$ of sets that has $V - B_0$ as a transversal. Let $\Delta[\mathcal{A}]$ be the bipartite graph associated with $\mathcal{A}$ where the latter equals $(A_j : j \in J)$. This graph contains a matching of $V - B_0$ onto J. If $j \in J$ and j is joined to v in this matching, relabel j as $\hat{v}$. Then, for each u in B_0, add a new vertex $\hat{u}$ to $\Delta[\mathcal{A}]$ joining this vertex to u and no other. The resulting graph is of the form $\hat{G}$ for some directed graph G. It follows, by Lemma 2.4.3, that N^* has $L(G, B_0)$ as its set of independent sets. □

Given a directed graph G and a subset B_0 of its vertex set, the matroid that has $L(G, B_0)$ as its collection of independent sets is called a *strict gammoid.* We shall use the same notation, $L(G, B_0)$, for this matroid as for its collection of independent sets. Such matroids were introduced by Mason (1972a). The following is an immediate consequence of the last theorem.

2.4.5 Corollary. *A matroid is a strict gammoid if and only if its dual matroid is transversal.* □

2.4.6 Example. Let G_1 and G_2 be the directed graphs shown in Figure 2.15(a). In each case, let $B_0 = \{1, 2, 3\}$. Then it is not difficult to check that $L(G_1, B_0)$ and $L(G_2, B_0)$ have the geometric representations shown in Figure 2.15(b). Note that $L(G_1, B_0) \cong \mathcal{W}^3$ (see Figure 2.1). □

Since every uniform matroid is transversal and has a uniform dual, every uniform matroid is a strict gammoid. We have already noted as a fact to be proved in Chapter 12 that every transversal matroid is representable over all sufficiently large fields. Since, by Corollary 2.2.9, the dual

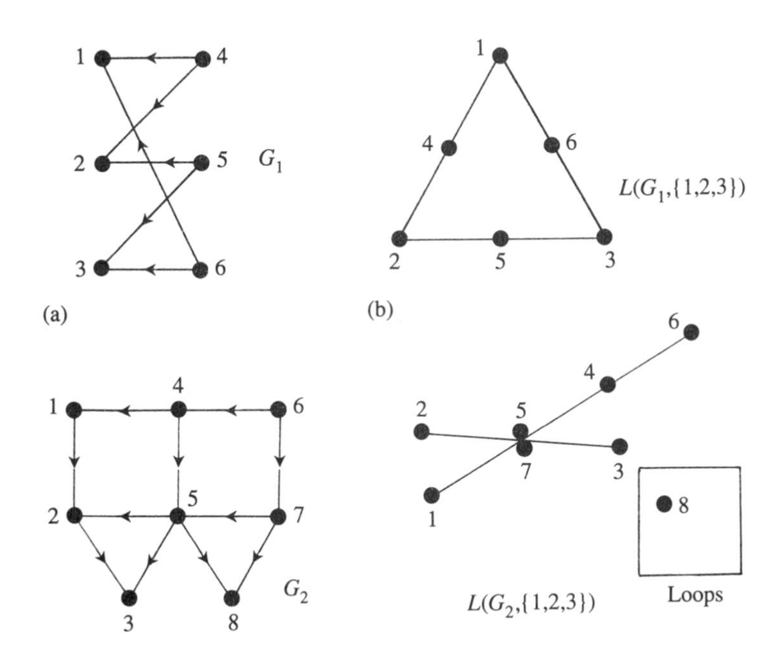

Fig. 2.15. G_1 and G_2 and the strict gammoids $L(G_i, \{1,2,3\})$.

of an F-representable matroid is F-representable, every strict gammoid is representable over all sufficiently large fields.

We showed in Example 1.6.3 that for the planar graph G in Figure 2.16, $M(G)$ is not transversal. Thus $M(G^*)$ is not a strict gammoid where G^* is a geometric dual of G (see Figure 2.16(b)). Note, however, that $M(G) = L(G_1, B_0)$ where G_1 is the directed graph shown in Figure 2.16(c) and $B_0 = \{1,5\}$. Thus $L(G_1, B_0)$ is a strict gammoid whose dual is not a strict gammoid. Hence $M(G^*)$ is a transversal matroid whose dual is not transversal.

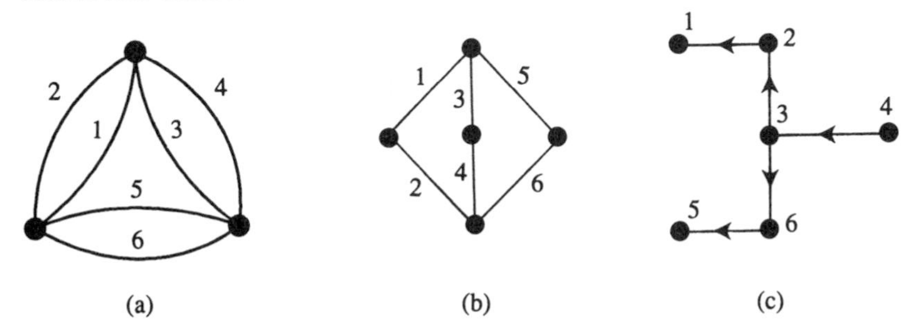

Fig. 2.16. (a) G. (b) G^*. (c) A digraph G_1 with $M(G) = L(G_1, \{1,5\})$.

Exercises

1. Prove that if M is a rank-r transversal matroid having no coloops, then every presentation of M has exactly r non-empty members.

2. Recall that a cyclic flat is a flat that is a union of circuits. Let $\mathcal{A} = (A_1, A_2, \ldots, A_m)$. Prove that if F is a cyclic flat of $M[\mathcal{A}]$ of rank k, then $|\{i : F \cap A_i \neq \emptyset\}| = k$.

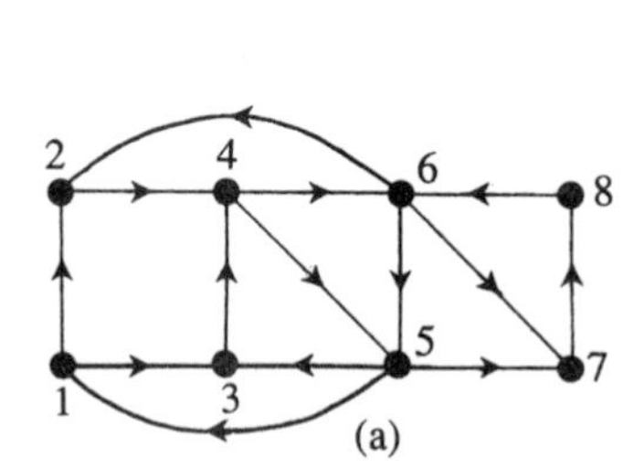

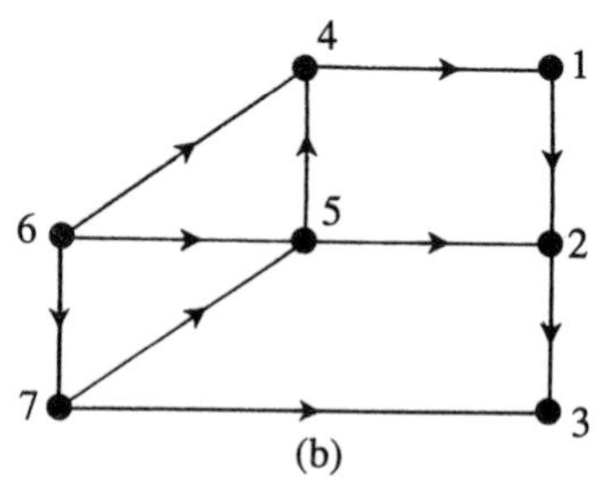

Fig. 2.17.

3. Let G be the directed graph shown in Figure 2.17(a).

   (i) Construct the corresponding bipartite graph $\hat{G}$.

   (ii) Let $V = \{1, 2, \ldots, 8\}$, $X = \{5, 7\}$, $Y = \{4, 2\}$, and the paths linking X to Y in G be 5134 and 7862. Construct the corresponding matching of $V - X$ to $\hat{V} - \hat{Y}$ in $\hat{G}$.

   (iii) Now take the matching $\{2\hat{1}, 3\hat{5}, 4\hat{3}, 5\hat{4}, 6\hat{8}, 8\hat{7}\}$ in $\hat{G}$ and construct the corresponding disjoint paths in G as in the proof of Lemma 2.4.3.

4. Let G be the directed graph shown in Figure 2.17(b) and let $M = L(G, B_0)$ where $B_0 = \{1, 2, 3\}$.

   (i) Find geometric representations for M and M^*.

   (ii) Give a presentation for the transversal matroid M^*.

   (iii) Reverse the directions on the arcs (5,4) and (7,5) of G and repeat (i) and (ii).

5. Let G_2 be the directed graph shown in Figure 2.15(a). Find $L(G_2, B_0)$ when $B_0 = \{2, 3, 8\}$.

6. Let G be the directed graph in Figure 2.12.

   (i) Find $L(G, B_0)$ when $B_0 = \{2, 5, 6\}$.

(ii) Show that there is no subset B'_0 of $V(G)$ such that $L(G, B'_0) \cong U_{3,6}$.

7. (i) Give a presentation for the transversal matroid $U_{r,n}$.

(ii) Find a directed graph G and a subset B_0 of $V(G)$ such that $L(G, B_0) \cong U_{n-r,n}$.

8. Show that each of the matroids P_6 and Q_6 in Figure 2.1 is a strict gammoid that is also transversal.

9.* (Bondy and Welsh 1971) Let $(A_1, A_2, \ldots, A_r)$ be a presentation of the rank-r transversal matroid M. Show that:

(i) If T is a transversal of $(A_2, A_3, \ldots, A_r)$ such that $A_1 \cap T$ has minimum cardinality, then $(A_1 - T, A_2, A_3, \ldots, A_r)$ is also a presentation of M.

(ii) M has distinct cocircuits $C_1^*, C_2^*, \ldots, C_r^*$ such that $(C_1^*, C_2^*, \ldots, C_r^*)$ is a presentation for M and $C_i^* \subseteq A_i$ for all i in $\{1, 2, \ldots, r\}$. Moreover, if $x \in C_j^*$ for some j in $\{1, 2, \ldots, r\}$, then $(C_1^*, C_2^*, \ldots, C_{j-1}^*, C_j^* - x, C_{j+1}^*, \ldots, C_r^*)$ is not a presentation for M.

3
Minors

We saw in Chapter 1 that, given a matroid M on a set E and a subset T of E, there is a matroid $M\backslash T$ on $E-T$ whose independent sets are those independent sets in M that are contained in $E-T$. In this chapter, we look more closely at this operation and its dual, we consider applying a sequence of these operations to a matroid, and we examine the effects of these operations on various classes of matroids.

3.1 Contraction

In this section, we introduce the operation of contraction as the dual of the operation of deletion. We then derive a definition of contraction that does not use duality and look at how contraction affects various special sets such as independent sets, circuits, and bases.

Let M be a matroid on E and T be a subset of E. The two fundamental operations on M that we have introduced so far are deletion and the taking of duals. The next definition combines these two operations. Let M/T, the *contraction of T from M*, be given by

3.1.1 $$M/T = (M^*\backslash T)^*.$$

Evidently M/T has ground set $E-T$. We shall sometimes write it as $M.(E-T)$ and call it the *contraction of M to $E-T$*.

Before presenting an example of this operation, we need to note a basic property of deletion for graphic matroids. If G is a graph and $T \subseteq E(G)$, then recall that $G\backslash T$ denotes the graph obtained from G by erasing or *deleting* the edges in T. Deletion for matroids extends this graph-theoretic operation, that is,

3.1.2 $$M(G\backslash T) = M(G)\backslash T.$$

This fact is used in the next example.

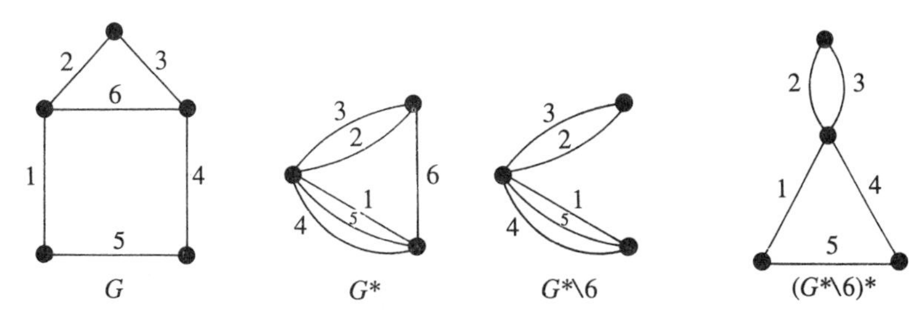

Fig. 3.1. $M(G)/6 = M(G/6)$.

3.1.3 Example. Let $M = M(G)$ where G is the planar graph shown in Figure 3.1. Then $M(G)/6 = (M^*(G)\backslash 6)^* = (M(G^*)\backslash 6)^*$ where G^* is a geometric dual of G and each edge of G^* has the same label as the corresponding edge of G. Thus, by 3.1.1 and 3.1.2, $M(G)/6 = (M(G^*\backslash 6))^* = M^*(G^*\backslash 6) = M((G^*\backslash 6)^*)$. But, from Figure 3.1, we see that $G^*\backslash 6$ has as a geometric dual the graph $G/6$ that is obtained from G by shrinking or contracting the edge 6 to a point. Therefore $M(G)/6 = M(G/6)$. This illustrates the general principle, to be proved in Section 3.2, that contraction for matroids generalizes the operation of contraction for graphs. In Section 3.3, we shall see that contraction of a non-loop element e from a matroid M corresponds geometrically to projecting the other elements of M from e onto a hyperplane avoiding e. □

In the next result, we shall need to recall from Proposition 2.1.9 that the rank function r^* of the dual M^* of M is given by

3.1.4
$$r^*(X) = |X| + r(E - X) - r(E).$$

Evidently if $T \subseteq E$, then the rank function of $M\backslash T$ is the restriction of r_M to the set of subsets of $E - T$, that is, for all $X \subseteq E - T$,

3.1.5
$$r_{M\backslash T}(X) = r_M(X).$$

3.1.6 Proposition. *If $T \subseteq E$, then, for all $X \subseteq E - T$,*

3.1.7
$$r_{M/T}(X) = r_M(X \cup T) - r_M(T).$$

Proof. By definition, $r_{M/T}(X) = r_{(M^*\backslash T)^*}(X)$. Thus,

$$\begin{aligned} r_{M/T}(X) &= |X| + r_{M^*\backslash T}(E - T - X) - r_{M^*\backslash T}(E - T), && \text{by (3.1.4)},\\ &= |X| + r^*(E - (T \cup X)) - r^*(E - T), && \text{by (3.1.5)},\\ &= |X| + [|E - (T \cup X)| + r_M(T \cup X) - r_M(E)] \\ &\quad - [|E - T| + r_M(T) - r_M(E)], && \text{by (3.1.4) again.} \end{aligned}$$

Hence, as $X \subseteq E - T$, $r_{M/T}(X) = r_M(T \cup X) - r_M(T)$, as required. □

The next three results determine the independent sets, bases, and circuits of M/T in terms of the corresponding sets for M. Recall that $M|T$ is $M\backslash(E-T)$.

3.1.8 Proposition. *Suppose that B_T is a basis for $M|T$. Then*

$$\begin{aligned}\mathcal{I}(M/T) &= \{I \subseteq E-T : I \cup B_T \in \mathcal{I}(M)\} \\ &= \{I \subseteq E-T : M|T \textit{ has a basis } B \textit{ such that } B \cup I \in \mathcal{I}(M)\}.\end{aligned}$$

Proof. Evidently $\{I \subseteq E-T : M|T$ has a basis B such that $B \cup I \in \mathcal{I}\}$ contains $\{I \subseteq E-T : I \cup B_T \in \mathcal{I}\}$. Now suppose that $B \cup I \in \mathcal{I}$ for some basis B of $M|T$. We shall show that $I \in \mathcal{I}(M/T)$. Clearly $I \cup B$ is a basis of $I \cup T$, so $r_M(I \cup B) = r_M(I \cup T)$. Therefore, as $r_{M/T}(I) = r_M(I \cup T) - r_M(T)$, it follows that

$$r_{M/T}(I) = r_M(I \cup B) - r_M(T) = |I \cup B| - |B|,$$

where the last step follows since $I \cup B \in \mathcal{I}$ and B is a basis of $M|T$. Hence $r_{M/T}(I) = |I|$, that is, $I \in \mathcal{I}(M/T)$. Therefore,

$$\{I \subseteq E-T : M|T \text{ has a basis } B \text{ such that } B \cup I \in \mathcal{I}\} \subseteq \mathcal{I}(M/T).$$

To complete the proof, we show that $\{I \subseteq E-T : I \cup B_T \in \mathcal{I}\}$ contains $\mathcal{I}(M/T)$. Thus assume $X \in \mathcal{I}(M/T)$. Then

$$\begin{aligned}|X| = r_{M/T}(X) &= r_M(X \cup T) - r_M(T) \\ &= r_M(X \cup B_T) - |B_T|,\end{aligned}$$

since B_T is a basis of $M|T$. Hence $|X \cup B_T| = r_M(X \cup B_T)$, so $X \cup B_T \in \mathcal{I}$ and the proposition follows. □

3.1.9 Corollary. *Let B_T be a basis of $M|T$. Then*

$$\begin{aligned}\mathcal{B}(M/T) &= \{B' \subseteq E-T : B' \cup B_T \in \mathcal{B}(M)\} \\ &= \{B' \subseteq E-T : M|T \textit{ has a basis } B \textit{ such} \\ &\qquad \textit{that } B' \cup B \in \mathcal{B}(M)\}.\end{aligned}$$ □

From Proposition 3.1.8, it is straightforward to determine the effect of contraction on uniform matroids.

3.1.10 Example. Let T be a t-element subset of $E = E(U_{m,n})$. Then

$$U_{m,n}/T \cong \begin{cases} U_{0,n-t} & \text{if } n \geq t \geq m, \\ U_{m-t,n-t} & \text{if } t < m. \end{cases}$$

For comparison, we observe that

$$U_{m,n}\backslash T \cong \begin{cases} U_{n-t,n-t} & \text{if } n \geq t \geq n-m, \\ U_{m,n-t} & \text{if } t < n-m. \end{cases}$$

3.1.11 Proposition. *The circuits of M/T consist of the minimal non-empty members of $\{C - T : C \in \mathcal{C}(M)\}$.*

Proof. Evidently we may assume that T is non-empty. Suppose that $C_1 \in \mathcal{C}(M/T)$. Then, if B_T is a basis of $M|T$, we have that $C_1 \cup B_T \notin \mathcal{I}(M)$, but $(C_1 - e) \cup B_T \in \mathcal{I}(M)$ for all e in C_1. As $C_1 \cup B_T$ is dependent in M, it contains a circuit D of M. Since $(C_1 - e) \cup B_T \in \mathcal{I}(M)$ for all e in C_1, we must have that $D \supseteq C_1$. Hence $C_1 = D - T$ where $D \in \mathcal{C}(M)$.

Now suppose that $C_2 - T$ is a minimal non-empty member of $\{C - T : C \in \mathcal{C}(M)\}$. Then $C_2 \cap T \subsetneqq C_2$. So $C_2 \cap T \in \mathcal{I}(M|T)$. Hence $C_2 \cap T$ is contained in a basis B_T of $M|T$. As $C_2 \cup B_T \supseteq C_2$, the set $C_2 \cup B_T$ is not in $\mathcal{I}(M)$. Hence $C_2 - T \notin \mathcal{I}(M/T)$. If $C_2 - T \notin \mathcal{C}(M/T)$, then $C_2 - T \supsetneqq C_3$ for some C_3 in $C(M/T)$. But, by the first paragraph, $C_3 = D - T$ for some D in $\mathcal{C}(M)$, and so the choice of $C_2 - T$ is contradicted. We conclude that $\mathcal{C}(M/T)$ contains the set $\mathcal{D}$ of minimal non-empty members of $\{C - T : C \in \mathcal{C}(M)\}$. It now follows by Lemma 2.1.19 that $\mathcal{C}(M/T)$ *equals* $\mathcal{D}$. □

The closure operator of M/T is not difficult to derive from 1.4.1 and Proposition 3.1.6 (Exercise 3).

3.1.12 Proposition. *For all $X \subseteq E - T$,*

$$\mathrm{cl}_{M/T}(X) = \mathrm{cl}_M(X \cup T) - T. \qquad \square$$

For comparison with the last four results, we note that

3.1.13 $\mathcal{I}(M\backslash T) = \{I \subseteq E - T : I \in \mathcal{I}(M)\}$;

3.1.14 $\mathcal{C}(M\backslash T) = \{C \subseteq E - T : C \in \mathcal{C}(M)\}$;

3.1.15 $\mathcal{B}(M\backslash T)$ *is the set of maximal members of* $\{B - T : B \in \mathcal{B}(M)\}$; *and*

3.1.16 $\mathrm{cl}_{M\backslash T}(X) = \mathrm{cl}_M(X) - T$.

Of these, 3.1.13 is just the definition; 3.1.14 follows immediately from 3.1.13; and 3.1.15 and 3.1.16 are straightforward and are left as exercises.

Using the above results and duality, we easily obtain the following.

3.1.17 $\mathcal{C}^*(M\backslash T)$ *is the set of minimal non-empty members of* $\{C^* - T \subseteq E - T : C^* \in \mathcal{C}^*(M)\}$;

3.1.18 $\mathcal{C}^*(M/T) = \{C^* \subseteq E - T : C^* \in \mathcal{C}^*(M)\}$;

3.1.19 $\mathcal{H}(M\backslash T)$ *is the set of maximal proper subsets of* $E - T$ *of the form* $H - T$ *where* $H \in \mathcal{H}(M)$;

3.1.20 $\mathcal{H}(M/T) = \{X \subseteq E - T : X \cup T \in \mathcal{H}(M)\}$.

Moreover, if $\mathcal{S}(M)$ denotes the set of spanning sets of M, then

3.1.21 $\mathcal{S}(M\backslash T) = \{X \subseteq E - T : M.T$ *has a basis* B *such that* $X \cup B \in \mathcal{S}(M)\}$;

3.1.22 $\mathcal{S}(M/T) = \{X \subseteq E - T : X \cup T \in \mathcal{S}(M)\} = \{Y - T : Y \in \mathcal{S}(M)\}$.

Since $M\backslash T$ and M/T are matroids on the same ground set, it is natural to try to determine when these matroids are equal. Evidently it is always true that

3.1.23 $\mathcal{I}(M\backslash T) \supseteq \mathcal{I}(M/T)$.

3.1.24 Proposition. *$M\backslash T = M/T$ if and only if $r(T) + r(E - T) = r(M)$.*

Proof. Suppose $M\backslash T = M/T$ and let B be a basis of $M\backslash T$. Then B is a basis of M/T and hence $B \dot\cup B_T$ is a basis of M for some basis B_T of $M|T$. Thus $r(M) = r(T) + r(E - T)$. Conversely, suppose that $r(M) = r(T) + r(E - T)$. By 3.1.23, to establish that $M\backslash T = M/T$, we need only show that $\mathcal{I}(M\backslash T) \subseteq \mathcal{I}(M/T)$. But if $I \in \mathcal{I}(M\backslash T)$, then I is a subset of a basis B of $E - T$ and B is contained in a basis $B \dot\cup B'$ of M. Evidently $r(M) = |B \dot\cup B'| = |B| + |B'| = r(E - T) + |B'|$. Since $r(M) = r(E - T) + r(T)$, it follows that $|B'| = r(T)$, that is, B' is a basis of $M|T$. Hence $B \in \mathcal{B}(M/T)$, so $I \in \mathcal{I}(M/T)$ and $M\backslash T = M/T$. □

If $T = \{e_1, e_2, \ldots, e_m\}$, then $M\backslash T$ and M/T will often be written as $M\backslash e_1, e_2, \ldots, e_m$ and $M/e_1, e_2, \ldots, e_m$, respectively.

3.1.25 Corollary. *$M\backslash e = M/e$ if and only if e is a loop or a coloop of M.*

Proof. This is left to the reader (Exercise 7). □

The existence of a non-empty proper subset T of $E(M)$ for which $M\backslash T = M/T$ has important structural significance for M and this will be examined in detail in the next chapter.

We close this section by observing that the operations of deletion and contraction commute both with each other and with themselves. This fact will be important since we shall want to consider matroids obtained from other matroids by a sequence of deletions and contractions.

3.1.26 Proposition. *Let T_1 and T_2 be disjoint subsets of $E(M)$. Then*

(i) $(M\backslash T_1)\backslash T_2 = M\backslash(T_1 \cup T_2) = (M\backslash T_2)\backslash T_1$;

(ii) $(M/T_1)/T_2 = M/(T_1 \cup T_2) = (M/T_2)/T_1$; *and*

(iii) $(M\backslash T_1)/T_2 = (M/T_2)\backslash T_1$.

Proof. Parts (i) and (ii) are left to the reader (Exercise 8). To prove (iii), we show that $(M/T_1)\backslash T_2$ and $(M\backslash T_2)/T_1$ have the same rank function. If $X \subseteq E - (T_1 \cup T_2)$, then

$$\begin{aligned} r_{(M/T_1)\backslash T_2}(X) &= r_{M/T_1}(X) \\ &= r_M(X \cup T_1) - r_M(T_1) \\ &= r_{M\backslash T_2}(X \cup T_1) - r_{M\backslash T_2}(T_1) \\ &= r_{(M\backslash T_2)/T_1}(X). \end{aligned}$$ □

In view of the last result, we may drop the parentheses in expressions such as $((M/X_1)\backslash X_2)\backslash X_3$ without introducing any ambiguity, it being understood here that X_1, X_2, and X_3 are disjoint. Another important consequence of the last result is that any sequence of deletions and contractions from M can be written in the form $M\backslash X/Y$ for some pair of disjoint sets X and Y, either of which may be empty. Matroids of this form are called *minors* of M. Such substructures were introduced by Tutte (1958) and many of the most celebrated results for matroids make reference to minors. We observe that, in view of Proposition 3.1.26(iii), a minor of M could equally well have been defined as any matroid of the form $M/Y\backslash X$ where X and Y are disjoint, possibly empty, subsets of $E(M)$. If $X \cup Y$ is non-empty, then we call $M/Y\backslash X$ a *proper minor* of M.

The minors of a matroid M are its fundamental substructures and we shall give many examples in later chapters of where a knowledge of the presence of certain matroids as minors of M has important implications for the properties of M. The matroid N_1 is called an *N-minor* of M if N_1 is a minor of M that is isomorphic to N. More generally, if $\mathcal{N}$ is a set of matroids, then N_1 is an *$\mathcal{N}$-minor* of M if N_1 is an N-minor of M for some N in $\mathcal{N}$.

We close this section with a basic property of minors, the proof of which is left to the reader (Exercise 11).

3.1.27 Proposition. *The matroid N is a minor of the matroid M if and only if N^* is a minor of M^*. More particularly,*

$$N = M\backslash X/Y \textit{ if and only if } N^* = M^*/X\backslash Y.$$ □

Exercises

1. Show that if $T \subseteq E(M)$, then

(i) $M\backslash T = (M^*/T)^*$;

(ii) $(M/T)^* = M^*\backslash T$; and

(iii) $M^*/T = (M\backslash T)^*$.

2. Let C be a circuit of the matroid M and e be an element of $E(M)$.

(i) Show that if $e \notin C$, then C is a union of circuits of M/e.

(ii) Show that if $e \in C$, then e is a loop of M, or $C - e$ is a circuit of M/e.

3. Show that, for all $X \subseteq E - T$, $\text{cl}_{M/T}(X) = \text{cl}_M(X \cup T) - T$.

4. Suppose that, for all elements f of a loopless matroid M, $r(M\backslash f) = r(M)$, but that, for some elements e and g, $r(M\backslash e, g) = r(M\backslash e/g)$. Show that $\{e, g\}$ is a cocircuit of M.

5. Let M be a matroid and T be a subset of $E(M)$. Show that:

(i) M/T has no loops if and only if T is a flat of M.

(ii) $\text{cl}(T) = T \cup \{e \in E(M) - T : e \text{ is a loop of } M/T\}$.

(iii) $\text{cl}^*(T) = T \cup \{e \in E(M) - T : e \text{ is a coloop of } M\backslash T\}$.

6. Let B_T be a basis of $M.T$. Show that

$$\mathcal{S}(M\backslash T) = \{X \subseteq E - T : X \cup B_T \in \mathcal{S}(M)\}.$$

7. Prove that $M\backslash e = M/e$ if and only if e is a loop or a coloop.

8. Prove (i) and (ii) of Proposition 3.1.26.

9. Prove that if $\{e, f, g\}$ is both a circuit and a cocircuit of the matroid M, then $M\backslash f/g = M\backslash g/f$.

10. Let M be a matroid, T be a subset of $E(M)$, and B_T be a basis of $M|T$. Prove directly that M/T is a matroid by showing that

$$\{B' \subseteq E - T : B' \cup B_T \in \mathcal{B}(M)\}$$

is the set of bases of a matroid on $E - T$.

11. Prove Proposition 3.1.27.

12. Let T be a subset of the ground set of a matroid M. Prove that the following statements are equivalent:

    (i) $M\backslash T = M/T$.

    (ii) $r(M\backslash T) \leq r(M/T)$.

    (iii) M has no circuits meeting both T and $E - T$.

13. Let $N = M\backslash X/Y$. Specify the rank function and the sets of bases, circuits, independent sets, hyperplanes, and cocircuits of N in terms of the corresponding objects in M.

3.2 Minors of certain classes of matroids

Some of the classes of matroids we have introduced have the property that all their minors are also in the class. We say that such classes are *closed under minors* or are *minor-closed.* Certain other classes are not minor-closed. We begin this section by noting some examples of minor-closed classes and end by showing that the classes of transversal matroids and strict gammoids are not minor-closed.

We noted in Example 3.1.10 that all deletions and all contractions of a uniform matroid are uniform. Hence the class of uniform matroids is closed under minors. This class has the further attractive property that it is closed under duality. We shall find that classes of matroids having this pair of properties are usually easier to work with than those without one or both of these properties.

It was noted in 3.1.2 that all deletions of graphic matroids are graphic. Next we consider contractions. If e is an edge of a graph G, then G/e is obtained from G by deleting e and identifying its ends. Repeating this process for all the edges in a subset T of $E(G)$ gives the graph G/T.

3.2.1 Proposition. *If G is a graph, then $M(G)/T = M(G/T)$ for all subsets T of $E(G)$.*

Proof. We shall show that, for any edge e of G,

$$M(G)/e = M(G/e). \tag{1}$$

The proposition then follows by a routine induction argument on $|T|$.

If e is a loop of G, then $G/e = G\backslash e$ and $M(G)/e = M(G)\backslash e$. The result follows in this case by 3.1.2. Now suppose that e is not a loop of G. Then, for a subset I of $E(G) - e$, it is not difficult to check that $I \cup e$ contains no cycle of G if and only if I contains no cycle of G/e. Hence $\mathcal{I}(M(G)/e) = \mathcal{I}(M(G/e))$ and (1) holds. □

On combining this proposition with 3.1.2, we deduce that the class of graphic matroids is minor-closed.

3.2.2 Corollary. *Every minor of a graphic matroid is graphic.* □

The concept of a minor is useful in graphs as well as in matroids. The graph H is a *minor* of the graph G if H is a contraction of some subgraph of G, or equivalently, if H can be obtained from G by a sequence of edge deletions, edge contractions, and deletions of isolated vertices. We leave the reader to check (Exercise 2(ii)) that if G is connected, then H is a minor of G if and only if $H = G\backslash X/Y$ for some disjoint subsets X and Y of $E(G)$. Evidently if H is a minor of G, then $M(H)$ is a minor of $M(G)$. However, the converse of this is not true (Exercise 2(iii)).

It was shown in Proposition 2.3.3 and Corollary 2.2.9 that the class of graphic matroids is not closed under duality whereas the class of F-representable matroids is. We shall show next that the class of F-representable matroids is also minor-closed.

Let A be a matrix over F and T be a subset of the set E of column labels of A. We shall denote by $A\backslash T$ the matrix obtained from A by deleting all the columns whose labels are in T. Evidently

3.2.3 $$M[A]\backslash T = M[A\backslash T].$$

Moreover:

3.2.4 Proposition. *Every minor of an F-representable matroid is F-representable.*

Proof. By 3.2.3, every deletion of an F-representable matroid is F-representable. Since the dual of an F-representable matroid is also F-representable, we deduce, from the definition of contraction, that every contraction of an F-representable matroid is F-representable. The required result follows immediately. □

In particular, the classes of binary and ternary matroids are minor-closed. Moreover, an easy modification of the last proof gives the following.

3.2.5 Proposition. *Every minor of a regular matroid is regular.* □

Although we know that a contraction of an F-representable matroid is F-representable, we do not yet know how to obtain an F-representation for the contraction. We shall describe this next. The term 'minor' in matroid theory is obviously taken from matrix theory. We have seen that deletion of elements corresponds to removing columns from the associated matrix. We shall now show that contraction essentially corresponds to removing rows.

Consider $M[A]$ and suppose that e is the label of a column of A. If this column is zero, then e is a loop of $M[A]$ and, by Corollary 3.1.25, $M[A]/e = M[A]\backslash e$. Hence, by 3.2.3, $M[A]/e$ is represented by $A\backslash e$. Now suppose that e is the label of a non-zero column of A. Then, by pivoting on a non-zero entry of e, we can transform A into a matrix A' in which the column labelled by e is a unit vector. In this case, A'/e will denote the matrix obtained from A' by deleting the row and column containing the unique non-zero entry in e.

3.2.6 Proposition. $M[A]/e = M[A']/e = M[A'/e]$.

Proof. Since A' is obtained from A by elementary row operations, $M[A'] = M[A]$ and so, $M[A']/e = M[A]/e$. To prove the equality of $M[A']/e$ and $M[A'/e]$, we first observe that, by using row and column swaps if necessary, we may assume that, in A', the unique non-zero entry of e is in row 1 and column 1. Let I be a subset of the ground set E of $M[A']$ such that $e \notin I$. Let $\{(\varepsilon_1, \underline{v}_1)^T, (\varepsilon_2, \underline{v}_2)^T, \ldots, (\varepsilon_k, \underline{v}_k)^T\}$ be the corresponding set of columns of A'. Then the set of columns labelled by $I \cup e$ is linearly independent if and only if the matrix in Figure 3.2 has rank $k+1$. The latter is true if and only if $[\underline{v}_1^T \underline{v}_2^T \cdots \underline{v}_k^T]$ has rank k, which in turn is true if and only if the columns of A'/e labelled by I are linearly independent. We conclude that $\mathcal{I}(M[A'/e]) = \mathcal{I}(M[A']/e)$. □

$$\begin{bmatrix} 1 & \varepsilon_1 & \varepsilon_2 & \cdots & \varepsilon_k \\ 0 & & & & \\ \vdots & \underline{v}_1^T & \underline{v}_2^T & \cdots & \underline{v}_k^T \\ 0 & & & & \end{bmatrix}$$

Fig. 3.2.

3.2.7 Example. Consider the vector matroid M of the following matrix A over $GF(3)$. We previously considered this matroid in Example 2.2.25.

$$A = \begin{array}{c} \begin{matrix} 1 & 2 & 3 & 4 & 5 & 6 & 7 & 8 \end{matrix} \\ \left[\begin{array}{c|cccc} & 0 & 1 & 1 & -1 \\ I_4 & 1 & 0 & 1 & 1 \\ & 1 & 1 & 0 & 1 \\ & -1 & 1 & 1 & 0 \end{array}\right] \end{array}$$

To find $M/8$, we transform A by adding row 3 to row 1, and subtracting row 3 from row 2. This gives the matrix A' shown below.

$$A' = \begin{bmatrix} 1 & 0 & 1 & 0 & 1 & -1 & 1 & 0 \\ 0 & 1 & -1 & 0 & 0 & -1 & 1 & 0 \\ 0 & 0 & 1 & 0 & 1 & 1 & 0 & 1 \\ 0 & 0 & 0 & 1 & -1 & 1 & 1 & 0 \end{bmatrix}$$

(columns labelled 1, 2, 3, 4, 5, 6, 7, 8)

Now delete row 3 and column 8 from A' to get the matrix

$$\begin{bmatrix} 1 & 0 & 1 & 0 & 1 & -1 & 1 \\ 0 & 1 & -1 & 0 & 0 & -1 & 1 \\ 0 & 0 & 0 & 1 & -1 & 1 & 1 \end{bmatrix}.$$

(columns labelled 1, 2, 3, 4, 5, 6, 7)

By Proposition 3.2.6, this matrix represents $M/8$. Of course, if we pivot on a different non-zero entry of column 8 of A, the above process will produce an alternative matrix representation for $M/8$. A geometric representation for $M/8$ is shown in Figure 3.3. □

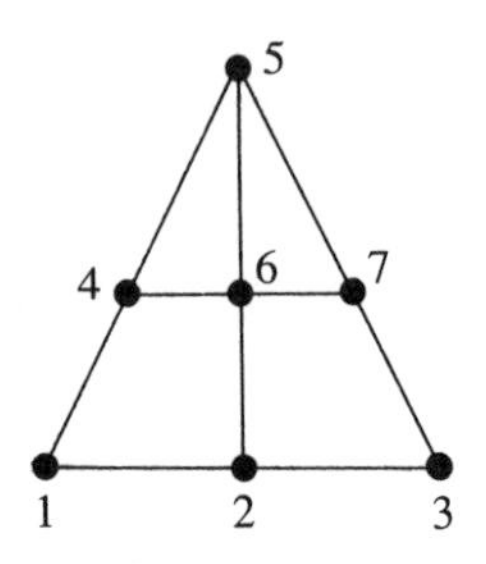

Fig. 3.3.

We have shown that the classes of uniform, graphic, and F-representable matroids are minor-closed. However, as we shall show next, the class of transversal matroids is not minor-closed.

We begin by noting that if $\mathcal{A}$ is a family of subsets of a set E and $T \subseteq E$, then

3.2.8 $$M[\mathcal{A}]\backslash T = M[\mathcal{A}\backslash T]$$

where $\mathcal{A}\backslash T$ is the family of sets obtained from $\mathcal{A}$ by deleting the elements of T from each member of $\mathcal{A}$. In contrast, a contraction of a transversal matroid need not be transversal.

3.2.9 Example. It was shown in Example 1.6.3 that, for the graphs G_1 and G_2 in Figure 1.20, $M(G_1)$ is transversal, whereas $M(G_2)$ is not. Since $G_2 = G_1/7$, $M(G_2) = M(G_1)/7$. Thus $M(G_1)$ is a transversal matroid with a non-transversal contraction. By duality, the class of strict gammoids is not closed under restriction. □

A *gammoid* is a matroid that is isomorphic to a restriction of a strict gammoid. It will follow from the next two propositions (Mason 1972a) that the class of gammoids is the smallest minor-closed class that contains all transversal matroids.

3.2.10 Proposition. *Every transversal matroid is a gammoid.*

Proof. Let $M = M[\mathcal{A}]$ where $\mathcal{A}$, a family of subsets of the set S, is indexed by the set J. Now let G be the directed graph that is formed from the bipartite graph $\Delta[\mathcal{A}]$ by orienting every edge from its endpoint in S to its endpoint in J. Then it is straightforward to check that the restriction to S of the strict gammoid $L(G, J)$ is precisely the matroid $M[\mathcal{A}]$. □

We now illustrate the construction used in the last proof.

3.2.11 Example. Let $S = \{x_1, x_2, \ldots, x_6\}$ and $\mathcal{A} = (A_1, A_2, A_3)$ where $A_1 = \{x_1, x_2, x_6\}$, $A_2 = \{x_2, x_3, x_4\}$, and $A_3 = \{x_4, x_5, x_6\}$. Then Figure 3.4(a) and (b) show $\Delta[\mathcal{A}]$ and the directed graph G constructed from it as above. The matroids $M[\mathcal{A}]$ and $L(G, \{1, 2, 3\})|S$, the equality of which was asserted in the last proof, are represented geometrically by Figure 3.4(c). □

We leave the reader to prove the next result and to deduce from it that the class of gammoids coincides with the class of transversal matroids and their contractions (Exercise 6).

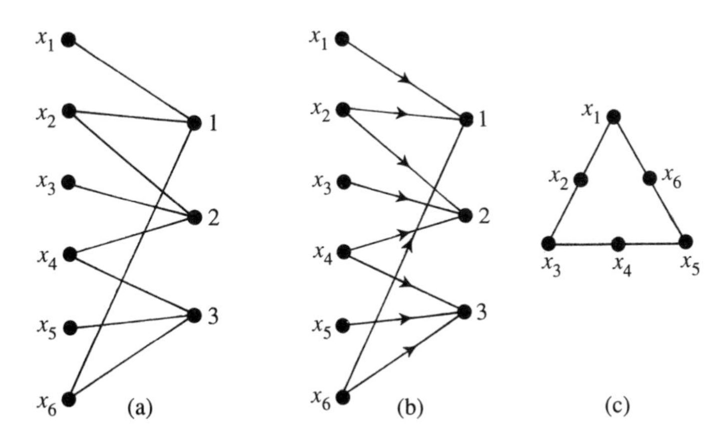

Fig. 3.4. Showing that every transversal matroid is a gammoid.

3.2.12 Proposition. *The class of gammoids is closed under minors and under duality.* □

Exercises

1. Find a graph G so that the bond matroid of G is isomorphic to $M(K_5)\backslash e$.

2. (i) Show that if e and f are edges of a graph G, then $G\backslash e/f = G/f\backslash e$.

   (ii) Show that if G is a connected graph, then the graph H is a minor of G if and only if, for some disjoint subsets X and Y of $E(G)$, $H = G\backslash X/Y$, or equivalently, $H = G/Y\backslash X$.

   (iii) Show that if G and H are graphs, then it is possible for $M(H)$ to be a minor of $M(G)$ without H being a minor of G.

3. Consider the wheel graph $\mathcal{W}_r$ in Exercise 10, Section 2.3. Show that if $r \geq 3$, then $M(\mathcal{W}_r)$ has $M(K_4)$ as a minor.

4. Let $[I_r|D]$ be a matrix over the field F. Suppose that e_1 labels the first column of I_r. By using the construction of a representation for the dual of a matroid, give an alternative derivation of a representation for $M[I_r|D]/e_1$.

5. Let T_8 and R_8 be the matroids in Exercise 8 of Section 2.2. Give geometric representations for each of $T_8/8$, $T_8/1$, $R_8/8$, and $R_8/1$.

6. Let $\mathcal{J}$ denote the class of gammoids. Prove the following:

   (i) $\mathcal{J}$ is minor-closed.

   (ii) $\mathcal{J}$ is closed under duality.

   (iii) $\mathcal{J}$ equals the class of transversal matroids and their contractions.

7. Consider the Petersen graph P_{10} shown in Figure 3.5.

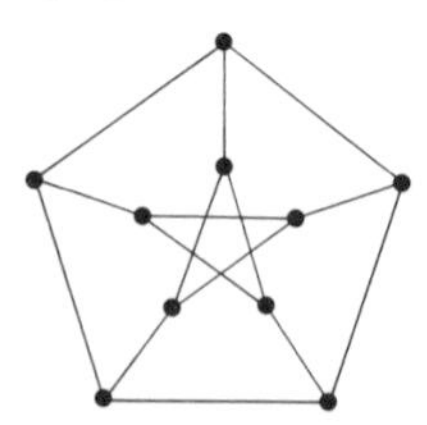

Fig. 3.5. The Petersen graph, P_{10}.

   (i) Argue as in Proposition 2.3.3 to show that $M^*(P_{10})$ is not graphic.

(ii) Show that P_{10} has K_5 as a minor but has no subgraph that is a subdivision of K_5.

(iii) Show that P_{10} has a subgraph that is a subdivision of $K_{3,3}$.

8. For each of the following three possibilities for M, find all the matroids having no minor isomorphic to M:

(i) $U_{0,1}$.

(ii) $U_{1,2}$.

(iii) $U_{0,1} \oplus U_{1,1}$ (see Exercise 7 of Section 1.1).

9.* (Bondy 1971; Piff 1972)

(i) Show that if $\{e, f\}$ is a circuit of a strict gammoid M, then $M \backslash e$ is a strict gammoid.

(ii) Deduce that if $\{e, f\}$ is a cocircuit of a transversal matroid N, then N/e is transversal.

3.3 Projection, flats, and the Scum Theorem

In this section we examine the effects of the operations of deletion and contraction on the flats of a matroid. In addition, we prove the Scum Theorem which, roughly stated, asserts that if a matroid has a certain minor, it has such a minor hanging from the top of its lattice of flats. We conclude by investigating how to obtain geometric representations for $M \backslash e$ and M/e from a geometric representation for M.

In 3.1.19 and 3.1.20, we specified the collections of hyperplanes of $M \backslash T$ and M/T. The next result extends this by determining all the flats of these matroids.

3.3.1 Proposition. *Let T be a subset of the ground set $E(M)$ of a matroid M and F be a subset of $E(M) - T$. Then*

(i) *F is a flat of M/T if and only if $F \cup T$ is a flat of M;*

(ii) *F is a flat of $M \backslash T$ if and only if M has a flat F' such that $F = F' - T$.*

Proof. This follows easily from 3.1.16 and Proposition 3.1.12 and is left as an exercise for the reader. □

We may use this result to relate the lattices of flats of M/T and $M \backslash T$ to that of M. If P is a partially ordered set and x and y are elements of

P such that $x \leq y$, recall that the *interval* $[x, y]$ is the poset induced on $\{z \in P : x \leq z \leq y\}$ by the partial order on P.

3.3.2 Proposition. *Let T be a subset of the ground set $E(M)$ of a matroid M. Then*

(i) *$\mathcal{L}(M/T)$ is isomorphic to the interval $[\text{cl}_M(T), E(M)]$ of $\mathcal{L}(M)$; and*

(ii) *if $E - T$ is a flat of M, then $\mathcal{L}(M \backslash T)$ is isomorphic to the interval $[\text{cl}_M(\emptyset), E - T]$ of $\mathcal{L}(M)$.*

Proof. The map $\phi : \mathcal{L}(M/T) \to [\text{cl}_M(T), E(M)]$ defined by $\phi(F) = F \cup T$ is easily shown to be an isomorphism. We leave the details of this and the proof of (ii) to the reader. □

The next two results are immediate consequences of the last proposition.

3.3.3 Corollary. *If T is a flat of the matroid M, then $\mathcal{L}(M/T)$ is isomorphic to the interval $[T, E(M)]$ of $\mathcal{L}(M)$.* □

3.3.4 Corollary. *If T_1 and $E - T_2$ are flats of M such that $T_1 \subseteq E - T_2$, then $\mathcal{L}(M/T_1 \backslash T_2)$ is isomorphic to the interval $[T_1, E - T_2]$ of $\mathcal{L}(M)$.* □

While Proposition 3.3.2 specified $\mathcal{L}(M/T)$ for all T, it determined $\mathcal{L}(M \backslash T)$ only when $E - T$ is a flat of M. If we drop this condition on $E - T$, then $\mathcal{L}(M \backslash T)$ need not be isomorphic to an interval of $\mathcal{L}(M)$. Indeed, the elements of $\mathcal{L}(M \backslash T)$, which are the flats of $M \backslash T$, correspond precisely to those flats F of M such that $\text{cl}_M(F - T) = F$.

We shall show that every minor of a matroid M can be written in the form $M \backslash I^*/I$ where I and I^* are independent and coindependent respectively. The proof of this will use the following lemma. Recall that $M.Z$ is $M/(E - Z)$.

3.3.5 Lemma. *Suppose that X and Y are disjoint subsets of the ground set of a matroid M. Let Y_1 be a basis of $M|Y$ and X_1 be a cobasis of $M.X$. Then, for every partition (X_2, Y_2) of $(X - X_1) \cup (Y - Y_1)$,*

$$M \backslash X/Y = M \backslash (X_1 \cup Y_2)/(Y_1 \cup X_2).$$

Proof. As Y_1 is a basis of $M|Y$, every element of $Y - Y_1$ is a loop of M/Y_1. Thus if (W_1, W_2) is a partition of $Y - Y_1$, then $M/Y = M/(Y_1 \cup W_1) \backslash W_2$. Moreover, X_1 is a basis of $M^*|X$, so if (Z_1, Z_2) is a partition of $X - X_1$, then $M^*/X = M^*/(X_1 \cup Z_1)\backslash Z_2$. Therefore $M\backslash X = M\backslash(X_1 \cup Z_1)/Z_2$. It is straightforward to check that

$$M\backslash X/Y = M\backslash(X_1 \cup Z_1 \cup W_2)/(Y_1 \cup W_1 \cup Z_2),$$

and so, as $(Z_1 \cup W_2, W_1 \cup Z_2)$ is a partition of $(X - X_1) \cup (Y - Y_1)$, the lemma follows. □

3.3.6 Proposition. *Let X and Y be disjoint subsets of the ground set of a matroid M. Let Y_1 be a basis of $M|Y$ and X_1 be a cobasis of $M.X$. Then $M\backslash X/Y = M\backslash[X_1 \cup (Y - Y_1)]/[Y_1 \cup (X - X_1)]$. Moreover, $X_1 \cup (Y - Y_1)$ and $Y_1 \cup (X - X_1)$ are coindependent and independent, respectively, in M.*

Proof. Take $Y_2 = Y - Y_1$ and $X_2 = X - X_1$ in the last lemma to get the first part. To prove the second part, note that, as every element of $X - X_1$ is a coloop of $M\backslash X_1$, the set $Y_1 \cup (X - X_1)$ is independent in $M\backslash X_1$ and hence in M. It follows, by duality, that $X_1 \cup (Y - Y_1)$ is coindependent in M. □

This proposition contains the essence of the next result.

3.3.7 The Scum Theorem. *(Higgs in Crapo and Rota 1970) Let M_1 be a simple minor of the simple matroid M. Then M has a flat F of rank $r(M) - r(M_1)$ such that there is a one-to-one order-preserving map from $\mathcal{L}(M_1)$ into the interval $[F, E(M)]$ of $\mathcal{L}(M)$.*

Proof. Let $M_1 = M\backslash X/Y$. By Proposition 3.3.6, $M_1 = M\backslash[X_1 \cup (Y - Y_1)]/[Y_1 \cup (X - X_1)]$ where $Y_1 \cup (X - X_1)$ is independent and $X_1 \cup (Y - Y_1)$ is coindependent in M. Thus $r(M_1) = r(M) - |Y_1 \cup (X - X_1)|$. Let $F = \mathrm{cl}_M(Y_1 \cup (X - X_1))$. Since M_1 is simple, $F \subseteq X \cup Y$. Moreover, $r(F) = r(M) - r(M_1)$. As every element of $F - [Y_1 \cup (X - X_1)]$ is a loop in $M/[Y_1 \cup (X - X_1)]$, Corollary 3.1.25 implies that $M_1 = M/F\backslash[(X \cup Y) - F]$. Now define $\phi : \mathcal{L}(M_1) \to [F, E(M)]$ by $\phi(Z) = \mathrm{cl}_M(Z \cup F)$ for all Z in $\mathcal{L}(M_1)$. It is straightforward to check that ϕ is one-to-one and order-preserving. □

We now turn our attention to obtaining geometric representations for $M\backslash e$ and M/e from a geometric representation for M. We have already implicitly used the fact that a geometric representation for $M\backslash e$ is obtained from one for M simply by deleting the point corresponding to e. Obtaining a geometric representation for the contraction of e is somewhat more complicated. We may assume that e is not a loop of M otherwise $M/e = M\backslash e$ and the geometric representation for M is as specified above. If P is the parallel class of M that contains e and P is non-trivial, then every element of $P - e$ becomes a loop of M/e. In order to determine the effect of contraction on the elements of $E(M) - P$, we shall now assume that $P = \{e\}$.

With these preliminaries, we are now ready to describe how to obtain a geometric representation for M/e. First note that all loops of M remain loops of M/e. Now let H be a hyperplane of M not containing e. We project every non-loop element of $E(M) - (H \cup e)$ from e onto H. Specifically, if f is such an element, we consider the line $\text{cl}(\{e, f\})$. If this line meets H in g, say, then f is mapped onto g, or equivalently, onto a point touching g. If $\text{cl}(\{e, f\})$ does not meet H, we add a new point on H. This point is the image of not only f but also of all other points in $\text{cl}(\{e, f\}) - e$. Thus all of the non-loops in the geometric representation for M/e are points of H consisting of the original non-loop points of H, together with the images of the non-loop elements of $E(M) - (H \cup e)$ under projection from e.

This projection from e has a predictable effect on a line X and a plane Y. Specifically, if $e \in X$, then the points of $X - e$ are all mapped onto the same point or, equivalently, onto mutually touching points; if $e \notin X$, the images of distinct non-touching points of X are distinct, non-touching, and collinear in M/e. If $e \in Y$, then the images of points of Y are collinear in M/e; if $e \notin Y$, the images of distinct non-touching points of Y are distinct, non-touching, and coplanar in M/e. The verification that this construction does indeed produce a geometric representation for M/e is left to the reader with the reminder that, since e is not a loop of M, a subset I of $E(M) - e$ is in $\mathcal{I}(M/e)$ if and only if $I \cup e \in \mathcal{I}(M)$.

We now illustrate this construction with two examples.

3.3.8 Example. Figure 3.6(a) shows a geometric representation for F_7. To obtain a geometric representation for $F_7/1$, we can project from 1 onto the line $\{3, 4, 5\}$. The result is shown in Figure 3.6(b). Note that the same diagram is obtained by projecting from 1 onto, for example, the line $\{2, 7, 5\}$. Moreover, if we relax the circuit–hyperplane $\{2, 4, 6\}$ in F_7 to get F_7^-, and then contract 1, we see that we get the same matroid that was obtained just by contracting 1 from F_7. This last observation

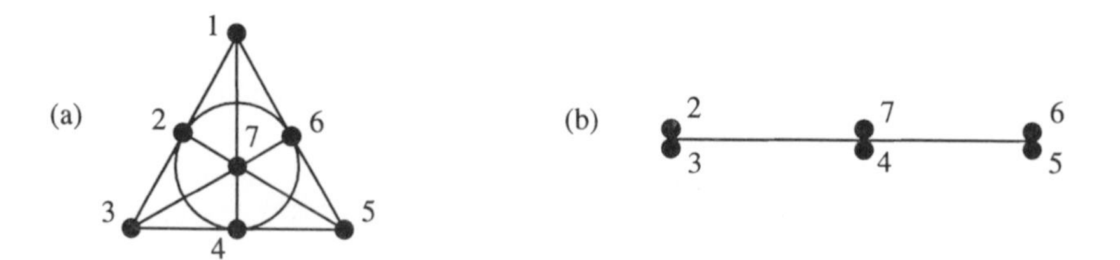

Fig. 3.6. (a) F_7. (b) $F_7/1$.

illustrates part of the following result (Kahn 1985), the proof of which is left to the reader (Exercise 9).

3.3.9 Proposition. *Let X be a circuit–hyperplane of the matroid M and M' be the matroid obtained by relaxing X.*

(i) *If $e \in E(M) - X$, then $M/e = M'/e$ and, provided M does not have e as a coloop, $M'\backslash e$ is obtained from $M\backslash e$ by relaxing the circuit–hyperplane X of the latter.*

Dually,

(ii) *if $f \in X$, then $M\backslash f = M'\backslash f$ and, provided M does not have f as a loop, M'/f is obtained from M/f by relaxing the circuit–hyperplane $X - f$ of the latter.* □

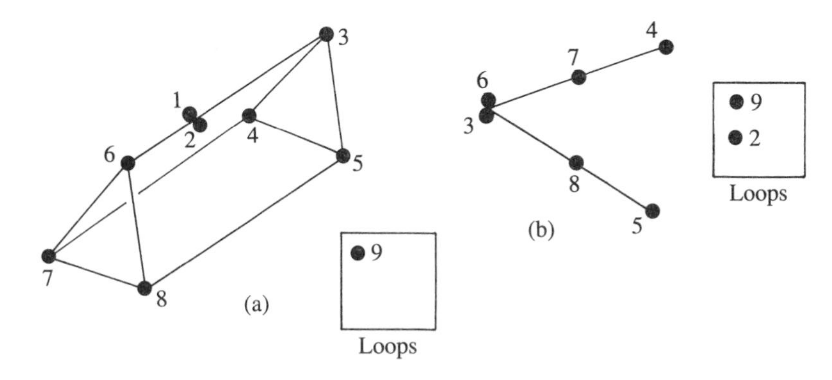

Fig. 3.7. (a) A matroid M. (b) The contraction $M/1$.

3.3.10 Example. Let M be the matroid for which a geometric representation is shown in Figure 3.7(a). To obtain a geometric representation for $M/1$, we project from 1 onto the plane $\{4, 5, 7, 8\}$. Evidently 2 must become a loop, while 3 and 6 become parallel. The resulting geometric representation for $M/1$ is shown in Figure 3.7(b) where, for example, 4, 7, and 6 are collinear in the contraction because 4, 7, 6, and 1 were coplanar in M. □

Exercises

1. Show that the matroid $M/1$ in Example 3.3.10 is both graphic and cographic by finding graphs G_1 and G_2 such that $M/1 \cong M(G_1) \cong M^*(G_2)$.

2. **(i)** Show that if C is a circuit of a matroid M and $e \in C$, then $M/C = M\backslash e/(C - e)$.

   (ii) State the dual of the result in (i).

3. For each of the matroids $M(K_4)$, $\mathcal{W}^3$, Q_6, and P_6 shown in Figure 2.1, find geometric representations for $M/1$ and $M/6$.

4. Give necessary and sufficient conditions, in terms of flats of M, for the set F to be a flat of $M\backslash X/Y$.

5. Give an example to show that if N is a restriction of a simple matroid M, then $\mathcal{L}(N)$ need not be a sublattice of $\mathcal{L}(M)$.

6. Prove that a matroid M has a minor isomorphic to $U_{2,n}$ if and only if M has a flat of rank $r(M) - 2$ that is contained in at least n distinct hyperplanes.

7. **(i)** Consider the matroid M for which a geometric representation is shown in Figure 1.17. Find geometric representations for $M/10$, $M/1\backslash 4$, and $M/3, 5\backslash 7$.

(ii) Consider the Vámos matroid V_8 (Figure 2.4). Give geometric representations for $V_8/1$ and $V_8/5$.

8. **(i)** Show that the non-Fano matroid F_7^- is non-binary by showing that it has a minor isomorphic to $U_{2,4}$.

(ii) Find a non-binary restriction N of F_7^- for which every proper restriction is binary.

9. Prove Proposition 3.3.9.

4
Connectivity

We have already met several examples of basic graph-theoretic notions that have matroid-theoretic analogues. In this chapter, following Whitney (1935), we shall extend the concept of 2-connectedness from graphs to matroids. Matroid extensions of k-connectedness for $k \geq 3$ will be looked at in Chapter 8. We begin our discussion of k-connectedness with $k = 2$ rather than $k = 1$ because, by Proposition 1.2.8, every graphic matroid is isomorphic to the cycle matroid of a connected graph. In particular, if G_1 is a disconnected graph, then there is a connected graph G_2 such that $M(G_1) \cong M(G_2)$. Hence there is no matroid concept that corresponds directly to the idea of connectedness, or 1-connectedness, for graphs.

4.1 Connectivity for graphs and matroids

The purpose of this section is to introduce a notion of 2-connectedness for matroids that extends the corresponding notion for graphs. We begin by recalling the definition of k-connectedness for graphs.

A subset X of the vertex set of a graph G is called a *vertex cut* if $G - X$ has more connected components than G, where $G - X$ is obtained from G by deleting the vertices in X and all incident edges. If a vertex cut X contains a single vertex v, then v is called a *cut-vertex* of G.

Consider the graphs G_1, G_2, and G_3 in Figure 4.1. None has a cut-vertex. However, both G_1 and G_2 have 2-element vertex cuts, whereas the smallest vertex cut in G_3 contains three elements. In general, if G is a connected graph that has at least one pair of distinct non-adjacent vertices, we define the *connectivity* $\kappa(G)$ of G to be the smallest j for which G has a j-element vertex cut. If G is connected, but has no pair of distinct non-adjacent vertices, we take $\kappa(G)$ to be $|V(G)| - 1$. Finally, if G is disconnected, we let $\kappa(G) = 0$. If k is a positive integer, G is said to be *k-connected* if $\kappa(G) \geq k$. Thus a graph with at least two vertices is 1-connected if and only if it is connected. For the graphs in Figure 4.1, it is easily seen that both $\kappa(G_1)$ and $\kappa(G_2)$ are 2, while $\kappa(G_3)$ is 3.

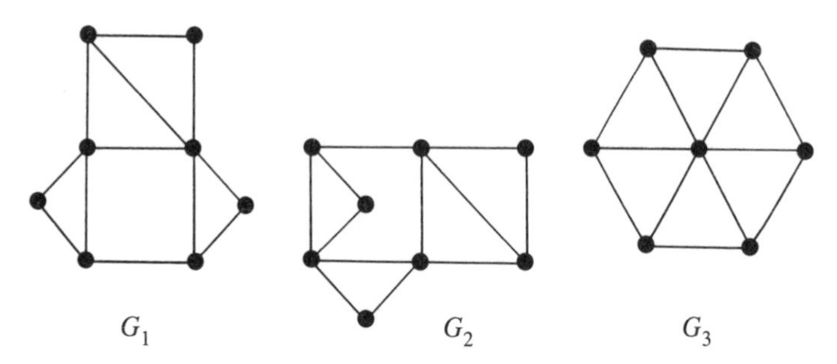

Fig. 4.1. $M(G_1)$ is isomorphic to $M(G_2)$ but $M(G_3)$ determines G_3.

The main result of the next chapter will characterize precisely when two graphs have isomorphic cycle matroids. Connectivity will be a key tool in that discussion, and the graphs in Figure 4.1 exemplify the core of that result. The reader is urged to show that $M(G_1) \cong M(G_2)$ while, if G_4 is a graph without isolated vertices for which $M(G_4) \cong M(G_3)$, then $G_4 \cong G_3$.

We have already noted that, for $k = 1$, the notion of k-connectedness does not extend to matroids. Now suppose that $k = 2$. The definition of 2-connectedness for graphs, relying as it does on deleting vertices, does not obviously extend to matroids. However, the following alternative characterization of 2-connectedness does.

4.1.1 Proposition. *Let G be a loopless graph without isolated vertices. If G has at least three vertices, then G is 2-connected if and only if, for every pair of distinct edges of G, there is a cycle containing both.*

This proposition can be proved by a direct graph-theoretic argument. Instead, we shall give a slightly different proof involving matroids, and we postpone this until later in the section.

We now indicate how the notion of 2-connectedness for graphs can be extended to matroids. For each element e of a matroid M, let $\gamma(e) = \{e\} \cup \{f\colon M$ has a circuit containing both e and $f\}$. Define the relation γ on E by $e\,\gamma\,f$ if and only if $e \in \gamma(f)$.

4.1.2 Proposition. *The relation γ is an equivalence relation on $E(M)$.*

Proof. Clearly γ is reflexive and symmetric. To show that it is transitive, suppose that $e \in \gamma(f)$ and $f \in \gamma(g)$ where e, f, and g are distinct. Then, as $f \in \gamma(g)$, $g \in \gamma(f)$. But e is also in $\gamma(f)$, so M has circuits C_1 and C_2 such that $e \in C_1$, $g \in C_2$, and $C_1 \cap C_2$ is non-empty. Suppose that, among all such pairs of circuits, $|C_1 \cup C_2|$ is minimal. We aim to show that M has a circuit containing both e and g. Assume that this is not so. Then $C_1 \neq C_2$. Now choose an element h from $C_1 \cap C_2$. The reader may find a Venn diagram useful in keeping track of the rest of the proof. By the strong circuit elimination axiom, 1.4.11, M has a circuit C_3 such that $e \in C_3 \subseteq (C_1 \cup C_2) - h$. Moreover, by assumption, $g \notin C_3$. As $C_3 \nsubseteq C_1$, there is an element i of $C_2 - C_1$ that is in C_3. Applying the strong circuit elimination axiom to C_2 and C_3, we find that M has a circuit C_4 such that $g \in C_4 \subseteq (C_2 \cup C_3) - i$. Since $C_4 \nsubseteq C_2$, the set $C_4 \cap (C_3 - C_2)$ is non-empty, hence $C_4 \cap C_1$ is non-empty. But $C_1 \cup C_4 \subseteq (C_1 \cup C_2) - i$ and so $|C_1 \cup C_4| < |C_1 \cup C_2|$. Therefore the pair (C_1, C_4) contradicts the choice of (C_1, C_2) because $e \in C_1$, $g \in C_4$, and $C_1 \cap C_4 \neq \emptyset$. □

4.1.3 Corollary. *If $T \subseteq E(M)$ and T is a γ-equivalence class, then, for all elements t of T, $\gamma(t) = T$.* □

The γ-equivalence classes are called the *(connected) components* of M. Clearly every loop of M is a component; so is every coloop. If $E(M)$ is a component of M, we call M *2-connected* or sometimes just *connected*; otherwise M is *disconnected*. The following is an immediate consequence of Corollary 4.1.3.

4.1.4 Proposition. *The matroid M is connected if and only if, for every pair of distinct elements of $E(M)$, there is a circuit containing both.* □

The term 'connected component' has now been defined here for both graphs and matroids. Although this may seem undesirable, it does conform with standard usage. To highlight the difference, consider the graphs in Figure 4.2. Evidently $M(G_1) \cong M(G_2) \cong M(G_3)$. But the graphs G_1, G_2, and G_3, have one, two, and three connected components, respectively. The connected components of the matroids $M(G_1)$, $M(G_2)$, and $M(G_3)$ are isomorphic to $M(G_{3,1})$, $M(G_{3,2})$, and $M(G_{3,3})$. The graphs $G_{3,1}$, $G_{3,2}$, and $G_{3,3}$ are the blocks of each of G_1, G_2, and G_3. In general, a *block* is a connected graph whose cycle matroid is connected. Clearly a loopless graph is a block if and only if it is connected and has no cut-vertices. Hence, by Proposition 4.1.1, a block with at least three vertices is 2-connected. A *block of a graph* is a subgraph that is a block and is maximal with this property.

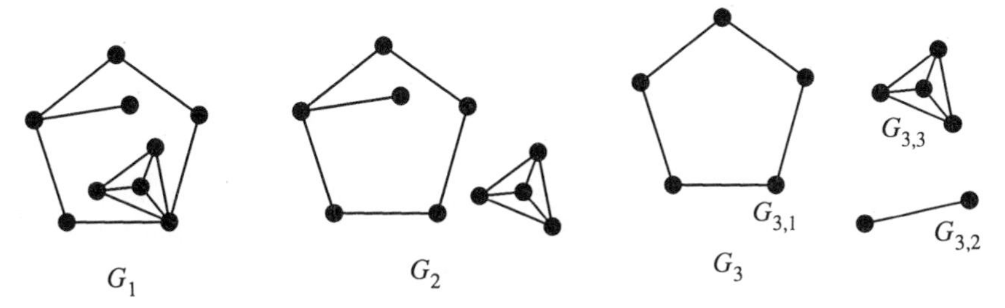

Fig. 4.2. Three different graphs with the same cycle matroid.

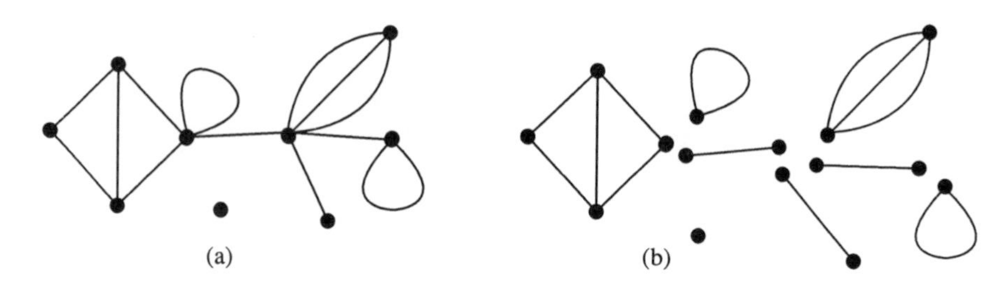

Fig. 4.3. A graph G. (b) The eight blocks of G.

4.1.5 Example. The graph G in Figure 4.3(a) has eight blocks, these being the eight components of the graph in Figure 4.3(b). □

We noted in Chapter 2 that the set of edges meeting at a vertex of a graph G is an edge cut of G but need not be a bond, a minimal edge cut. The last example provides another illustration of this phenomenon. The next result notes that this phenomenon cannot occur in 2-connected graphs. Its proof is left to the reader.

4.1.6 Proposition. *Let G be a connected loopless graph having at least three vertices. Then G is 2-connected if and only if, for every vertex v of G, the set of edges meeting at v is a bond.* □

It is not difficult to determine when a small matroid is connected.

4.1.7 Example. Table 4.1 lists all of the connected matroids on at most four elements. The graph $\mathcal{W}_2$ is shown in Figure 4.4. The reader is encouraged to check the accuracy of the table. □

In general, the uniform matroid $U_{m,n}$ is disconnected if and only if $n \geq 2$ and m is 0 or n.

Table 4.1. All connected matroids on at most four elements.

Number of elements	Connected matroids
0	$U_{0,0}$
1	$U_{0,1}, U_{1,1}$
2	$U_{1,2}$
3	$U_{1,3}, U_{2,3}$
4	$U_{1,4}, U_{2,4}, U_{3,4}, M(\mathcal{W}_2)$

Fig. 4.4. $\mathcal{W}_2$.

To conclude this section, we prove Proposition 4.1.1 by verifying the following restatement of it.

4.1.8 Proposition. *Let G be a loopless graph without isolated vertices. If G has at least three vertices, then $M(G)$ is a connected matroid if and only if G is a 2-connected graph.*

Proof. It is straightforward to show, using Proposition 4.1.4, that G is 2-connected if $M(G)$ is connected. We leave this as an exercise.

Now suppose that G is 2-connected and let G_1 be the subgraph of G induced by some arbitrarily chosen component of $M(G)$. Then, as G is loopless, $|V(G_1)| \geq 2$. We may assume that $G_1 \neq G$, otherwise the required result holds. Therefore G has an edge xy with exactly one endpoint, say x, in G_1. Now x is not a cut-vertex of G, so $G - x$ contains a path P_{yz} that joins y to some vertex z of $G_1 - x$ but is otherwise disjoint from $G_1 - x$. Because G_1 is certainly connected, it contains a path P_{zx} joining z and x. The union of P_{yz}, P_{zx}, and the edge xy is a cycle in G that contains xy together with at least one edge of G_1. But $M(G_1)$ is a component of $M(G)$, and hence we obtain the contradiction that $xy \in E(G_1)$. □

Exercises

1. Show that a matroid M is connected if and only if, for every pair of distinct elements of $E(M)$, there is a hyperplane avoiding both.

2. Show that every component of a loopless matroid is a flat.

3. Find all connected matroids on a 5-element set, giving geometric representations for each.

4. If X is a subset of a matroid M and $M|X$ is connected, find necessary and sufficient conditions for $M|(\text{cl}(X))$ to be connected.

5. Show that a matroid M with at least two elements is connected if and only if $\widetilde{M}$ is connected and M has no loops.

6. Let X and Y be subsets of a matroid M such that both $M|(\text{cl}(X))$ and $M|(\text{cl}(Y))$ are connected and $X \cap Y$ is non-empty. Show that $M|(\text{cl}(X \cup Y))$ is connected.

7. Let A be an $r \times n$ matrix over the field F. Prove that $M[A]$ is connected if and only if there is no partition of the set of columns of A into non-empty sets X_1 and X_2 such that $\langle X_1 \rangle \cap \langle X_2 \rangle = \{\underline{0}\}$.

8. (Harary 1969, Theorem 3.3) Let G be a loopless graph without isolated vertices and suppose that $|V(G)| \geq 3$. Prove that the following statements are equivalent:

   (a) G is 2-connected.

   (b) If u and v are vertices of G, there is a cycle meeting both.

   (c) If v is a vertex and e is an edge of G, there is a cycle containing e and meeting v.

   (d) If e and f are edges of G, there is a cycle containing both.

4.2 Properties of matroid connectivity

In this section we indicate some of the many alternative characterizations of matroid connectivity. The first of these relates to the problem considered in the preceding chapter of determining those subsets T of $E(M)$ for which $M \backslash T = M/T$.

A subset X of a matroid M is called a *separator* of M if X is a union of components of M. Thus both E and $\emptyset$ are separators of M. All other separators are called *non-trivial separators.* Clearly, if X is a separator of M, so is $E - X$. The following result characterizes the separators of a matroid.

4.2.1 Proposition. *Let T be a subset of the ground set E of a matroid M. Then T is a separator of M if and only if*

4.2.2 $$r(T) + r(E-T) = r(M).$$

Proof. Suppose that T is a separator of M. Let B_T and B_{E-T} be bases of $M|T$ and $M|(E-T)$, respectively, and let $B = B_T \cup B_{E-T}$. If B is dependent, it contains a circuit C. Since C cannot meet both T and $E-T$, it is contained in B_T or B_{E-T}; a contradiction. Thus B is independent in M. As B_T and B_{E-T} are maximal independent subsets of T and $E-T$, respectively, B is a maximal independent subset of M. Hence 4.2.2 holds. The proof of the converse is similar and is left to the reader. □

We note here that, by the semimodularity of r, for all subsets T of M,

4.2.3 $$r(T) + r(E-T) \geq r(M).$$

The last proposition characterized when equality occurs in 4.2.3. On combining that result with Proposition 3.1.24, we obtain the following.

4.2.4 **Corollary.** *If T is a subset of a matroid M, then $M \backslash T = M/T$ if and only if T is a separator of M.* □

One of the attractive features of the class of connected matroids is that it is closed under duality. To prove this we shall use the next two results.

4.2.5 **Lemma.** *If T is a subset of a matroid M, then*

$$r(T) + r(E - T) - r(M) = r(T) + r^*(T) - |T|.$$

Proof. This follows easily from the formula for $r^*(T)$ (Proposition 2.1.9) and is left as an exercise. □

4.2.6 **Proposition.** *A matroid M is disconnected if and only if, for some proper non-empty subset T of M,*

4.2.7
$$r(T) + r^*(T) - |T| = 0.$$

Proof. By the preceding lemma, $r(T) + r^*(T) - |T| = 0$ if and only if $r(T) + r(E - T) - r(M) = 0$. By Proposition 4.2.1, this holds for a non-empty proper subset T of M if and only if M is disconnected. □

4.2.8 **Corollary.** *M is connected if and only if M^* is connected.*

Proof. Equation 4.2.7 is self-dual. □

We can obtain an alternative proof of the last corollary directly from the definition of connectedness by using the following result.

4.2.9 **Proposition.** *If x and y are distinct elements of a circuit C of a matroid M, then M has a cocircuit C^* such that $C \cap C^* = \{x, y\}$.*

Proof. As $C - x$ is independent in M, it is contained in a basis B that contains y but avoids x. Now $E(M) - B$ is a basis of M^* that meets C in x. The fundamental cocircuit $C^*(y, E(M) - B)$ of y with respect to the cobasis $E(M) - B$ of M is contained in $y \cup (E(M) - B)$. Since it has the element y in common with the cocircuit C of M^*, Proposition 2.1.11 implies that it must contain at least one other element of C. As $C \cap (y \cup (E(M) - B)) = \{x, y\}$, it follows that $x \in C^*(y, E(M) - B)$. Hence $C^*(y, E(M) - B) \cap C = \{x, y\}$, so $C^*(y, E(M) - B)$ is the required cocircuit of M. □

It is of interest to know how the components of a minor of M are related to the components of M. The following result is obtained by combining Proposition 4.1.4 and Corollary 4.2.8.

4.2.10 Proposition. *Every connected component of $M\backslash X/Y$ is contained in a connected component of M.* □

The next result characterizes disconnected matroids in terms of their collections of independent sets. Its straightforward proof is omitted.

4.2.11 Proposition. *The matroid M is disconnected if and only if, for some proper non-empty subset T of $E(M)$, $\mathcal{I}(M) = \{I_1 \cup I_2 : I_1 \in \mathcal{I}(M|T), I_2 \in \mathcal{I}(M|(E-T))\}$.* □

By the last result, every disconnected matroid M has the property that it can be determined from two disjoint restrictions. We now describe how to form a new matroid from two arbitrary matroids on disjoint sets.

4.2.12 Proposition. *Let M_1 and M_2 be matroids on disjoint sets E_1 and E_2. Let $E = E_1 \cup E_2$ and $\mathcal{I} = \{I_1 \cup I_2 : I_1 \in \mathcal{I}(M_1), I_2 \in \mathcal{I}(M_2)\}$. Then $(E, \mathcal{I})$ is a matroid.*

Proof. This is routine and was stated as Exercise 7 of Section 1.1. □

The matroid $(E, \mathcal{I})$ in the last proposition is called the *direct sum* or *1-sum* of M_1 and M_2 and is denoted $M_1 \oplus M_2$. More generally, for n matroids, $M_1, M_2, \ldots, M_n$, on disjoint sets, $E_1, E_2, \ldots, E_n$, the direct sum $M_1 \oplus M_2 \oplus \cdots \oplus M_n$ is the pair $(E, \mathcal{I})$ where $E = E_1 \cup E_2 \cup \cdots \cup E_n$ and $\mathcal{I} = \{I_1 \cup I_2 \cup \cdots \cup I_n : I_i \in \mathcal{I}(M_i)$ for all i in $\{1, 2, \ldots, n\}\}$. It is not difficult to extend Proposition 4.2.12 to show that $M_1 \oplus M_2 \oplus \cdots \oplus M_n$ is indeed a matroid. We call $M_1, M_2, \ldots, M_n$ the *direct sum components* of $M_1 \oplus M_2 \oplus \cdots \oplus M_n$. Note that, whenever we write $M_1 \oplus M_2 \oplus \cdots \oplus M_n$, it will be implicit that $M_1, M_2, \ldots, M_n$ have disjoint ground sets.

The next result is a straightforward extension of Proposition 4.2.11.

4.2.13 Corollary. *If $T_1, T_2, \ldots, T_k$ are the connected components of a matroid M, then $M = M|T_1 \oplus M|T_2 \oplus \cdots \oplus M|T_k$. Moreover, if $M_1 \oplus M_2 \oplus \cdots \oplus M_n = N_1 \oplus N_2 \oplus \cdots \oplus N_m$, where each of $M_1, M_2, \ldots, M_n$, $N_1, N_2, \ldots, N_m$ is connected and non-empty, then $m = n$, and there is a permutation σ of $\{1, 2, \ldots, n\}$ such that, for all i in $\{1, 2, \ldots, n\}$, $M_i = N_{\sigma(i)}$.* □

4.2.14 Example. The rank-4 matroid for which a geometric representation is shown in Figure 4.5 is isomorphic to $U_{2,4} \oplus U_{2,4}$. It is easy to see that this matroid is not uniform. □

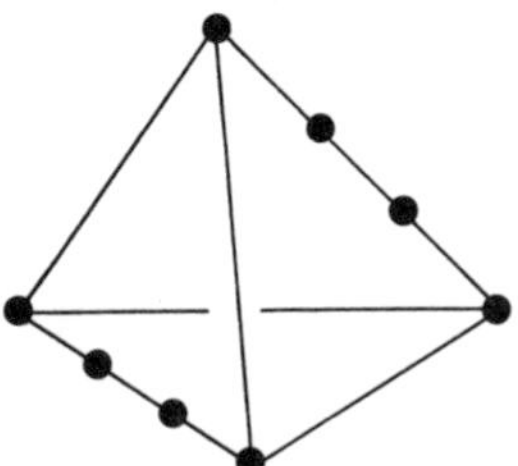

Fig. 4.5. $U_{2,4} \oplus U_{2,4}$.

In contrast to the last example, we have the following result.

4.2.15 Proposition. *The classes of F-representable, graphic, cographic, transversal, and regular matroids are all closed under the operation of direct sum.*

Proof. This is left to the reader (Exercise 6). □

We conclude this section by noting a number of basic properties of the direct sum. These are stated for $M_1 \oplus M_2$ but can easily be extended to $M_1 \oplus M_2 \oplus \cdots \oplus M_n$. The straightforward proofs of these facts are left to the reader.

4.2.16 $\mathcal{C}(M_1 \oplus M_2) = \mathcal{C}(M_1) \cup \mathcal{C}(M_2)$.

4.2.17 *If $X \subseteq E(M_1 \oplus M_2)$, then*

$$r_{M_1 \oplus M_2}(X) = r_{M_1}(X \cap E(M_1)) + r_{M_2}(X \cap E(M_2)).$$

4.2.18 $\mathcal{B}(M_1 \oplus M_2) = \{B_1 \cup B_2 : B_1 \in \mathcal{B}(M_1),\ B_2 \in \mathcal{B}(M_2)\}$.

4.2.19 $\mathcal{H}(M_1 \oplus M_2) = \{H_1 \cup E(M_2) : H_1 \in \mathcal{H}(M_1)\}$
$\cup \{E(M_1) \cup H_2 : H_2 \in \mathcal{H}(M_2)\}$.

4.2.20 *F is a flat of $M_1 \oplus M_2$ if and only if $F \cap E(M_1)$ and $F \cap E(M_2)$ are flats of M_1 and M_2, respectively.*

4.2.21 $\mathcal{S}(M_1 \oplus M_2) = \{S_1 \cup S_2 : S_1 \in \mathcal{S}(M_1),\ S_2 \in \mathcal{S}(M_2)\}$.

4.2.22 $\mathcal{C}^*(M_1 \oplus M_2) = \mathcal{C}^*(M_1) \cup \mathcal{C}^*(M_2)$.

4.2.23 *If $X \subseteq E(M_1 \oplus M_2)$, then*

$$(M_1 \oplus M_2)\backslash X = [M_1\backslash(X \cap E(M_1))] \oplus [M_2\backslash(X \cap E(M_2))],$$

and

$$(M_1 \oplus M_2)/X = [M_1/(X \cap E(M_1))] \oplus [M_2/(X \cap E(M_2))].$$

Finally, we note another attractive feature of duality.

4.2.24 Proposition. $(M_1 \oplus M_2)^* = M_1^* \oplus M_2^*$.

Proof.
$$\begin{aligned} \mathcal{C}((M_1 \oplus M_2)^*) &= \mathcal{C}^*(M_1 \oplus M_2) \\ &= \mathcal{C}^*(M_1) \cup \mathcal{C}^*(M_2), \text{ by 4.2.22;} \\ &= \mathcal{C}(M_1^*) \cup \mathcal{C}(M_2^*) \\ &= \mathcal{C}(M_1^* \oplus M_2^*), \text{ by 4.2.16.} \quad \square \end{aligned}$$

Exercises

1. Let T be a subset of the ground set of a matroid M. Prove that the following statements are equivalent:

   (a) T is a separator of M.

   (b) $M^*\backslash T = M^*/T$.

   (c) Every hyperplane of M contains T or $E - T$.

   (d) T is a separator of M^*.

2. Show that if $\{e, f\}$ is both a circuit and a cocircuit of the matroid M, then $\{e, f\}$ is a component of M.

3. Let M be a matroid and r^* be the rank function of M^*. If X is a subset of $E(M)$, show that $r^*(X)$ equals $r((M|X)^*)$, the rank of $(M|X)^*$, if and only if X is a separator of M.

4. Find all disconnected paving matroids.

5. Find all matroids with at most three circuits and show that all such matroids are both graphic and cographic.

6. (i) Suppose that A_1 and A_2 are F-representations of the matroids M_1 and M_2. Show that $\left[\begin{smallmatrix} A_1 & 0 \\ 0 & A_2 \end{smallmatrix}\right]$ is an F-representation of $M_1 \oplus M_2$.

   (ii) Complete the proof of Proposition 4.2.15.

7. Prove directly that, in a graph G, every pair of distinct edges is in a cycle if and only if every pair of distinct edges is in a bond.

8. (i) Prove that every minor of a paving matroid is paving.

   (ii) Prove that a matroid is paving if and only if it has no minor isomorphic to $U_{2,2} \oplus U_{0,1}$.

4.3 More properties of connectivity

Many matroid properties are such that a matroid M has the property if and only if all of its connected components have the property. This fact, which was exemplified in Proposition 4.2.15, means that in many arguments one is able to restrict attention to connected matroids. In this section, we present some more properties of connected matroids. The first result (Tutte 1966b) is particularly useful in induction arguments.

4.3.1 Theorem. *Let e be an element of a connected matroid M. Then $M \backslash e$ or M/e is connected.*

Proof. Suppose $M \backslash e$ is not connected and let E_1 be one of its components. If $x \in E_1$ and $y \in E(M \backslash e) - E_1$, then M has a circuit C_{xy} containing $\{x, y\}$. Now C_{xy} is not a circuit of $M \backslash e$ since x and y are in different components of $M \backslash e$. Hence $e \in C_{xy}$ and so $C_{xy} - e$ is a circuit of M/e containing $\{x, y\}$. But x and y were arbitrarily chosen in E_1 and $E(M) - (E_1 \cup e)$, respectively. Using this and the fact that the components of M/e are $\gamma_{M/e}$- equivalence classes, it is not difficult to deduce that $E(M) - e$ is a component of M/e; that is, M/e is connected. □

The next result, due to Lehman (1964), asserts that a connected matroid is uniquely determined by the set of circuits containing any fixed element. An extension of this result for binary matroids will be noted in Exercise 9 of Section 9.3.

4.3.2 Theorem. *Let e be an element of a connected matroid M, and $\mathcal{C}_e$ be the set of circuits of M containing e. Then the circuits of M not containing e are the minimal sets of the form*

$$(C_1 \cup C_2) - \bigcap\{C : C \in \mathcal{C}_e,\ C \subseteq C_1 \cup C_2\}$$

where C_1 and C_2 are distinct members of $\mathcal{C}_e$.

Proof. Let C_1 and C_2 be distinct members of $\mathcal{C}_e$. We shall denote $\{C : C \in \mathcal{C}_e,\ C \subseteq C_1 \cup C_2\}$ by $\mathcal{C}_e(C_1 \cup C_2)$. We begin by showing that

$$(C_1 \cup C_2) - \bigcap\{C : C \in \mathcal{C}_e(C_1 \cup C_2)\} \text{ is dependent.} \tag{1}$$

By circuit elimination, M has a circuit C_3 such that $C_3 \subseteq (C_1 \cup C_2) - e$. To prove (1), we shall show that

$$C_3 \subseteq (C_1 \cup C_2) - \bigcap\{C : C \in \mathcal{C}_e(C_1 \cup C_2)\}. \tag{2}$$

Choose f in $C_3 \cap C_1$. Then, by the strong circuit elimination axiom, there is a circuit C_4 such that $e \in C_4 \subseteq (C_1 \cup C_3) - f$. Evidently

$C_4 \in \mathcal{C}_e(C_1 \cup C_2)$ and hence $f \notin \bigcap\{C : C \in \mathcal{C}_e(C_1 \cup C_2)\}$. It follows that $C_3 \cap C_1$ does not meet $\bigcap\{C : C \in \mathcal{C}_e(C_1 \cup C_2)\}$. Similarly, the last set avoids $C_3 \cap C_2$. Thus (2) holds and hence so does (1).

Now let D be an arbitrary circuit of M avoiding e. As M is connected, it has a circuit that contains e and meets D. Choose such a circuit C_1 so that $C_1 \cup D$ is minimal. As $e \in C_1 - D$, the strong circuit elimination axiom implies that M has a circuit C_2 that is distinct from C_1 such that $e \in C_2 \subseteq C_1 \cup D$. We shall show that

$$D = (C_1 \cup C_2) - \bigcap\{C : C \in \mathcal{C}_e(C_1 \cup C_2)\}. \tag{3}$$

If $C' \in \mathcal{C}_e(C_1 \cup C_2)$, then $C' \subseteq C_1 \cup C_2 \subseteq C_1 \cup D$. Since C' is not properly contained in C_1, it must meet D. As C' contains e, the choice of C_1 implies that $C' \cup D = C_1 \cup D$. Hence, as C' was arbitrarily chosen in $\mathcal{C}_e(C_1 \cup C_2)$,

$$(\bigcap\{C : C \in \mathcal{C}_e(C_1 \cup C_2)\}) \cup D = C_1 \cup D \supseteq C_1 \cup C_2.$$

Therefore

$$D \supseteq (C_1 \cup C_2) - \bigcap\{C : C \in \mathcal{C}_e(C_1 \cup C_2)\}. \tag{4}$$

But, by (1), the right-hand side of (4) contains a circuit. Hence equality must hold in (4), that is, (3) holds.

It is now not difficult to complete the proof. For example, one can use Lemma 2.1.19. □

By Corollary 4.2.8 and the definition of connectivity, it follows that:

4.3.3 Corollary. *M is connected if and only if, for every pair of distinct elements of M, there is a circuit and a cocircuit containing both.* □

We now present two variants of this. The first is an immediate consequence of Proposition 4.1.4 and Corollary 4.2.8. The second exemplifies the use of Theorem 4.3.1 in induction arguments.

4.3.4 Corollary. *A matroid M is connected if and only if, for every pair of distinct elements, there is a circuit or a cocircuit containing both.* □

4.3.5 Proposition. *Let M be a matroid having at least three elements. Then M is connected if and only if, for all 3-element subsets X of $E(M)$, there is a circuit or a cocircuit containing X.*

Proof. If the condition on 3-element subsets of $E(M)$ holds, then M is certainly connected. Conversely, suppose that M is connected. We shall argue by induction on $|E(M)|$ to prove that the specified condition on 3-element subsets holds. If $|E(M)| = 3$, then, as M is connected, it follows from Table 4.1 (p. 126) that M is isomorphic to $U_{1,3}$ or $U_{2,3}$. Hence, in this case, the result holds. Now assume the result holds for $|E(M)| < n$. Let $|E(M)| = n \geq 4$ and let X be an arbitrary 3-element subset of $E(M)$. Choose an element e from $E(M) - X$. By Theorem 4.3.1, either (i) $M\backslash e$ is connected, or (ii) M/e is connected. Assume that (i) occurs. Then, by the induction assumption, X is contained in a circuit C or a cocircuit C^* of $M\backslash e$. In the first case, C is a circuit of M containing X. In the second case, C^* or $C^* \cup e$ is a cocircuit of M containing X. Thus if (i) occurs, then the result holds for M. But if (ii) occurs, then, by Corollary 4.2.8, $(M/e)^*$ is connected, that is, $M^*\backslash e$ is connected. Applying the argument used in case (i) with M^* in place of M, we get that M^* has a circuit or a cocircuit containing X. Hence so does M and the result holds in case (ii). We conclude, by induction, that the proposition holds. □

We close this chapter with an extension of Theorem 4.3.1 and with a corollary of this extension. This corollary is indicative of the sort of results that have received a lot of recent research attention. In Chapter 11, we shall consider a number of other results of this type in the context of higher connectivity for matroids.

4.3.6 Proposition. *(Brylawski 1972; Seymour 1977a) Let N be a connected minor of a connected matroid M and suppose that $e \in E(M) - E(N)$. Then at least one of $M\backslash e$ and M/e is connected and has N as a minor.*

Proof. We may assume that N is a minor of $M\backslash e$; otherwise, in the argument that follows, we replace M and N by their duals. If $M\backslash e$ is connected, the proposition is proved. Thus we may suppose that $M\backslash e$ is disconnected. It follows, by Theorem 4.3.1, that M/e is connected. To complete the proof, we shall show that N is a minor of M/e.

As N is a connected minor of $M\backslash e$, Proposition 4.2.10 implies that N is a minor of some component E_1 of $M\backslash e$. Let $E_2 = E(M\backslash e) - E_1$. Then N is a minor of $M\backslash e\backslash E_2$. Moreover, by Proposition 4.2.1 and the fact that M is connected, we deduce that

$$r(E_1) + r(E_2) = r(M\backslash e) \leq r(M) \tag{5}$$

and

$$r(E_1 \cup e) + r(E_2) > r(M). \tag{6}$$

By (5), (6), and (R2)$'$, $r(E_1 \cup e) = r(E_1) + 1$, that is, e is a coloop of $M\backslash E_2$. Thus $M\backslash e\backslash E_2 = (M\backslash E_2)\backslash e = (M\backslash E_2)/e = M/e\backslash E_2$. Since N is a minor of $M\backslash e\backslash E_2$, it follows that N is also a minor of $M/e\backslash E_2$ and hence, as required, N is a minor of M/e. □

4.3.7 **Corollary.** *Let N be a connected minor of a connected matroid M. Then there is a sequence $M_0, M_1, M_2, \ldots, M_n$ of connected matroids such that $M_0 = N$, $M_n = M$, and, for each i in $\{0, 1, \ldots, n-1\}$, M_i is a single-element deletion or a single-element contraction of M_{i+1}.* □

Exercises

1. Give an example of a connected matroid M having distinct elements e_1, e_2, and e_3 such that all of $M\backslash e_1$, M/e_1, $M\backslash e_2$, and M/e_3 are connected, but M/e_2 and $M\backslash e_3$ are disconnected.

2. Let e and f be distinct elements of a connected matroid M.

   (i) Show that $M\backslash e, f$, $M\backslash e/f$, $M\backslash f/e$, or $M/e, f$ is connected.

   (ii) Give examples to show that every set of three of these four matroids can be disconnected.

3. Let M_1 and M_2 be matroids on the same set E and suppose that e is an element of E that is a loop of neither M_1 nor M_2.

   (i) Show that the set of bases of M_1 containing e equals the set of bases of M_2 containing e if and only if $M_1/e = M_2/e$.

   (ii) Give an example to show that a connected matroid is not uniquely determined by the set of bases containing a fixed element e.

4. Let M be a connected matroid having at least $2n-1$ elements. Show that $E(M)$ has a subset $\{e_1, e_2, \ldots, e_n\}$ such that either $M\backslash e_i$ is connected for all i or M/e_i is connected for all i.

5. Let e be an element of a connected matroid M and $\mathcal{H}_e$ be the set of hyperplanes of M avoiding e. Specify all the hyperplanes of M in terms of the members of $\mathcal{H}_e$.

6. (Dirac 1967; Plummer 1968) Let G be a 2-connected graph such that, for all edges e, the graph $G\backslash e$ is not 2-connected. Prove that G has a vertex of degree two. (Hint: Argue by induction on the number of edges of G.)

7. Let M be a connected matroid such that, for every element e, $M\backslash e$ is disconnected.

   (i) (Murty 1974) Prove that M has a cocircuit of size two.

   (ii) Show that every circuit of M is a flat.

(iii) Give an example of a connected matroid N in which every circuit is a flat such that $|E(N)| \geq 3$ and $N\backslash x$ is connected for every element x of N.

We shall consider several extensions of (i) in Section 10.2.

8. Prove that the following are equivalent for a matroid M:

(a) M has no two intersecting cocircuits.

(b) M is a direct sum of matroids of rank at most one.

(c) M has no circuit of size exceeding two.

9. (i) Show that there are connected matroids with arbitrarily many elements in which some 4-element set is in neither a circuit nor a cocircuit.

(ii)° Characterize those connected matroids in which every 4-element subset is contained in a circuit or a cocircuit.

10. Let M be a connected matroid, C be a circuit of M, and e be an element of M. Prove that:

(i) M has circuits C_1 and C_2 such that $e \in C_1 \cap C_2$ and C is a subset of $C_1 \cup C_2$.

(ii)* (Mason and Oxley 1980) If f is an element of M, then M has circuits C_e and C_f such that $e \in C_e$, $f \in C_f$, and $C \subseteq C_e \cup C_f$.

11. (Lovász, Schrijver, and Seymour, private communication) Let M be a connected matroid whose largest circuit has m elements and whose largest cocircuit has n elements. Prove that $|E(M)| < 2^{m+n}$.

5
Graphic matroids

We have already seen that graphic matroids form a fundamental class of matroids. Moreover, numerous operations and results for graphs provide the inspiration or motivation for corresponding operations and results for matroids. In this chapter, we examine graphic matroids in more detail. In particular, we shall present several proofs delayed from Chapters 1 and 2 including proofs that a graphic matroid is representable over every field, and that a cographic matroid $M^*(G)$ is graphic only if G is planar. The main result of the chapter is Whitney's 2-Isomorphism Theorem which establishes necessary and sufficient conditions for two graphs to have isomorphic cycle matroids.

5.1 Representability

In this section, we shall prove that every graphic matroid is representable over every field and that every graphic matroid is regular. The proofs of these results use the following construction for a given graph G. Form a directed graph $D(G)$ from G by arbitrarily assigning a direction to each edge. Let $A_{D(G)}$ denote the incidence matrix of $D(G)$, that is, $A_{D(G)}$ is the matrix $[a_{ij}]$ whose rows and columns are indexed by the vertices and arcs, respectively, of $D(G)$ where

$$a_{ij} = \begin{cases} 1, & \text{if vertex } i \text{ is the tail of non-loop arc } j; \\ \text{-1}, & \text{if vertex } i \text{ is the head of non-loop arc } j; \\ 0, & \text{otherwise.} \end{cases}$$

5.1.1 **Example.** Let G be the graph shown in Figure 5.1(a) and $D(G)$ be as shown in Figure 5.1(b). Then $A_{D(G)}$ is as given in Figure 5.2. □

5.1.2 **Proposition.** *If G is a graph, then $M(G)$ is representable over every field.*

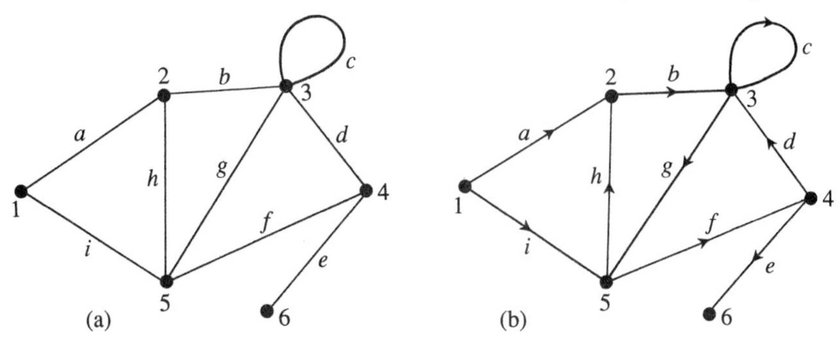

Fig. 5.1. (a) A graph G. (b) An orientation $D(G)$ of G.

Proof. Let $D(G)$ be an arbitrary orientation of G and form $A_{D(G)}$ as described above. Now let F be an arbitrary field and $A'_{D(G)}$ denote the matrix obtained from $A_{D(G)}$ by reducing each entry modulo F. This leaves $A_{D(G)}$ unchanged unless F has characteristic two, in which case each negative entry changes sign. We shall show that $A'_{D(G)}$ represents $M(G)$ over F.

Let C be a cycle of G. If C is a loop e, then the corresponding column of $A_{D(G)}$ is zero, hence C is a circuit of $M[A'_{D(G)}]$. Now suppose that C is not a loop and let its edges, in cyclic order, be $e_1, e_2, \ldots, e_k$. Let $\underline{v}(e_1), \underline{v}(e_2), \ldots, \underline{v}(e_k)$ be the corresponding columns of $A'_{D(G)}$. Now cyclically traverse C beginning with e_1 and, for each e_j, let α_j be 1 or -1 according to whether the direction of traversal agrees or disagrees with the direction of e_j. Then $\alpha_1 \underline{v}(e_1), \alpha_2 \underline{v}(e_2), \ldots, \alpha_k \underline{v}(e_k)$ are the incidence vectors of a reorientation of C in which every arc is oriented in the direction of traversal. It follows that, in this reorientation of C, every vertex of C is the head of exactly one arc and the tail of exactly one arc. Thus $\sum_{j=1}^{k} \alpha_j \underline{v}(e_j) = \underline{0}$. Hence C is dependent in $M[A'_{D(G)}]$.

Now suppose that $\{f_1, f_2, \ldots, f_m\}$ is a circuit of $M[A'_{D(G)}]$. If $m = 1$, then f_1 is a loop in G, so $\{f_1\}$ is a circuit in $M(G)$. Assume, then, that $m > 1$ and consider the columns $\underline{v}(f_1), \underline{v}(f_2), \ldots, \underline{v}(f_m)$ of $A'_{D(G)}$. Evidently, for some non-zero members $\varepsilon_1, \varepsilon_2, \ldots, \varepsilon_m$ of F, $\sum_{i=1}^{m} \varepsilon_i \underline{v}(f_i) = \underline{0}$.

	a	b	c	d	e	f	g	h	i
1	1	0	0	0	0	0	0	0	1
2	−1	1	0	0	0	0	0	−1	0
3	0	−1	0	−1	0	0	1	0	0
4	0	0	0	1	1	−1	0	0	0
5	0	0	0	0	0	1	−1	1	−1
6	0	0	0	0	−1	0	0	0	0

Fig. 5.2. $A_D(G)$.

Thus every row of the matrix $[\underline{v}(f_1)\ \underline{v}(f_2)\ \cdots\ \underline{v}(f_m)]$ that contains at least one non-zero entry contains at least two such entries. But the rows of this matrix correspond to vertices of G. Hence, in the subgraph G_1 of G induced by $\{f_1, f_2, \ldots, f_m\}$, every vertex has degree at least two. Thus G_1 must contain a cycle (Exercise 2, Section 1.6) and therefore $\{f_1, f_2, \ldots, f_m\}$ contains a circuit of $M(G)$.

We have shown that every circuit of $M(G)$ is dependent in $M[A'_{D(G)}]$, and that every circuit of $M[A'_{D(G)}]$ is dependent in $M(G)$. It follows by Lemma 2.1.19 that $\mathcal{C}(M(G)) = \mathcal{C}(M[A'_{D(G)}])$, and therefore $M(G) = M[A'_{D(G)}]$. □

We shall use $A_{D(G)}$ again in the next result.

5.1.3 Proposition. *If G is a graph, then $M(G)$ is regular.*

Proof. Throughout this proof, we shall view $A_{D(G)}$ as a matrix over $\mathbb{R}$. To verify the proposition, it will be shown, by induction on $|E(G)|$, that $A_{D(G)}$ is totally unimodular. This is clearly true if $|E(G)| = 1$. Assume it true for $|E(G)| < n$ and let $|E(G)| = n$. Take an $m \times m$ submatrix X of $A_{D(G)}$. We want to show that $\det X \in \{0, 1, -1\}$. If $m < n$, then X is a submatrix of $A_{D(G\backslash e)}$ for some edge e of G, so, by the induction assumption, $\det X \in \{0, 1, -1\}$. Hence we may assume that $m = n$. We may also suppose that every row of X has at least one non-zero entry, otherwise $\det X = 0$.

Now suppose that the ith row of X has exactly one non-zero entry x_{ij}. Then x_{ij} is ± 1 and $|\det X| = |\det(X_{ij})|$ where X_{ij} is the matrix obtained by deleting row i and column j from X. But X_{ij} is a square submatrix of $A_{D(G/e_j)}$ where e_j is the edge of G corresponding to the jth column of X. Hence, by the induction assumption, $\det(X_{ij}) \in \{0, 1, -1\}$ and so $\det X \in \{0, 1, -1\}$.

We may now assume that every row of X has at least two non-zero entries. Thus X has at least $2n$ non-zero entries. But each column of X has at most two non-zero entries. Hence X has exactly $2n$ non-zero entries, two per row and two per column. Thus X is the vertex-arc incidence matrix for a subdigraph of $D(G)$. Let G_X be the subgraph of G corresponding to this subdigraph. Evidently every vertex of G_X has degree two, so G_X contains a cycle. But, by the proof of the last result, X represents $M(G_X)$ over $\mathbb{R}$. Therefore the columns of X are linearly dependent and so $\det X = 0$. □

On combining Propositions 2.2.22 and 5.1.3, we immediately obtain the following.

5.1.4 Corollary. *If G is a graph, then $M^*(G)$ is regular.* □

We remark that, although $M(G)$ has rank at most $|V(G)| - 1$, the matrix $A_{D(G)}$ has exactly $|V(G)|$ rows. Next we describe a direct procedure for obtaining an F-representation A for $M(G)$ that has exactly $r(M(G))$ rows. Let $r(M(G)) = r$ where $r \geq 1$ and let $\{e_1, e_2, \ldots, e_r\}$ be a basis B of $M(G)$. Take the first r columns of A to be the matrix I_r labelling the columns of this submatrix by $e_1, e_2, \ldots, e_r$ and letting the corresponding column vectors be $\underline{v}_1, \underline{v}_2, \ldots, \underline{v}_r$. Now take an arbitrary orientation $D(G)$ of G. For each edge e of $E(G) - B$, recall that $C(e, B)$ is the unique circuit of $M(G)$ contained in $B \cup e$. Evidently $C(e, B)$ is a cycle of G; we call it the *fundamental cycle* of e with respect to B. Suppose that its edges are, in cyclic order, $e, e_{i_1}, e_{i_2}, \ldots, e_{i_k}$. Traverse this cycle in the direction determined by e letting α_{i_j} be 1 or -1 according to whether the direction of traversal agrees or disagrees with the direction of e_j. Take, as the column of A corresponding to e, the vector $\sum_{j=1}^{k} \alpha_{i_j} \underline{v}_j$ (modulo F). The proof that the matrix A is indeed an F-representation for $M(G)$ is based on the fact that A can be obtained from $A_{D(G)}$ by a sequence of the operations 2.2.1–2.2.6 (see Exercise 4).

5.1.5 Example. For $r \geq 2$, let $\mathcal{W}_r$ denote the graph shown in Figure 5.3(a).

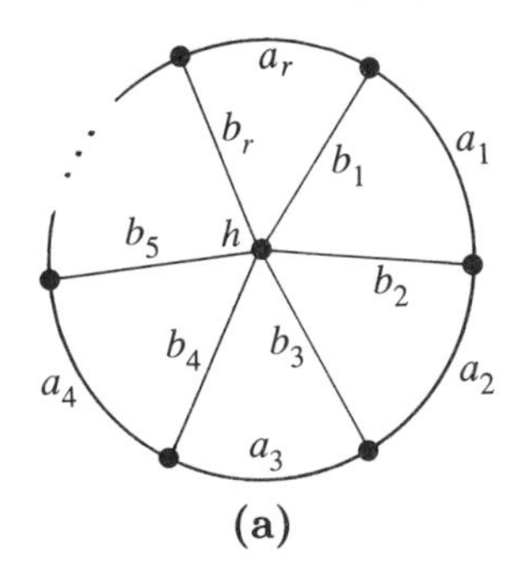

(a)

$$
\begin{array}{cccc|ccccccc}
b_1 & b_2 & \cdots & b_r & a_1 & a_2 & a_3 & \cdots & a_{r-1} & a_r \\
 & & & & 1 & 0 & 0 & \cdots & 0 & -1 \\
 & & & & -1 & 1 & 0 & \cdots & 0 & 0 \\
 & & & & 0 & -1 & 1 & \cdots & 0 & 0 \\
 & & I_r & & 0 & 0 & -1 & \cdots & 0 & 0 \\
 & & & & \vdots & \vdots & \vdots & \ddots & \vdots & \vdots \\
 & & & & 0 & 0 & 0 & \cdots & 1 & 0 \\
 & & & & 0 & 0 & 0 & \cdots & -1 & 1
\end{array}
$$

(b)

Fig. 5.3. (a) $\mathcal{W}_r$. (b) A representation for $M(\mathcal{W}_r)$.

This graph is called the *r-spoked wheel.* The edges $b_1, b_2, \ldots, b_r$ are its *spokes*, the vertex h is its *hub*, and the cycle with edge set $\{a_1, a_2, \ldots, a_r\}$ is its *rim*. Let $\{b_1, b_2, \ldots, b_r\}$ be the distinguished basis of $M(\mathcal{W}_r)$ and orient all these edges away from the hub. Then, orienting the rim clockwise, we obtain the representation for $M(\mathcal{W}_r)$ in Figure 5.3(b) where, if F has characteristic two, all negative entries change sign. Now suppose r is even and we reverse the orientations on each of $b_2, b_4, \ldots, b_r$ and each of $a_2, a_4, \ldots, a_r$. Then, we obtain the same representation as in Figure 5.3(b) except that all negative entries have their signs changed. □

5.1.6 Example. Orient the edges of K_5 as shown in Figure 5.4(a) and let the distinguished basis of $M(K_5)$ be $\{1, 2, 3, 4\}$. Then the matrix in Figure 5.4(b) represents $M(K_5)$ over all fields. □

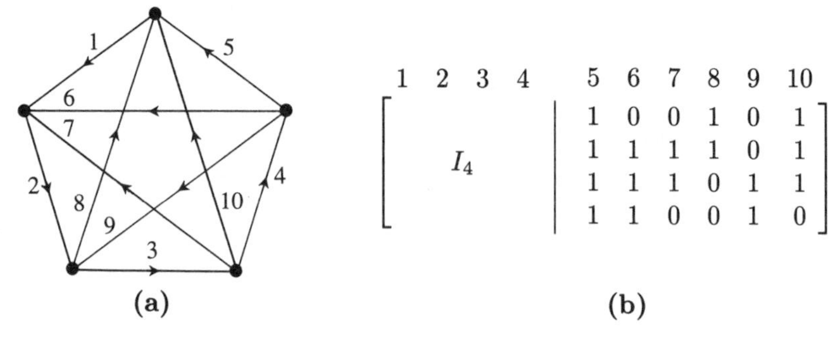

Fig. 5.4. (a) An orientation of K_5. (b) A representation for $M(K_5)$.

Exercises

1. Show that if G is a graph, then $M^*(G) \oplus M(G)$ is representable over every field.

2. (i) Construct F-representations for $M(K_{3,3})$ by using both of the methods described in this section.

   (ii) Use these representations to find representations for $M^*(K_{3,3})$.

3. Let A be an $m \times n$ matrix over the field F such that every column contains at most two non-zero entries.

   (i) Is $M[A]$ graphic?

   (ii) Is $M[A]$ graphic if $F = GF(2)$?

4. Let A be the matrix constructed following 5.1.4.

   (i) Show that $G[B]$ has a vertex of degree one.

(ii) Suppose that e_1 is the edge v_1v_2 of $G[B]$ where v_1 has degree one in $G[B]$. Replace row v_2 of $A_{D(G)}$ by the sum of rows v_1 and v_2 and then delete row v_1 and column e_1 from the resulting matrix. Show that the matrix thus obtained is the incidence matrix of the orientation induced on G/e_1 by $D(G)$.

(iii) Argue by induction on r using (i) and (ii) to show that A can be obtained from $A_{D(G)}$ by a sequence of the operations 2.2.1–2.2.6 where the only element of F by which a row or column is multiplied is -1.

(iv) Deduce that A is an F-representation for $M(G)$.

(v) Show that, when viewed over $\mathbb{R}$, A is totally unimodular.

5.2 Duality in graphic matroids

In Section 2.3, we proved that if G is a planar graph, then $M^*(G)$ is graphic. In this section we shall prove the converse of this result. The main tool in this proof will be Kuratowski's Theorem (see 2.3.8). We shall also use the notion of the abstract dual of a graph, which we now define.

If G' is a geometric dual of the planar graph G, then there is a bijection ψ from $E(G)$ to $E(G')$ such that X is a cycle in G if and only if $\psi(X)$ is a bond in G'. Suppose now that G' is an arbitrary graph for which such a bijection exists. Then G' is called an *abstract dual* of G. Evidently G' is an abstract dual of G if and only if $M(G) \cong M^*(G')$. Every geometric dual of a graph is clearly also an abstract dual. However, the converse of this statement is not true (see Exercise 2). The concept of an abstract dual will facilitate the proof of the next theorem, the main result of this section. This proof will use the following result, an easy consequence of duality (Exercise 1).

5.2.1 Lemma. *If G' is an abstract dual of G, then G is an abstract dual of G'.* □

5.2.2 Theorem. *The following statements are equivalent for a graph G:*

(i) *G is planar.*

(ii) *$M^*(G)$ is graphic.*

(iii) *G has an abstract dual.*

Proof. The fact that (i) implies (ii) is the content of Theorem 2.3.4. To see that (ii) implies (iii), note that if $M^*(G) \cong M(G')$ for some graph G', then, by Lemma 5.2.1, G' is an abstract dual of G. It remains to show that (iii) implies (i) and, following Parsons (1971), we break the proof of this into several steps, the first two of which are easy preliminaries designed to allow us to use Kuratowski's Theorem.

5.2.3 Lemma. *If G_2 is a subgraph of the graph G_1 and G_1 has an abstract dual, then so does G_2.*

Proof. The general result will follow if we can show that it holds in the special case in which G_2 is obtained by deleting a single edge, say x, from G_1. Let G_1' be an abstract dual of G_1 and let x' be the edge of G_1' corresponding to x. Then G_1'/x' is an abstract dual of G_2, for the cycles of G_2 are just the cycles of G_1 not containing x, and the bonds of G_1'/x' are just the bonds of G_1' not containing x'. □

5.2.4 Lemma. *If G_1 has an abstract dual G_1' and G_1 is a subdivision of G_2, then G_2 has an abstract dual.*

Proof. We shall prove the lemma in the case in which G_1 is obtained from G_2 by replacing an edge e by a path of length two having edges f and g. The general result will then follow without difficulty. Let f' and g' be the edges of G_1' corresponding to f and g, respectively. Then f' and g' are parallel in G_1' unless e is a cut-edge in G_2, in which case f' and g' are loops in G_1'. In either case, it is easily checked that $G_1' \backslash g'$ is an abstract dual of G_2. □

To complete the proof of the theorem, suppose that G is a non-planar graph having an abstract dual. Then, by Kuratowski's Theorem (2.3.8), G has a subgraph G_1 that is a subdivision of K_5 or $K_{3,3}$. But, by Lemma 5.2.3, G_1 has an abstract dual, and hence, by Lemma 5.2.4, K_5 or $K_{3,3}$ has an abstract dual. Thus $M(K_5)$ or $M(K_{3,3})$ is cographic, so $M^*(K_5)$ or $M^*(K_{3,3})$ is graphic. This contradiction to Proposition 2.3.3 completes the proof of Theorem 5.2.2. □

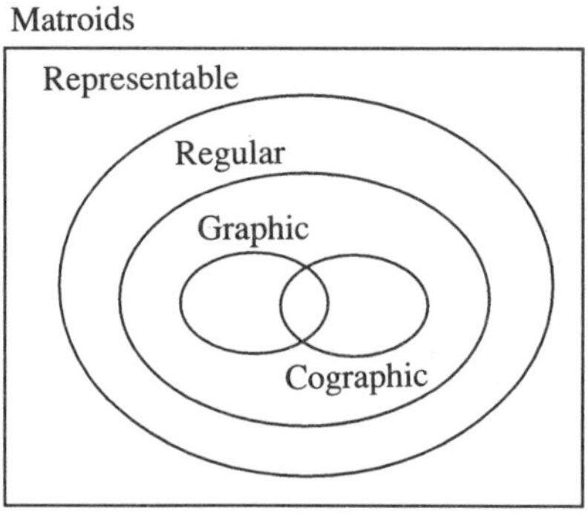

Fig. 5.5. Relationships between certain classes of matroids.

The Venn diagram in Figure 5.5 summarizes the relationships that have so far been proved between various classes of matroids. We remark that, although all the inclusions shown are indeed proper, we have not yet proved all of them to be so. Theorem 5.2.2 tells us that the intersection of the classes of graphic and cographic matroids is precisely the class of matroids that are isomorphic to the cycle matroids of planar graphs. We shall call such matroids *planar* with the warning that, for some authors, a 'planar matroid' is a rank-3 matroid.

By Proposition 5.1.2 and duality, all graphic and cographic matroids are representable over every field. We shall see in Chapter 6 that the class of regular matroids coincides precisely with the class of matroids that are representable over every field. This result will provide an alternative proof of Proposition 5.1.3 since the latter then becomes an immediate consequence of Proposition 5.1.2.

The classes of graphic and cographic matroids are important subclasses of the class of regular matroids. The union of the first two classes is not equal to the third for, by Proposition 2.3.3, the regular matroid $M(K_5) \oplus M^*(K_5)$ is neither graphic nor cographic. However, a difficult result of Seymour (1980b) shows that all regular matroids can be built up from graphic matroids, cographic matroids, and one other 10-element matroid. A precise statement of this important result will be given in Chapter 13.

We noted in Corollary 3.2.2 that the class of graphic matroids is minor-closed. By duality, so too is the class of cographic matroids. On combining these facts, we immediately deduce the following:

5.2.5 Corollary. *The class of planar matroids is minor-closed.* □

For any minor-closed class $\mathcal{N}$ of matroids, it is natural to seek a list of the minor-minimal matroids not in $\mathcal{N}$, that is, those matroids M such that $M \notin \mathcal{N}$ but every proper minor of M is in $\mathcal{N}$. Indeed, as we shall see in several later chapters, almost all of the most celebrated theorems in matroid theory are excluded-minor characterizations of various classes. In Chapter 13, we shall give such a characterization of the class of planar matroids. For the moment, we note that it is straightforward to use Kuratowski's Theorem to obtain the following characterization of the members of this class within the class of graphic matroids.

5.2.6 Proposition. *The following statements are equivalent for a graph G:*

(i) *G is a planar graph.*

(ii) *$M(G)$ is a planar matroid.*

(iii) *$M(G)$ has no minor isomorphic to $M(K_5)$ or $M(K_{3,3})$.* □

Exercises

1. Prove that if G' is an abstract dual of G, then G is an abstract dual of G'.

2. (D. R. Woodall, in Welsh 1976, Section 6.4) Construct a graph without isolated vertices that is an abstract dual of the graph G in Figure 5.6 but is not a geometric dual of G.

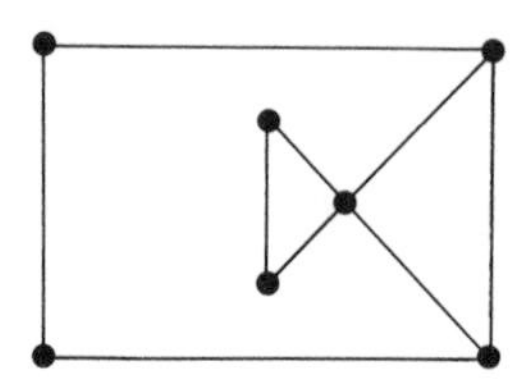

Fig. 5.6. A graph with an abstract dual that is not a geometric dual.

3. (i) Give examples to show that all the inclusions indicated in Figure 5.5 are proper.

   (ii) Indicate precisely how the class of binary matroids fits into Figure 5.5.

4. Show that if the graph G has a subgraph that is a subdivision of the graph H, then $M(G)$ has a minor isomorphic to $M(H)$.

5. (Whitney 1932a; Ore 1967, Section 3.3) The graph H is a *Whitney dual* of the graph G if there is a bijection $\psi : E(G) \to E(H)$ such that, for all subsets Y of $E(G)$,

$$r(M(H)) - r(M(H)\backslash\psi(Y)) = |Y| - r(M(G)|Y).$$

   (i) Show that if H is a Whitney dual of G, then G is a Whitney dual of H.

   (ii) Determine the relationship between Whitney duals and geometric and abstract duals.

   (iii)* Prove that a graph is planar if and only if it has a Whitney dual.

6. (Wagner 1937a,b; Harary and Tutte 1965) Prove that a graph is planar if and only if it has no minor isomorphic to K_5 or $K_{3,3}$.

7. Let G be a connected plane graph.

(i) Use matroid duality to show that G satisfies Euler's polyhedron formula,

$$|V(G)| - |E(G)| + |F(G)| = 2,$$

where $F(G)$ is the set of faces of G.

(ii) Prove that if G is simple and $|V(G)| \geq 3$, then

$$|E(G)| \leq 3|V(G)| - 6.$$

(iii) Characterize the graphs for which equality is attained in (ii).

(iv) Show that if G is simple, then it has a vertex of degree at most five.

(v) Give an example of a simple plane graph in which every vertex has degree five.

5.3 Whitney's 2-Isomorphism Theorem

In this section, we shall prove Whitney's theorem characterizing when two graphs have isomorphic cycle matroids. Our proof, due to Truemper (1980), is considerably shorter than the original. A different short proof has been given by Wagner (1985).

If G is a graph, then, evidently, adding vertices to G with no incident edges will not alter the cycle matroid of the graph. For this reason, *throughout this section, we shall consider only graphs having no isolated vertices.* The following operations on a graph G have appeared, at least implicitly, in earlier chapters. Examples of these operations are shown in Figures 5.7 and 5.8:

(a) *Vertex identification.* Let v and v' be vertices of distinct components of G. We modify G by identifying v and v' as a new vertex $\overline{v}$.

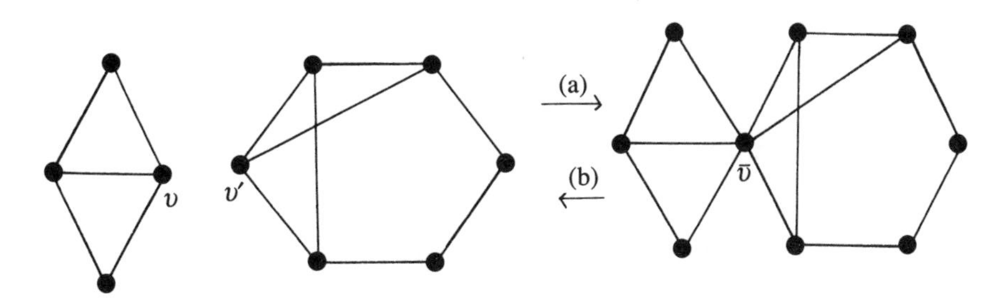

Fig. 5.7. (a) Vertex identification. (b) Vertex cleaving.

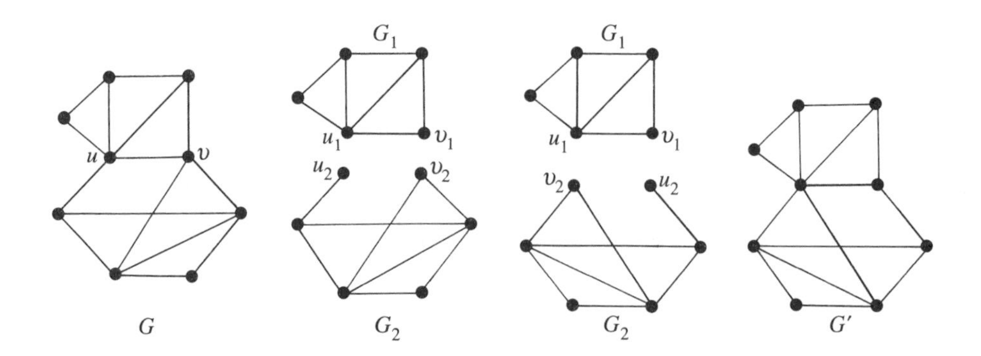

Fig. 5.8. Twisting about $\{u, v\}$.

(b) *Vertex cleaving.* This is the reverse operation of vertex identification so that a graph can only be cleft at a cut-vertex.

(c) *Twisting.* Suppose that the graph G is obtained from the disjoint graphs G_1 and G_2 by identifying the vertices u_1 of G_1 and u_2 of G_2 as the vertex u of G, and identifying the vertices v_1 of G_1 and v_2 of G_2 as the vertex v of G. In a *twisting* G' of G *about* $\{u, v\}$, we identify, instead, u_1 with v_2 and v_1 with u_2. We call G_1 and G_2 the *pieces* of the twisting.

The graph G is 2-*isomorphic* to the graph H if H can be transformed into a graph isomorphic to G by a sequence of operations of types (a), (b), and (c). Evidently 2-isomorphism is an equivalence relation. Since none of the three operations (a) – (c) alters the edge sets of the cycles of a graph, if G is 2-isomorphic to H, then $M(G) \cong M(H)$. Most of the rest of this section will be devoted to proving the converse of this, the main result of the chapter.

5.3.1 Whitney's 2-Isomorphism Theorem. *(Whitney 1933) Let G and H be graphs having no isolated vertices. Then $M(G)$ and $M(H)$ are isomorphic if and only if G and H are 2-isomorphic.*

The proof of this theorem will use the next four lemmas. The first of these establishes the required result in the case that G is 3-connected. Its proof is due to Edmonds (see Truemper 1980) and Greene (1971).

5.3.2 Lemma. *Let G and H be loopless graphs without isolated vertices. Suppose that $\psi : E(G) \to E(H)$ is an isomorphism from $M(G)$ to $M(H)$. If G is 3-connected, then ψ induces an isomorphism between the graphs G and H.*

Proof. First note that if J is a loopless block, a hyperplane of $M(J)$ can only be connected if the complementary cocircuit is a vertex bond. Moreover, if $M(J)$ has exactly $|V(J)|$ connected hyperplanes, then, up to relabelling, $M(J)$ uniquely determines the vertex–edge incidence matrix of J, and hence uniquely determines J itself.

As G is 3-connected, the complement in $E(G)$ of a vertex bond is a connected hyperplane of $M(G)$. Thus $M(G)$ has exactly $|V(G)|$ connected hyperplanes. But $M(G) \cong M(H)$, and $M(G)$ is connected and loopless. Hence H is a loopless block. Therefore, as $|V(G)| - 1 = r(M(G)) = r(M(H)) = |V(H)| - 1$, we have $|V(G)| = |V(H)|$. Thus $M(H)$ has precisely $|V(H)|$ connected hyperplanes and so $G \cong H$. □

For those 2-connected graphs that are not 3-connected, the proof of 5.3.1 will use Tutte's characterization (1966a) of the structure of such graphs. To state this result, we shall need another definition.

Suppose $k \geq 2$. A connected graph G is called a *generalized cycle* with *parts* $G_1, G_2, \ldots, G_k$ if the following conditions hold:

(i) Each G_i is a connected subgraph of G having a non-empty edge set, and, if $k = 2$, both G_1 and G_2 have at least three vertices.

(ii) The edge sets of $G_1, G_2, \ldots, G_k$ partition the edge set of G, and each G_i shares exactly two vertices, its *contact vertices*, with $\bigcup_{j \neq i} G_j$.

(iii) If each G_i is replaced by an edge joining its contact vertices, the resulting graph is a cycle.

5.3.3 Example. The graph G shown in Figure 5.9 can be viewed as a generalized cycle with six parts, namely its blocks. It can also be viewed as a generalized cycle with two, three, four, or five parts. □

5.3.4 Lemma. *Let G be a block having at least four vertices and suppose that G is not 3-connected. Then G has a representation as a generalized cycle, each part of which is a block.*

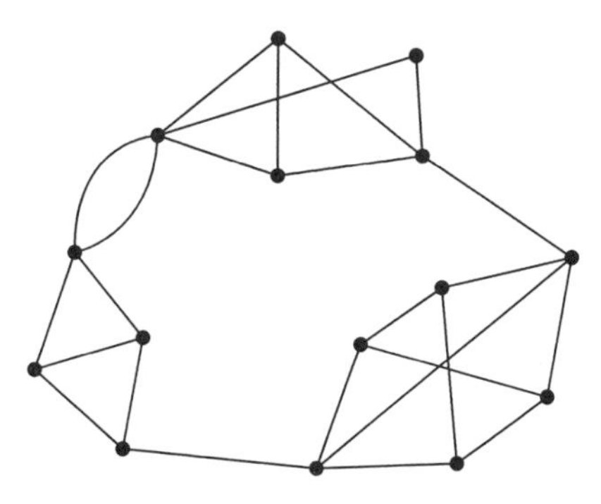

Fig. 5.9. A generalized cycle.

Proof. Suppose that $\{u, v\}$ is a vertex cut of G. Choose a component H_1 of $G - \{u, v\}$ and let $H_2 = G - \{u, v\} - V(H_1)$. For $i = 1, 2$, let G_i be the subgraph of G induced by $V(H_i) \cup \{u, v\}$, and let G_2' be the graph obtained from G_2 by deleting those edges that join u and v. Then G is a generalized cycle with parts G_1 and G_2'. If G_1 is not a block, it is the union of two connected subgraphs $G_{1,1}$ and $G_{1,2}$ that have at least two vertices each, but have only one common vertex, say x. As G is a block, x is distinct from both u and v, and one of u and v is in $G_{1,1}$ while the other is in $G_{1,2}$. Therefore, G is a generalized cycle with parts $G_{1,1}$, $G_{1,2}$, and G_2'. If necessary, we repeat the foregoing process finitely many times until a representation of G is obtained in which each part is a block. □

The core of the proof of Whitney's Theorem is contained in the next result.

5.3.5 Lemma. *Suppose that G is a block having a representation as a generalized cycle, the parts, $G_1, G_2, \ldots, G_k$, of which are blocks. Let H be a graph for which there is an isomorphism θ from $M(G)$ to $M(H)$. Then H is a generalized cycle with parts $H_1, H_2, \ldots, H_k$, where, for all i, H_i is the subgraph of H induced by $\theta(E(G_i))$.*

Proof. For an arbitrary element j of $\{1, 2, \ldots, k\}$, let $G_{-j} = \bigcup_{i \neq j} G_i$ and $H_{-j} = \bigcup_{i \neq j} H_i$. The main part of the proof of this lemma involves showing that

$$|V(H_j) \cap V(H_{-j})| = 2. \qquad (1)$$

Now, since $G, G_1, G_2, \ldots, G_k$ are blocks, all of $H, H_1, H_2, \ldots, H_k$ are blocks. As G has a cycle meeting all of $E(G_1), E(G_2), \ldots, E(G_k)$, the graph H_{-j} has an edge e having an endpoint x in H_j. Choose an edge f in H_j. Then, since H is a block, it has a cycle C_1 containing both e and f. Starting at x, move along C_1 using the edge e first and stop at the first vertex y of H_j that is encountered. Let P be the path from x to y thus traversed. Since C_1 contains the edge f of H_j, such a vertex y exists, and $y \in (V(H_j) \cap V(H_{-j})) - x$. Hence $|V(H_j) \cap V(H_{-j})| \geq 2$. We shall complete the proof of (1) by showing that

$$V(H_j) \cap V(H_{-j}) = \{x, y\}. \qquad (2)$$

Let u be one of the contact vertices of G_j. As G_j is a block, the set E_u of edges of G_j meeting u is a bond in G_j. Hence $\theta(E_u)$ is a bond in H_j, so $H_j \backslash \theta(E_u)$ has two components. Moreover:

5.3.6 Lemma. *Every cycle of H that contains an edge of H_j and an edge of H_{-j} also contains an edge of $\theta(E_u)$.*

Proof. This is an immediate consequence of the fact that every cycle of G that contains an edge of G_j and an edge of G_{-j} also contains an edge of E_u. □

If x and y are joined by a path in $H_j \backslash \theta(E_u)$, then, on combining this path with P, we obtain a cycle. Since $E(P) \subseteq E(H_{-j})$, this cycle contains edges of both H_j and H_{-j}. But, as $\theta(E_u) \subseteq E(H_j)$, the cycle avoids $\theta(E_u)$. This contradiction to Lemma 5.3.6 implies that one of the two components of $H_j \backslash \theta(E_u)$, say H_j^x, contains x, while the other, H_j^y, contains y.

Now assume that equation (2) does not hold. Then there is a vertex z in $(V(H_j) \cap V(H_{-j})) - \{x, y\}$. Let g and h be edges of H_{-j} and H_j, respectively, meeting z. Since H is a block, it has a cycle C_2 that contains both g and h. Starting at z and using the edge g first, move along C_2 stopping at the first vertex w of $V(H_j) \cup V(P)$ that is encountered. Clearly $w \neq z$. Let Q be the path from z to w thus traversed.

We now distinguish three cases, showing that each leads to the conclusion that H has a cycle C_3 meeting H_j and H_{-j} but containing no edge of $\theta(E_u)$. Since this contradicts Lemma 5.3.6, it will follow that (2) holds and hence so does (1). The cases are as follows: (i) w is a vertex of P; (ii) $w \in V(H_j^x) - V(P)$; and (iii) $w \in V(H_j^y) - V(P)$.

We shall describe the construction of C_3 in each case assuming that $z \in H_j^x$. If $z \in H_j^y$, then C_3 is obtained similarly. Recall that $E(P)$ is a subset of $E(H_{-j})$. In case (i), C_3 is obtained by combining Q, that part of P from w to x, and an arbitrary path from x to z in H_j^x. In case (ii), we obtain C_3 by combining Q with an arbitrary path from w to z in H_j^x. Finally, in case (iii), C_3 is formed by combining P and Q with a path from x to z in H_j^x and a path from y to w in H_j^y.

We conclude that (1) holds. Therefore, as H is a block, it is a generalized cycle having parts $H_1, H_2, \ldots, H_k$. □

With these preliminaries, we are now ready to prove Whitney's 2-Isomorphism Theorem.

Proof of Theorem 5.3.1. By the remarks prior to the theorem statement, it suffices to show that if there is a bijection θ from $E(G)$ onto $E(H)$ that induces an isomorphism between $M(G)$ and $M(H)$, then G and H are 2-isomorphic. Let G^+ and H^+ be the graphs that are formed from the disjoint unions of the blocks of G and H, respectively. Because $M(G^+) \cong M(G) \cong M(H) \cong M(H^+)$, it follows that there is a bijection from the set of blocks of G^+ to the set of blocks of H^+ so that corresponding blocks have isomorphic cycle matroids. Now G and G^+ are 2-isomorphic, as are H and H^+. Therefore, the required result will follow if we can show that G^+ and H^+ are 2-isomorphic. This will be done by proving by induction on $|V(G)|$ that if G is a block and H is a graph for which $M(G)$ and $M(H)$ are isomorphic, then H can be transformed into a graph isomorphic to G by a sequence of twistings. We shall therefore assume, for the remainder of the proof, that G is a block. Clearly, if $|V(G)| \leq 4$, the required result holds. Now suppose that the result holds for $|V(G)| < n$ and let $|V(G)| = n \geq 5$. Then G is 2-connected.

By Lemma 5.3.2, if G is 3-connected, then $G \cong H$ and the required result certainly holds. It follows that we may assume that G is not 3-connected. Therefore, by Lemma 5.3.4, G has a representation as a generalized cycle in which each of the parts $G_1, G_2, \ldots, G_k$ is a block. Hence, by Lemma 5.3.5, H has a representation as a generalized cycle, the parts of which are $H_1, H_2, \ldots, H_k$ where $H_i = H[\theta(E(G_i))]$.

For each i in $\{1, 2, \ldots, k\}$, consider the graphs $G_i + e_i$ and $H_i + f_i$ that are obtained from G_i and H_i by adding edges e_i and f_i so that, in each case, the new edge joins the contact vertices. Now C is a cycle of G meeting both $E(G_i)$ and $E(G_{-i})$ if and only if $\theta(C)$ is a cycle of H that meets $\theta(H_i)$ in $\theta(E(G_i) \cap C)$ and also meets $\theta(H_{-i})$. From this, it follows that the isomorphism between $M(G_i)$ and $M(H_i)$ can be extended to an isomorphism between $M(G_i + e_i)$ and $M(H_i + f_i)$ by mapping e_i to f_i. Now $|V(G_i + e_i)| = |V(G_i)| < |V(G)|$. Hence, by the induction assumption, $H_i + f_i$ can be transformed by a sequence of twistings into a graph isomorphic to $G_i + e_i$. Since every 2-element vertex cut in $H_i + f_i$ is also a vertex cut in H, one can perform the same sequence of twistings in H. After this has been done for all i in $\{1, 2, \ldots, k\}$, one obtains a generalized cycle H' with parts $H'_1, H'_2, \ldots, H'_k$. Moreover, $M(H') \cong M(G)$ and, for each i, there is an isomorphism θ_i from G_i to H'_i under which the contact vertices of G_i are mapped to the contact vertices of H'_i.

Suppose that the cyclic order of the parts of G is $G_1, G_2, \ldots, G_k$, while the cyclic order of the parts of H' is $H'_1, H'_{\sigma(2)}, \ldots, H'_{\sigma(k)}$ for some permutation σ of $\{2, 3, \ldots, k\}$. Then it is clear that, by a sequence of twistings, H' can be transformed into a generalized cycle in which the parts, in cyclic order, are $H'_1, H'_2, \ldots, H'_k$. If the resulting graph H'' is still not isomorphic to G, this last remaining difficulty can be overcome by twisting some of the parts of H'' about their contact vertices. □

We note that in transforming H into a graph isomorphic to G, no attempt was made to minimize the number of twistings. We defer to the exercises consideration of Truemper's (1980) extension of Whitney's Theorem which determines a best-possible bound on the minimum number of twistings necessary.

To close this section, we note some further properties of planar graphs. First we observe that a graph is planar if and only if it can be embedded on the surface of a sphere so that distinct edges do not meet, except possibly at their endpoints. To see this, suppose that we have such an

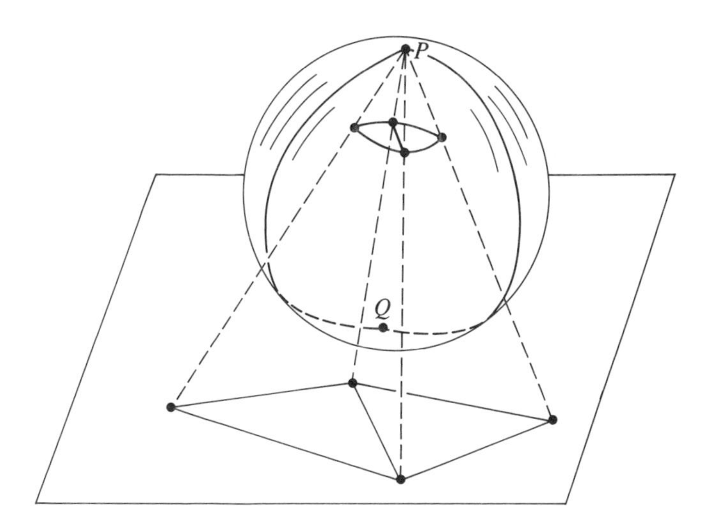

Fig. 5.10. Stereographic projection.

embedding G' of a graph G. Choose a point P on the sphere so that P is not a vertex of G' and is not on any edge. Let Q be the point on the sphere diametrically opposite P. Now project G' from P onto the tangent plane to the sphere at Q. This operation, which is known as *stereographic projection*, gives a plane graph G'' isomorphic to G. Figure 5.10 shows an example of the operation. To establish that every planar graph can be embedded on the sphere so that distinct edges do not cross, one needs only to reverse the above construction.

Using this idea of stereographic projection and the 2-Isomorphism Theorem, one can prove the following proposition (Whitney 1932b). An outline of this proof is given in Exercise 4. In Chapter 8, various links between graph and matroid connectivity will be established and these will be used to give a shorter proof of the proposition and to prove another of Whitney's results, namely that a 3-connected simple planar graph has a unique embedding on the sphere.

5.3.7 Proposition. *Let G be a 3-connected loopless planar graph. If G_1 and G_2 are abstract duals of G each having no isolated vertices, then $G_1 \cong G_2$.* □

Exercises

1. (i) From among the five graphs in Figure 5.11, determine which have isomorphic cycle matroids.

   (ii) For each of these five graphs G, find the number of non-isomorphic graphs H for which $M(H) \cong M(G)$.

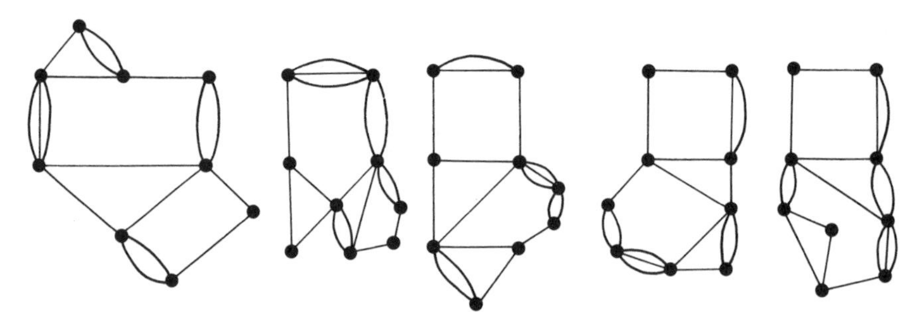

Fig. 5.11.

2. (Ore 1967, Theorem 3.4.1) Let G be a planar block and G^* be a graph without isolated vertices. Prove that G^* is an abstract dual of G if and only if it is a geometric dual of G.

3. Let G be a 3-connected simple plane graph. Prove that the following statements are equivalent for a cycle C of G:

(i) C is the boundary of a face of G.

(ii) $M(G/C)$ is connected.

(iii) C is a vertex bond in every abstract dual of G.

4. **(i)** Show that, for any edge e of a planar graph G, there is a planar embedding of G in which e is on the boundary of the infinite face.

(ii) Use (i) and the 2-Isomorphism Theorem to prove Proposition 5.3.7.

5.* (Truemper 1980) Let G be a 2-connected graph having n vertices and H be a graph that is 2-isomorphic to G. Show that:

(i) G can be transformed into a graph that is isomorphic to H by a sequence of at most $n-2$ twistings.

(ii) For any integer m, there are such graphs G and H for which $n \geq m$ and at least $n-2$ twistings are needed to transform G into a graph that is isomorphic to H.

5.4 Series–parallel networks

The operations of joining electrical components in series and in parallel are fundamental in electrical network theory. In this section we shall investigate the corresponding operations for graphs. We shall show how, by

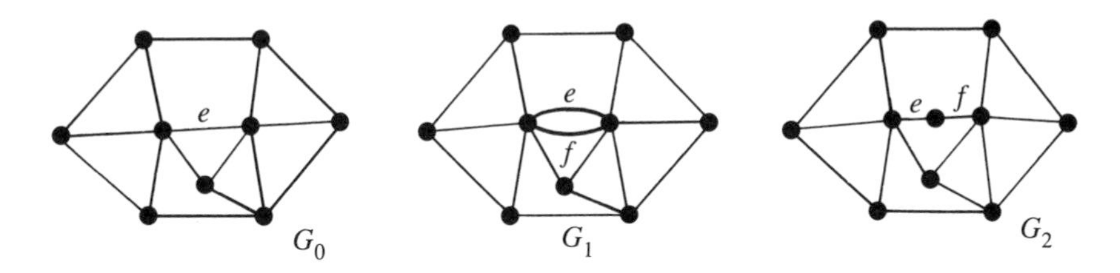

Fig. 5.12. Parallel extension and series extension.

beginning with a graph with just one edge and using these operations, we can build up a class of planar graphs with the property that all abstract duals of members of the class are also in the class. The graph operations considered here have been successfully extended to matroids and, in Chapter 7, a detailed discussion will be given of these matroid operations.

We begin with some examples. The graph G_1 in Figure 5.12 is obtained from G_0 by adding the edge f *in parallel* with the edge e, whereas, in G_2, the edge f has been added *in series* with e. We call G_1 a parallel extension of G_0, and G_2 a series extension of G_0. For an arbitrary graph G, these operations are defined as follows: G' is a *parallel extension* of G or, equivalently, G is a *parallel deletion* of G' if G' has a two-edge cycle $\{e, f\}$ such that $G' \backslash f = G$. If, instead, $\{e, f\}$ is a two-edge bond of G' and $G'/f = G$, then G' is a *series extension* of G, and G is a *series contraction* of G'. Thus, for example, replacing a cut-edge by a path of length two is not a series extension. The example in Figure 5.13 shows that not every series extension consists of replacing an edge by a path of length two.

The graph-theoretic operations of series and parallel extension generalize immediately to matroids. In particular, if $M \backslash f = N$ and f is in a 2-circuit of M, then M is called a *parallel extension* of N, and N is called a *parallel deletion* of M. If, instead, $M/f = N$ and f is in a 2-cocircuit of M, then M is called a *series extension* of N, and N is called a *series contraction* of M. Clearly M is a parallel extension of N if and only if M^* is a series extension of N^*. A *series class* of M is a parallel class of M^*; it is *non-trivial* if it contains at least two elements.

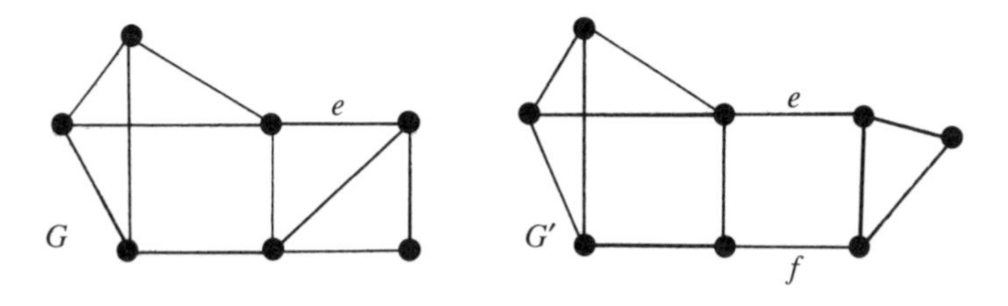

Fig. 5.13. A series extension that is not a subdivision.

If the graph G has a subgraph that is a subdivision of the graph H, and H is 2-connected, then H can be obtained from G by a sequence of edge deletions, series contractions, and deletions of isolated vertices. Motivated by this and the form of such graph-theoretic results as Kuratowski's Theorem, Bixby (1977) defined the matroid N to be a *series minor* of M if N can be obtained from M by a sequence of deletions and series contractions. If, instead, N can be obtained from M by a sequence of contractions and parallel deletions, then N is a *parallel minor* of M. Clearly M_1 is a parallel minor of M_2 if and only if M_1^* is a series minor of M_2^*.

We noted in Proposition 3.1.26 that the operations of deletion and contraction commute. This is not the case for deletion and series contraction. For example, suppose that $\{x, y, z\}$ is a cocircuit of M but that y is not in a 2-element cocircuit of M. Then $M\backslash x$ has $\{y, z\}$ as a cocircuit, so $M\backslash x/y$ is a series minor of M. However, M/y is not a series contraction of M. In spite of the failure of these operations to commute, we do have the following result.

5.4.1 Lemma. *If $N = M/y\backslash x$ and $\{y, z\}$ is a cocircuit of M, then N is either a series contraction or a deletion of $M\backslash x$.*

Proof. If $\{y, z\}$ is a cocircuit of $M\backslash x$, then evidently N is a series contraction of $M\backslash x$. Thus we may assume that $\{y, z\}$ is not a cocircuit of $M\backslash x$. It follows easily that $\{x, y\}$ must be a cocircuit of M. Therefore y is a coloop of $M\backslash x$. Hence $N = M\backslash x\backslash y$. □

The next result is a straightforward consequence of this lemma.

5.4.2 Proposition. *Let M and N be matroids. If N is a series minor of M, then $N = M\backslash X/Y$ where every element in Y is in series with an element of $M\backslash X$ not in Y.*

Proof. This is left to the reader (Exercise 4). □

The notion of a series minor of a matroid was motivated by reference to Kuratowski's Theorem. The following variant of that theorem follows easily on combining the theorem with Proposition 5.2.6 (Bixby 1977).

5.4.3 Proposition. *Let G be a graph. Then G is planar if and only if $M(G)$ has no series minor isomorphic to $M(K_5)$ or $M(K_{3,3})$.* □

Wagner (1937a) showed that one can also characterize planar graphs in terms of excluded parallel minors. The proof of his result is more difficult than the proof of the last result and is omitted. The graphs $K'_{3,3}$, $K''_{3,3}$, and $K'''_{3,3}$ are shown in Figure 5.14.

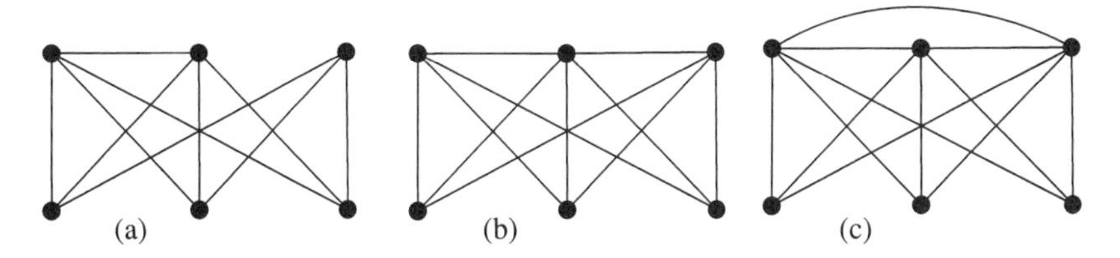

Fig. 5.14. (a) $K'_{3,3}$. (b) $K''_{3,3}$. (c) $K'''_{3,3}$.

5.4.4 Proposition. *A graph G is planar if and only if its cycle matroid $M(G)$ has no parallel minor isomorphic to $M(K_5)$, $M(K_{3,3})$, $M(K'_{3,3})$, $M(K''_{3,3})$, or $M(K'''_{3,3})$.* □

Returning to the operations of series and parallel extension for graphs, we remark that these operations are special cases of the operations of series and parallel connection of graphs, which will be more closely examined in Section 7.1. Examples of these operations are illustrated in Figure 5.15. The graphs G and H are, respectively, the series and parallel connections of G_1 and G_2 relative to the directed edges e_1 and e_2. Evidently if G_2 were the graph consisting of a pair of parallel edges, then the resulting series and parallel connections would be just a series and a parallel extension of G_1.

Notice that, although both the graphs G_1 and G_2 in Figure 5.15 are 3-connected, neither G nor H is. Furthermore, $G\backslash e$ and H/e are not even 2-connected. These observations foreshadow several of the important properties of series and parallel connections of matroids. These matroid operations, which are of basic structural importance, will be looked at in detail in Chapter 7.

Electrical networks that can be built up by adding resistors in series and in parallel arise frequently in electrical engineering. An attractive feature of such a network is that its joint resistance can easily be calculated (see Duffin 1965). We now consider the graphs corresponding to such networks. A graph G without isolated vertices is called a *series–parallel network* if it can be obtained from either of the two connected single-edge graphs by a sequence of operations each of which is either a series or a parallel extension. Thus, as is easily checked, the graphs G_1 and G_2 in Figure 5.16 are series–parallel networks, while G_3 is not. Kuratowski's Theorem characterized planar graphs by a list of excluded subgraphs, that is, a list of graphs which cannot occur as subgraphs. The main result of this section is an excluded-subgraph characterization of series–parallel networks. Some readers may wish to try to derive this result themselves. In that case, a detailed consideration of G_3 may prove helpful.

The definition of series–parallel networks strongly suggests that inductive arguments can be successfully applied in determining their properties.

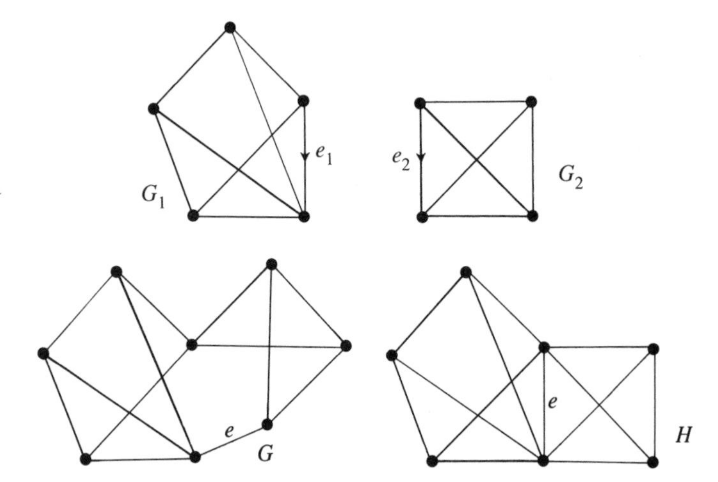

Fig. 5.15. Series and parallel connection of graphs.

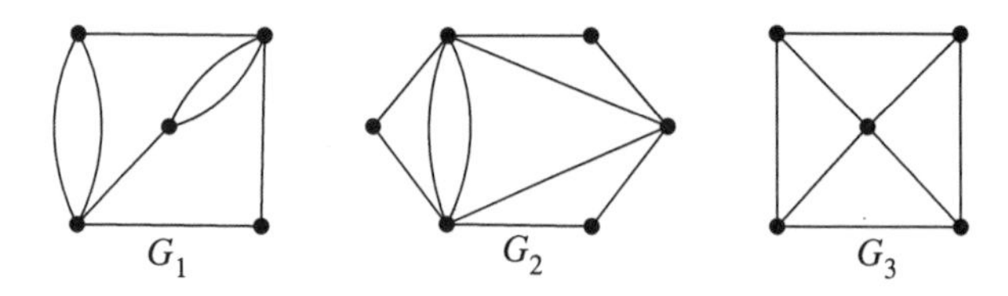

Fig. 5.16. G_1 and G_2 are series–parallel networks; G_3 is not.

Indeed, each of the following properties of a series–parallel network G is easily obtained using such an argument.

5.4.5 *G is a planar graph.*

5.4.6 *G is a block.*

5.4.7 *If G has at least three edges, one of which is e, then either $G\backslash e$ or G/e is not a block.*

The next result is essentially an extension of Lemma 5.2.4.

5.4.8 Lemma. *Let G be a planar graph and H be an abstract dual of G. Then, corresponding to every series extension G' of G, there is a parallel extension H' of H which is an abstract dual of G'. Moreover, every parallel extension of H is an abstract dual of a series extension of G.*

Proof. As H is an abstract dual of G, there is a bijection ψ from $E(G)$ onto $E(H)$ with the property that X is a cycle in G if and only if $\psi(X)$ is a bond in H. Now suppose that the graph G' is obtained by adding a new edge f in series with the edge e of G, and let H' be the graph obtained from H by adding a new edge g in parallel with $\psi(e)$. It is routine to verify that H' is an abstract dual of G', and the remainder of the proof is left as an exercise. □

From this lemma we immediately obtain the following.

5.4.9 Proposition. *Let G be a graph having no isolated vertices. If G is an abstract dual of a series–parallel network, then G is a series–parallel network.* □

The next theorem (Dirac 1952a; Ádám 1957; Duffin 1965) is the main result of this section, and most of the remainder of the section will be devoted to proving it.

5.4.10 Theorem. *A graph G with at least one edge is a series–parallel network if and only if it is a block having no subgraph that is a subdivision of K_4.*

Proof. If G is a series–parallel network, a straightforward induction argument shows that G is a block with no subgraph that is a subdivision of K_4. The converse of this is also proved by induction on $|E(G)|$. It clearly holds for $|E(G)| \leq 2$. Assume it true for $|E(G)| < n$ and let $|E(G)| = n \geq 3$. If G has two edges in series or two edges in parallel, then the result follows easily by the induction assumption. We may therefore assume that G is simple with every vertex having degree at least three. The next lemma, due to Dirac (1952a), completes the proof of the theorem by establishing the contradiction that G has a subgraph that is a subdivision of K_4.

A *chordal path* of a cycle in a graph is a path that is edge-disjoint from the cycle and that joins two non-neighbouring vertices of the cycle. An *internal vertex* of such a path is a vertex of the path other than one of its end-vertices.

5.4.11 Lemma. *A simple 2-connected graph G in which the degree of every vertex is at least three has a subgraph that is a subdivision of K_4.*

Proof. From among the cycles of G, choose one, say C, which has the greatest number of edges. Let $a_1, a_2, \ldots, a_m$ be the vertices of C in cyclic order. We shall show next that every vertex of C meets a chordal path of C. Since every vertex of G has degree at least three, for each i in $\{1, 2, \ldots, m\}$, there is an edge e_i that meets a_i but is not in C. As G is 2-connected, e_i is the first edge of a path P_i in G that joins a_i to some other vertex a_j of C but is otherwise disjoint from C. If this path is not a chordal path of C, then a_i and a_j are neighbouring vertices of C. But then, either P_i has more than one edge, in which case the choice of C

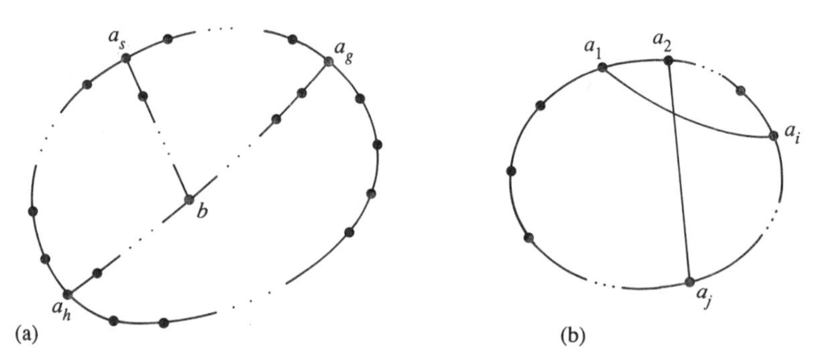

Fig. 5.17. Chordal paths of C giving subdivisions of K_4.

is contradicted, or P_i has exactly one edge, and the simplicity of G is contradicted. We conclude that every vertex of C does indeed meet a chordal path of C.

We now distinguish two cases:

(i) C has two chordal paths joining different pairs of vertices and having a common internal vertex.

(ii) Whenever two chordal paths of C have a common internal vertex, they have the same end-vertices.

In the first case, let R_1 and R_2 be chordal paths of C that have a common internal vertex but do not join the same pair of vertices. Suppose R_1 joins a_g and a_h, and R_2 joins a_s and a_t where a_s, a_g, and a_h are distinct, although a_t may equal a_g or a_h. Beginning at a_s, follow R_2 towards a_t until the first time a vertex b of R_1 is encountered. Then the subgraph of G induced by C, the chordal path R_1, and that part of the chordal path R_2 joining a_s and b is a subdivision of K_4 (see Figure 5.17(a)).

In case (ii), for every pair $\{a_i, a_j\}$ of distinct vertices of C that are joined by at least one chordal path, arbitrarily choose such a path P_{ij}. Let G' be the subgraph of G induced by C and the set of chosen paths P_{ij}. By assumption, no two chordal paths of C in G' share a common internal vertex. Moreover, each such chordal path cuts C into two paths. Take a shortest such path P in C and suppose that it joins a_1 and a_i. Evidently $i \neq 2$. Also, since C has a chordal path that meets a_2, there is a chordal path in G' joining a_2 to a_j for some j not in $\{1, 2, 3\}$. As a shortest path joining a_2 and a_j in C cannot be shorter than P, we must have that $j > i$. It follows that the subgraph of G' induced by the union of C, the chordal path joining a_1 and a_i, and the chordal path joining a_2 and a_j is a subdivision of K_4 (see Figure 5.17(b)).

This completes the proof of the lemma and thereby finishes the proof of Theorem 5.4.10. □

The next result essentially restates Theorem 5.4.10 as an excluded-minor result. The proof is left as an exercise. In Chapter 11, we shall give an excluded-minor characterization of series–parallel networks within the class of all matroids.

5.4.12 **Corollary.** *The following statements are equivalent for a graph G having at least one edge:*

(i) *G is a series–parallel network.*

(ii) *$M(G)$ is connected and has no series minor isomorphic to $M(K_4)$.*

(iii) *$M(G)$ is connected and has no parallel minor isomorphic to $M(K_4)$.*

(iv) *$M(G)$ is connected and has no minor isomorphic to $M(K_4)$.* □

Exercises

1. For each of the graphs in Figures 5.11, 5.12, 5.13, and 5.18, determine whether the graph is a series–parallel network and, if not, find the least number of edges that must be deleted to give such a graph.

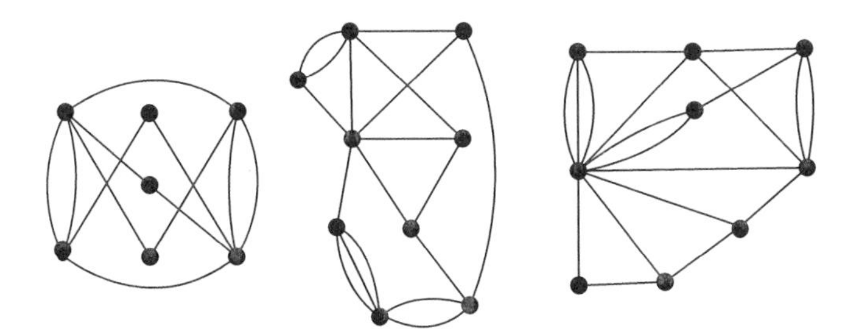

Fig. 5.18.

2. Let G be a series–parallel network having no parallel edges. If C is a cycle, determine, in terms of $|C|$, the maximum number of chordal paths that C can have.

3. Show that if the graph G has a subgraph that is a subdivision of the graph H, then $M(G)$ has $M(H)$ as a series minor.

4. Prove Proposition 5.4.2.

5. Suppose that the matroid M has no series minor isomorphic to the matroid N and has no parallel minor isomorphic to N. Prove or disprove that M has no N-minor.

6. Show that a graph is a series–parallel network if and only if it is connected having a single edge, or it can be obtained from a cycle with two edges by a sequence of subdivisions and parallel extensions.

7. (Dirac 1960) Show that a simple n-vertex series–parallel network has at most $2n - 3$ edges and characterize the graphs for which equality holds in this bound.

8. (Oxley 1984a) Prove that a block with at least one edge is a series–parallel network if and only if it does not have a cycle and a bond that meet in exactly four edges.

9. (Duffin 1965) Prove that if a 2-connected graph G has a planar embedding in which every vertex lies on the boundary of the infinite face, then G is a series–parallel network. Is the converse of this true?

10.* (Duffin 1965) Let G be a block and assign a direction to every edge of G. Prove that G is a series–parallel network if and only if, whenever a cycle C of G meets a pair of edges, x and y, that are oppositely directed relative to C, then x and y are oppositely directed relative to every cycle containing them both.

6 Representable matroids

Ever since Whitney (1935) used graphic and vector matroids as the two fundamental examples of matroids, there have been numerous basic problems associated with these classes, one of the most prominent of which has been to characterize the classes. In the case of vector matroids, this problem is usually specialized to that of characterizing the class of F-representable matroids for some particular field F. Since the classes of graphic and F-representable matroids are minor-closed, one of the most commonly sought ways to characterize these classes has been via a list of minor-minimal matroids that are not in the class. In Section 6.6, we shall present Tutte's excluded-minor characterization of the class of graphic matroids, while, in Section 6.5, we state the excluded-minor characterizations of the classes of binary and ternary matroids. The last two results will be proved in Chapters 9 and 10, respectively. The fields $GF(2)$ and $GF(3)$ are the only fields for which the sets of excluded minors are known. Indeed, they are the only fields for which the sets of excluded minors are known to be finite.

There are two main goals of this long chapter. The first is to provide an overview of the basic questions associated with matroid representability. The second is to indicate how one actually goes about constructing representations. In view of the length of this chapter, perhaps a word of guidance to the reader is in order. The key ideas are presented in Sections 6.1 and 6.3–6.6. The other four sections discuss topics that are less central: Section 6.2 looks at affine geometries, a class of highly symmetric structures that are closely linked to the projective geometries of Section 6.1; Section 6.7 discusses algebraic matroids, a class of matroids that properly contains the class of representable matroids and arises from algebraic dependence over a field; Section 6.8 focuses on characteristic sets, its main idea being concerned with how one can capture geometrically certain algebraic properties of a field; and Section 6.9 examines modularity, a special property of flats that is important in several contexts including matroid constructions.

6.1 Projective geometries

By definition, a matroid M is representable over a field F if and only if $M \cong M[A]$ for some matrix A over F. An element e of $M[A]$ is a loop of this matroid if and only if the corresponding column of A is the zero vector. Two elements f and g are parallel in $M[A]$ if and only if the corresponding columns of A are non-zero and are scalar multiples of each other. It follows from these observations that a matroid M is F-representable if and only if its associated simple matroid $\widetilde{M}$ is F-representable. Hence, when we discuss representability questions, we shall usually concentrate on simple matroids.

This section introduces projective geometries and examines their properties. These objects arise frequently in mathematics and are extremely natural to consider in matroid theory, their position among representable matroids being analogous to that of complete graphs in graph theory. Let V be a vector space over a field F. The *projective geometry* $PG(V)$ consists of a set of *points*, a disjoint set of *lines*, and an *incidence relation* between points and lines. The points and lines are precisely the 1- and 2-dimensional subspaces of V, and incidence is determined by set inclusion. Evidently the construction of $PG(V)$ from V is analogous to the construction of the simple matroid $\widetilde{M}$ from a matroid M. Indeed, if V is finite and we view it as a matroid, these constructions are identical. An alternative way to construct $PG(V)$ from V is to first delete the zero vector and then, from each 1-dimensional subspace, delete all but one of the remaining elements. The elements that are left become the points of the projective geometry. Frequently these representatives for the 1-dimensional subspaces are chosen according to some pattern. For example, if $V = V(n+1, F)$, we often choose from each 1-dimensional subspace that vector whose first non-zero coordinate is a one. When $PG(V)$ is constructed in this way, the lines are the intersections of the 2-dimensional subspaces of V with the chosen set of points.

We now look briefly at the geometric structure of projective geometries. It is not difficult to check that $PG(V)$ has the following properties:

(i) Any two distinct points, a and b, are on exactly one line ab.

(ii) Every line contains at least three points.

(iii) If a, b, c, and d are four distinct points, no three of which are collinear, and if the line ab intersects the line cd, then the line ac intersects the line bd.

In general, suppose P and L are disjoint sets of points and lines, respectively, and ι is an incidence relation such that (i)–(iii) hold. Then the triple (P, L, ι) is called a *projective space*. In view of (i) and (ii), we can interpret lines as subsets of P. When this is done, it is natural to indicate incidence by using set-theoretic notation. Two projective spaces (P, L, ι) and (P', L', ι') are *isomorphic* if there is a bijection $\zeta : P \to P'$ such that a subset X of P is in L if and only if $\zeta(X)$ is in L'. As we shall see in just a moment, most finite projective spaces are actually isomorphic to projective geometries.

A *subspace* of a projective space (P, L, ι) is a subset P_1 of P such that if a and b are distinct elements of P_1, then all points on the line ab are in P_1. Thus, for example, $\emptyset$ and P are subspaces, as are $\{p\}$ and X for all points p and all lines X. A subspace is a *hyperplane* if the only subspace properly containing it is P. The subspaces of (P, L, ι), ordered by inclusion, form a partially ordered set having a zero. The (*projective*) *dimension* of a subspace is one less than its height in this poset provided this height is finite; otherwise the dimension of the subspace is ∞. Thus $\emptyset$ has dimension -1, while points and lines have dimensions 0 and 1, respectively. If $V = V(n+1, F)$, then $PG(V)$ has dimension n and it is customary to denote this projective geometry by $PG(n, F)$.

We have noted that every projective geometry is a projective space. The next result is a partial converse to this for finite projective spaces. When F is the finite field $GF(q)$, we shall frequently write $PG(n, q)$ for $PG(n, F)$.

6.1.1 Theorem. *Every finite projective space of dimension other than two is isomorphic to $PG(n, q)$ for some integer n exceeding -2 and some prime power q.* □

In fact, a natural extension of this result holds for infinite projective spaces. The reader is referred to Dembowski's book (1968, p. 27) for a discussion of this result and for several references for proofs of Theorem 6.1.1. A word of warning here concerning terminology: this is not standardized. For example, Dembowski uses 'projective geometry' for what we have called a 'projective space'; and he distinguishes $PG(n, q)$ by calling it a 'finite desarguesian projective geometry'.

A projective space of dimension 2 is universally called a *projective plane*. Figure 1.11 shows the projective planes $PG(2, 2)$ and $PG(2, 3)$, both of which are certainly projective geometries. It is known, however, that there are finite projective planes that are not isomorphic to any projective geometry $PG(2, q)$ (see, for example, Hughes and Piper 1973). Clearly any finite projective plane gives rise to a rank-3 paving matroid on the set of points. Using this fact and Theorem 6.1.1, we deduce that every finite projective space can be regarded as a matroid.

We now know that the projective geometries $PG(n, F)$ can be viewed algebraically, using their derivation from $V(n+1, F)$, or geometrically, using the idea of a projective space. Moreover, for $n \leq 3$, we can draw a diagram representing $PG(n,q)$, this diagram being just a geometric representation of the corresponding matroid; that is, the points, lines, and planes of the projective space coincide with points, lines, and planes of the matroid. Next we indicate the fundamental importance of projective geometries in any discussion of matroid representability. Let M be an n-element rank-r matroid. Then, by definition, M is F-representable if and only if M is isomorphic to $M[A]$ for some $m \times n$ matrix A over F with $m \geq r$. Associated with this isomorphism, there is a natural map ψ_A from $E(M)$ into $V(m,F)$.

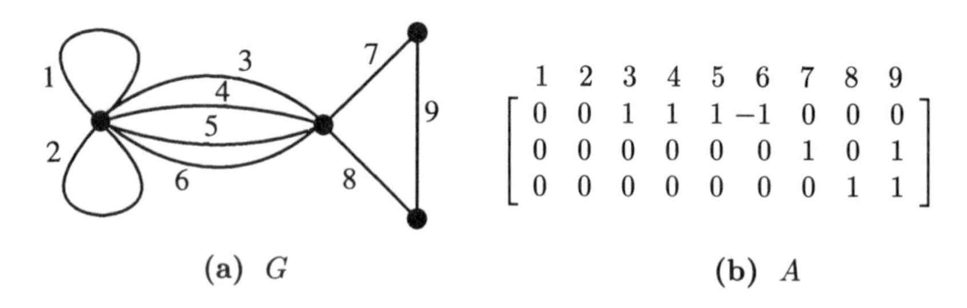

(a) G (b) A

Fig. 6.1. A graph G and a matrix A representing $M(G)$ over $\mathbb{R}$.

6.1.2 **Example.** Let $E = \{1,2,...,9\}$ and $M = M(G)$ where G is the graph shown in Figure 6.1(a). Let A be the matrix over $\mathbb{R}$ shown in Figure 6.1(b). Then under the identity map on E, it is easily checked that $M(G) \cong M[A]$. The corresponding map ψ_A from E into $V(3,\mathbb{R})$ is $\psi_A(1) = \psi_A(2) = (0,0,0)^T$, $\psi_A(3) = \psi_A(4) = \psi_A(5) = (1,0,0)^T$, $\psi_A(6) = (-1,0,0)^T$, $\psi_A(7) = (0,1,0)^T$, $\psi_A(8) = (0,0,1)^T$, and $\psi_A(9) = (0,1,1)^T$. □

It is clear that, in general, a rank-r matroid M is F-representable if and only if, for some $m \geq r$, there is a function $\psi : E(M) \to V(m,F)$ such that $r_M(X) = \dim\langle\psi(X)\rangle$ for all $X \subseteq E(M)$. Such a map ψ will be called a *coordinatization* of M *over* F or an F-*coordinatization* of M. Clearly if ψ is a coordinatization of M, then $r = \dim\langle\psi(E(M))\rangle$. Hence the rank-$r$ matroid M is F-representable if and only if M has a coordinatization ϕ over F such that $\phi : E(M) \to V(r,F)$.

It was noted at the outset of this section that a matroid is F-representable if and only if the associated simple matroid is F-representable. Suppose, then, that the matroid M is simple and that $\psi : E(M) \to V(m,F)$ is a coordinatization of M over F. Then ψ is one-to-one. Moreover, $\psi(E(M))$ does not contain $\underline{0}$; nor does it contain more than one element of any 1-dimensional subspace of $V(m,F)$. Hence $\psi(E(M))$ can be viewed as a subset of $PG(m-1,F)$. Indeed, for any finite subset S of $PG(m-1,F)$, there is a matroid induced on S by linear independence over F. We shall denote this matroid by $PG(m-1, F)|S$. Summarizing the above discussion, we have the following theorem, the main result of this section.

6.1.3 Theorem. *Let M be a simple rank-r matroid and F be a field. The following statements are equivalent:*

(i) *M is F-representable.*

(ii) *$PG(r-1, F)$ has a finite subset T such that $M \cong PG(r-1, F)|T$.*

(iii) *For some $m \geq r$, there is a finite subset S of $PG(m-1, F)$ such that $M \cong PG(m-1, F)|S$.* □

In view of this result, we see that a study of F-representable simple matroids is a study of the restrictions of projective geometries over F. Just as every n-vertex simple graph can be obtained from K_n by deleting edges, so too can every rank-r simple F-representable matroid be obtained from $PG(r-1, F)$ by deleting elements. We remark here that in a geometric representation of $PG(r-1, F)|S$, the points of the representation are just points of the projective space; the lines and planes of the representation come from intersecting S with lines and planes of the projective space that contain, respectively, at least two points and at least three non-collinear points of S.

Since projective geometries are so fundamental in a discussion of representability, the remainder of this section will discuss some properties of these structures. We begin by noting some statistics for $PG(r-1, q)$. Featured prominently here are the *Gaussian coefficients* $\left[{m \atop k}\right]_q$, which are defined for all integers m and k with $0 \leq k \leq m$ by

$$\begin{aligned} \left[{m \atop k}\right]_q &= \frac{(q^m-1)(q^m-q)\cdots(q^m-q^{k-1})}{(q^k-1)(q^k-q)\cdots(q^k-q^{k-1})} \\ &= \frac{(q^m-1)(q^{m-1}-1)\cdots(q^{m-k+1}-1)}{(q^k-1)(q^{k-1}-1)\cdots(q-1)}. \end{aligned}$$

Note that $\left[{m \atop 0}\right]_q = 1$ for all $m \geq 0$ since, by convention, an empty product is 1.

We have seen that $PG(r-1, q)$ can be looked at in several different ways. For most of the rest of this book, we shall view it as the simple matroid associated with $V(r, q)$, where the latter is also being thought of as a matroid here. Evidently the matroid $PG(r-1, q)$ has rank r. Moreover, every flat of $PG(r-1, q)$ corresponds to a flat of $V(r, q)$. Since every rank-k flat of $V(r, q)$ is isomorphic to $V(k, q)$, every rank-k flat of $PG(r-1, q)$ is isomorphic to $PG(k-1, q)$. The next result specifies the number of such flats. From this, we get immediately that $PG(r-1, q)$ has exactly $(q^r-1)/(q-1)$ elements.

6.1.4 Proposition. *Let k be a non-negative integer.*

(i) *The number of rank-k flats in $PG(r-1,q)$ is $\left[{r \atop k}\right]_q$.*

(ii) *The number of k-element independent sets in $PG(r-1,q)$ is $\frac{1}{k!}\left(\frac{q^r-1}{q-1}\right)\left(\frac{q^r-q}{q-1}\right)\cdots\left(\frac{q^r-q^{k-1}}{q-1}\right)$.*

(iii) *The number of k-element circuits in $PG(r-1,q)$ is 0 for $k<3$ and is $\frac{1}{k!(q-1)}(q^r-1)(q^r-q)\cdots(q^r-q^{k-2})$ for $k\geq 3$.*

Proof. We shall use the following result.

6.1.5 Lemma. *The number of ordered k-tuples $(\underline{v}_1,\underline{v}_2,\ldots,\underline{v}_k)$ of distinct members of $V(r,q)$ such that $\{\underline{v}_1,\underline{v}_2,\ldots,\underline{v}_k\}$ is linearly independent is $(q^r-1)(q^r-q)\cdots(q^r-q^{k-1})$.*

Proof. Evidently $V(r,q)$ has q^r members. As $\underline{v}_1$ must be non-zero, there are q^r-1 choices for it. Once distinct vectors $\underline{v}_1,\underline{v}_2,\ldots,\underline{v}_j$ have been chosen so that $\{\underline{v}_1,\underline{v}_2,\ldots,\underline{v}_j\}$ is linearly independent, $\underline{v}_{j+1}$ can be any of the q^r elements of $V(r,q)$ except for the q^j elements of $\langle\underline{v}_1,\underline{v}_2,\ldots,\underline{v}_j\rangle$. Hence there are q^r-q^j choices for $\underline{v}_{j+1}$. □

From this lemma, we deduce that the number of k-element independent sets in $V(r,q)$ is

$$\frac{1}{k!}(q^r-1)(q^r-q)\cdots(q^r-q^{k-1}). \tag{1}$$

Since every non-zero member of $V(r,q)$ is in a parallel class of size $q-1$, (ii) follows.

The number of rank-k flats of $PG(r-1,q)$ is the same as the number n_k of rank-k flats of $V(r,q)$. But every rank-k flat of $V(r,q)$ is isomorphic to $V(k,q)$ and so the total number of k-element independent sets in $V(r,q)$ is the product of n_k and the number of k-element independent sets in $V(k,q)$. It follows from (1) that

$$n_k=\frac{(1/k!)(q^r-1)(q^r-q)\cdots(q^r-q^{k-1})}{(1/k!)(q^k-1)(q^k-q)\cdots(q^k-q^{k-1})}=\left[{r \atop k}\right]_q,$$

that is, (i) holds.

To prove (iii), we shall use the following lemma whose proof is left as an exercise.

6.1.6 Lemma. *If $r\geq 2$ and B is a basis of $PG(r-1,q)$, then there are precisely $(q-1)^{r-1}$ elements x of $PG(r-1,q)$ such that $B\cup x$ is a circuit.* □

Let $\mathcal{C}_{r,k}$ denote the set of k-element circuits of $PG(r-1,q)$. Evidently $|\mathcal{C}_{r,k}| = 0$ for $k < 3$, so suppose $k \geq 3$. Then, as every k-element circuit is in a unique flat of rank $k-1$,

$$|\mathcal{C}_{r,k}| = \begin{bmatrix} r \\ k-1 \end{bmatrix}_q |\mathcal{C}_{k-1,k}|. \tag{2}$$

Now, in $PG(k-2,q)$, consider the set of ordered pairs (B, C) where B is a basis and C is a circuit containing B. By counting the number of such pairs in two different ways, first over circuits and then over bases, we get, using Lemma 6.1.6, that

$$k|\mathcal{C}_{k-1,k}| = (q-1)^{k-2}|\mathcal{B}(PG(k-2,q))|.$$

Thus, by (ii),

$$\begin{aligned} |\mathcal{C}_{k-1,k}| &= \frac{(q-1)^{k-2}}{k(k-1)!}\left(\frac{q^{k-1}-1}{q-1}\right)\left(\frac{q^{k-1}-q}{q-1}\right)\cdots\left(\frac{q^{k-1}-q^{k-2}}{q-1}\right) \\ &= \frac{1}{k!(q-1)}\left(q^{k-1}-1\right)\left(q^{k-1}-q\right)\cdots\left(q^{k-1}-q^{k-2}\right). \end{aligned}$$

Therefore, by (2), for $k \geq 3$,

$$|\mathcal{C}_{r,k}| = \frac{1}{k!(q-1)}\left(q^r-1\right)\left(q^r-q\right)\cdots\left(q^r-q^{k-2}\right). \qquad \square$$

6.1.7 Corollary. *A simple rank-r matroid M that is representable over $GF(q)$ has at most $\frac{q^r-1}{q-1}$ elements. Moreover, if $|E(M)| = \frac{q^r-1}{q-1}$, then M is isomorphic to $PG(r-1,q)$.*

Proof. By Theorem 6.1.3, as M is representable over $GF(q)$ and has rank r, there is a subset S of $E(PG(r-1,q))$ such that M is isomorphic to $PG(r-1,q)|S$. Since, by Proposition 6.1.4 (i), $PG(r-1,q)$ has $\frac{q^r-1}{q-1}$ elements, both parts of the result now follow easily. $\square$

The lattice of flats of $PG(r-1,q)$ is particularly well-behaved. Clearly it is isomorphic to the lattice of flats of $V(r,q)$. Using this fact and Corollary 3.3.4, it is straightforward to prove the following.

6.1.8 Proposition. *Let X and Y be flats of $PG(r-1,q)$ with $X \subseteq Y$ and $r(Y)-r(X) = k$. Then the interval $[X,Y]$ of $\mathcal{L}(PG(r-1,q))$ is isomorphic to $\mathcal{L}(PG(k-1,q))$.* $\square$

The next corollary follows immediately from this proposition and the fact that every line of $PG(r-1,q)$ contains exactly $q+1$ points.

6.1.9 Corollary. *If $r \geq 2$ and $e \in PG(r-1,q)$, then $PG(r-1,q)/e$ is the matroid that is obtained from $PG(r-2,q)$ by replacing each element by q elements in parallel.* □

From elementary properties of $V(r,q)$, it is not difficult to see that the lattice obtained from $\mathcal{L}(PG(r-1,q))$ by reversing all order relations is actually isomorphic to $\mathcal{L}(PG(r-1,q))$. Informally, $\mathcal{L}(PG(r-1,q))$ looks the same upside down as it does right side up. A geometric lattice with this property is an example of a modular lattice, an arbitrary lattice $\mathcal{L}$ being *modular* if it satisfies the Jordan–Dedekind chain condition and, for all x, y in $\mathcal{L}$,

$$h(x) + h(y) = h(x \vee y) + h(x \wedge y).$$

Modularity of lattices can be defined in several other equivalent ways (see, for example, Birkhoff 1967). In Section 6.9, we shall examine modularity in matroids. For the moment, we note that the fact that $\mathcal{L}(PG(r-1,q))$ is modular follows from the familiar vector space identity,

$$\dim U + \dim W = \dim(U+W) + \dim(U \cap W).$$

In Figure 6.2, we summarize several basic statistics associated with $\mathcal{L}(PG(r-1,q))$ including the numbers of flats of each rank, the number of rank-$(j-1)$ flats contained in each rank-j flat, and the number of rank-$(j+1)$ flats containing each rank-j flat. These numbers are not difficult to derive using Proposition 6.1.4.

It was stated without proof in Chapter 2 that the Vámos matroid V_8 (Vámos 1968) is not representable. We can now show this. The proof is based on the same idea that was used to establish that $\mathcal{L}(PG(r-1,q))$ is modular. A geometric representation for V_8 is shown in Figure 6.3. As we shall see, the crucial feature of V_8 which prevents it from being representable is that the points 5, 6, 7, and 8 are not coplanar.

6.1.10 Proposition. *The Vámos matroid V_8 is not representable.*

Proof. Assume that V_8 is representable over the field F. Then there is a map $\psi : E(V_8) \to V(4,F)$ such that $r(X) = \dim\langle\psi(X)\rangle$ for all $X \subseteq E(V_8)$. If $\{x_1, x_2, \ldots, x_k\} \subseteq E(V_8)$, then $W(x_1, x_2, \ldots, x_k)$ will denote the subspace $\langle\psi(\{x_1, x_2, \ldots, x_k\})\rangle$. Now

$$\begin{aligned}\dim(W(5,6) \cap W(1,2,3,4)) &= \dim W(5,6) + \dim W(1,2,3,4) \\ &\quad - \dim(W(5,6) + W(1,2,3,4)) \\ &= 2 + 3 - 4 = 1.\end{aligned}$$

Thus $W(5,6) \cap W(1,2,3,4) = \langle \underline{v} \rangle$ for some non-zero $\underline{v}$ in $V(4,F)$.

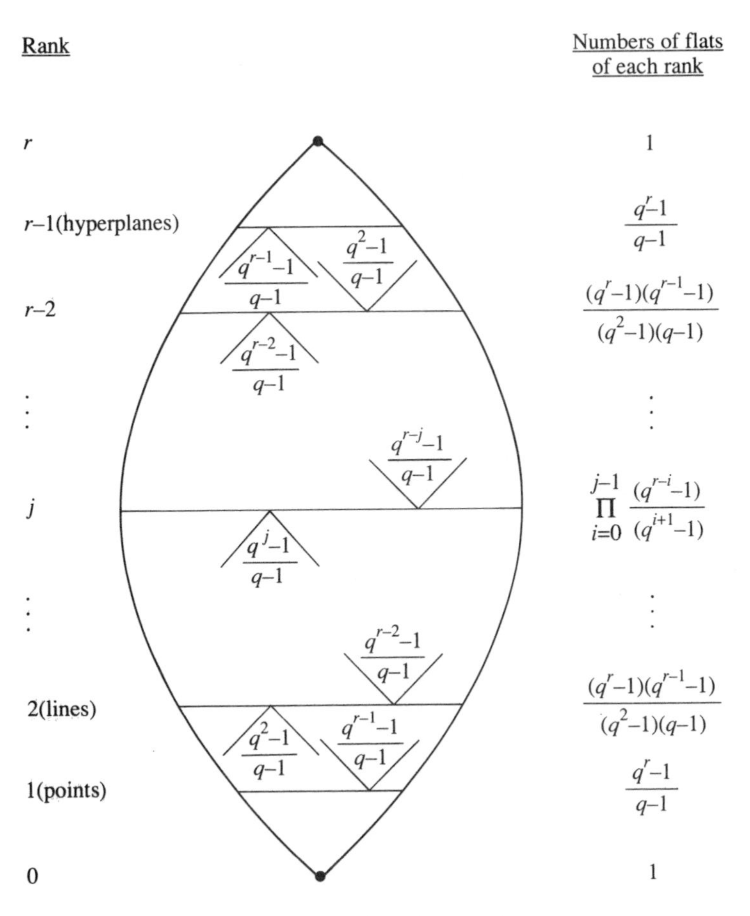

Fig. 6.2. $\mathcal{L}(PG(r-1,q))$.

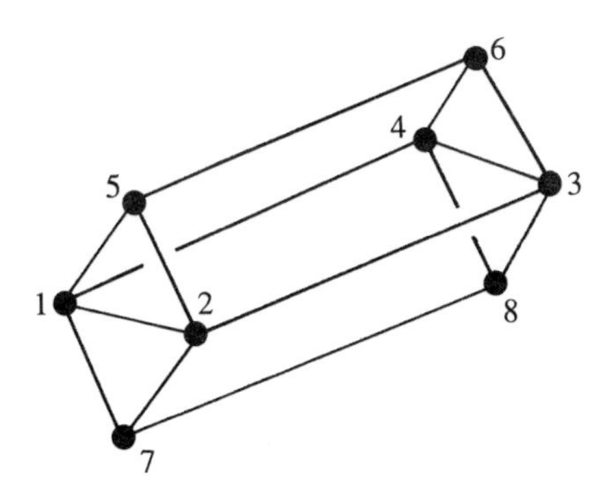

Fig. 6.3. The Vámos matroid, V_8.

Again

$$\begin{aligned}\dim(W(1,4,5,6)\cap W(1,2,3,4)) &= \dim W(1,4,5,6)+\dim W(1,2,3,4)\\ &\quad -\dim(W(1,4,5,6)+W(1,2,3,4))\\ &= 3+3-4=2.\end{aligned}$$

But $W(1,4)$ is a 2-dimensional subspace of $W(1,4,5,6)\cap W(1,2,3,4)$ and therefore

$$W(1,4)=W(1,4,5,6)\cap W(1,2,3,4)\supseteq W(5,6)\cap W(1,2,3,4)=\langle \underline{v}\rangle.$$

By symmetry, $\langle \underline{v}\rangle \subseteq W(2,3)$.

Now

$$\dim(W(1,4)\ \cap\ W(2,3))=2+2-3=1.$$

But $\langle \underline{v}\rangle$ is a subspace of $W(1,4)\cap W(2,3)$, so $\langle \underline{v}\rangle = W(1,4)\cap W(2,3)$. Geometrically, if V_8 is F-representable, then we can add a point p to the diagram so that its relationship to $\{1,2,3,4,5,6\}$ is as shown in Figure 6.4(a). Of course, p corresponds to the subspace $\langle \underline{v}\rangle$ of $V(4,F)$. By symmetry again, as $W(1,4)\cap W(2,3)=W(5,6)\cap W(1,2,3,4)$, it follows that $W(1,4)\cap W(2,3)=W(7,8)\cap W(1,2,3,4)$. Hence $\langle \underline{v}\rangle\subseteq W(7,8)$, and so $\langle \underline{v}\rangle \subseteq W(5,6)\cap W(7,8)$. Geometrically, this says that p is on both the line containing 5 and 6 and the line containing 7 and 8, so $\{5,6,7,8\}$ is dependent (see Figure 6.4(b)); or, in terms of subspaces:

$$\begin{aligned}\dim W(5,6,7,8) &= \dim(W(5,6)+W(7,8))\\ &= 2+2-\dim(W(5,6)\cap W(7,8))\leq 3.\end{aligned}$$

This is a contradiction since $r(\{5,6,7,8\})=4$. We conclude that V_8 is not F-representable. □

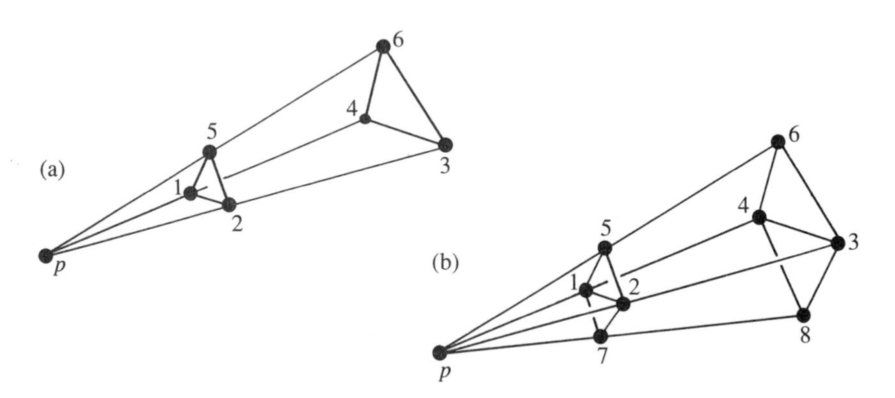

Fig. 6.4. Showing that the Vámos matroid is non-representable.

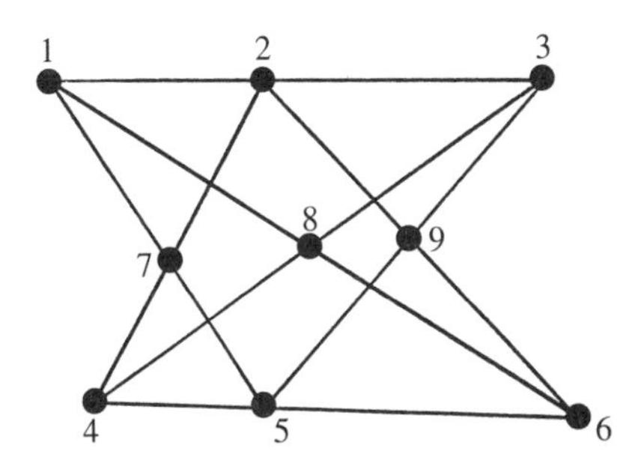

Fig. 6.5. The non-Pappus matroid.

The next result establishes an assertion made in Chapter 1 about the non-representability of the non-Pappus matroid.

6.1.11 Proposition. *If F is a field, then the non-Pappus matroid is not F-representable.*

Proof. Assume, to the contrary, that the non-Pappus matroid N is F-representable for some field F. Then, by Theorem 6.1.3, there is a 9-element subset S of $PG(2,F)$ such that the matroid $PG(2,F)|S$ has the geometric representation shown in Figure 6.5. In particular, 7, 8, and 9 are not collinear in this diagram, that is, $\{7,8,9\}$ is a basis of N. But the next result precludes the occurrence of such a configuration in $PG(2,F)$, thereby completing the proof of the proposition. □

6.1.12 Pappus's Theorem. *Let $\{1,2,3\}$ and $\{4,5,6\}$ be triples of distinct points which lie on the lines L and L', respectively, of $PG(2,F)$ such that none of these six points is on both L and L'. Let 7, 8, and 9 be the points of intersection of the pairs of lines, 15 and 24, 16 and 34, and 26 and 35, respectively. Then 7, 8, and 9 are collinear.*

Proof. This can be found, for example, in Hughes and Piper (1973, Theorem 2.6). □

We remark here that the pairs of lines whose intersections give the points 7, 8, and 9 in the last theorem are actually opposite sides of the twisted hexagon whose vertices are, in cyclic order, 1, 5, 3, 4, 2, and 6.

If we modify Figure 6.5 to make 7, 8, and 9 collinear, the resulting configuration is called the *Pappus configuration.* As noted in Chapter 1, the corresponding matroid is called the Pappus matroid.

We now look briefly at projective planes, projective spaces of dimension two. From Figure 6.2, we have that $PG(2,q)$ has $\frac{q^3-1}{q-1}$ or q^2+q+1 points and the same number of lines. Moreover, each line contains $\frac{q^2-1}{q-1}$

or $q+1$ points, and each point is on $q+1$ lines. The projective planes that are not isomorphic to $PG(2,q)$ for any q also have a very uniform structure. In particular, if π is a finite projective plane, then one can show that all lines of π contain the same number of points. If this number is $m+1$, then π is said to have *order* m. It is not difficult to prove:

6.1.13 Proposition. *If the projective plane π has order m, then*

(i) *every point of π is on exactly $m+1$ lines;*

(ii) *π has exactly m^2+m+1 points; and*

(iii) *π has exactly m^2+m+1 lines.* □

Clearly the plane $PG(2,q)$ has order q. It is known that there are unique planes of orders 2, 3, 4, 5, 7, and 8, and it follows from the theorem of Bruck and Ryser stated below that there is no plane of order 6. Moreover, there are at least four non-isomorphic planes of order 9 and at least two non-isomorphic planes of order m for all integers m exceeding 8 that are of the form p^k for p a prime and k an integer exceeding 1. Among the many interesting unsolved questions associated with projective planes are the following, both of which are extremely difficult:

(i) If p is a prime exceeding seven, is $PG(2,p)$ the only projective plane of order p?

(ii) Is there a finite projective plane whose order is not a power of a prime?

The following important result of Bruck and Ryser (1949) eliminates many potential orders for projective planes.

6.1.14 Proposition. *Let m be an integer of the form $4k+1$ or $4k+2$ where $k \geq 1$. If there is a projective plane of order m, then there are integers a and b such that $m = a^2+b^2$.*

Proof. This can be found in Hall (1986). □

From a matroid-theoretic point of view, projective planes are of particular interest in connection with the problem of matroid reconstruction, an analogue of the well-known reconstruction problem for graphs. Let M_1 and M_2 be non-isomorphic projective planes of order m. Then every single-element contraction of each of M_1 and M_2 is isomorphic to the matroid obtained from $U_{2,m+1}$ by replacing each element by m parallel elements. Equivalently, the multiset that consists of the isomorphism classes of all of the single-element deletions of M_1^* is identical to the corresponding multiset for M_2^*; yet $M_1^* \not\cong M_2^*$. This shows that we may not be able to uniquely determine a matroid, even if we know all of its single-element deletions up to isomorphism. For further discussion of this interesting topic of matroid reconstruction, the reader is referred to Chapter 14 and the work of Brylawski (1974, 1975a).

To close this section, we comment briefly on a more general type of representability than we have so far considered. Let A be an $m \times n$ matrix over a division ring R. Let E be the set of column labels of A and $\mathcal{I}$ be the set of subsets I of E for which the multiset of columns labelled by I is a linearly independent set in the left vector space of m-tuples over R. Then the same argument that was used in Proposition 1.1.1 to show that $(E, \mathcal{I})$ is a matroid when R is a field can also be used to establish that $(E, \mathcal{I})$ is a matroid when R is an arbitrary division ring. The reader who is unfamiliar with left vector spaces should consult Jacobson's *Lectures on Abstract Algebra, Volume* II (1953) where most of the treatment of linear algebra is in terms of one-sided vector spaces.

As before, we shall denote the matroid obtained from the matrix A by $M[A]$. An arbitrary matroid that is isomorphic to $M[A]$ is said to be *representable over* R. A good reference for matroids representable over division rings is Ingleton (1971a). It is noted there that the non-Pappus matroid, while it is not representable over any field, is representable over some division ring. This raises the question as to whether there are any matroids that are not representable over any division ring. Ingleton gives several examples of such matroids including the 10-point rank-3 non-Desargues matroid for which a geometric representation is shown in Figure 6.6. The fact that this diagram does actually represent a matroid follows by 1.5.8;

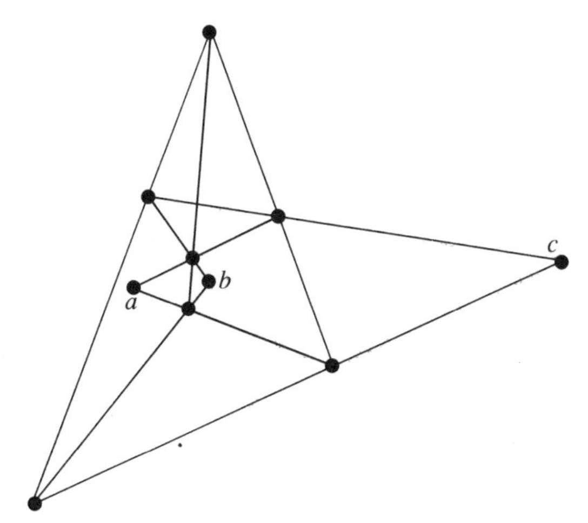

Fig. 6.6. The non-Desargues matroid.

the fact that this matroid is not representable over any division ring follows by Desargues' Theorem in projective geometry (see, for example, Dembowski 1968, p. 26) which asserts that if the configuration of points and lines in Figure 6.6 occurs in a projective space over a division ring, then the points a, b, and c must be collinear.

Exercises

1. What is the maximum number of bases in a simple rank-r matroid that is representable over $GF(q)$, and which matroids attain this maximum?

2. Prove that the Gaussian coefficients have the following properties:

   (i) $\begin{bmatrix} m \\ k \end{bmatrix}_q = \begin{bmatrix} m \\ m-k \end{bmatrix}_q$.

   (ii) $q^{k(m-k)} \le \begin{bmatrix} m \\ k \end{bmatrix}_q \le q^{k(m-k+1)}$.

   (iii) $\begin{bmatrix} m \\ k \end{bmatrix}_q = \begin{bmatrix} m-1 \\ k \end{bmatrix}_q + q^{m-k} \begin{bmatrix} m-1 \\ k-1 \end{bmatrix}_q$.

3. Show that $PG(1,q) \cong U_{2,q+1}$.

4. Show that $\begin{bmatrix} r-t \\ s-u \end{bmatrix}_q \begin{bmatrix} t \\ u \end{bmatrix}_q q^{(s-u)(t-u)}$ is the number of rank-s subspaces of $PG(r-1,q)$ whose intersection with a fixed rank-t subspace has rank u.

5. Let I be a k-element independent set in $PG(r-1,q)$. Identify the matroid $PG(r-1,q)/I$.

6. Let M be a simple rank-r matroid that is representable over $GF(q)$ and has exactly $\begin{bmatrix} r \\ k \end{bmatrix}_q$ flats of rank k for some k with $0 \le k \le r$. For which values of k must it be true that $M \cong PG(r-1,q)$?

7. (Ingleton 1971a) Show that the matroid for which a geometric representation is shown in Figure 6.7 is not representable over any division ring.

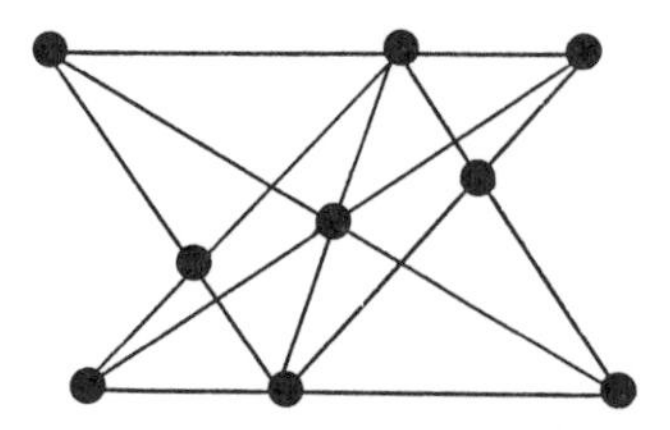

Fig. 6.7. A matroid that is not representable over any division ring.

8.[*] (Ingleton 1971a) Let V_1, V_2, V_3, and V_4 be subspaces of a vector space V.

(i) Show that

$$\begin{aligned}\dim(V_1 \cap V_2 \cap V_3) &\geq \dim(V_1 \cap V_2) + \dim V_3 - \dim(V_1 + V_3) \\ &- \dim(V_2 + V_3) + \dim(V_1 + V_2 + V_3).\end{aligned}$$

(ii) Show that

$$\begin{aligned}\dim(V_1 \cap V_2 \cap V_3 \cap V_4) &\geq \dim(V_1 \cap V_2 \cap V_3) \\ &+ \dim(V_1 \cap V_2 \cap V_4) - \dim(V_1 \cap V_2).\end{aligned}$$

(iii) Deduce from (i) and (ii) that

$$\begin{aligned}\dim(V_1 \cap V_2 \cap V_3 \cap V_4) &\geq \dim(V_1 \cap V_2) + \dim V_3 + \dim V_4 \\ &- \dim(V_1 + V_3) - \dim(V_2 + V_3) \\ &- \dim(V_1 + V_4) - \dim(V_2 + V_4) \\ &+ \dim(V_1 + V_2 + V_3) \\ &+ \dim(V_1 + V_2 + V_4).\end{aligned}$$

(iv) Prove that

$$\begin{aligned}\dim V_1 + \dim V_2 &+ \dim(V_1 + V_2 + V_3) + \dim(V_1 + V_2 + V_4) \\ &\qquad + \dim(V_3 + V_4) \\ \leq \dim(V_1 + V_2) &+ \dim(V_1 + V_3) + \dim(V_1 + V_4) \\ &+ \dim(V_2 + V_3) + \dim(V_2 + V_4).\end{aligned}$$

(v) Deduce from (iv) that if M is a representable matroid and X_1, X_2, X_3, and X_4 are subsets of $E(M)$, then

$$\begin{aligned}r(X_1) + r(X_2) &+ r(X_1 \cup X_2 \cup X_3) \\ &+ r(X_1 \cup X_2 \cup X_4) + r(X_3 \cup X_4) \\ &\leq r(X_1 \cup X_2) + r(X_1 \cup X_3) + r(X_1 \cup X_4) \\ &+ r(X_2 \cup X_3) + r(X_2 \cup X_4).\end{aligned}$$

(vi) Deduce from (v) that V_8 is non-representable.

(vii)[°] Are there any other inequalities, similar to but independent of that noted in (v), which hold for all representable matroids?

6.2 Affine geometries

Affine matroids were introduced in Section 1.5. In this section, we introduce affine geometries and indicate their relationship to both affine matroids and projective geometries. Finite affine geometries form an important class of highly symmetric matroids and play a fundamental role in geometry. However, they are not as central to a discussion of matroid representability as projective geometries. This section is independent of the remaining sections of this chapter and hence could be omitted by a reader whose main interest is in representability.

Let n be an integer exceeding -2. The *affine geometry* $AG(n, F)$ is obtained from $PG(n, F)$ by deleting from the latter all the points of a hyperplane. Equivalently, we can construct $AG(n, F)$ directly from $V(n+1, F)$ as follows. First delete a hyperplane H of $V(n+1, F)$. Then, for each 1-dimensional subspace X of $V(n+1, F)$ that is not contained in H, choose a single representative from $X - H$. These representatives are the points of $AG(n, F)$. Since, for any two hyperplanes H_1 and H_2 of $V(n+1, F)$, there is a non-singular linear transformation of $V(n+1, F)$ mapping H_1 onto H_2, it does not matter which hyperplane H of $V(n+1, F)$ is deleted in the above construction; that is, $AG(n, F)$ is well-defined. Now every hyperplane of $V(n+1, F)$ is the kernel of a linear functional, the latter being a linear transformation from $V(n+1, F)$ onto $V(1, F)$. In particular, as $V(n+1, F) = \{(x_1, x_2, \ldots, x_{n+1}) : x_i \in F$ for all $i\}$, the set of vectors $(x_1, x_2, \ldots, x_{n+1})$ with $x_1 = 0$ is a hyperplane of $V(n+1, F)$. If this is the hyperplane H that is deleted, then all the remaining vectors have non-zero first coordinate. If we pick, as the representative for each 1-dimensional subspace, that vector whose first non-zero coordinate is a one, then we see that the point set of $AG(n, F)$ can be viewed as $\{(1, x_2, x_3, \ldots, x_{n+1}) : x_i \in F$ for all $i\}$. Evidently if S is the finite subset $\{(1, x_{i2}, x_{i3}, \ldots, x_{i(n+1)}) : 1 \le i \le m\}$ of $AG(n, F)$, then there is a matroid $AG(n, F)|S$ induced on the points of S by linear dependence over F. This matroid is precisely the affine matroid on $\{(x_{i2}, x_{i3}, \ldots, x_{i(n+1)}) : 1 \le i \le m\}$ identified in Section 1.5. It is now not difficult to see that a matroid M is affine over F if and only if M has no loops and there is an integer $n \ge -2$ and a finite subset S of $AG(n, F)$ such that $\widetilde{M} \cong AG(n, F)|S$.

We now turn attention to the affine geometry $AG(r-1, q)$, that is, $AG(r-1, GF(q))$, viewing this as a matroid. Clearly this matroid has q^{r-1} elements and rank r. Moreover, all its rank-k flats are isomorphic to $AG(k-1, q)$.

6.2.1 Example. The affine matroid $AG(3, 2)$ was looked at several times in Chapters 1 and 2. With $V(4, 2) = \{(x_1, x_2, x_3, x_4)^T : x_i \in \{0, 1\}$ for all $i\}$ we have, from above, that $AG(3, 2)$ is represented over $GF(2)$ by the

$$\begin{bmatrix} 1 & 1 & 1 & 1 & 1 & 1 & 1 & 1 \\ 1 & 0 & 0 & 1 & 1 & 0 & 1 & 0 \\ 0 & 1 & 0 & 1 & 0 & 1 & 1 & 0 \\ 0 & 0 & 1 & 0 & 1 & 1 & 1 & 0 \end{bmatrix} \qquad \left[\begin{array}{c|cccc} & 0 & 1 & 1 & 1 \\ I_4 & 1 & 0 & 1 & 1 \\ & 1 & 1 & 0 & 1 \\ & 1 & 1 & 1 & 0 \end{array}\right]$$

(a) (b)

Fig. 6.8. Two $GF(2)$-representations for $AG(3,2)$.

matrix in Figure 6.8(a), this representation being obtained from $V(4,2)$ by deleting all the points of the hyperplane $x_1 = 0$. Another representation for $AG(3,2)$ that was noted following Example 2.2.25 is given in Figure 6.8(b). This second representation is obtained from $V(4,2)$ by deleting all the points of the hyperplane $x_1 + x_2 + x_3 + x_4 = 0$. □

6.2.2 Example. Consider $AG(2,2)$ and $AG(2,3)$. Viewed as the matroids induced on $V(2,2)$ and $V(2,3)$ by affine dependence, these affine geometries have the geometric representations shown in Figure 6.9. Notice that each point has been labelled by the corresponding element of $V(2,q)$, and dependence among these points must be determined affinely rather than linearly. To obtain linear representations for these matroids, recall, from Section 1.5, that if a point has affine coordinates (v_1, v_2), we take its corresponding linear coordinates to be $(1, v_1, v_2)$.

Notice here that $AG(2,2) \cong U_{3,4}$. Moreover, comparing Figure 6.9 with Figure 1.11, we can see, in these two cases, how $AG(2,q)$ has been obtained from $PG(2,q)$ by deleting a line. □

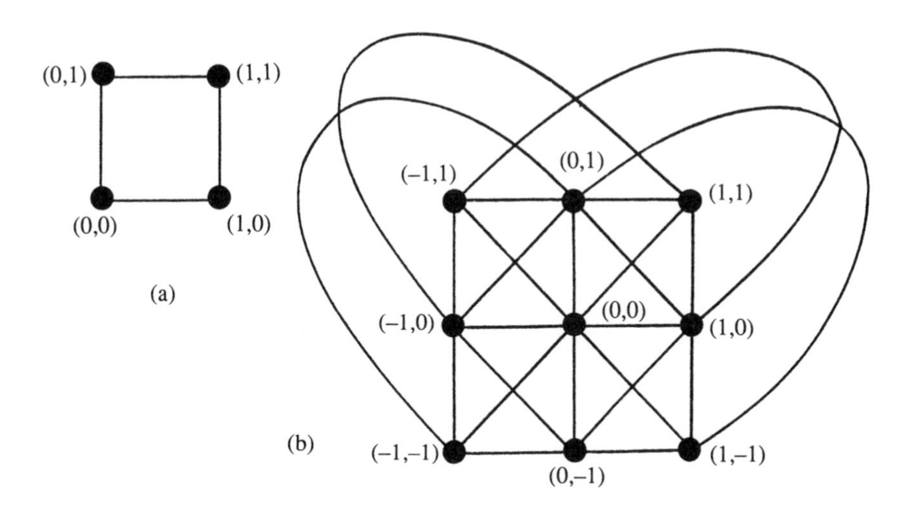

Fig. 6.9. (a) $AG(2,2)$ and (b) $AG(2,3)$.

The next result is the analogue of Proposition 6.1.4 for $AG(r-1,q)$.

6.2.3 Proposition. *Let k be a non-negative integer.*

(i) *The number of rank-k flats of $AG(r-1,q)$ is*

$$q^{r-k}\begin{bmatrix} r-1 \\ k-1 \end{bmatrix}_q = q^{r-k}\frac{(q^{r-1}-1)(q^{r-1}-q)\cdots(q^{r-1}-q^{k-2})}{(q^{k-1}-1)(q^{k-1}-q)\cdots(q^{k-1}-q^{k-2})}.$$

(ii) *The number of k-element independent sets of $AG(r-1,q)$ is*

$$\frac{1}{k!}q^{r-1}(q^{r-1}-1)(q^{r-1}-q)\cdots(q^{r-1}-q^{k-2}).$$

(iii) *$AG(r-1,q)$ has no circuits with fewer than three elements. For $k \geq 3$, its number of k-element circuits is*

$$\frac{1}{k!}q^{r-2}(q^{r-1}-1)(q^{r-1}-q)\cdots(q^{r-1}-q^{k-3})\left[(q-1)^{k-1}-(-1)^{k-1}\right].$$

Proof. $AG(r-1,q)$ is obtained from $PG(r-1,q)$ by deleting the points of a hyperplane H of the latter. Thus every flat of $AG(r-1,q)$ is of the form $X-H$ where X is a flat of $PG(r-1,q)$. It follows without difficulty using the modularity of $\mathcal{L}(PG(r-1,q))$ that the rank-k flats of $AG(r-1,q)$ are all the non-empty sets of the form $X-H$ where X is a rank-k flat of $PG(r-1,q)$. Thus the number of rank-k flats of $AG(r-1,q)$ is the difference between the number of such flats in $PG(r-1,q)$ and the number in a hyperplane of $PG(r-1,q)$. Since every hyperplane of $PG(r-1,q)$ is isomorphic to $PG(r-2,q)$, we deduce from Proposition 6.1.4(i) that the number of rank-k flats in $AG(r-1,q)$ is $\left[{r \atop k}\right]_q - \left[{r-1 \atop k}\right]_q$, which simplifies to $q^{r-k}\left[{r-1 \atop k-1}\right]_q$.

To prove (ii), we shall first count the number of ordered k-tuples $(\underline{v}_1,\underline{v}_2,\ldots,\underline{v}_k)$ of distinct elements of $\{(1,x_2,\ldots,x_r) : x_i \in GF(q)$ for all $i\}$ such that $\{\underline{v}_1,\underline{v}_2,\ldots,\underline{v}_k\}$ is linearly independent. Evidently there are q^{r-1} choices for $\underline{v}_1$. If distinct vectors $\underline{v}_1,\underline{v}_2,\ldots,\underline{v}_j$ have been chosen so that $\{\underline{v}_1,\underline{v}_2,\ldots,\underline{v}_j\}$ is independent, then $\{\underline{v}_1,\underline{v}_2,\ldots,\underline{v}_j\}$ spans a rank-j flat of $AG(r-1,q)$ and this flat is isomorphic to $AG(j-1,q)$. Hence there are $q^{r-1}-q^{j-1}$ choices for $\underline{v}_{j+1}$. We conclude that the number of k-element independent sets of $AG(r-1,q)$ is

$$\frac{1}{k!}q^{r-1}(q^{r-1}-1)(q^{r-1}-q)\cdots(q^{r-1}-q^{k-2}).$$

The proof of (iii) mimics the proof of (iii) of Proposition 6.1.4. It uses the following lemma in which, for convenience, $AG(r-1,q)$ is viewed as a restriction of $V(r,q)$.

6.2.4 Lemma. *If $r \geq 2$ and B is a basis of $AG(r-1,q)$, then the number of elements $\underline{x}$ of $AG(r-1,q)$ such that $B \cup \underline{x}$ is a circuit equals*

$$q^{-1}[(q-1)^r - (-1)^r].$$

Proof. Let $B = \{(1, \underline{v}_i) : 1 \leq i \leq r\}$ and $\underline{x} = (1, \underline{w})$. Then $B \cup \underline{x}$ is a circuit of $AG(r-1,q)$ if and only if there are non-zero elements $\alpha_1, \alpha_2, \ldots, \alpha_r$ of $GF(q)$ such that

$$\sum_{i=1}^{r} \alpha_i (1, \underline{v}_i) = (1, \underline{w}).$$

Thus, from considering the first coordinates here, we get that, for $r \geq 2$, the number of circuits of $AG(r-1,q)$ containing B equals the number $m(r)$ of r-tuples $(\alpha_1, \alpha_2, \ldots, \alpha_r)$ of non-zero elements of $GF(q)$ such that $\alpha_1 + \alpha_2 + \cdots + \alpha_r = 1$. Clearly $m(r)$ is defined for all $r \geq 1$, and $m(1) = 1$. Moreover, for $r \geq 2$,

$$\begin{aligned}
m(r) &= \Big|\Big\{(\alpha_1, \alpha_2, \ldots, \alpha_r) : \sum_{i=1}^{r} \alpha_i = 1 \text{ and } \alpha_i \in GF(q) - \{0\} \\
&\qquad\qquad \text{for } 1 \leq i \leq r\Big\}\Big| \\
&= \Big|\Big\{(\alpha_1, \alpha_2, \ldots, \alpha_{r-1}, \alpha_r) : \sum_{i=1}^{r-1} \alpha_i \neq 1 \text{ and } \alpha_r = 1 - \sum_{i=1}^{r-1} \alpha_i, \\
&\qquad\qquad \text{where } \alpha_i \in GF(q) - \{0\} \text{ for } 1 \leq i \leq r-1\Big\}\Big| \\
&= |\{(\alpha_1, \alpha_2, \ldots, \alpha_{r-1}) : \alpha_i \in GF(q) - \{0\} \quad \text{for } 1 \leq i \leq r-1\}| \\
&\quad - \Big|\Big\{(\alpha_1, \alpha_2, \ldots, \alpha_{r-1}) : \sum_{i=1}^{r-1} \alpha_i = 1 \text{ and } \alpha_i \in GF(q) - \{0\} \\
&\qquad\qquad \text{for } 1 \leq i \leq r-1\Big\}\Big|.
\end{aligned}$$

Thus

$$m(r) = (q-1)^{r-1} - m(r-1).$$

An easy induction argument now completes the proof of the lemma. □

Let $\mathcal{C}'_{r,k}$ denote the set of k-element circuits in $AG(r-1,q)$. Clearly $|\mathcal{C}'_{r,k}| = 0$ for $k < 3$, while, for $k \geq 3$, $|\mathcal{C}'_{r,k}|$ is the product of $|\mathcal{C}'_{k-1,k}|$ and the number of rank-$(k-1)$ flats in $AG(r-1,q)$. The rest of the argument closely follows the proof of (iii) of Proposition 6.1.4 and the details are left to the reader. □

We now consider $\mathcal{L}(AG(r-1,q))$. As the next result shows, this lattice is somewhat like $\mathcal{L}(PG(r-1,q))$ in that every interval of it that does not include the zero of the lattice is isomorphic to $\mathcal{L}(PG(k-1,q))$ for some k.

6.2.5 Proposition. *Let X and Y be flats of $AG(r-1,q)$ with $X \subseteq Y$ and $r(Y) - r(X) = k$. Then*

$$[X,Y] \cong \begin{cases} \mathcal{L}(AG(k-1,q)), & \text{if } X = \emptyset; \\ \mathcal{L}(PG(k-1,q)), & \text{otherwise.} \end{cases}$$

Proof. As noted earlier, every rank-k flat of $AG(r-1,q)$ is isomorphic to $AG(k-1,q)$. This proves the result when $X = \emptyset$. Now suppose $X \neq \emptyset$. We shall establish the result in the case that $Y = E(AG(r-1,q))$. The general result will then follow since, by Proposition 6.1.8, every interval in the lattice of flats of a projective geometry is also the lattice of flats of a projective geometry.

Clearly X has rank $r-k$. Let m denote the number of rank-$(r-k+1)$ flats of $AG(r-1,q)$ containing X. Now, for all x in $E(AG(r-1,q)) - X$, the set $X \cup x$ is contained in a unique rank-$(r-k+1)$ flat. As each rank-$(r-k+1)$ flat containing X has q^{r-k} elements, each such flat has $q^{r-k} - q^{r-k-1}$ elements not in X. Therefore

$$m = \left(\frac{1}{q^{r-k} - q^{r-k-1}}\right) |E(AG(r-1,q)) - X| = \frac{q^{r-1} - q^{r-k-1}}{q^{r-k} - q^{r-k-1}} = \frac{q^k - 1}{q - 1}.$$

Thus the simple matroid M associated with $AG(r-1,q)/X$ has $\frac{q^k-1}{q-1}$ elements. But this matroid has rank k and is representable over $GF(q)$. Hence, by Corollary 6.1.7, $M \cong PG(k-1,q)$ and so $\mathcal{L}(AG(r-1,q)/X) \cong \mathcal{L}(PG(k-1,q))$. □

Using Propositions 6.2.3 and 6.2.5, and arguing as in their proofs, it is not difficult to check that the various numbers shown in Figure 6.10 are correct.

Every line of $AG(r-1,q)$ has exactly q elements. Using this, together with Proposition 6.2.5, we immediately obtain the following:

6.2.6 Corollary. *Let e be an element of $AG(r-1,q)$. Then $AG(r-1,q)/e$ is isomorphic to the matroid obtained from $PG(r-2,q)$ by replacing every element by $q-1$ elements in parallel.* □

In the final result of this section, the matroid $AG(r-1,q)$ will be identified with $PG(r-1,q)\backslash H$ where H is a hyperplane of $PG(r-1,q)$.

We then consider the effect of adding an element of H to $AG(r-1,q)$ and contracting out this added element. This operation is an example of the quotient operation, which we shall examine in more detail in the next chapter.

6.2.7 Proposition. *Suppose $e \in H$ and $M = PG(r-1,q)\backslash(H-e)/e$. Then M is isomorphic to the matroid that is obtained from $AG(r-2,q)$ by replacing every element by q elements in parallel.*

Proof. This is not difficult and is left to the reader (Exercise 5). □

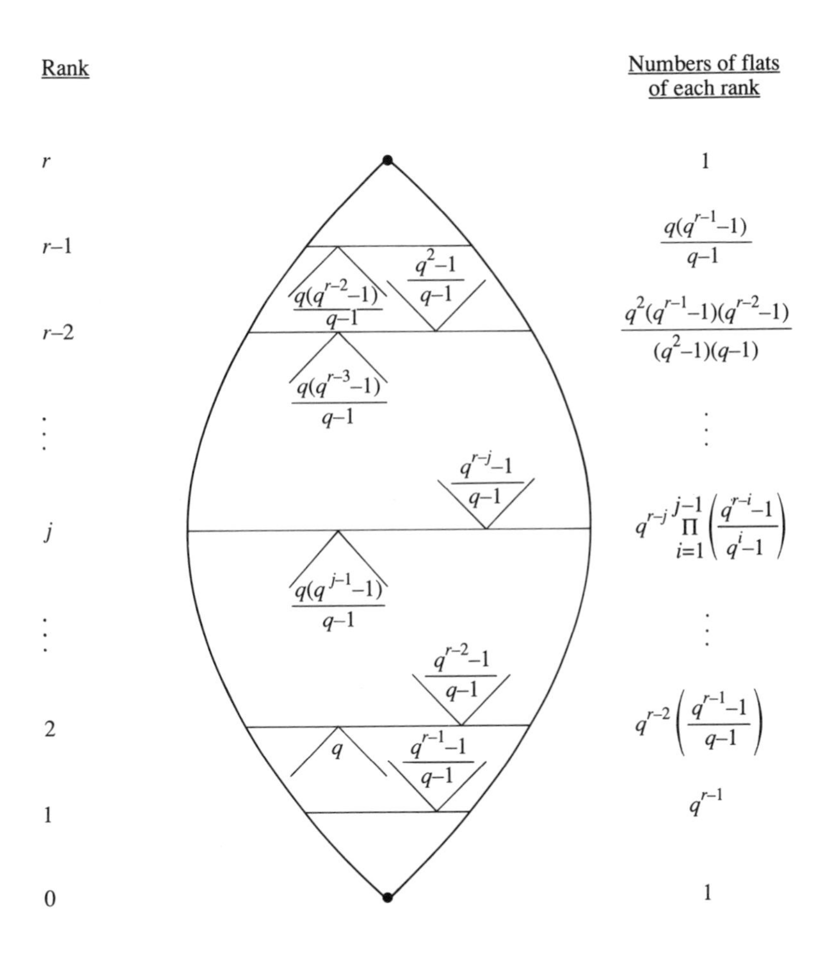

Fig. 6.10. $\mathcal{L}(AG(r-1,q))$.

Brylawski and Kelly (1980, p. 80) have noted that the last result corresponds intuitively to the fact that photographs from an infinite distance give perfect perspective.

Exercises

1. Show that $AG(1,q) \cong U_{2,q}$.

2. Show that $M(K_4)$ is not affine over $GF(2)$ or $GF(3)$ but that it is affine over all fields with more than three elements.

3. Show that $AG(r-1,q)$ has the following properties:

(i) The intersection of two distinct hyperplanes is either empty or has rank $r-2$.

(ii) There is a partition of the ground set into q sets, each of which is a hyperplane.

4. Let X and Y be flats of $AG(n,q)$ of ranks s and t, respectively. If $s \leq m \leq t$, find the number of rank-m flats Z for which $X \subseteq Z \subseteq Y$.

5. Prove Proposition 6.2.7.

6. Which of the affine geometries $AG(n,q)$ are self-dual matroids?

7. Let M be a binary matroid. Show that M is affine over $GF(2)$ if and only if $E(M)$ is a disjoint union of cocircuits.

8. Recall that if W is a subspace of a vector space V, then a *coset* of W is a subset of V of the form $\{\underline{v}+\underline{w} : \underline{w} \in W\}$ where $\underline{v}$ is some fixed member of V. View $AG(n,q)$ as the matroid on $V(n,q)$ whose independent sets are the affinely independent subsets of $V(n,q)$.

(i) Show that, for all $k \geq 1$, a set X is a rank-k flat of $AG(n,q)$ if and only if X is a coset of a $(k-1)$-dimensional subspace of $V(n,q)$.

(ii) Show that H is a hyperplane of $AG(n,q)$ if and only if there is a non-zero vector $\underline{w}$ of $V(n,q)$ and an element c of $GF(q)$ such that $H = \{\underline{v} \in V(n,q) : \underline{v} \cdot \underline{w} = c\}$.

9.* (Oxley 1978a) Let M be a $GF(q)$-representable matroid whose ground set is a disjoint union of cocircuits. Prove that M is affine over $GF(q)$.

6.3 Different matroid representations

This section will answer the question as to when two representations of a matroid are genuinely different, one of the key questions associated with matroid representability. Let M be an F-representable rank-r matroid. Roughly speaking, two representations of M are equivalent if one is the image of the other under an automorphism of $PG(r-1, F)$ where an *automorphism* of this projective geometry is a permutation of its set of points that maps lines onto lines. These automorphisms are often referred to as *collineations*. It is not difficult to check that, under an automorphism of $PG(r-1, F)$, every subspace is mapped onto one of the same dimension (Exercise 2).

Before we formally define equivalence of representations, let us recall from Section 2.2 that if A is a matrix over the field F, then the matroid $M[A]$, whose ground set is the set of column labels of A, is unchanged if one performs any of the following operations on A.

6.3.1 *Interchange two rows.*

6.3.2 *Multiply a row by a non-zero member of F*

6.3.3 *Replace a row by the sum of that row and another.*

6.3.4 *Interchange two columns (moving their labels with the columns).*

6.3.5 *Multiply a non-zero column by a non-zero member of F.*

6.3.6 *Replace each entry of the matrix by its image under some automorphism of F.*

These six operations are fundamental to the definition of equivalence of representations which we shall describe next. Let M be a rank-r matroid on the set $\{e_1, e_2, \ldots, e_n\}$ where $r \geq 1$. Suppose that A_1 and A_2 are $r \times n$ matrices over the field F with the columns of each being labelled, in some order, by $e_1, e_2, \ldots, e_n$. Assume that the identity map on $\{e_1, e_2, \ldots, e_n\}$ is an isomorphism between M and $M[A_i]$ for $i = 1, 2$. We define A_1 and A_2 to be *equivalent representations* of M if $r \leq 2$, or $r \geq 3$ and A_2 can be obtained from A_1 by a sequence of the operations 6.3.1–6.3.6. The reason why there is a dichotomy here depending on r will become clear later when we indicate the link between this definition and the automorphisms of $PG(r-1, F)$.

Two remarks are in order here. Firstly, we note that equivalence of representations for rank-0 matroids has not been defined. In this trivial case, the representation problem is completely solved: every such matroid is representable over every field, the representations consisting of all of the zero matrices with the appropriate number of columns. Secondly, operations 6.3.1–6.3.6 are but six of seven matrix operations that were noted in Section 2.2 as not altering the associated vector matroid. The seventh operation, deletion of a zero row, was not incorporated into the definition of equivalence because we want equivalent representations to have the same size.

We now examine more closely what it means for two representations to be equivalent with a view to relating this definition to projective space automorphisms. The straightforward proof of the next result is left to the reader. The operation of *column scaling* referred to in this result is operation 6.3.5. Similarly, operation 6.3.2 is called *row scaling*.

6.3.7 Lemma. *Let A_1 and A_2 be F-representations of a matroid of rank at least three. Then A_1 and A_2 are equivalent if and only if A_2 can be obtained from A_1 by the following sequence of operations: a single column permutation; a sequence of elementary row operations; a single application of an automorphism of F to all the entries of the matrix; and a sequence of column scalings.* □

Focusing for the moment on the sequence of elementary row operations that are needed here, we recall the following well-known result from linear algebra (see, for example, Hungerford 1974).

6.3.8 Lemma. *Let A and A' be $r \times n$ matrices over the field F. Then the following statements are equivalent:*

(a) *A' can be obtained from A by a sequence of elementary row operations.*

(b) *A' can be obtained by multiplying A on the left by a non-singular $r \times r$ matrix.*

(c) *There is a non-singular linear transformation τ on $V(r, F)$ such that, for all i in $\{1, 2, \ldots, n\}$, the* ith *column of A' is the image under τ of the* ith *column of A.* □

In the sequence of operations used in Lemma 6.3.7 to obtain A_2 from A_1, the sequence of elementary row operations was followed by the application of an F-automorphism to all the entries of the matrix. In general, a permutation σ of $V(r, F)$ is called a *semilinear transformation* if there is a non-singular linear transformation τ of $V(r, F)$ and an automorphism α of F such that if $\underline{v} \in V(r, F)$ and $\tau(\underline{v}) = (w_1, w_2, \ldots, w_r)$, then $\sigma(\underline{v}) = (\alpha(w_1), \alpha(w_2), \ldots, \alpha(w_r))$. It is not difficult to characterize semilinear transformations in terms of a weakened linearity condition (see Exercise 1).

For matroids of rank at least three, we now reformulate the definition of equivalence of representations in terms of semilinear transformations.

6.3.9 Proposition. *Let M be a rank-r matroid on the set $\{e_1, e_2, \dots, e_n\}$ and suppose that $r \geq 3$. Let A_1 and A_2 be $r \times n$ matrices over the field F such that, for $i = 1, 2$, the map that takes e_j to the jth column $\underline{v}_j^{(i)}$ of A_i is an isomorphism from M to $M[A_i]$. Then A_1 and A_2 are equivalent representations of M if and only if there is a semilinear transformation σ of $V(r, F)$ and a sequence $c_1, c_2, \dots, c_n$ of non-zero elements of F such that $\underline{v}_j^{(2)} = c_j \sigma(\underline{v}_j^{(1)})$ for all j in $\{1, 2, \dots, n\}$.*

Proof. This follows easily from the last two lemmas and the preceding discussion. The details are left to the reader. □

The following result, sometimes known as the Fundamental Theorem of Projective Geometry, indicates the importance of semilinear transformations.

6.3.10 Theorem. *If $r \geq 3$, then every automorphism of $PG(r-1, F)$ is induced by a semilinear transformation of $V(r, F)$.*

Proof. This can be found in Artin (1957) and Baer (1952). □

We are now in a position to indicate the precise relationship between equivalence of representations and projective space automorphisms, thereby justifying the definition of the former. Let A_1 and A_2 be $r \times n$ matrices representing the rank-r matroid M over the field F and suppose that $r \geq 1$. Let ψ_1 and ψ_2 be the F-coordinatizations ψ_{A_1} and ψ_{A_2} that correspond to A_1 and A_2, respectively (see Example 6.1.2). Then $\psi_i : E(M) \to V(r, F)$ for $i = 1, 2$. Moreover, ψ_i maps every loop of M to $\underline{0}$. Now let $T_i = \{\langle \psi_i(e) \rangle$: e is a non-loop element of $M\}$. Define $\theta : T_1 \to T_2$ by $\theta(\langle \psi_1(e) \rangle) = \langle \psi_2(e) \rangle$ for all non-loop elements e of M. Then θ is well-defined and can be viewed as a map between two subsets of $PG(r-1, F)$.

6.3.11 Corollary. *The matrices A_1 and A_2 are equivalent representations of M if and only if θ is the restriction to T_1 of an automorphism of $PG(r-1, F)$.*

Proof. If $r \geq 3$, the result follows immediately on combining Proposition 6.3.9 and Theorem 6.3.10. If $r \leq 2$, then every permutation of the point set of $PG(r-1, F)$ is an automorphism of the projective geometry and again the result is immediate. □

The next example illustrates the use of the last corollary to show that two particular representations are not equivalent.

6.3.12 Example. Consider the following matrices over $\mathbb{R}$:

$$A_1 = \begin{array}{c} \begin{matrix} 1 & 2 & 3 & \;\; 4 & 5 & 6 \end{matrix} \\ \left[\begin{array}{ccc|ccc} & & & 0 & 1 & 1 \\ & I_3 & & 1 & 2 & 1 \\ & & & 2 & 2 & 1 \end{array}\right] \end{array}, \qquad A_2 = \begin{array}{c} \begin{matrix} 1 & 2 & 3 & \;\; 4 & 5 & 6 \end{matrix} \\ \left[\begin{array}{ccc|ccc} & & & 0 & 1 & 1 \\ & I_3 & & 1 & 2 & 1/2 \\ & & & 2 & 2 & 1/2 \end{array}\right] \end{array}.$$

We shall view the geometric representations for $M[A_1]$ and $M[A_2]$ shown in Figure 6.11 as subsets of $PG(2, \mathbb{R})$ where certain additional points have been included in these diagrams. Clearly the identity map on the set $\{1, 2, \ldots, 6\}$ is an isomorphism between $M[A_1]$ and $M[A_2]$, so A_1 and A_2 are both representations for $M[A_1]$. However, they are not equivalent representations for, as we can see with the aid of the point 7, which corresponds to $(1, 1, 2)^T$, there is no collineation of $PG(2, \mathbb{R})$ that acts as the identity on $\{1, 2, \ldots, 6\}$. □

A rank-r matroid M on an n-element set is called *uniquely F-representable* if all of the $r \times n$ matrices representing M over F are equivalent. As we shall see in Chapter 10, a crucial step in the proof of the excluded-minor characterization of ternary matroids is the fact that such matroids are uniquely $GF(3)$-representable. The field $GF(3)$, like all the fields $GF(p)$ for p prime, has no non-identity automorphisms. To see this, recall that the automorphism group of the field $GF(p^k)$ is $\{x \to x^{p^i} : i \in \{0, 1, 2, \ldots, k-1\}\}$ (see, for example, Hungerford 1974). This group is generated by the map $x \to x^p$ and is cyclic of order k. In particular, if k is 1, the automorphism group is trivial. Hence, every semilinear transformation of $V(r, p)$ is actually just a non-singular linear transformation. Thus, in the case of both binary and ternary matroids, we shall not need to worry at all about operation 6.3.6. Indeed, for all

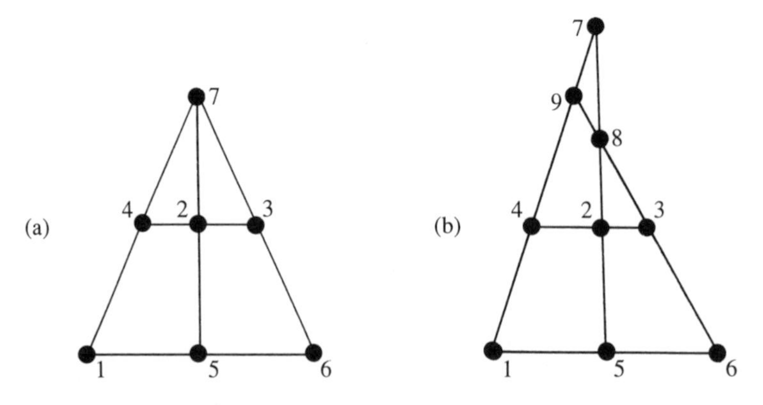

Fig. 6.11. Different embeddings of $\{1, 2, \ldots, 6\}$ in $PG(2, \mathbb{R})$.

prime fields $GF(p)$, we have the following characterization of equivalent representations (Brylawski and Lucas 1976). This comes from combining 6.3.7–6.3.10 with the observation that multiplying a matrix on the right by a non-singular diagonal matrix is equivalent to performing a sequence of column scalings.

6.3.13 Proposition. *Suppose that $r \geq 3$ and let A_1 and A_2 be $r \times n$ matrices over a field F with the columns of each matrix being labelled, in order, by $e_1, e_2, \ldots, e_n$. Suppose that F has no non-identity automorphisms. Then A_1 and A_2 are equivalent representations of a matroid on $\{e_1, e_2, \ldots, e_n\}$ if and only if there is a non-singular $r \times r$ matrix X and a non-singular $n \times n$ diagonal matrix Y such that $A_2 = XA_1Y$.* □

Exercises

1. Consider the set S of semilinear transformations of $V(r, F)$.

   (i) Show that the members of S are precisely the permutations η of $V(r, F)$ for which there is an automorphism α of F such that
   $$\eta(\underline{x} + \underline{y}) = \eta(\underline{x}) + \eta(\underline{y}) \text{ for all } \underline{x},\ \underline{y} \text{ in } V(r, F),$$
   and $\eta(c\underline{x}) = \alpha(c)\eta(\underline{x})$ for all $\underline{x}$ in $V(r, F)$ and all c in F.

   (ii) Prove that S forms a group under composition.

2. Let ζ be a permutation of the set of points of $PG(r-1, F)$. Show that:

   (i) if ζ is an automorphism of $PG(r-1, F)$ and X is a subspace of $PG(r-1, F)$, then both $\zeta(X)$ and $\zeta^{-1}(X)$ are subspaces having the same dimension as X;

   (ii) ζ is an automorphism of $PG(r-1, F)$ if and only if it maps subspaces onto subspaces.

3. Show that Proposition 6.3.13 can fail if the restriction that $r \geq 3$ is dropped.

4. Let the elements of $GF(4)$ be 0, 1, ω, and $\omega + 1$ where $\omega^2 = \omega + 1$. Show that the following matrices are equivalent $GF(4)$-representations for Q_6, this matroid being as shown in Figure 2.1:

$$\begin{array}{c} \begin{array}{cccccc} 6 & 3 & 5 & 2 & 4 & 1 \end{array} \\ \left[\begin{array}{c|ccc} & 1 & 0 & 1 \\ I_3 & 1 & 1 & \omega \\ & \omega+1 & 1 & \omega+1 \end{array}\right] \end{array}, \quad \begin{array}{c} \begin{array}{cccccc} 1 & 4 & 3 & 6 & 2 & 5 \end{array} \\ \left[\begin{array}{c|ccc} & 1 & 1 & 0 \\ I_3 & \omega & 0 & 1 \\ & 1 & \omega & 1 \end{array}\right] \end{array}.$$

5. (i) Show that the matroid in Figure 6.11(a) is a ternary affine matroid by exhibiting it as a restriction of $AG(2,3)$.

(ii) Show that the matroid in Figure 6.11(b) is non-ternary by finding a non-ternary proper minor of it.

6. (i) Show that the following matrices are equivalent $GF(2)$-representations of the same matroid, S_8:

$$\begin{array}{cccccccc} 1 & 2 & 3 & 4 & 5 & 6 & 7 & 8 \end{array} \qquad \begin{array}{cccccccc} 4 & 7 & 3 & 1 & 5 & 6 & 2 & 8 \end{array}$$
$$\left[\begin{array}{c|cccc} & 0 & 1 & 1 & 0 \\ I_4 & 1 & 0 & 1 & 1 \\ & 1 & 1 & 0 & 1 \\ & 1 & 1 & 1 & 0 \end{array}\right], \quad \left[\begin{array}{c|cccc} & 0 & 1 & 1 & 1 \\ I_4 & 1 & 0 & 1 & 1 \\ & 1 & 1 & 0 & 1 \\ & 1 & 1 & 1 & 1 \end{array}\right].$$

(ii) Show that S_8 is self-dual.

(iii) Show that S_8 has a unique element e such that $S_8/e \cong F_7$ and a unique element f such that $S_8\backslash f \cong F_7^*$.

6.4 Constructing representations for matroids

In this section we describe explicitly how to construct representations for matroids, and we illustrate this technique on a number of examples. In general, finding whether a matroid is representable over some field will require a lot of computation. The question of the complexity of this problem will not be addressed here, although in Section 9.3, we shall prove a result of Seymour (1981a) that there is no polynomially bounded algorithm to decide whether or not a given matroid is binary.

Recall from Chapter 2 that every $r \times n$ matrix that is an F-representation of a rank-r matroid M is equivalent to a standard representative matrix $[I_r|D]$ for M. In view of this fact, we shall now concentrate on matrices of this type. Given such a matrix, let its columns be labelled, in order, $e_1, e_2, \ldots, e_n$. Let B be the basis $\{e_1, e_2, \ldots, e_r\}$ of M. For all i in $\{1, 2, \ldots, r\}$, the unique non-zero entry in column i of $[I_r|D]$ is in row i. Thus it is natural to label the rows of $[I_r|D]$ by $e_1, e_2, \ldots, e_r$. Hence D has its rows labelled by $e_1, e_2, \ldots, e_r$ and its columns labelled by $e_{r+1}, e_{r+2}, \ldots, e_n$. Evidently, for all k in $\{r+1, r+2, \ldots, n\}$, the fundamental circuit $C(e_k, B)$ is $e_k \cup \{e_i : e_i \in B$ and D has a non-zero entry in row e_i and column $e_k\}$. Thus, if we let $D^\#$ be the matrix obtained from D by replacing each non-zero entry of D by a 1, then the columns of $D^\#$ are precisely the incidence vectors of the sets $C(e_k, B) - e_k$ where the rows and columns of $D^\#$ inherit their labels from D. This matrix $D^\#$ is called the *B-fundamental-circuit incidence matrix* of M

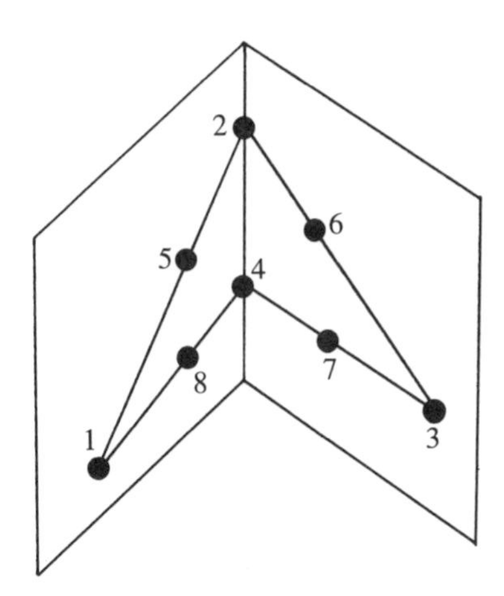

Fig. 6.12. $\mathcal{W}^4$.

(Brylawski and Lucas 1976). In addition $[I_r|D^\#]$ is sometimes called a *partial representation* for M (Truemper 1984a). This name is justified by the following result which is easily proved using fundamental circuits.

6.4.1 Proposition. *Suppose that $[I_r|D_1]$ and $[I_r|D_2]$ are matrices over the field F with the columns of each labelled, in order, by $e_1, e_2, \ldots, e_n$. Assume that the identity map on $\{e_1, e_2, \ldots, e_n\}$ is an isomorphism of $M[I_r|D_1]$ and $M[I_r|D_2]$. Then $D_1^\# = D_2^\#$.* □

Note that an immediate consequence of this is that a binary matroid is uniquely $GF(2)$-representable. In fact, as we shall see in Chapter 10, if F is an arbitrary field and M is an F-representable matroid that is also binary, then M is uniquely F-representable.

Partial representations share many of the properties of representations. For example, using Proposition 2.1.11, it is straightforward to prove the following analogue of Theorem 2.2.8. Some other properties of partial representations will be looked at in the exercises.

6.4.2 Proposition. *Let M be a matroid on the set E and B be a basis of M. If X is the B-fundamental-circuit incidence matrix of M, then X^T is the $(E-B)$-fundamental-circuit incidence matrix of M^*.* □

6.4.3 Example. Let M be the matroid $\mathcal{W}^4$ for which a geometric representation is shown in Figure 6.12 and let $B = \{1, 2, 3, 4\}$. Then B is a basis for $\mathcal{W}^4$. The B-fundamental-circuit incidence matrix of M and the $(E-B)$-fundamental-circuit incidence matrix of M^* are the matrices X and Y in Figure 6.13. Evidently $Y = X^T$.

$$X = \begin{array}{c} \\ 1 \\ 2 \\ 3 \\ 4 \end{array} \begin{array}{c} \begin{array}{cccc} 5 & 6 & 7 & 8 \end{array} \\ \left[\begin{array}{cccc} 1 & 0 & 0 & 1 \\ 1 & 1 & 0 & 0 \\ 0 & 1 & 1 & 0 \\ 0 & 0 & 1 & 1 \end{array}\right] \end{array} \qquad Y = \begin{array}{c} \\ 5 \\ 6 \\ 7 \\ 8 \end{array} \begin{array}{c} \begin{array}{cccc} 1 & 2 & 3 & 4 \end{array} \\ \left[\begin{array}{cccc} 1 & 1 & 0 & 0 \\ 0 & 1 & 1 & 0 \\ 0 & 0 & 1 & 1 \\ 1 & 0 & 0 & 1 \end{array}\right] \end{array}$$

Fig. 6.13. Partial representations for $\mathcal{W}^4$ and its dual.

We remark that if we modify $\mathcal{W}^4$ by requiring that 5, 6, 7, and 8 be coplanar, then the matroid we obtain is graphic. We leave it to the reader to identify exactly which graphic matroid it is. □

The Fano and non-Fano matroids, F_7 and F_7^-, are of fundamental importance in matroid theory. Figure 6.14(a) gives geometric representations for both F_7 and F_7^-. In the former, $\{4, 5, 6\}$ is a line; in the latter, it is not. For each of these matroids, we aim to determine the fields over which it is representable. As a first step towards this, let $M \in \{F_7, F_7^-\}$ and let $B = \{1, 2, 3\}$. Then B is a basis of M, and the B-fundamental-circuit incidence matrix X of M is as in Figure 6.14(b). In Example 2.2.10, it was noted, though not proved, that $[I_3|X]$ can represent F_7 or F_7^- depending on the field over which it is viewed. This follows immediately from the next result, the proof of which *will* be given.

6.4.4 Lemma. *Let $A = [I_3|X]$ where X is the matrix in Figure 6.14(b) viewed over the field F. If the characteristic of F is two, then $M[A] = F_7$, while if the characteristic of F is not two, then $M[A] = F_7^-$.*

Proof. Let M be F_7 if F has characteristic two, and F_7^- otherwise. To prove that $M = M[A]$, we first perform the routine check that each 3-point line of M is a circuit of $M[A]$. It then remains to show that $M[A]$ has no other circuits of size less than four. To achieve this, we note that, as each of the unbroken 3-point lines in Figure 6.14(a) is certainly a circuit

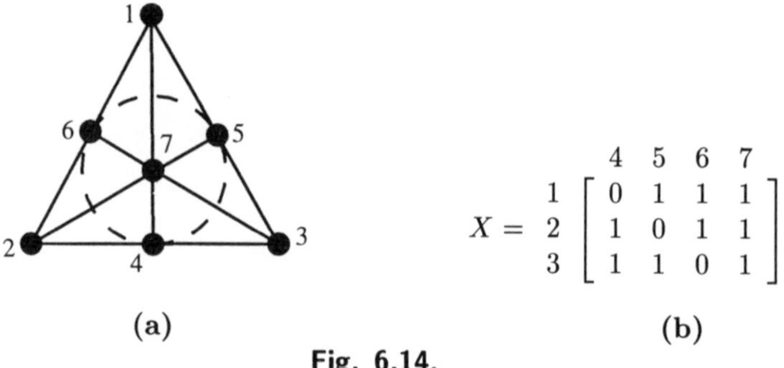

$$X = \begin{array}{c} \\ 1 \\ 2 \\ 3 \end{array} \begin{array}{c} \begin{array}{cccc} 4 & 5 & 6 & 7 \end{array} \\ \left[\begin{array}{cccc} 0 & 1 & 1 & 1 \\ 1 & 0 & 1 & 1 \\ 1 & 1 & 0 & 1 \end{array}\right] \end{array}$$

Fig. 6.14.

of $M[A]$, and $M[A]$ has rank three, there are very few other possibilities for circuits of $M[A]$ of size less than four. For instance, if $\{4, 6, 7\}$ is a circuit, then 1, 4, 6, and 7 are all collinear. But 3 is collinear with 6 and 7, and 2 is collinear with 1 and 6, so 1, 4, 6, 7, 3, and 2 are collinear. Hence, as 5 is collinear with 1 and 3, all seven elements are collinear; that is, $M[A]$ has collapsed to rank two; a contradiction. Using this idea, it is not difficult to show that if F has characteristic two, then $M[A]$ has no other circuits of size less than four and so $M[A] = F_7$. Similarly, if F has characteristic other than two, then the only other possible circuit of $M[A]$ of size less than four is $\{4, 5, 6\}$. But the 3×3 submatrix of X whose columns are 4, 5, and 6 has determinant 2. Thus $\{4, 5, 6\}$ is independent in $M[A]$, and hence, $M[A] = F_7^-$. □

Later in the section, we shall complete the determination of all the fields over which each of F_7 and F_7^- is representable. In the last example, we used determinants to decide about the independence of a set of vectors. The next result uses this elementary idea again to get an extension of Proposition 6.4.1. The proof follows Brylawski and Lucas (1976).

6.4.5 Proposition. *Let $[I_r|D_1]$ and $[I_r|D_2]$ be $r \times n$ matrices over the fields F and F', respectively, with the columns of each matrix being labelled, in order, by $e_1, e_2, \ldots, e_n$. Then the identity map on $\{e_1, e_2, \ldots, e_n\}$ is an isomorphism from $M[I_r|D_1]$ to $M[I_r|D_2]$ if and only if, whenever D_1' and D_2' are corresponding square submatrices of D_1 and D_2 respectively,* $\det D_1' = 0$ *exactly when* $\det D_2' = 0$.

Proof. We shall prove the result in one direction. The proof for the converse can be obtained by reversing this argument. Assume that the identity map on $\{e_1, e_2, \ldots, e_n\}$ is an isomorphism from $M[I_r|D_1]$ to $M[I_r|D_2]$. Let D_1' and D_2' be the square submatrices of D_1 and D_2 whose rows and columns are labelled by $\{e_{i_1}, e_{i_2}, \ldots, e_{i_k}\}$ and $\{e_{j_1}, e_{j_2}, \ldots, e_{j_k}\}$, respectively. Then the following statements are equivalent:

(i) $\det D_1' \neq 0$;

(ii) $(\{e_1, e_2, \ldots, e_r\} - \{e_{i_1}, e_{i_2}, \ldots, e_{i_k}\}) \cup \{e_{j_1}, e_{j_2}, \ldots, e_{j_k}\}$ is a basis of $M[I_r|D_1]$;

(iii) $(\{e_1, e_2, \ldots, e_r\} - \{e_{i_1}, e_{i_2}, \ldots, e_{i_k}\}) \cup \{e_{j_1}, e_{j_2}, \ldots, e_{j_k}\}$ is a basis of $M[I_r|D_2]$;

(iv) $\det D_2' \neq 0$. □

Next we describe how to construct representations. Let M be a rank-r matroid and B be the basis $\{e_1, e_2, \ldots, e_r\}$ for M. Let X be the B-fundamental-circuit incidence matrix of M, the columns of this matrix being labelled by $e_{r+1}, e_{r+2}, \ldots, e_n$. We know from Proposition 6.4.1 that if $[I_r|D]$ is an F-representation of M with columns labelled, in order, $e_1, e_2, \ldots, e_n$, then $[I_r|D^{\#}] = [I_r|X]$. Thus the task of finding an F-representation for M can be viewed as being one of finding the specific elements of F that correspond to the non-zero elements of $D^{\#}$. By column scaling, we know that the first non-zero entry in each column of D can be taken to be one. Moreover, by row scaling, the first non-zero entry in each row can be taken to be one. The next theorem (Brylawski and Lucas 1976) gives a best-possible lower bound on the number of entries of D whose values can be predetermined by row and column scaling. Before stating it, we shall require some preliminaries.

Consider the matrix $D^{\#}$. Its rows are indexed by $e_1, e_2, \ldots, e_r$ and its columns by $e_{r+1}, e_{r+2}, \ldots, e_n$. Let $G(D^{\#})$ denote the *associated* simple *bipartite graph*, that is, $G(D^{\#})$ has vertex classes $\{e_1, e_2, ..., e_r\}$ and $\{e_{r+1}, e_{r+2}, \ldots, e_n\}$, and two vertices e_i and e_j are adjacent if and only if the entry in row e_i and column e_j of $D^{\#}$ is 1.

6.4.6 Example. If $D^{\#}$ is the matrix in Figure 6.15(a), then $G(D^{\#})$ is as shown in Figure 6.15(b) where, for example, e_2 and e_5 are adjacent because the entry in the row indexed by e_2 and the column indexed by e_5 is 1. Similarly, e_3 and e_8 are non-adjacent because the (e_3, e_8)-entry is 0. □

6.4.7 Theorem. *Let the $r \times n$ matrix $[I_r|D_1]$ be an F-representation for the matroid M. Let $\{b_1, b_2, \ldots, b_k\}$ be a basis of the cycle matroid of $G(D_1^{\#})$. Then $k = n - \omega(G(D_1^{\#}))$. Moreover, if $(\theta_1, \theta_2, \ldots, \theta_k)$ is an ordered k-tuple of non-zero elements of F, then M has an F-representation $[I_r|D_2]$ that is equivalent to $[I_r|D_1]$ such that, for each i in $\{1, 2, \ldots, k\}$, the entry of D_2 corresponding to b_i is θ_i. Indeed, $[I_r|D_2]$ can be obtained from $[I_r|D_1]$ by a sequence of row and column scalings.*

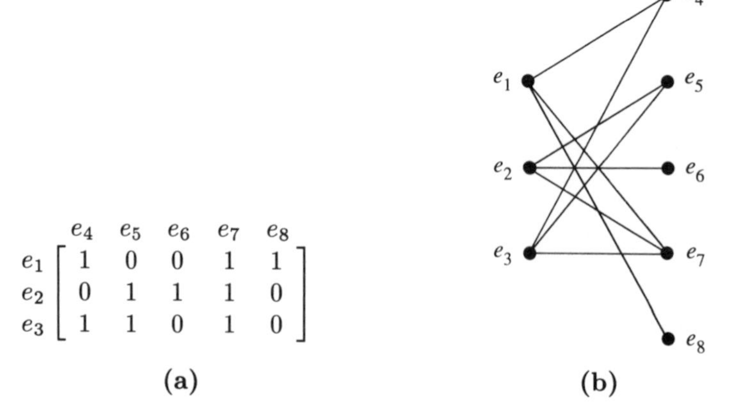

Fig. 6.15. (a) A matrix and (b) its associated bipartite graph.

Before proving this, consider Example 6.4.6 again. The boxed entries in $D^{\#}$ below correspond to a basis of the cycle matroid of $G(D^{\#})$.

$$\begin{array}{c} \\ e_1 \\ e_2 \\ e_3 \end{array} \begin{array}{c} \begin{array}{ccccc} e_4 & e_5 & e_6 & e_7 & e_8 \end{array} \\ \left[\begin{array}{ccccc} \boxed{1} & 0 & 0 & \boxed{1} & \boxed{1} \\ 0 & \boxed{1} & \boxed{1} & \boxed{1} & 0 \\ \boxed{1} & 1 & 0 & 1 & 0 \end{array}\right] \end{array}$$

Hence if $[I_3|D]$ is an F-representation of a matroid M, then, by Theorem 6.4.7, we can independently assign arbitrary non-zero elements of F to each of these boxed entries. In particular, we can assume that, in D, each of these entries is 1. In this case, then, there are only two non-zero entries of D whose values we are not free to arbitrarily assign.

Proof of Theorem 6.4.7. The fact that $k = n - \omega(G(D_1^{\#}))$ is immediate from 1.3.7. To prove the second part, suppose that G_1 is a forest that is a subgraph of $G(D_1^{\#})$ and, for each edge x of G_1, let $\theta(x)$ be a non-zero element of F. We shall show, by induction on $|E(G_1)|$, that, by a sequence of row and column scalings of $[I_r|D_1]$, we can obtain a matrix $[I_r|D_2]$ such that, for all edges x of G_1, the entry in D_2 corresponding to x is $\theta(x)$. This is trivially true for $|E(G_1)| = 0$. Assume it true for $|E(G_1)| < m$ and let $|E(G_1)| = m \geq 1$. As G_1 is a forest with at least one edge, it has a vertex v of degree one. Let y be the edge of G_1 incident with v. On applying the induction assumption to $G_1 \backslash y$, it follows that, by appropriate row and column scaling of $[I_r|D_1]$, we obtain a matrix $[I_r|D_2]$ such that, for all edges x of $G_1 \backslash y$, the entry of D_2 corresponding to x is $\theta(x)$.

Now the vertex v of $G(D_1^{\#})$ corresponds to a row or a column of D_2. In the former case, as y is the only edge of G_1 meeting v, none of the entries of D_2 corresponding to edges of $G_1 \backslash y$ is in this row. We can therefore multiply this row by an appropriate non-zero scalar t to make the y-entry equal to $\theta(y)$ without changing any of the $(G_1 \backslash y)$-entries. This multiplication may alter the entry in row v of I_r. But multiplying the corresponding column of I_r by $1/t$ will fix this without affecting any other entries.

If the vertex v corresponds to a column of D_2, then, by taking an appropriate non-zero scalar multiple of this column, we can make the y-entry equal to $\theta(y)$ without altering any of the $(G_1 \backslash y)$-entries. The required result now follows by induction. □

The last result enables us to complete our determination of the fields over which F_7 is representable and the fields over which F_7^- is representable.

6.4.8 Proposition. *Let F be a field. Then*

(i) *F_7 is F-representable if and only if the characteristic of F is two; and*

(ii) *F_7^- is F-representable if and only if the characteristic of F is not two.*

By Lemma 6.4.4, to prove this result it suffices to establish the following.

6.4.9 Lemma. *If $M \in \{F_7, F_7^-\}$ and M is F-representable, then $M = M[I_3|X]$ where X is the matrix in Figure 6.14(b).*

Proof. The matrix X is the B-fundamental-circuit incidence matrix for M where B is the basis $\{1,2,3\}$ of M. Thus if $[I_3|D]$ is an F-representation for M, then

$$D^{\#} = X = \begin{array}{c} \\ 1 \\ 2 \\ 3 \end{array}\begin{array}{c} \begin{array}{cccc} 4 & 5 & 6 & 7 \end{array} \\ \begin{bmatrix} 0 & \boxed{1} & \boxed{1} & \boxed{1} \\ \boxed{1} & 0 & 1 & \boxed{1} \\ 1 & 1 & 0 & \boxed{1} \end{bmatrix} \end{array}$$

where the boxed entries correspond to a basis for $M(G(D^{\#}))$. Hence there are non-zero elements a, b, and c of F such that $M = M[I_3|D_1]$ where D_1 is as shown in Figure 6.16(a). Since $\{1,4,7\}$ is a circuit of M, the submatrix of $[I_3|D_1]$ shown in Figure 6.16(b) has determinant 0. Thus $a = 1$. Similarly, as $\{2,5,7\}$ and $\{3,6,7\}$ are circuits of M, both b and c equal one. We conclude that, as required, $D_1 = X$. □

$$D_1 = \begin{array}{c} \\ 1 \\ 2 \\ 3 \end{array}\begin{array}{c} \begin{array}{cccc} 4 & 5 & 6 & 7 \end{array} \\ \begin{bmatrix} 0 & 1 & 1 & 1 \\ 1 & 0 & c & 1 \\ a & b & 0 & 1 \end{bmatrix} \end{array}$$

(a)

$$\begin{array}{c} \begin{array}{ccc} 1 & 4 & 7 \end{array} \\ \begin{bmatrix} 1 & 0 & 1 \\ 0 & 1 & 1 \\ 0 & a & 1 \end{bmatrix} \end{array}$$

(b)

Fig. 6.16.

We showed in Proposition 6.1.10 that the Vámos matroid V_8 is not representable. In fact, it is a consequence of the following result of Fournier (1971) that V_8 is a smallest non-representable matroid.

6.4.10 Proposition. *Let M be a matroid having fewer than eight elements. Then M is representable.* □

The Vámos matroid is not the only non-representable 8-element matroid, though as Fournier (1971) also showed, all such matroids have rank four.

6.4.11 Example. (Blackburn, Crapo, and Higgs 1973) Consider the following representation for $AG(3,2)$:

$$\begin{array}{cccc|cccc} 1 & 2 & 3 & 4 & 5 & 6 & 7 & 8 \\ \end{array}$$
$$\left[\begin{array}{c|cccc} & 0 & 1 & 1 & 1 \\ I_4 & 1 & 0 & 1 & 1 \\ & 1 & 1 & 0 & 1 \\ & 1 & 1 & 1 & 0 \end{array}\right].$$

Evidently $\{1,2,3,8\}$ is a circuit of this matroid. From the geometric representation of $AG(3,2)$ in Figure 1.14, we know that every circuit of $AG(3,2)$ is also a hyperplane. Let M' be the matroid obtained from $AG(3,2)$ by relaxing $\{1,2,3,8\}$. By Proposition 3.3.9(ii), $M'\backslash 8 = AG(3,2)\backslash 8$. Clearly the last matroid is isomorphic to F_7^*. Hence if M' is F-representable, then F has characteristic two. But, by Proposition 3.3.9(ii), $M'/1$ can be obtained from $AG(3,2)/1$ by relaxing the circuit–hyperplane $\{2,3,8\}$ of the latter. Moreover, by Proposition 6.2.5, $AG(3,2)/1 \cong F_7$. Thus $M'/1 \cong F_7^-$. Hence M' is not representable over any characteristic-two fields. We conclude that M' is not representable. Since M' has thirteen 4-circuits, but V_8 has only five such circuits, M' is a non-representable matroid that is different from V_8. □

In Example 2.2.25, we considered the vector matroid of the following matrix over $GF(3)$:

$$\begin{array}{cccc|cccc} 1 & 2 & 3 & 4 & 5 & 6 & 7 & 8 \\ \end{array}$$
$$\left[\begin{array}{c|cccc} & 0 & 1 & 1 & -1 \\ I_4 & 1 & 0 & 1 & 1 \\ & 1 & 1 & 0 & 1 \\ & -1 & 1 & 1 & 0 \end{array}\right].$$

It was shown there that this matroid, call it P_8, is self-dual. It is not difficult to check that P_8 has no circuits of size less than four and that its 4-circuits are $\{1,2,3,8\}$, $\{1,2,4,7\}$, $\{1,3,4,6\}$, $\{2,3,4,5\}$, $\{1,4,5,8\}$, $\{2,3,6,7\}$, $\{1,5,6,7\}$, $\{2,5,6,8\}$, $\{3,5,7,8\}$, and $\{4,6,7,8\}$. The next result (Oxley 1986a) uses Theorem 6.4.7 to show that P_8 has another special property. Part of the proof of this shows that P_8 has a transitive automorphism group. In general, for a matroid M, an *automorphism* is a permutation σ of $E(M)$ such that $r(X) = r(\sigma(X))$ for all $X \subseteq E(M)$. The set of automorphisms of M forms a group under composition. This automorphism group is *transitive* if, for every two elements x and y of M, there is an automorphism that maps x to y.

6.4.12 Proposition. *Let k be an integer exceeding one. Then P_8 is a minor-minimal matroid that is not representable over $GF(2^k)$.*

$$A = \left[\begin{array}{c|cccc} & 5 & 6 & 7 & 8 \\ & 1 & 0 & 1 & 1 \\ I_3 & 1 & 1 & 0 & 1 \\ & -1 & 1 & 1 & 0 \end{array}\right]$$

(columns of I_3 labelled 2 3 4)

(a)

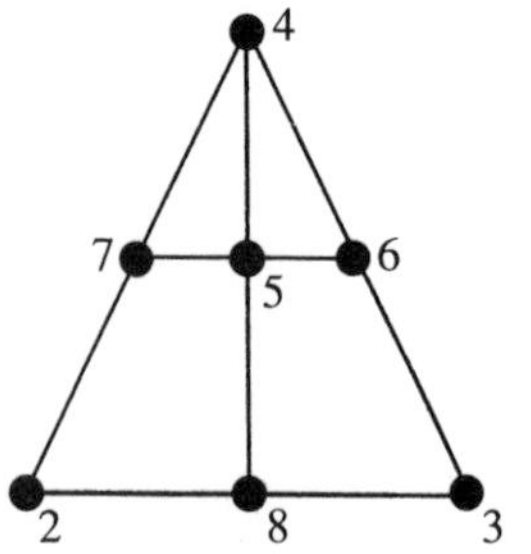

(b)

Fig. 6.17. Matrix and geometric representations for $P_8/1$.

Proof. From the list of 4-circuits of P_8, it is easy to check that the permutations $\sigma_1 = (1,8,4,5)(2,7,3,6)$ and $\sigma_2 = (1,2,4,3)(5,6,8,7)$ are both automorphisms of P_8. By using products of these two automorphisms, it is easy to show that the automorphism group of P_8 is transitive. Therefore, for all e in $E(P_8)$, we have $P_8/e \cong P_8/1$. The last matroid is represented over $GF(3)$ by the matrix A in Figure 6.17(a).

Consider the matroid P_7 for which a geometric representation is shown in Figure 6.17(b). By showing that each 3-point line of P_7 is a circuit in $M[A]$ and that the latter has no 1- or 2-circuits, we get that $P_7 \cong M[A]$.

6.4.13 Lemma. *The matroid P_7 is representable over a field F if and only if $|F| \geq 3$.*

Proof. If $B = \{2,3,4\}$, then the B-fundamental-circuit incidence matrix of P_7 is

$$X = \begin{array}{c} \\ 2 \\ 3 \\ 4 \end{array}\begin{array}{c} \begin{array}{cccc} 5 & 6 & 7 & 8 \end{array} \\ \left[\begin{array}{cccc} \boxed{1} & 0 & \boxed{1} & \boxed{1} \\ \boxed{1} & \boxed{1} & 0 & 1 \\ 1 & \boxed{1} & 1 & 0 \end{array}\right] \end{array}.$$

The boxed entries here correspond to a basis of $M(G(X))$ where $G(X)$ is the bipartite graph associated with X. Suppose that P_7 is F-representable for some field F. Then, by Theorem 6.4.7, F has non-zero elements a, b, and c such that $P_7 \cong M[I_3|D]$ where D is as shown in Figure 6.18(a).

$$D = \begin{array}{c} \\ 2 \\ 3 \\ 4 \end{array}\begin{array}{c} \begin{array}{cccc} 5 & 6 & 7 & 8 \end{array} \\ \left[\begin{array}{cccc} 1 & 0 & 1 & 1 \\ 1 & 1 & 0 & c \\ a & 1 & b & 0 \end{array}\right] \end{array} \qquad \begin{array}{c} \begin{array}{ccc} 4 & 5 & 8 \end{array} \\ \left[\begin{array}{ccc} 0 & 1 & 1 \\ 0 & 1 & c \\ 1 & a & 0 \end{array}\right] \end{array}$$

(a) (b)

Fig. 6.18.

Since $\{4,5,8\}$ is a circuit of P_7, the submatrix of $[I_3|D]$ shown in Figure 6.18(b) has determinant 0. Thus $c = 1$. A similar argument, using the circuit $\{5,6,7\}$ of P_7, shows that $b = a - 1$. As $b \neq 0$, it follows that $a \neq 1$. So, in particular, P_7 is non-binary. We now have that D is

$$\begin{array}{c} \\ 2 \\ 3 \\ 4 \end{array}\begin{array}{c} \begin{array}{cccc} 5 & 6 & 7 & 8 \end{array} \\ \left[\begin{array}{cccc} 1 & 0 & 1 & 1 \\ 1 & 1 & 0 & 1 \\ a & 1 & a-1 & 0 \end{array}\right] \end{array}.$$

Then, regardless of the particular field F, it is easy to check that, provided $a \notin \{0,1\}$, the matroid $M[I_3|D]$ is simple and has among its circuits all 3-point lines of P_7. It follows that $P_7 \cong M[I_3|D]$. Moreover, since a can certainly be chosen in $F - \{0,1\}$ provided $|F| \geq 3$, we conclude that P_7 is representable over all fields other than $GF(2)$. □

Since P_8 is self-dual and all of its single-element contractions are isomorphic to P_7, all of its single-element deletions are isomorphic to P_7^*. Hence all proper minors of P_8 are representable over all the fields $GF(2^k)$ for $k > 1$. The following result will complete the proof of Proposition 6.4.12, though, like Lemma 6.4.13, it is actually stronger than is needed for the proof of that proposition.

6.4.14 Lemma. *The matroid P_8 is representable over a field F if and only if the characteristic of F is not two.*

Proof. Let $[I_r|D]$ be an F-representation for P_8. By Theorem 6.4.7, we can assume that D is as shown in Figure 6.19(a) where a, b, c, d, and e are non-zero elements of F. As $\{1,4,5,8\}$ is a circuit of P_8, the submatrix of $[I_r|D]$ shown in Figure 6.19(b) has determinant 0. Hence

$$a = e. \tag{1}$$

Similarly, as $\{2,3,6,7\}$ is a circuit,

$$c = 1; \tag{2}$$

$$\begin{array}{c} \\ 1 \\ 2 \\ 3 \\ 4 \end{array}\begin{array}{c} \begin{array}{cccc} 5 & 6 & 7 & 8 \end{array} \\ \left[\begin{array}{cccc} 0 & 1 & 1 & d \\ 1 & 0 & 1 & 1 \\ a & 1 & 0 & e \\ b & 1 & c & 0 \end{array}\right] \end{array} \qquad \begin{array}{c} \begin{array}{cccc} 1 & 4 & 5 & 8 \end{array} \\ \left[\begin{array}{cccc} 1 & 0 & 0 & d \\ 0 & 0 & 1 & 1 \\ 0 & 0 & a & e \\ 0 & 1 & b & 0 \end{array}\right] \end{array}$$

(a) (b)

Fig. 6.19.

and, from the circuits $\{1,5,6,7\}$, $\{2,5,6,8\}$, $\{3,5,7,8\}$, and $\{4,6,7,8\}$, respectively, we deduce that

$$c + a - b = 0, \tag{3}$$

$$ad + be - bd = 0, \tag{4}$$

$$b + dc - db = 0, \tag{5}$$

and

$$e + 1 - d = 0. \tag{6}$$

Using (1) and (2) to eliminate e and c from (3)–(6), we get

$$a = b - 1, \tag{7}$$

$$ad + ab - bd = 0, \tag{8}$$

$$b + d - db = 0, \tag{9}$$

and

$$a + 1 = d. \tag{10}$$

From (7) and (10), $b = d$. Substituting b for d in (8) and (9) and simplifying, we get

$$b(2a - b) = 0$$

and

$$b(2 - b) = 0.$$

As $b \neq 0$, it follows that $b = 2$ and $a = 1$. Thus if F has characteristic two, then $b = 0$. Therefore P_8 is not representable over fields of characteristic two. Moreover, if F does not have characteristic two and $[I_r|D]$ is an F-representation for P_8, then D is

$$\begin{array}{c} \\ 1 \\ 2 \\ 3 \\ 4 \end{array} \begin{array}{c} \begin{array}{cccc} 5 & 6 & 7 & 8 \end{array} \\ \left[\begin{array}{cccc} 0 & 1 & 1 & 2 \\ 1 & 0 & 1 & 1 \\ 1 & 1 & 0 & 1 \\ 2 & 1 & 1 & 0 \end{array}\right] \end{array}.$$

To complete the proof of the lemma, it remains to show that, if D is as specified here, then $[I_4|D]$ represents P_8 over every field whose characteristic is not two. Evidently $M[I_4|D]$ has no 1- or 2-circuits. If C is a 3- or 4-circuit of $M[I_4|D]$, then, as $\{1,2,3,4\}$ is a basis, C meets $\{5,6,7,8\}$. Moreover, as det $D = -4 \neq 0$, $\{5,6,7,8\}$ is a basis and hence C meets $\{1,2,3,4\}$. Thus, if $|C| = 3$, then $|C \cap \{1,2,3,4\}|$ is 1 or 2 and it is easy to check that neither of these occurs. Therefore $M[I_4|D]$ has no 3-circuits. By the construction of D, each of the ten 4-circuits of P_8 is dependent in $M[I_4|D]$. Hence each is a circuit. Suppose that C is a 4-circuit of $M[I_4|D]$ that is not one of these ten. As $M[I_4|D]$ is self-dual and has no circuits of size less than four, it has no cocircuits of size less than four. From above, $|C \cap \{1,2,3,4\}|$ is 1, 2, or 3. In the first case, since every 3-element subset of $\{5,6,7,8\}$ is in some 4-circuit of P_8, the closure of C in $M[I_4|D]$ has size at least five; that is, $M[I_4|D]$ has a cocircuit of size less than four. Similar arguments show that neither of the other two cases can occur. We conclude that $M[I_4|D] = P_8$. □

Exercises

1. Let X be a matrix all of whose entries are in $\{0,1\}$. Show that the associated bipartite graph $G(X)$ has a cycle if and only if X has a submatrix in which every row and every column contains at least two ones.

2. Show that P_8 has no minor isomorphic to $M(K_4)$.

3. Show that $AG(3,2)$ is representable only over fields of characteristic two.

4. Suppose that the columns of the matrix $[I_r|X]$ are labelled, in order, $e_1, e_2, \ldots, e_n$ and let $[I_r|X]$ be a partial representation for the matroid M. Show that:

(i) If $r+1 \leq i \leq n$, then the matrix obtained from $[I_r|X]$ by deleting column i is a partial representation for $M \backslash e_i$.

(ii) If $1 \leq i \leq r$, then the matrix obtained from $[I_r|X]$ by deleting row i and column i is a partial representation for M/e_i.

5. Let M be the 8-element rank-4 matroid shown in Figure 6.20. Show that M is a non-representable matroid that is not isomorphic to either V_8 or the matroid M' in Example 6.4.11.

6. By attempting to construct a representation, show directly that the non-Pappus matroid is non-representable.

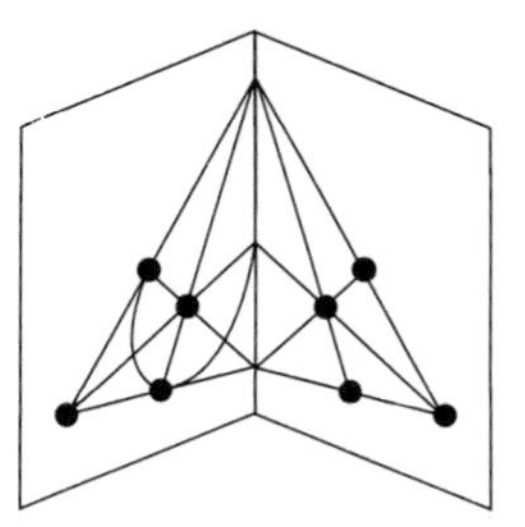

Fig. 6.20. A non-representable matroid, F_8.

7. Determine all fields over which the matroid Q_6 in Figure 2.1 is representable.

8. (i) Show that $\mathcal{W}^4$ is F-representable if and only if $|F| \geq 3$.

   (ii) Show that $U_{3,6}$ is F-representable if and only if $|F| \geq 4$.

9. (Brylawski and Kelly 1980)

   (i) Show that $AG(2,3)$ is F-representable if and only if F contains a root of the equation $x^2 - x + 1 = 0$.

   (ii) Using quadratic residues, deduce from (i) that $AG(2,3)$ is representable over $GF(q)$ if and only if q is a square, q is a power of 3, or $q \equiv 1 \pmod 3$.

10. Show that $AG(2,4)$ is F-representable if and only if F has $GF(4)$ as a subfield.

6.5 Representability over finite fields

This section will be primarily concerned with surveying various results on matroid representability over finite fields. The proofs of the main results, the excluded-minor characterizations of binary and ternary matroids, will be presented in Chapters 9 and 10 respectively.

Probably the most popular approach to finding necessary and sufficient conditions for a matroid to be F-representable has been to find the minimal obstructions to F-representability. Since a matroid M is F-representable if and only if all its minors are F-representable, one way to characterize the class of F-representable matroids is by listing the minor-minimal matroids that are not F-representable. These matroids, which are called the *excluded* or *forbidden minors* for F-representability, are just the non-F-representable matroids for which every proper minor is F-representable. Finding the complete set of excluded minors for representability over a particular field is a notoriously difficult problem and has, in fact, only been solved for the 2-element and 3-element fields. Nevertheless, we can still make some general observations about excluded minors. For example, since the class of F-representable matroids is closed under duality, we have:

6.5.1 Lemma. *If the matroid M is an excluded minor for F-representability, then so is its dual M^*.* □

Excluded minors for F-representability fall loosely into two categories: those that are excluded because the field is too small, and those that are excluded for structural reasons, be they geometric or algebraic. A class of matroids of the first type is the class of rank-two uniform matroids. The straightforward proof of the next result is left to the reader (see Exercise 5 of Section 1.2).

6.5.2 Proposition. *Let F be a field and k be an integer exceeding one. Then $U_{2,k}$ is F-representable if and only if $|F| \geq k-1$.* □

From this proposition and Lemma 6.5.1, it follows that, for all finite fields $GF(q)$, some of the excluded minors for $GF(q)$-representability are excluded because of the size of the field.

6.5.3 Corollary. *The matroids $U_{2,q+2}$ and $U_{q,q+2}$ are excluded minors for $GF(q)$-representability.* □

In fact, Tutte (1958) established that $U_{2,4}$ is the unique excluded minor for $GF(2)$-representability, that is:

6.5.4 Theorem. *A matroid is binary if and only if it has no $U_{2,4}$-minor.* □

The proof of this result will be given in Chapter 9 along with a number of other characterizations of binary matroids.

The next two results show that the Fano and non-Fano matroids and their duals are examples of structural excluded minors for representability.

6.5.5 Proposition. *Let F be a field of characteristic other than two. Then F_7 and F_7^* are excluded minors for F-representability.*

Proof. By Proposition 6.4.8, F_7 is not F-representable. In Example 1.5.6, we noted that $F_7 \backslash e \cong M(K_4)$ for all elements e. Hence, by Proposition 5.1.2, every single-element deletion of F_7 is F-representable. On the other hand, by extending the argument in Example 3.3.8 and using Proposition 5.1.2 again, we get that every single-element contraction of F_7 is graphic and hence is F-representable. We conclude that F_7 is an excluded-minor for F-representability. By Lemma 6.5.1, so is F_7^*. □

The matroid P_8 in the next result was defined after Example 6.4.11.

6.5.6 Proposition. *Let F be a field of characteristic two. If $F \neq GF(2)$, then P_8, F_7^-, and $(F_7^-)^*$ are excluded minors for F-representability.*

Proof. The assertion for P_8 is a restatement of Proposition 6.4.12. The proofs for F_7^- and $(F_7^-)^*$ are similar to the last proof and are left as exercises. □

The reason why $GF(2)$ is an exception in the last proposition is that each of P_8, F_7^-, and $(F_7^-)^*$ has a $U_{2,4}$-minor. Hence none of these three matroids is minor-minimal with the property of being non-representable over $GF(2)$.

On combining Corollary 6.5.3 and Proposition 6.5.5, we get that $U_{2,5}$, $U_{3,5}$, F_7, and F_7^* are excluded minors for $GF(3)$-representability. The fact that they are the only such excluded minors was announced in 1971 by Ralph Reid but was never published by him. The first published proofs of this result were due independently to Bixby (1979) and Seymour (1979). In Chapter 10, we shall present a proof that is due to Kahn and Seymour (1988).

6.5.7 Theorem. *A matroid is ternary if and only if it has no minor isomorphic to any of the matroids $U_{2,5}$, $U_{3,5}$, F_7, and F_7^*.* □

Given the excluded-minor characterizations of the classes of binary and ternary matroids stated above, the next natural class to consider is the class of *quaternary matroids*, that is, matroids representable over $GF(4)$. By Corollary 6.5.3 and Proposition 6.5.6, all of the matroids $U_{2,6}$, $U_{4,6}$, F_7^-, $(F_7^-)^*$, and P_8 are excluded minors for $GF(4)$-representability. Another such matroid is P_6, a geometric representation of which is shown in Figure 6.21. It is easy to check that every proper minor of the self-dual matroid P_6 is $GF(4)$-representable. However, P_6 itself is not $GF(4)$-representable because the field is too small.

6.5.8 Proposition. *The matroid P_6 is F-representable if and only if $|F| \geq 5$.*

Proof. Suppose that P_6 is $GF(q)$-representable. Then P_6 is a restriction of $PG(2,q)$. Let L be the line of $PG(2,q)$ containing 1, 2, and 4. Then the three lines of $PG(2,q)$ spanned by $\{3,5\}$, $\{3,6\}$, and $\{5,6\}$ meet L in distinct points of $PG(2,q)$. Moreover, none of these points of intersection is in $\{1,2,4\}$. Thus L contains at least six points, so $q \geq 5$.

The proof of the converse is left to the reader. □

The following is unsolved.

6.5.9 Problem. *Is $\{U_{2,6}, U_{4,6}, F_7^-, (F_7^-)^*, P_6, P_8\}$ the complete set of excluded minors for $GF(4)$-representability?* □

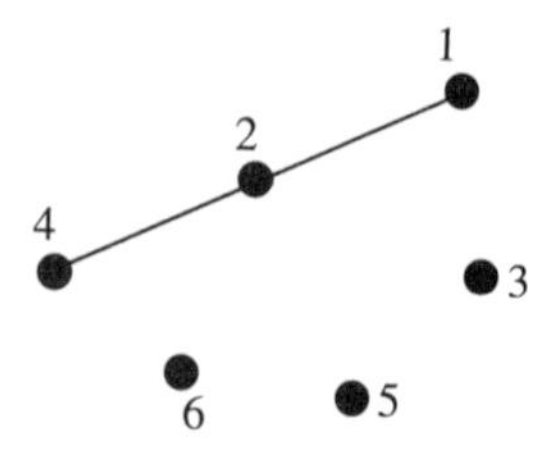

Fig. 6.21. P_6.

Before Oxley (1986a) had proved Proposition 6.4.12, Kahn and Seymour (private communication) had conjectured that the first five matroids in this list form a complete set. The discovery of P_8 has lead to the feeling that there may be other such matroids, although none is yet known.

For the fields $GF(q)$ with $q \geq 5$, there do not seem to have been any conjectures made as to what a complete list of excluded minors for $GF(q)$-representability might be. When $q = 5$, we know that $U_{2,7}$, $U_{5,7}$, F_7, and F_7^* are excluded minors. Moreover, it is not difficult to check that V_8 is an excluded minor for $GF(5)$-representability. Since the non-Pappus matroid has no $U_{2,7}$-minor but is not $GF(5)$-representable, some rank-3 restriction M of it will also be an excluded minor; of course, M^* will also be an excluded minor. We leave the reader to check that the matroid in the next example is also an excluded minor for $GF(5)$-representability.

6.5.10 Example. Let Q_8 be the rank-4 paving matroid whose circuits of size less than five are the faces of the cube in Figure 6.22 together with exactly five of the six diagonal planes such as $\{1, 2, 7, 8\}$. One easily checks that Q_8 is self-dual. □

No attempt has been made here to find a potentially complete list of excluded minors for $GF(5)$-representability. Indeed, as we shall soon see, there are two more uniform matroids that belong on such a list (see also Exercise 5). However, the matroids listed do provide some indication of how the complexity of finding the complete set of excluded minors for $GF(q)$-representability grows with q. For large q, finding such a complete list appears impossible with currently available methods. Instead, the focus of research for $q \geq 5$ has been on resolving the following conjecture of Rota (1971), one of the major unsolved problems in matroid theory.

6.5.11 Conjecture. *For all prime powers q, the complete set of excluded minors for $GF(q)$-representability is finite.* □

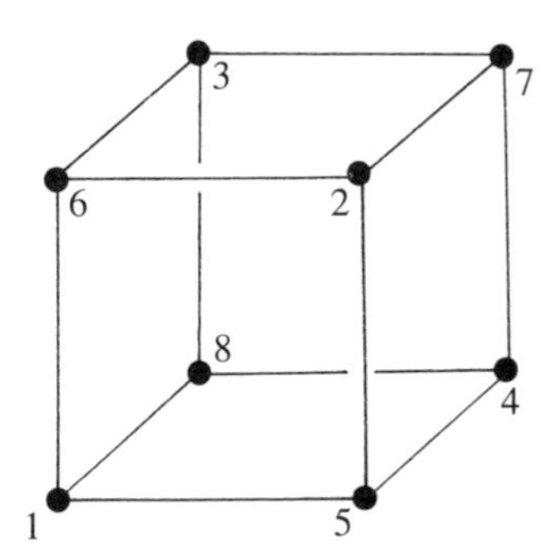

Fig. 6.22. 4-circuits of Q_8: all faces and five diagonal planes.

Table 6.1. $U_{r,n}$ is $GF(q)$-representable if and only if $n \leq k$.

r	k	Restriction on q
3	$q+1$	q odd
	$q+2$	q even
4	5	$q \leq 3$
	$q+1$	$q \geq 4$
5	6	$q \leq 4$
	$q+1$	$q \geq 5$

Theorems 6.5.4 and 6.5.7 prove this conjecture when q is 2 or 3. No other case of the conjecture has yet been solved. As a contrast to the above conjecture for finite fields, we note that the set of excluded minors for $\mathbb{R}$-representability is infinite (Exercise 6).

We now turn to another unsolved problem that concerns minors which are excluded for size reasons.

6.5.12 Problem. *Find all the fields over which the uniform matroid $U_{r,n}$ is representable.* □

If $r \leq 1$, then evidently $U_{r,n}$ is F-representable for every field F. Moreover, Proposition 6.5.2 solves 6.5.12 for $r = 2$. For larger values of r, the problem has received considerable attention from projective geometers who use the term *k-arc* for a set S of points of $PG(r, q)$ such that $PG(r, q)|S \cong U_{r+1,k}$ (see, for example, Fenton and Vámos 1982). For all r in $\{3, 4, 5\}$, the matroid $U_{r,n}$ is $GF(q)$-representable if and only if n does not exceed the value of k specified in Table 6.1. This table is obtained by combining results of Bose (1947), Bush (1952), Segre (1955), Casse (1969), and Gulati and Kounias (1970). For a more detailed consideration of k-arcs and their properties, the reader is referred to Hirschfeld's (1983) survey paper.

For r exceeding five, although partial results are known, there is no complete answer. For example, Thas (1968) proved a result which implies that, if q is odd and exceeds $(4r-1)^2$, then $U_{r,n}$ is $GF(q)$-representable if and only if $n \leq q+1$. Again, we refer to Hirschfeld (1983) for this and other related results. It follows immediately from Table 6.1 that $U_{3,7}$ and $U_{4,7}$ are excluded minors for $GF(5)$-representability. In fact, for all q in $\{3, 5, 7, 9, 11\}$, each of $U_{2,q+2}, U_{3,q+2}, \ldots, U_{q,q+2}$ is an excluded minor for $GF(q)$-representability. It is natural to ask whether the last statement remains true for all odd prime powers q, though this problem would appear to be very difficult.

To close this section, we note that there are various known necessary and sufficient conditions for representability. Among these are results due to Vámos (1971a) and White (1980a). In general, these conditions amount to giving an algebraic reformulation of what it means for a matroid to be representable. For a survey of these results and how they have been used, we refer the reader to White (1987b) where a proof can also be found of the following proposition of Tutte (1958), the first result of this kind.

6.5.13 Proposition. *Let M be a matroid and F be a field. A necessary and sufficient condition for M to be F-representable is that, for all hyperplanes H of M, there is a function $f_H : E(M) \to F$ such that*

(i) $H = \{x \in E(M) : f_H(x) = 0\}$; *and*

(ii) *if H_1, H_2, and H_3 are distinct hyperplanes of M for which the rank of $H_1 \cap H_2 \cap H_3$ is $r(M) - 2$, then there are non-zero elements c_1, c_2, and c_3 of F such that*

$$c_1 f_{H_1} + c_2 f_{H_2} + c_3 f_{H_3} = 0.$$

□

Exercises

1. Let F be a field. Show that every excluded minor for F-representability is connected and simple.

2. Find all fields F for which

   (i) Q_8 is F-representable;

   (ii) Q_8 is an excluded minor for F-representability.

3. (Brylawski 1975d) Show that:

   (i) The matroid M in Figure 6.23(a) is an excluded minor for $GF(5)$-representability.

   (ii) For all $m \geq 5$, the matroid N in Figure 6.23(b) is F-representable if and only if $|F| \geq m$.

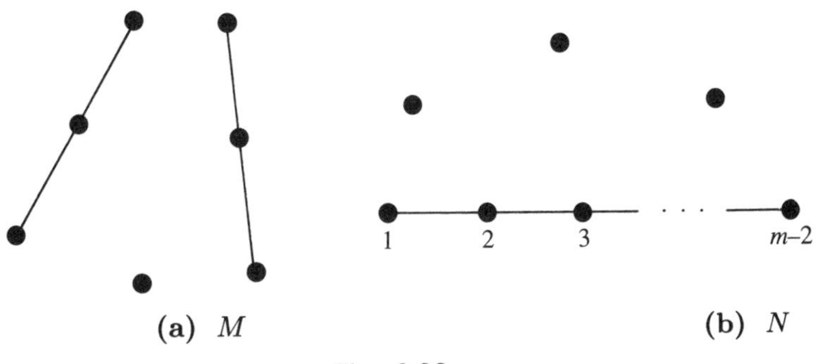

Fig. 6.23.

4. Find all restrictions of the non-Pappus matroid that are excluded minors for $GF(5)$-representability.

5. Let T_8 and R_8 be as in Exercise 8 of Section 2.2. Show that:

   (i) R_8 is F-representable if and only if the characteristic of F is not two.

   (ii) T_8 is F-representable if and only if F has characteristic three.

   (iii) R_8 has a transitive automorphism group.

   (iv) T_8 is an excluded minor for F-representability for all fields F whose characteristic is not two or three.

   (v) Q_8 is obtained from R_8 by relaxing a circuit–hyperplane.

6. (Lazarson 1958) Suppose p is a prime number exceeding two. Let L_p denote the vector matroid of the matrix $[I_{p+1}|J_{p+1} - I_{p+1}]$ over $GF(p)$ where J_n is the $n \times n$ matrix of all ones. Then $L_3 = T_8$. Let the columns of $[I_{p+1}|J_{p+1} - I_{p+1}]$ be labelled by $x_1, x_2, \ldots, x_{p+1}$, $y_1, y_2, \ldots, y_{p+1}$. Show the following:

   (i) The non-spanning circuits of L_p are $\{y_1, y_2, \ldots, y_{p+1}\}$ plus all the sets of the form $\{x_i, y_i, x_j, y_j\}$ for $1 \leq i < j \leq p+1$.

   (ii) L_p is representable over a field F if and only if F has characteristic p.

   (iii) $L_p \backslash y_{p+1}$ is represented over $\mathbb{Q}$ and over $GF(p')$ for all $p' \geq p$ by the matrix obtained from $[I_{p+1}|J_{p+1} - I_{p+1}]$ by deleting the last column.

   (iv) L_p / y_{p+1} is represented over $\mathbb{Q}$ and over $GF(p')$ for all $p' \geq p$ by the matrix

$$
\begin{array}{c}
\begin{array}{cccc|cccccc}
x_2 & x_3 & \cdots & x_{p+1} & x_1 & y_1 & y_2 & y_3 & \cdots & y_p
\end{array}\\
\left[\begin{array}{c|cccccc}
 & 1 & 0 & 0 & 0 & \cdots & 0 \\
 & 1 & 1 & p-1 & 0 & \cdots & 0 \\
 & 1 & 1 & 0 & p-1 & \cdots & 0 \\
I_p & 1 & 1 & 0 & 0 & \cdots & 0 \\
 & \vdots & \vdots & \vdots & \vdots & & \vdots \\
 & 1 & 1 & 0 & 0 & \cdots & p-1 \\
 & 0 & 1 & 1 & 1 & \cdots & 1
\end{array}\right].
\end{array}
$$

   (v) If F is a field of characteristic p', then L_p is an excluded minor for F-representability if $p' > p$, or $p' = 0$.

(vi) For all fields F of characteristic zero, the set of excluded minors for F-representability is infinite.

6.6 Regular matroids

In the last section, we were concerned with matroids that are representable over some fixed finite field. We now consider the class of matroids that are representable over every field. By Proposition 5.1.2 and duality, this class includes the classes of graphic and cographic matroids. In this section, we shall show that it coincides with the class of regular matroids. In addition, we shall state Tutte's important excluded-minor theorems for the classes of regular, graphic, and cographic matroids.

The proof that the class of regular matroids equals the class of matroids representable over every field will use a modification of the pivot operation described in Section 2.2. Suppose that, in a standard representative matrix $[I_r|D]$, a pivot is performed on the non-zero element d_{st} of D where $D = [d_{ij}]$. Then the resulting matrix need no longer have the form $[I_r|X]$. However, one obtains a matrix in this form simply by interchanging columns s and $r+t$ of the transformed matrix. Thus, in order to maintain the form of the matrix, whenever a pivot is performed in $[I_r|D]$ on an element of D, we shall always follow this by the appropriate column interchange; that is, we absorb the natural column interchange into the pivoting operation.

6.6.1 Example. Let A be the matrix over $\mathbb{Q}$ shown in Figure 6.24(a). Pivoting on the element in the top right corner, we get the matrix shown in Figure 6.24(b). We remark that $M[A]$ can be obtained from $AG(3,2)$ by a sequence of circuit–hyperplane relaxations. We leave it to the reader to determine exactly how many such relaxations are needed (Exercise 1). □

$$\begin{array}{cccc|cccc} 1 & 2 & 3 & 4 & 5 & 6 & 7 & 8 \\ \hline & & & & 0 & 1 & 1 & 1 \\ & I_4 & & & 1 & 0 & 1 & 1 \\ & & & & 1 & 1 & 0 & 1 \\ & & & & 1 & 1 & 1 & 0 \end{array} \qquad \begin{array}{cccc|cccc} 8 & 2 & 3 & 4 & 5 & 6 & 7 & 1 \\ \hline & & & & 0 & 1 & 1 & 1 \\ & I_4 & & & 1 & -1 & 0 & -1 \\ & & & & 1 & 0 & -1 & -1 \\ & & & & 1 & 1 & 1 & 0 \end{array}$$

(a) (b)

Fig. 6.24. Pivoting on the first entry of column 8 of (a) gives (b).

The next theorem characterizes regular matroids in terms of representability (Tutte 1965). The following lemma plays an important role in the proof of that theorem.

6.6.2 Lemma. *Let $[d_{ij}]$ be a matrix D_1 all of whose entries are in $\{0, 1, -1\}$. Suppose $[I_r|D_1]$ is an $\mathbb{R}$-representation of a binary matroid M. Assume* *that* $[I_r|D_2]$ *is obtained from* $[I_r|D_1]$ *by pivoting on a non-zero entry* d_{st} *of* D_1. *Then every entry of* D_2 *is in* $\{0, 1, -1\}$.

Proof. It is straightforward to check that all the entries in row s of D_2 and all those in column t are in $\{0, 1, -1\}$. Now suppose $j \neq t$ and $i \neq s$. Then the pivot replaces d_{ij} by $d_{ij} - (d_{it}/d_{st})d_{sj}$, which equals $(1/d_{st})(d_{st}d_{ij} - d_{it}d_{sj})$. As all the entries of D_1 are in $\{0, 1, -1\}$ and $d_{st} \neq 0$, the last quantity is in $\{0, 1, -1\}$ unless $|d_{st}d_{ij} - d_{it}d_{sj}| = 2$. Hence assume that this equation holds. Then the matrix $\begin{bmatrix} d_{st} & d_{sj} \\ d_{it} & d_{ij} \end{bmatrix}$, or an appropriate row or column permutation thereof, is a submatrix D_1' of D_1 whose determinant is ± 2. But, when viewed over $GF(2)$, the matrix $[I_r|D_1]$ represents M. Thus, by Proposition 6.4.5, as $\det D_1'$ is 0 over $GF(2)$, it must also be 0 over $\mathbb{R}$; a contradiction. □

6.6.3 Theorem. *The following statements are equivalent for a matroid* M:

(i) M *is regular.*

(ii) M *is representable over every field.*

(iii) M *is binary and, for some field* F *of characteristic other than two,* M *is* F*-representable.*

Before proving this theorem we make two important observations. Firstly, an immediate consequence of it is that the class of regular matroids is the intersection of the classes of binary and ternary matroids. Secondly, from the first part of the proof, it will follow that if $[I_r|D]$ is a totally unimodular matrix that represents the matroid M over $\mathbb{R}$, then, when $[I_r|D]$ is viewed over an arbitrary field F, it is an F-representation for M.

If C is a cycle of a graph G, a *diagonal* of C is an edge of G that joins two distinct vertices of C but is not in C.

Proof of Theorem 6.6.3. Evidently we may assume that $r(M) > 0$. Now suppose that (i) holds. Then it follows by Lemma 2.2.21 that M has a totally unimodular representation of the form $[I_r|D]$. Let F be an arbitrary field and X be a set of r columns of $[I_r|D]$. Then X is a basis of M if and only if the determinant over $\mathbb{R}$ of the submatrix corresponding to X is a non-zero member of $\{0, 1, -1\}$. But the latter holds if and only if this determinant, when evaluated over F, is non-zero. In turn, this holds if and only if X is a basis of the vector matroid on $[I_r|D]$ when the latter is viewed as a matrix over F. We conclude that $[I_r|D]$ represents M over F, so (i) implies (ii). Since (ii) clearly implies (iii), it remains to show that (iii) implies (i). The following proof of this is due to Brylawski (1975c) (see also White 1987b).

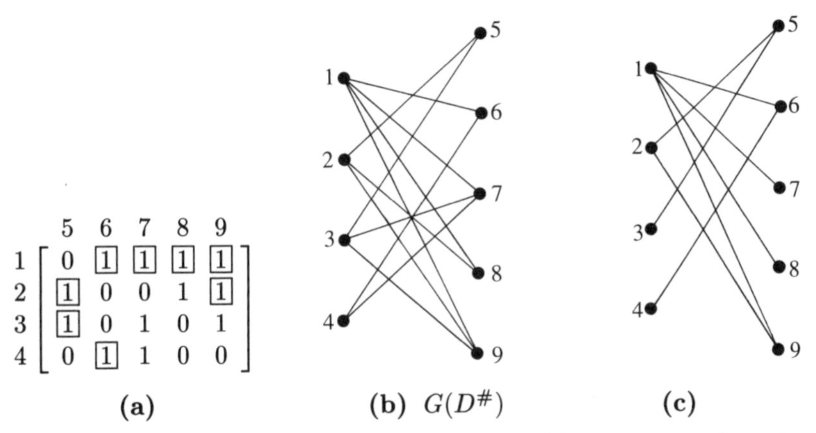

Fig. 6.25. The bipartite graphs associated with (a) and its boxed entries.

Assume that (iii) holds and let $[I_r|D]$ be an F-representation for M. By Theorem 6.4.7, if B_D is a basis of the cycle matroid of $G(D^{\#})$, then, we may assume that every entry in D corresponding to an edge of this basis is equal to 1. Now, let d be any other non-zero entry of D, and e_d be the corresponding edge of $G(D^{\#})$. Then there is a unique cycle C_d in $G(D^{\#})$ that contains e_d and is contained in $B_D \cup e_d$. Of course, C_d is the fundamental cycle of e_d with respect to B_D. We shall argue by induction on $|C_d|$ to show that $d \in \{1, -1\}$.

Before continuing with the proof, we consider an example. Assume $D^{\#}$ is the matrix shown in Figure 6.25(a) with the boxed entries corresponding to the edges in B_D. Then (b) and (c) of Figure 6.25 show $G(D^{\#})$ and the subgraph of $G(D^{\#})$ induced by B_D. Now, for some non-zero elements d_1, d_2, d_3, and d_4 of F, the matrix D is as shown in Figure 6.26(a). Thus C_{d_1} and C_{d_2} are as shown in Figure 6.26(b) and (c). The submatrices of D whose rows and columns are indexed by the vertices of C_{d_1} and C_{d_2} are

$$\begin{array}{c} \\ 1 \\ 2 \end{array}\!\!\begin{array}{c} \begin{array}{cc} 8 & 9 \end{array} \\ \begin{bmatrix} 1 & 1 \\ d_1 & 1 \end{bmatrix} \end{array} \quad \text{and} \quad \begin{array}{c} \\ 1 \\ 2 \\ 3 \end{array}\!\!\begin{array}{c} \begin{array}{ccc} 5 & 7 & 9 \end{array} \\ \begin{bmatrix} 0 & 1 & 1 \\ 1 & 0 & 1 \\ 1 & d_2 & d_3 \end{bmatrix} \end{array},$$

respectively. We call these submatrices D_{d_1} and D_{d_2}.

Returning to the proof that (iii) implies (i), we have that, in general, for some $k \geq 2$, there are exactly k rows and exactly k columns of D that contain entries corresponding to edges of C_d. Let D_d be the $k \times k$ submatrix of D induced by these sets of k rows and k columns. In D_d, each row and each column contains exactly two entries that correspond to edges of C_d. Now either (a) D_d contains some other non-zero entries, or (b) D_d contains no other non-zero entries. In case (a), for every other non-zero entry d' of D_d, the corresponding edge $e_{d'}$ of $G(D^{\#})$ is a diagonal of C_d.

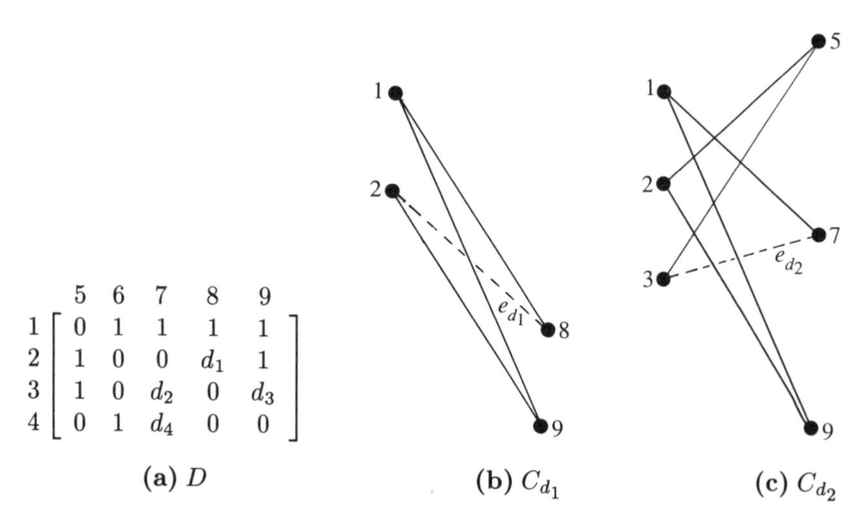

(a) D (b) C_{d_1} (c) C_{d_2}

Fig. 6.26. The fundamental cycles of d_1 and d_2 in $G(D^{\#})$.

Hence $|C_{d'}| < |C_d|$ and so, by the induction assumption, $|d'| = 1$. Thus every entry of D_d, except possibly d, is in $\{0, 1, -1\}$. Now let $G(D^{\#}_d)$ be the subgraph of $G(D^{\#})$ induced by the vertices of C_d and, among the cycles of $G(D^{\#}_d)$ containing e_d, let C'_d be one of shortest length. The submatrix D'_d of D_d induced by the edges of C'_d has j rows and j columns for some $j \leq k$. Moreover, each row and each column of D'_d contains exactly two entries corresponding to edges of C'_d and has no other non-zero entries. Thus, in case (a), D has a $j \times j$ submatrix D'_d having d as an entry such that each row and column contains exactly two non-zero entries and all these entries, except possibly d, are in $\{1, -1\}$. But this also holds in case (b) for, in that case, we take $j = k$ and $D'_d = D_d$.

Evaluating $\det(D'_d)$ using the summation formula 2.2.11, we see that there are exactly two non-zero terms in this summation and $\det(D'_d) = \pm 1 \pm d$. Since M is binary, $[I_r|D^{\#}]$ is a $GF(2)$-representation for M. Hence, over $GF(2)$, d is 1. Therefore $\det(D'_d)$ is 0 when evaluated over $GF(2)$. Thus, by Proposition 6.4.5, $\det(D'_d)$ is also 0 when evaluated over F, and so $d \in \{1, -1\}$. This completes the induction argument that every non-zero entry of D is in $\{1, -1\}$.

We know that $[I_r|D]$ represents M over F. To complete the proof we shall show that, when viewed over $\mathbb{R}$, $[I_r|D]$ is a totally unimodular representation of M. To establish that $[I_r|D]$ represents M over $\mathbb{R}$, we note that by Proposition 6.4.5 it suffices to show that, for every square submatrix D' of D, the determinant of D' is 0 over F if and only if it is 0 over $\mathbb{R}$. Clearly if $\det D'$ is 0 over $\mathbb{R}$, it is 0 over F. Moreover, the converse of this will follow if we can show that if $\det D'$ is non-zero over

$\mathbb{R}$, then it is in $\{1,-1\}$. But establishing this will also verify that M is regular and thereby complete the proof of the theorem.

Now suppose that $\det D' \neq 0$. Then, by Lemma 6.6.2, we can pivot in D on elements of D' until each column of D' is a unit vector. Since these pivots can only affect the determinant of D' by changing its sign, $\det D' \in \{1,-1\}$, and the theorem is proved. □

It follows by the last theorem, Corollary 6.5.3, and the proof of Proposition 6.5.5 that the excluded minors for regularity must include $U_{2,4}$, F_7, and F_7^*. The next result (Tutte 1958) establishes that this list is complete. We shall give two different proofs of this result, one in Chapter 10 and the other in Chapter 13.

6.6.4 Theorem. *A matroid is regular if and only if it has no minor isomorphic to any of the matroids $U_{2,4}$, F_7, and F_7^*.* □

In contrast to this excluded-minor characterization, Seymour (1980b) has given a constructive characterization of regular matroids showing that every such matroid can be built up from graphic matroids, cographic matroids, and a certain 10-element self-dual matroid R_{10}. The precise statement of this result will be given in Chapter 13. In that chapter, we shall also prove the following excluded-minor characterizations of the classes of graphic and cographic matroids. These results, which are again due to Tutte (1959), may be viewed as matroid generalizations of Kuratowski's Theorem (2.3.8).

6.6.5 Theorem. (i) *A matroid is graphic if and only if it has no minor isomorphic to any of the matroids $U_{2,4}$, F_7, F_7^*, $M^*(K_5)$, and $M^*(K_{3,3})$.*

(ii) *A matroid is cographic if and only if it has no minor isomorphic to any of the matroids $U_{2,4}$, F_7, F_7^*, $M(K_5)$, and $M(K_{3,3})$.* □

6.6.6 Corollary. *The following statements are equivalent for a matroid M.*

(i) *$M \cong M(G)$ for some planar graph G.*

(ii) *M is both graphic and cographic.*

(iii) *M has no minor isomorphic to any of the matroids $U_{2,4}$, F_7, F_7^*, $M(K_5)$, $M^*(K_5)$, $M(K_{3,3})$, and $M^*(K_{3,3})$.* □

The results of this section enable us to update the Venn diagram in Figure 5.5 (see Figure 6.27). Moreover, by using various examples described earlier, the reader should have no difficulty in showing that each of the inclusions indicated in this diagram is proper (Exercise 2).

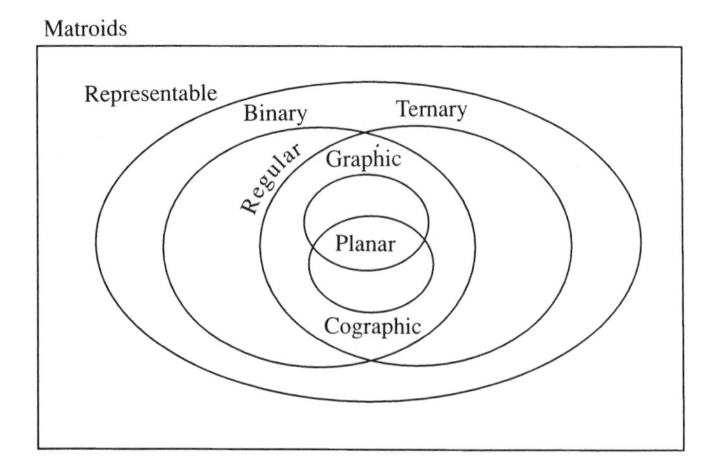

Fig. 6.27. Relationships between various classes of matroids.

Exercises

1. (i) Show that the matroid $M[A]$ in Example 6.6.1 is isomorphic to neither R_8 nor T_8 from Exercise 8 of Section 2.2.

   (ii) How many circuit–hyperplane relaxations are needed to get $M[A]$ from $AG(3,2)$?

2. Show that each of the inclusions in Figure 6.27 is proper.

3. A simple rank-r matroid M is *maximal regular* if M is regular and there is no simple rank-r regular matroid N of which M is a proper restriction. Show that:

   (i) $M(K_n)$ is maximal regular for all $n \geq 2$.

   (ii) $M^*(K_{3,3})$ is maximal regular.

4. Let M be a matroid and A be a matrix $[I_r|D]$ such that for every field F, when A is viewed over F, it is an F-representation for M.

   (i) Prove that, for every square submatrix X of A, the determinant of X, calculated over F, is 0, 1, or -1.

   (ii) Deduce that M is regular.

   (iii) Use this, in conjunction with Proposition 5.1.2, to give a short proof of Proposition 5.1.3.

5.* (Lindström 1987a) Use Proposition 6.5.13 to prove that a matroid M is both binary and ternary if and only if M is F-representable for every field F.

6.7 Algebraic matroids

The focus of this chapter thus far has been on matroids that arise via linear dependence. There is another type of dependence that occurs in algebra, specifically in the study of fields. In this section we shall see how this notion of algebraic dependence also gives rise to a class of matroids. This observation was first made by van der Waerden (1937) when he noted several features common to both linear and algebraic dependence. Soon after, the properties of these algebraic matroids were investigated by Mac Lane (1938). Despite the proximity of this work to Whitney's (1935) founding paper in matroid theory, algebraic matroids have received far less attention over the years than representable matroids. Undoubtedly, an important reason for this is that algebraic matroids are not nearly as easy to deal with as representable matroids. Probably the most striking illustration of this fact is that it is still not known whether or not the class of algebraic matroids is closed under duality.

Before giving the precise definition of an algebraic matroid, we recall some of the basics of field theory. While a detailed development of this theory can be found in most graduate algebra texts, we shall not assume this background, attempting to make the treatment here as self-contained as possible. Much of our discussion will follow van der Waerden (1937). If F is a field, then $F[x_1, x_2, \ldots, x_n]$ and $F(x_1, x_2, \ldots, x_n)$ will denote, respectively, the ring of polynomials over F in the n indeterminates $x_1, x_2, \ldots, x_n$, and the associated field of rational functions in $x_1, x_2, \ldots, x_n$. Suppose F is a subfield of the field K. Then K is called an *extension field* of F. So, for instance, $GF(4)$ and $GF(8)$ are extension fields of $GF(2)$, although $GF(8)$ is not an extension field of $GF(4)$. The extension field K is a vector space over the field F. Clearly the dimension of this F-vector space K may be finite or infinite. An element u of K is said to be *algebraic* over F if u is a root of some non-zero polynomial in $F[x]$. If u is not algebraic over F, it is *transcendental* over F.

Now let $\{t_1, t_2, \ldots, t_n\}$ be a subset of K. Then $F(t_1, t_2, \ldots, t_n)$ denotes the subfield of K generated by $\{t_1, t_2, \ldots, t_n\}$, that is, $F(t_1, t_2, \ldots, t_n)$ consists of all elements of the form $h(t_1, t_2, \ldots, t_n)/k(t_1, t_2, \ldots, t_n)$ where h and k are in $F[x_1, x_2, \ldots, x_n]$. An element s of K is *algebraically dependent* on $\{t_1, t_2, \ldots, t_n\}$ over F if s is algebraic over $F(t_1, t_2, \ldots, t_n)$. The latter occurs if and only if s is a root of an equation of the form

$$a_0(t_1, t_2, \ldots, t_n)x^m + a_1(t_1, t_2, \ldots, t_n)x^{m-1} + \cdots + a_m(t_1, t_2, \ldots, t_n) = 0$$

where, for all i, $a_i(t_1, t_2, \ldots, t_n)$ is a polynomial in $t_1, t_2, \ldots, t_n$ with coefficients in F, and at least one of these polynomials is non-zero.

A finite subset T of K is *algebraically dependent* over F if, for some t in T, the element t is algebraically dependent on $T - t$. If T is not

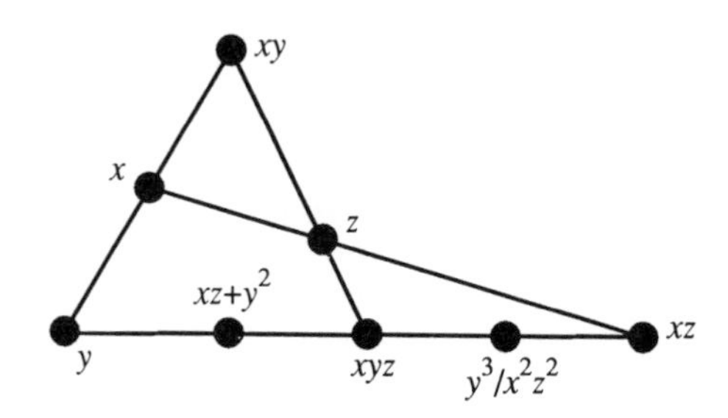

Fig. 6.28. A geometric representation of an algebraic matroid.

algebraically dependent over F, it is called *algebraically independent* over F. Reference to the particular field F is often omitted when one is discussing algebraic dependence.

We show next that the abstract notion of dependence that is captured in the definition of a matroid has algebraic dependence as a special case.

6.7.1 Theorem. *Suppose that K is an extension field of a field F and E is a finite subset of K. Then the collection $\mathcal{I}$ of subsets of E that are algebraically independent over F is the set of independent sets of a matroid on E.*

Before proving this theorem, we note the following.

6.7.2 Example. Let $F = GF(2)$ and $K = F(x,y,z)$ where x, y, and z are indeterminates. If $E = \{x, y, z, xy, xz, xyz, xz + y^2, y^3/x^2z^2\}$, then it is not difficult to check that Figure 6.28 gives a geometric representation for the matroid $(E, \mathcal{I})$ whose existence is asserted by the theorem. □

Following van der Waerden (1937), we shall use a sequence of three elementary lemmas to prove Theorem 6.7.1.

6.7.3 Lemma. *If s is algebraically dependent on $\{t_1, t_2, \ldots, t_n\}$ but not on $\{t_1, t_2, \ldots, t_{n-1}\}$, then t_n is algebraically dependent on $\{t_1, t_2, \ldots, t_{n-1}, s\}$.*

Proof. In $F(t_1, t_2, \ldots, t_{n-1})$, the element s is algebraically dependent on $\{t_n\}$. Thus, for some polynomials $a_0(t_n), a_1(t_n), \ldots, a_m(t_n)$ that have coefficients in $F(t_1, t_2, \ldots, t_{n-1})$ and are not all zero,

$$a_0(t_n)s^m + a_1(t_n)s^{m-1} + \cdots + a_m(t_n) = 0.$$

Grouping the terms in this equation by powers of t_n, we get the equation

$$b_0(s)t_n^k + b_1(s)t_n^{k-1} + \cdots + b_k(s) = 0 \tag{1}$$

where each of $b_0(s), b_1(s), \ldots, b_k(s)$ is a polynomial having coefficients in $F(t_1, t_2, \ldots, t_{n-1})$. By hypothesis, the element s is not algebraic over $F(t_1, t_2, \ldots, t_{n-1})$. Therefore each of $b_0(s), b_1(s), \ldots, b_k(s)$ is either identically zero in s or is non-zero. But if all of these polynomials are identically zero in s, then all of $a_0(t_n), a_1(t_n), \ldots, a_m(t_n)$ are zero; a contradiction. Thus $b_i(s)$ is non-zero for some i. Hence, by (1), t_n is algebraically dependent on $\{s\}$ over $F(t_1, t_2, \ldots, t_{n-1})$, so t_n is algebraically dependent on $\{t_1, t_2, \ldots, t_{n-1}, s\}$ over F. □

6.7.4 Lemma. *Suppose that s is algebraic over $F(t)$ and t is algebraic over F. Then s is algebraic over F.*

Proof. As t is algebraic over F and s is algebraic over $F(t)$, the vector spaces $F(t)$ and $F(t)(s)$ are finite dimensional over F and $F(t)$, respectively. From this it follows easily that, for some positive integer k, the vector space $F(t)(s)$ has dimension k over F. Therefore $\{1, s, s^2, \ldots, s^k\}$ cannot be linearly independent over F and so s is algebraic over F. □

6.7.5 Lemma. *If s is algebraically dependent on $\{t_1, t_2, \ldots, t_n\}$ and each t_i is algebraically dependent on $\{u_1, u_2, \ldots, u_k\}$, then s is algebraically dependent on $\{u_1, u_2, \ldots, u_k\}$.*

Proof. We argue by induction on n. If $n = 1$, the result is immediate by the last lemma. Assume the result true for $n < m$ and let $n = m$. Now $F(t_1, t_2, \ldots, t_m) = F(t_m)(t_1, t_2, \ldots, t_{m-1})$. Thus s is algebraic over $F(t_m)(t_1, t_2, \ldots, t_{m-1})$. Moreover, for all i in $\{1, 2, \ldots, m-1\}$, t_i is algebraic over $F(u_1, u_2, \ldots, u_k)$ and hence over $F(t_m)(u_1, u_2, \ldots, u_k)$. Thus, by the induction assumption, s is algebraic over $F(t_m)(u_1, u_2, \ldots, u_k)$. As the last field equals $F(u_1, u_2, \ldots, u_k, t_m)$, and t_m is algebraic over $F(u_1, u_2, \ldots, u_k)$, it follows, by Lemma 6.7.4 again, that s is algebraic over $F(u_1, u_2, \ldots, u_k)$. Hence, by induction, the lemma holds. □

The next result is an easy consequence of the last three lemmas and its proof is left to the reader.

6.7.6 Corollary. *Let T and U be finite subsets of an extension field K of a field F and suppose that U is algebraically independent over F.*

(i) *If every element of T is algebraic over F, then U is algebraically independent over $F(T)$.*

(ii) *$T \cup U$ is algebraically independent over F if and only if T is algebraically independent over $F(U)$.* □

With these preliminaries, we are now ready to prove that algebraic dependence does indeed give rise to a matroid.

Proof of Theorem 6.7.1. Evidently $\mathcal{I}$ satisfies (I1) and (I2). Assume that $\mathcal{I}$ does not satisfy (I3) and let I_1 and I_2 be algebraically independent subsets of E with $|I_1| < |I_2|$ such that, for every element e of $I_2 - I_1$, the set $I_1 \cup e$ is algebraically dependent. It follows easily by Lemma 6.7.3 that each such e depends algebraically on some subset of I_1. Hence every element of I_2 depends algebraically on I_1. Now, from among the minimum-cardinality subsets of $I_1 \cup I_2$ on which every element of I_2 depends algebraically, choose one, say I_1', such that $I_1' - I_2$ is as small as possible. Evidently $|I_1'| \le |I_1|$. As $|I_2| > |I_1|$, the set $I_2 - I_1'$ is non-empty. Let e be an element of this set. Then e depends algebraically on I_1'. Let I_1'' be a minimal subset of I_1' on which e is algebraically dependent. As $I_1'' \cup e$ is not contained in the algebraically independent set I_2, there is an element f in $I_1'' - I_2$. Now e depends algebraically on I_1'' but not on $I_1'' - f$. Thus, by Lemma 6.7.3, f depends algebraically on $(I_1'' - f) \cup e$ and hence on $(I_1' - f) \cup e$. Therefore every element of I_1' depends algebraically on $(I_1' - f) \cup e$. But every element of I_2 depends algebraically on I_1'. Hence, by Lemma 6.7.5, every element of I_2 depends algebraically on $(I_1' - f) \cup e$. Since $|((I_1' - f) \cup e) - I_2| < |I_1' - I_2|$, the choice of I_1' is contradicted. We conclude that $\mathcal{I}$ satisfies (I3) and hence that $(E, \mathcal{I})$ is a matroid. □

Let M be an arbitrary matroid and let $(E, \mathcal{I})$ be the matroid in the last theorem. Suppose that there is a map φ from $E(M)$ into E such that a subset T of $E(M)$ is independent in M if and only if $|\varphi(T)| = |T|$ and $\varphi(T)$ is independent in $(E, \mathcal{I})$. Then the matroid M is said to be *algebraic over F* or *algebraically representable over F*, and the map φ is called an *algebraic representation* of M over F. An *algebraic matroid* is one that is algebraic over some field. To emphasize the distinction between algebraic representability and the type of representability discussed earlier, we shall sometimes refer to an F-representable matroid as being *linearly representable* over F. Before giving some examples of algebraic matroids, we prove a characterization of the independent sets of such a matroid. This result will facilitate the determination of the collection of independent sets from a list of the elements.

6.7.7 Proposition. *Let $\{t_1, t_2, \dots, t_n\}$ be a non-empty subset of an extension field K of a field F. Then $\{t_1, t_2, \dots, t_n\}$ is algebraically independent over F if and only if there is no non-zero polynomial $f(x_1, x_2, \dots, x_n)$ with coefficients in F such that $f(t_1, t_2, \dots, t_n) = 0$.*

Proof. If $\{t_1, t_2, \ldots, t_n\}$ is algebraically dependent over F, then some t_i is algebraically dependent on $\{t_1, t_2, \ldots, t_{i-1}, t_{i+1}, \ldots, t_n\}$. Therefore there is a non-zero member f of $F[x_1, x_2, \ldots, x_n]$ such that $f(t_1, t_2, \ldots, t_n) = 0$.

Now suppose that $\{t_1, t_2, \ldots, t_n\}$ is algebraically independent. We shall argue by induction on n to show that there is no non-zero member f of $F[x_1, x_2, \ldots, x_n]$ such that $f(t_1, t_2, \ldots, t_n) = 0$. This is immediate if $n = 1$. Assume it true for $n < m$ and suppose that $n = m \geq 2$. Let f be a member of $F[x_1, x_2, \ldots, x_m]$ for which $f(t_1, t_2, \ldots, t_m) = 0$. Group the terms in the polynomial f according to the power of t_m involved. Since $\{t_1, t_2, \ldots, t_m\}$ is algebraically independent, it follows that, for all i, the coefficient $f_i(t_1, t_2, \ldots, t_{m-1})$ of t_m^i is zero. But $\{t_1, t_2, \ldots, t_{m-1}\}$ is also algebraically independent and therefore, by the induction assumption, each f_i is identically zero. Hence f itself is identically zero and the proposition follows by induction. □

6.7.8 Example. We noted in Proposition 6.5.2 that the uniform matroid $U_{2,n}$ is F-representable if and only if $|F| \geq n - 1$. In contrast, $U_{2,n}$ is algebraically representable over F for all fields F. To see this, let E be the subset $\{xy, x^2y, \ldots, x^ny\}$ of $F(x, y)$ where x and y are indeterminates. Then it is straightforward to check that if $\mathcal{I}$ is the set of algebraically independent subsets of E over F, then $(E, \mathcal{I}) \cong U_{2,n}$. Similarly, if $U_{3,n}$ has ground set $\{1, 2, \ldots, n\}$ and φ is the map from this set into $F(x, y, z)$ defined by $\varphi(i) = xy^iz^{i^2}$, then one can show that φ is an algebraic representation for $U_{3,n}$ over F. Indeed, as we shall see in Chapter 12, all transversal matroids are algebraic over all fields, hence so are all uniform matroids. □

Let E be a finite subset of the extension field K of the field F and let M be the matroid induced on E by algebraic dependence over F. Then an element t of E is a loop of M if and only if t is algebraic over F. Equivalently, t is transcendental over F if and only if $\{t\}$ is algebraically independent over F. More generally, suppose that $\{t_1, t_2, \ldots, t_n\}$ is algebraically independent over F. Then $t_1, t_2, \ldots, t_n$ are also called *independent transcendentals* over F. Now let $x_1, x_2, \ldots, x_n$ be indeterminates over F. Then it is not difficult to see that the polynomial rings $F[t_1, t_2, \ldots, t_n]$ and $F[x_1, x_2, \ldots, x_n]$ are isomorphic and hence that the quotient fields $F(t_1, t_2, \ldots, t_n)$ and $F(x_1, x_2, \ldots, x_n)$ are also isomorphic. Thus, algebraically, the n independent transcendentals $t_1, t_2, \ldots, t_n$ behave just like the n indeterminates $x_1, x_2, \ldots, x_n$.

We saw in Example 6.7.8 that a matroid can be algebraic over a field but be non-representable over that field. The next example (Ingleton 1971a) is of an algebraic matroid that is not linearly representable over any field.

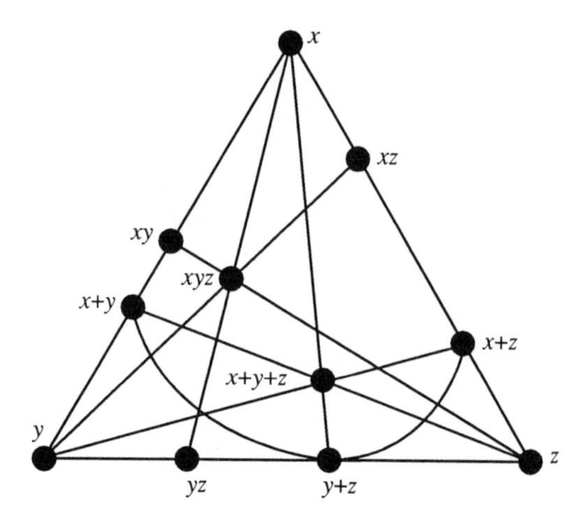

Fig. 6.29. An algebraic matroid that is not linearly representable.

6.7.9 Example. Let E be the subset $\{x, y, z, xy, xz, yz, x+y, x+z, y+z, xyz, x+y+z\}$ of $GF(2)(x, y, z)$ where x, y, and z are independent transcendentals. Let M be the matroid induced on E by algebraic dependence over $GF(2)$. Then Figure 6.29 is a geometric representation of M. To see this, note first that $\{x+y, y+z, x+z\}$ is algebraically dependent because, over $GF(2)$,

$$(x+y)+(y+z)+(x+z)=0.$$

The algebraic equations that correspond to each of the other 3-element circuits shown are easily derived. Therefore Figure 6.29 is indeed a geometric representation of M unless some 3-element subset of $\{xy, xz, yz, x+y+z\}$ is a circuit. But it is straightforward to show that there is no non-zero polynomial f in $GF(2)[x_1, x_2, x_3]$ such that $f(xy, xz, yz) = 0$. Hence $\{xy, xz, yz\}$ is not a circuit. By symmetry, either all or none of the remaining 3-element subsets of $\{xy, xz, yz, x+y+z\}$ is a circuit. But in the former case, circuit elimination implies that $\{xy, xz, yz\}$ is a circuit. We conclude that no 3-element subset of $\{xy, xz, yz, x+y+z\}$ is a circuit and this completes the proof that Figure 6.29 is a geometric representation of M.

Now M is certainly algebraic. But since it clearly has both F_7 and F_7^- as restrictions, it is not linearly representable. □

We show next that the class of matroids that are linearly representable over a field F is a subset of the class of matroids that are algebraic over F. The proof of this follows Piff (1969) (see also Welsh 1976).

6.7.10 Proposition. *If a matroid M is linearly representable over a field F, then M is algebraic over F.*

Proof. Let $r(M) = r$. Evidently we may assume that $r > 0$. Let $\psi : E(M) \to V(r, F)$ be an F-coordinatization of M where the members of $V(r, F)$ are viewed as column vectors. Let $\{e_1, e_2, \ldots, e_r\}$ be a basis of M, and, for all i in $\{1, 2, \ldots, r\}$, let $\psi(e_i) = \underline{v}_i$. Then $\{\underline{v}_1, \underline{v}_2, \ldots, \underline{v}_r\}$ is a basis for $V(r, F)$. Now let $t_1, t_2, \ldots, t_r$ be independent transcendentals over F. Then, for every element e of $E(M)$, the vector $\psi(e)$ can be uniquely written in the form $\sum_{i=1}^{r} a_i \underline{v}_i$ where $a_1, a_2, \ldots, a_r \in F$. For each such e, let $\phi(e)$ be $\sum_{i=1}^{r} a_i t_i$. Then ϕ is a well-defined function from $E(M)$ into $F(t_1, t_2, \ldots, t_r)$. To complete the proof of the proposition, we shall show that ϕ is an algebraic representation of M over F.

First let $\{g_1, g_2, \ldots, g_r\}$ be an arbitrary basis of M. Then, for some non-singular $r \times r$ matrix A over F,

$$(\psi(g_1)\, \psi(g_2) \cdots \psi(g_r)) = (\underline{v}_1\, \underline{v}_2 \cdots \underline{v}_r)A.$$

Thus $(\phi(g_1), \phi(g_2), \ldots, \phi(g_r)) = (t_1, t_2, \ldots, t_r)A$, and so,

$$(\phi(g_1), \phi(g_2), \ldots, \phi(g_r))A^{-1} = (t_1, t_2, \ldots, t_r).$$

Hence each of $t_1, t_2, \ldots, t_r$ is algebraically dependent on $\{\phi(g_1), \phi(g_2), \ldots, \phi(g_r)\}$. But $\{t_1, t_2, \ldots, t_r\}$ is a basis for the algebraic matroid on $\phi(E(M))$ and therefore so too is $\{\phi(g_1), \phi(g_2), \ldots, \phi(g_r)\}$.

Now let $\{c_1, c_2, \ldots, c_k\}$ be a circuit C of M. Then $\psi(C)$ is a minimal linearly dependent subset of $V(r, F)$. Hence, for some non-zero elements $a_1, a_2, \ldots, a_k$ of F, $\sum_{i=1}^{k} a_i \psi(c_i) = \underline{0}$. Thus $\sum_{i=1}^{k} a_i \phi(c_i) = 0$, so $\phi(C)$ is algebraically dependent over F. We conclude that ϕ is indeed an algebraic representation for M over F. □

Example 6.7.8 shows that the converse of Proposition 6.7.10 fails for all finite fields. However, for fields of characteristic zero, we have the following result whose proof we omit (see, for example, Lang 1965, Chapter 10, Proposition 10).

6.7.11 Proposition. *If a matroid M is algebraic over a field F of characteristic zero, then M is linearly representable over $F(T)$ for some finite set T of transcendentals over F.* □

We know from Example 6.7.9 and Proposition 6.7.10 that the class of algebraic matroids is large. Indeed, Ingleton (1971a) asked whether this class coincides with the class of all matroids. Subsequently, he and Main (1975) proved that the Vámos matroid is non-algebraic. Their result was used by Lindström (1984a, 1986b) to prove that the matroid for which a geometric representation is shown in Figure 6.30(a) is also non-algebraic. By generalizing the construction of the Vámos matroid, Lindström (1988a) defined a class of non-algebraic matroids consisting of

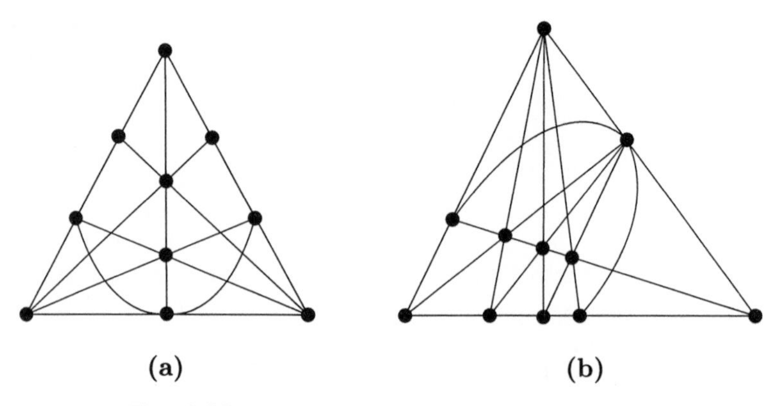

Fig. 6.30. Two non-algebraic matroids.

one matroid of each rank exceeding three. In addition, he found (1987c) an infinite class of rank-3 non-algebraic matroids, each having the property that all of its proper minors are algebraic. A geometric representation for the first member of this class is shown in Figure 6.30(b).

Next we shall consider the effect on algebraic matroids of the three fundamental operations of deletion, contraction, and the taking of duals. Evidently if a matroid M is algebraic over F and $T \subseteq E(M)$, then $M|T$ is algebraic over F. The behaviour of M under contraction is more difficult to determine and relies on the following result of Lindström (1989) which proves a conjecture of Piff (1972). We remark that Lindström's paper notes that two earlier papers (Lindström 1987b, Shameeva 1985) claiming to prove Piff's conjecture were, in fact, incomplete.

6.7.12 Proposition. *If a matroid M is algebraic over an extension field $F(t)$ of a field F, then M is algebraic over F.* □

The real substance of the last result lies in the case when t is transcendental over F, for Piff (1972) showed that the result is straightforward when t is algebraic over F (Exercise 3). From the last proposition, Lindström (1989) was able to deduce that the class of algebraic matroids is closed under contraction.

6.7.13 Proposition. *If a matroid M is algebraic over a field F and $T \subseteq E(M)$, then $M.T$ is algebraic over F.*

Proof. By Proposition 6.7.12, it suffices to show that $M.T$ is algebraic over $F(U)$ for some finite set U. Our proof of this follows Welsh (1976). Let $\phi : E(M) \to K$ be an algebraic representation of M over F. Then K is an extension field of $F(\phi(E(M) - T))$. Denote by ϕ' the restriction of the map ϕ to the set T. Let I be a subset of T and B be a basis of

$E(M) - T$. The following sequence of equivalent statements shows that ϕ' is an algebraic representation of $M.T$ over $F(\phi(E(M) - T))$ where the verification that (iii) is equivalent to (iv) uses Corollary 6.7.6(ii), while the equivalence of (iv) and (v) follows by Corollary 6.7.6(i):

(i) I is independent in $M.T$.

(ii) $I \cup B$ is independent in M.

(iii) $\phi(I \cup B)$ is algebraically independent over F, and $|\phi(I \cup B)| = |I \cup B|$.

(iv) $\phi(I)$ is algebraically independent over $F(\phi(B))$, and $\phi(I)| = |I|$.

(v) $\phi(I)$ is algebraically independent over $F(\phi(E(M) - T))$, and $|\phi(I)| = |I|$. □

6.7.14 Corollary. *If a matroid M is algebraic over a field F, then every minor of M is algebraic over F.* □

In view of this corollary and the characterizations of binary and ternary matroids noted in Section 6.5, it is natural to try to determine the excluded minors for algebraic representability over certain specific fields. However, as Lemos (1988) has noted, one can use one of Lindström's (1987c) classes of non-algebraic matroids or another class of matroids considered by Gordon (1984) to show that, for every field F, the set of such excluded minors is infinite.

Probably the most basic unsolved problem in the study of algebraic matroids is the following.

6.7.15 Problem. *Is the dual of an algebraic matroid also algebraic?* □

Lindström (1985b) has shown that to answer this question it suffices to solve the following problem, although he notes that this problem also seems very difficult.

6.7.16 Problem. *If a matroid is algebraic over a field of a certain characteristic, is its dual also algebraic over a field of this characteristic?* □

For a discussion of various other unsolved problems related to algebraic matroids, the reader is referred to Lindström (1988b).

Exercises

1. Show that all graphic matroids are algebraic over every field.
2. By giving an explicit algebraic representation, show that $U_{r,n}$ is algebraic over all fields for all r and n with $0 \le r \le n$.

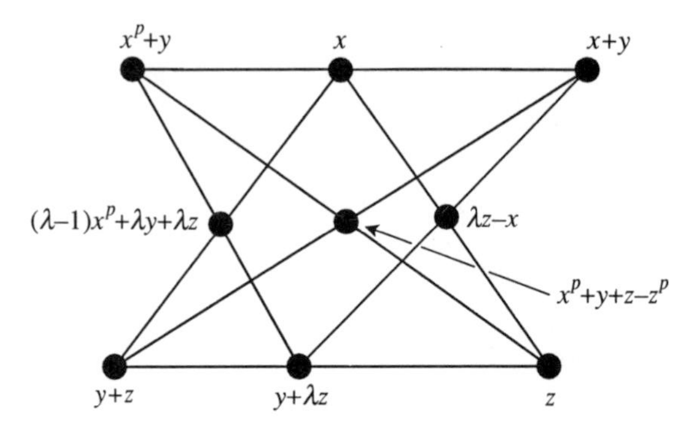

Fig. 6.31. The non-Pappus matroid is algebraic over $GF(p^2)$.

3. Show that if M is algebraic over $F(t)$ where t is algebraic over F, then M is algebraic over F.

4. Suppose that $M \cong M(G)$ for some graph G and let F be a field. Describe how to obtain explicit algebraic representations for M and M^* over F.

5. (Lindström 1988b) Prove that if M is algebraic over a field F, then M is algebraic over the prime field of F. Deduce that if $n \in \mathbb{Z}^+$ and M is algebraic over all fields with at least n elements, then M is algebraic over all fields.

6. (Lindström 1983, 1986a) Let p be a prime number and λ be an element of $GF(p^2)$ such that $\lambda^p \neq \lambda$.

(i) Show that Figure 6.31 depicts an algebraic representation for the non-Pappus matroid over $GF(p^2)$.

(ii) Deduce that the non-Pappus matroid is algebraic over $GF(p)$ for all primes p.

7. **(i)*** (Lindström 1985d) Prove that if a matroid is linearly representable over $\mathbb{Q}$, then it is algebraic over all fields.

(ii) Deduce that a matroid which is algebraic over some field of characteristic 0 is algebraic over all fields.

8. **(i)*** (Lindström 1985c) Prove that the Fano matroid is algebraic only over fields of characteristic two.

(ii) Show that the non-Fano matroid is algebraic over all fields.

9.* (Lindström 1985a) Show that the non-Desargues matroid (see Figure 6.6) is non-algebraic.

10.* (Ingleton and Main 1975) Let $\{1,1',2,2',3,3',4,4'\}$ be a set E of elements of an extension field K of a field F and let M be the matroid induced on E by algebraic dependence over F. Suppose that every subset of E with fewer than four elements is independent, while every subset of E with more than four elements has rank 4.

(i) Show that if $\{1,1',2,2'\}$, $\{1,1',3,3'\}$, $\{1,1',4,4'\}$, $\{2,2',3,3'\}$, and $\{2,2',4,4'\}$ are dependent, then $\{3,3',4,4'\}$ is also dependent.

(ii) Deduce that the Vámos matroid, Q_8 (6.5.10), and the relaxation of $AG(3,2)$ are non-algebraic.

6.8 Characteristic sets

For a matroid M, the *characteristic set* $\mathcal{K}(M)$ of M is the set $\{k : M$ is representable over some field of characteristic $k\}$. This notion was introduced by Ingleton (1971a), although his definition of characteristic set was somewhat broader, since it was in terms of representability over division rings. We have seen a number of examples of where a matroid M fails to be representable over some prime field $GF(p)$ simply because the field is too small. In such circumstances, p is not excluded from $\mathcal{K}(M)$ since M will be representable over some suitable extension of $GF(p)$. In general, one can view the fact that p is not in $\mathcal{K}(M)$ as a reflection of some non-trivial structural property of M.

If P is the set of prime numbers, then $\mathcal{K}(M) \subseteq P \cup \{0\}$ for all matroids M. The following question was asked in the division-ring case by Ingleton (1971a).

6.8.1 Problem. *For which subsets T of $P \cup \{0\}$, is there a matroid M having T as its characteristic set?* □

This question has now been completely answered and, in this section, we survey the results that lead to the answer along with various results about characteristic sets of certain particular classes of matroids.

Part of the motivation for Problem 6.8.1 derived from the following result of Rado (1957).

6.8.2 Proposition. *If $0 \in \mathcal{K}(M)$, then $\mathcal{K}(M)$ contains all sufficiently large primes.* □

Vámos (1971b) showed that the converse of this proposition is also true. Indeed, he established the following stronger result.

6.8.3 Proposition. *If $\mathcal{K}(M)$ is infinite, then $0 \in \mathcal{K}(M)$.* □

On combining the last two results, we deduce that, if T is the characteristic set of some matroid, then either

(i) $0 \in T$ and $P - T$ is infinite; or

(ii) $0 \notin T$ and T is finite.

The question as to which sets satisfying (i) do actually occur as characteristic sets was resolved by R. Reid (see Brylawski and Kelly 1980, pp. 101–2) who discovered the following.

6.8.4 Example. Let k be an integer exceeding one and α be a primitive kth root of unity in $\mathbb{C}$, that is, α is an element of $\mathbb{C}$ for which $\alpha^k = 1$ and $1, \alpha, \alpha^2, \ldots, \alpha^{k-1}$ are distinct. Let A_k be the following matrix over $\mathbb{C}$:

$$\begin{array}{c} \begin{array}{ccc|ccccccccccc} e_1 & e_2 & e_3 & e_4 & e_5 & a_0 & b_0 & a_1 & b_1 & \cdots & a_{k-2} & b_{k-2} \end{array} \\ \left[\begin{array}{c|ccccccccc} & \boxed{1} & \boxed{1} & \boxed{1} & 0 & \boxed{1} & 0 & \cdots & \boxed{1} & 0 \\ I_3 & 1-\alpha & 1 & \boxed{1} & \boxed{1} & 1 & \boxed{1} & \cdots & 1 & \boxed{1} \\ & 1 & 0 & \boxed{1} & 1 & f_1(\alpha) & f_1(\alpha) & \cdots & f_{k-2}(\alpha) & f_{k-2}(\alpha) \end{array}\right] \end{array}$$

where, for all i in $\{1, 2, \ldots, k-1\}$, $f_i(\alpha) = \alpha^i + \alpha^{i-1} + \cdots + 1$. Then it is not difficult to check that Figure 6.32 is a geometric representation for $M[A_k]$. Evidently $0 \in \mathcal{K}(M[A_k])$. In what follows, we shall be viewing A_k as a matrix over fields other than $\mathbb{C}$. We emphasize here that, throughout this discussion, $M[A_k]$ will always denote the vector matroid of the *complex* matrix A_k.

6.8.5 Lemma. *Let p be a prime number. Then $p \in \mathcal{K}(M[A_k])$ if and only if k is not divisible by p.*

Proof. Let F be a field. If $M[A_k]$ is F-representable, then, by Theorem 6.4.7, we may assume that $M[A_k]$ has an F-representation $[I_3|X]$ so that, in X, each entry that corresponds to a boxed entry of A_k equals 1. Since $\{e_3, e_5, a_0, a_1, a_2, \ldots, a_{k-2}\}$ is a line of $M[A_k]$, it follows that the second entries in the e_5-, a_1-, a_2-, ..., and a_{k-2}-columns of X are all 1. The circuit $\{e_2, a_0, e_4\}$ implies that the third entry in the e_4-column of X is a 1, while the circuits $\{e_1, a_i, b_i\}$, for i in $\{0, 1, \ldots, k-2\}$, imply that the third entries of the a_i- and b_i-columns of X are equal. Thus, for some non-zero elements $x_0, x_1, \ldots, x_{k-2}$ of F, the matrix X is

$$\begin{array}{c} \begin{array}{ccccccccccc} e_4 & e_5 & a_0 & b_0 & a_1 & b_1 & a_2 & b_2 & \cdots & a_{k-2} & b_{k-2} \end{array} \\ \left[\begin{array}{ccccccccccc} 1 & 1 & 1 & 0 & 1 & 0 & 1 & 0 & \cdots & 1 & 0 \\ x_0 & 1 & 1 & 1 & 1 & 1 & 1 & 1 & \cdots & 1 & 0 \\ 1 & 0 & 1 & 1 & x_1 & x_1 & x_2 & x_2 & \cdots & x_{k-2} & x_{k-2} \end{array}\right]. \end{array}$$

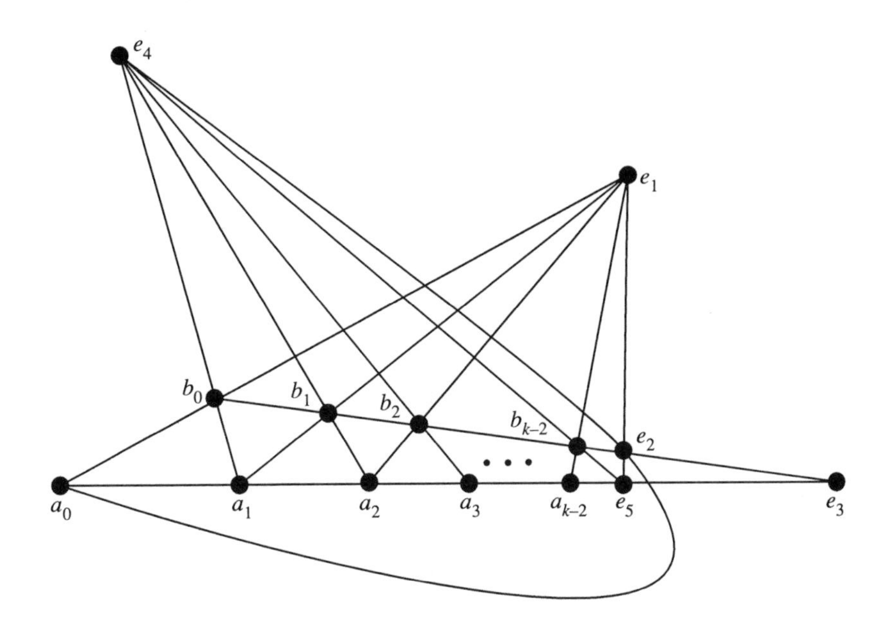

Fig. 6.32. $M[A_k]$ is $GF(p)$-representable if and only if $p \nmid k$.

Now writing $1 - x$ for x_0, we deduce from the circuit $\{e_4, b_0, a_1\}$ of $M[A_k]$ that $x_1 = 1+x$. In general, since $\{e_4, b_i, a_{i+1}\}$ is a circuit of $M[A_k]$ for all i in $\{1, 2, \ldots, k-3\}$, we get that $x_{i+1} = xx_i + 1$. It follows that $x_{i+1} = f_{i+1}(x)$. Finally, as $\{e_4, b_{k-2}, e_5\}$ is a circuit, $xx_{k-2} + 1 = 0$, that is, $f_{k-1}(x) = 0$. Hence $(x-1)f_{k-1}(x) = 0$, so $x^k - 1 = 0$.

As x_0 is non-zero, $x \neq 1$. Moreover, as none of $x_1, x_2, \ldots, x_{k-2}$ is 0, none of $f_1(x), f_2(x), \ldots, f_{k-2}(x)$ is 0, that is, $x^2-1, x^3-1, \ldots, x^{k-1}-1$ are all non-zero. We conclude that if $M[A_k]$ is F-representable, then F contains x, a primitive kth root of unity, and an F-representation for $M[A_k]$ is $[I_3|X]$ where X is

$$\begin{array}{ccc|ccccccccc}
e_1 & e_2 & e_3 & e_4 & e_5 & a_0 & b_0 & a_1 & b_1 & \cdots & a_{k-2} & b_{k-2} \\
 & & & 1 & 1 & 1 & 0 & 1 & 0 & \cdots & 1 & 0 \\
 & I_3 & & 1-x & 1 & 1 & 1 & 1 & 1 & \cdots & 1 & 1 \\
 & & & 1 & 0 & 1 & 1 & f_1(x) & f_1(x) & \cdots & f_{k-2}(x) & f_{k-2}(x)
\end{array}.$$

But, as every line of $M[A_k]$ is also a line of $M[I_3|X]$, it follows without difficulty that $[I_3|X]$ is indeed an F-representation for $M[A_k]$. Thus $M[A_k]$ is F-representable if and only if, when A_k is viewed as a matrix over F, it is an F-representation for $M[A_k]$ where α is a primitive kth root of unity in F. Now suppose F has characteristic p. Then, viewed over F, every entry in A_k is in $GF(p)(\alpha)$. Hence $M[A_k]$ is F-representable if and only if it is represented over $GF(p)(\alpha)$ by A_k. The result now follows by using the following standard result from field theory.

6.8.6 Lemma. *The algebraic closure of $GF(p)$ contains a primitive kth root of unity if and only if k is not divisible by p.*

Proof. This follows, for example, from the remarks on page 295 of Hungerford (1974). □

6.8.7 Proposition. *If $0 \in T$ and $P - T$ is finite, then there is a matroid M for which $\mathcal{K}(M) = T$.*

Proof. Let $k = \prod_{p \in P-T} p$ and $M = M[A_k]$. The result follows immediately from Lemma 6.8.5. □

The following result of Kahn (1982) completes the answer to Problem 6.8.1.

6.8.8 Proposition. *If T is a finite set of prime numbers, then there is a matroid whose characteristic set is T.* □

A sketch of the main ideas in the proof of the last result can be found in White (1987b). This proof uses several transcendental field extensions in its construction of a matroid with characteristic set T, so the matroids formed are, in general, not representable over the corresponding prime fields. By extending the ideas in Example 6.8.4 and some other unpublished work of R. Reid, Brylawski (1982b) constructed large numbers of rank-3 matroids that are only representable over a small number of prime fields. For example, one such matroid is representable over $GF(23)$ and $GF(59)$ but over no other characteristics.

In closing this section, we note that, in contrast to the classes of matroids with small characteristic sets just considered, many of the matroids M we have already met have characteristic set equal to $P \cup \{0\}$. This is true, for example, if M is regular, and so, in particular, if M is graphic or cographic. Moreover, it follows from a result of Piff and Welsh (1970), to be proved in Chapter 12, that if M is transversal, and hence if M is uniform, then $\mathcal{K}(M) = P \cup \{0\}$.

Exercises

1. With A_k as in Example 6.8.4, show that $M[A_2] \cong F_7^-$.
2. Let M be a matroid that is representable over $\mathbb{Q}$. Show that:

   (i) M is representable over all fields of characteristic zero;

(ii) there is an integer $n(M)$ such that M is representable over $GF(p)$ for all primes p exceeding $n(M)$.

3. (Reid in Brylawski and Kelly 1980) Let p be a prime number exceeding two and M_p be the vector matroid of the following matrix over $GF(p)$:

$$\left[\begin{array}{c|cccccccccc} & 1 & 1 & 0 & 0 & \cdots & 0 & 1 & 1 & \cdots & 1 \\ I_3 & 1 & 1 & 1 & 1 & \cdots & 1 & 0 & 0 & \cdots & 0 \\ & 1 & 0 & 1 & 2 & \cdots & p-1 & 1 & 2 & \cdots & p-1 \end{array}\right].$$

(i) Give a geometric representation for M_p.

(ii) Show that $M[A_p]$ is isomorphic to a relaxation of M_p.

(iii) Prove that M_p has characteristic set $\{p\}$.

(iv*) Delete the third element from M_p. Prove that the resulting matroid is an excluded minor for $\mathbb{R}$-representability.

4. (Brylawski 1982b) Let p be an odd prime number and let $m = \lfloor \log_2(p+1) \rfloor$. For $0 \leq i \leq m$, define x_i to be $\lfloor (p+1)/2^{m-i+1} \rfloor$. Let N_p be the vector matroid of the following matrix over $GF(p)$:

$$\begin{array}{c|ccccccccccc} e_1\ e_2\ e_3 & e_4 & e_5 & e_6 & f_1 & g_1 & f_2 & g_2 & \cdots & f_m & g_m \\ \hline & 1 & 1 & 1 & 1 & 0 & 1 & 0 & \cdots & 1 & 0 \\ I_3 & 1 & 1 & 0 & 2 & 1 & 2 & 1 & \cdots & 2 & 1 \\ & 1 & 0 & 1 & x_1 & x_1 & x_2 & x_2 & \cdots & x_m & x_m \end{array}.$$

(i) Show that N_{13} is the vector matroid of the following matrix over $GF(13)$:

$$\left[\begin{array}{c|ccccccccc} & 1 & 1 & 1 & 1 & 0 & 1 & 0 & 1 & 0 \\ I_3 & 1 & 1 & 0 & 2 & 1 & 2 & 1 & 2 & 1 \\ & 1 & 0 & 1 & 1 & 1 & 3 & 3 & 7 & 7 \end{array}\right].$$

(ii) Suppose that $y_1 y_2 \cdots y_{m+1}$ is $p+1$ written in binary, that is, $p+1 = y_1 2^m + y_2 2^{m-1} + \cdots + y_{m+1} 2^0$. Show that x_i is the number which, when written in binary, is $y_1 y_2 \cdots y_i$.

(iii*) Show that N_p has characteristic set $\{p\}$.

(iv) Suppose that $p \geq 5$. Prove that $AG(2,p)$ has characteristic set $\{p\}$ by showing that it has N_p as a restriction.

6.9 Modularity

Let X and Y be flats in a matroid M. Then (X, Y) is a *modular pair of flats* if

$$r(X) + r(Y) = r(X \cup Y) + r(X \cap Y).$$

If Z is a flat of M such that (Z, Y) is a modular pair for all flats Y, then Z is called a *modular flat* of M. Evidently $E(M)$, $\mathrm{cl}(\emptyset)$, and all rank-1 flats of M are modular flats; so are all separators of M. In this section, we shall study some of the properties of modular flats. Such flats are, in general, relatively well-behaved and the property of modularity will be particularly important in certain matroid constructions that will be considered in Chapters 7 and 12. We noted in Section 6.1 that the projective geometries $PG(n, q)$ have the property that all their flats are modular. Another important result of this section will note that there are relatively few other matroids with this property.

Because modularity is a property of flats, it is basically a lattice property. Indeed, it is clear that a flat X is modular in a matroid M if and only if, in the simple matroid $\widetilde{M}$ associated with M, the flat corresponding to X is modular. For this reason, much of the present discussion of modularity will concentrate on simple matroids. The reader should have little difficulty deriving the corresponding results for non-simple matroids.

We begin by looking at *modular matroids*, those matroids in which every flat is modular. Every free matroid is easily seen to be modular, so too is every finite projective plane. Since, by Theorem 6.1.1, a finite projective space is either a finite projective plane or a projective geometry, $PG(n, q)$, we deduce that every finite projective space is a modular matroid. Moreover, it is easy to show (Exercise 1) that the direct sum of two modular matroids is also modular. Thus, by taking direct sums of free matroids and finite projective spaces, one can construct more examples of modular matroids. It follows from the next result that every simple modular matroid is obtainable in this way.

6.9.1 Proposition. *A matroid M is modular if and only if, for every connected component N of M, the simple matroid associated with N is either a free matroid or a finite projective space.*

Proof. A proof, in the context of geometric lattices, can be found in Birkhoff (1967, pp. 90–93). □

A flat Y in a matroid M is a *complement* of the flat X if $X \cap Y = \mathrm{cl}(\emptyset)$ and $\mathrm{cl}(X \cup Y) = E(M)$. Thus Y is a complement of X if and only if, in $\mathcal{L}(M)$, the meet and join of X and Y are the zero and one, respectively, of the lattice.

The remaining results in this section are taken from Brylawski (1975b). The first such result gives two alternative characterizations of modular flats which appear to be progressively weaker than the definition.

6.9.2 Proposition. *The following statements are equivalent for a flat X in a matroid M:*

(i) *X is modular.*

(ii) *$r(X) + r(Y) = r(X \cup Y)$ for all flats Y that meet X in* $\mathrm{cl}(\emptyset)$.

(iii) *$r(X) + r(Y) = r(M)$ for all complements Y of X.*

Proof. Evidently (i) implies (ii), and (ii) implies (iii). We shall show that (iii) implies (ii) and omit the similar proof that (ii) implies (i).

Assume that (iii) holds, but (ii) does not, and let Y be a flat for which $X \cap Y = \mathrm{cl}(\emptyset)$ and $r(X) + r(Y) \neq r(X \cup Y)$. Because (iii) holds, Y is not a complement of X, so $r(X \cup Y) < r(M)$. Let Z be a basis for $M/(X \cup Y)$. We show next that $\mathrm{cl}(Y \cup Z)$ is a complement of X. Clearly $\mathrm{cl}(X \cup \mathrm{cl}(Y \cup Z)) = E(M)$. Now suppose that a is a non-loop element of $X \cap \mathrm{cl}(Y \cup Z)$. Then $a \notin Y$ since $X \cap Y = \mathrm{cl}(\emptyset)$. Thus if B is a basis of Y, then $B \cup a$ is independent in $M|(X \cup Y)$. But Z is independent in $M/(X \cup Y)$. Therefore $Z \cup B \cup a$ is independent in M, a contradiction to the fact that $a \in \mathrm{cl}(Y \cup Z)$. Hence $X \cap \mathrm{cl}(Y \cup Z) = \mathrm{cl}(\emptyset)$, so $\mathrm{cl}(Y \cup Z)$ is indeed a complement of X. By (iii),

$$\begin{aligned} r(M) = r(X) + r(\mathrm{cl}(Y \cup Z)) &= r(X) + r(Y) + |Z| \\ &= r(X) + r(Y) + r(M) - r(X \cup Y). \end{aligned}$$

Therefore $r(X \cup Y) = r(X) + r(Y)$. This contradiction completes the proof that (iii) implies (ii). □

Recall that a *line* in a matroid is a rank-2 flat.

6.9.3 Corollary. *In a simple matroid, a hyperplane is modular if and only if it intersects every line.*

Proof. This is immediate from the last proposition. □

6.9.4 Example. Consider the matroids $\mathcal{W}^3$ and M for which geometric representations are shown in Figure 6.33. From the last corollary, we deduce that $\mathcal{W}^3$ has no modular hyperplanes, whereas M has exactly two modular hyperplanes, $\{2,3,4,8\}$ and $\{5,6,7,8\}$. □

The next result enables us to identify certain particular flats as being modular. Its straightforward proof is left to the reader (Exercise 6).

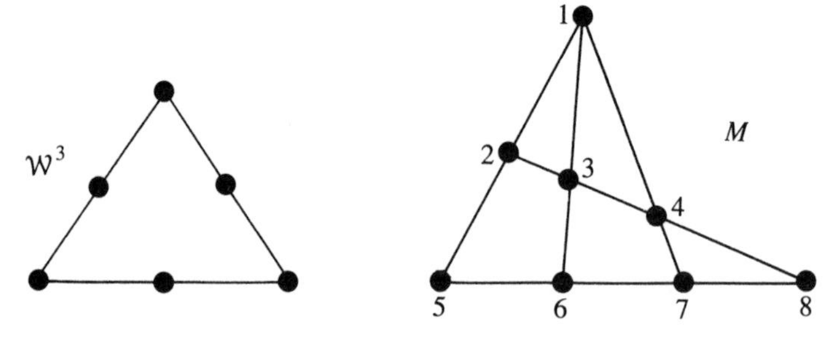

Fig. 6.33. M has two modular lines; $\mathcal{W}^3$ has none.

6.9.5 Proposition. *Let X be a modular flat in a matroid M and T be a subset of $E(M)$ containing X. Then X is a modular flat of $M|T$.* □

Using this result, the reader can easily prove the following.

6.9.6 Corollary. *If M is a simple matroid representable over $GF(q)$ and X is a subset of $E(M)$ such that $M|X \cong PG(k-1, q)$ for some k, then X is a modular flat of M.* □

From this corollary, we deduce that 3-point and 4-point lines are always modular flats in simple binary and simple ternary matroids, respectively. In particular, a 3-circuit is always a modular flat in a simple regular matroid.

The proofs of the next two results are based on the characterization of modular flats in (iii) of Proposition 6.9.2. Each adds further support to the earlier assertion that modular flats are relatively well-behaved.

6.9.7 Proposition. *Let X be a modular flat in a matroid M and Y be a modular flat in $M|X$. Then Y is a modular flat in M.*

Proof. Clearly Y is a flat of M. Now let Z be a complement of Y in M. Then $\text{cl}(Y \cup Z) = E(M)$. Thus, as $X \supseteq Y$, it follows that $\text{cl}(X \cup Z) = E(M)$. Moreover, $X \cap Z \subseteq \text{cl}((X \cap Z) \cup Y) \subseteq X$, so $X \cap Z \subseteq \text{cl}((X \cap Z) \cup Y) \cap Z \subseteq X \cap Z$, that is,

$$\text{cl}((X \cap Z) \cup Y) \cap Z = X \cap Z.$$

Therefore, since $Y \cap Z = \text{cl}(\emptyset)$, the Hasse diagram in Figure 6.34 represents a sublattice of $\mathcal{L}(M)$ where, possibly, $\text{cl}((X \cap Z) \cup Y) = X$.

As X is a modular flat and $\text{cl}(X \cup Z) = E(M)$,

$$r(M) = r(X \cup Z) = r(X) + r(Z) - r(X \cap Z). \tag{1}$$

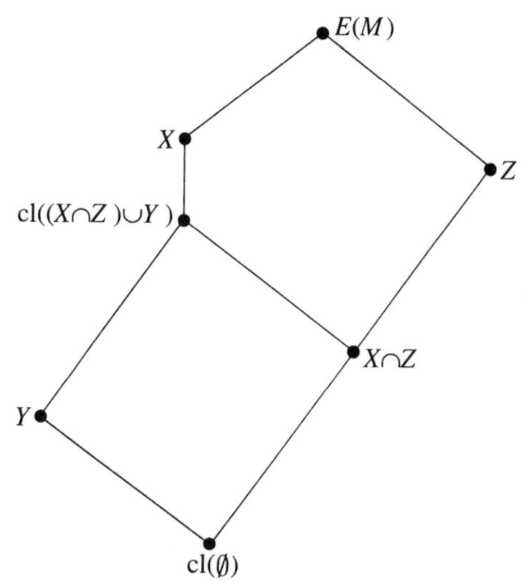

Fig. 6.34. A sublattice of $\mathcal{L}(M)$.

Thus

$$\begin{aligned}
r(M) &\geq r(\text{cl}((X \cap Z) \cup Y)) + r(Z) - r(X \cap Z) \\
&= r(\text{cl}((X \cap Z) \cup Y)) + r(Z) - r(\text{cl}((X \cap Z) \cap Y) \cap Z) \\
&\geq r(\text{cl}((X \cap Z) \cup Y \cup Z)) \\
&\geq r(Y \cup Z) = r(M).
\end{aligned}$$

Clearly equality must hold throughout the last sequence of inequalities and therefore

$$r(M) = r(\text{cl}((X \cap Z) \cup Y)) + r(Z) - r(X \cap Z). \tag{2}$$

Comparing (1) and (2), we deduce, since $\text{cl}((X \cap Z) \cup Y) \subseteq X$, that $\text{cl}((X \cap Z) \cup Y) = X$. Thus, in the Hasse diagram in Figure 6.34, the vertices corresponding to $\text{cl}((X \cap Z) \cup Y)$ and X coincide. This diagram now suggests how to finish the proof, for Y is modular in $M|X$, so

$$\begin{aligned}
r(Y) + r(X \cap Z) &= r((X \cap Z) \cup Y) + r((X \cap Z) \cap Y) \\
&= r(X) + r(\text{cl}(\emptyset)).
\end{aligned}$$

Substituting for $r(X \cap Z)$ from (1) and simplifying, we get that

$$r(Y) + r(Z) = r(M),$$

and the proposition follows by Proposition 6.9.2(iii). □

6.9.8 Corollary. *If X and Y are modular flats in a matroid M, then $X \cap Y$ is a modular flat.*

Proof. By the last result, it suffices to show that $X \cap Y$ is modular in $M|X$. This can be done using an argument similar to the last one. We leave it to the reader to complete the details. □

Brylawski (1975b) calls the next result the *modular short-circuit axiom.* It provides a useful characterization of modular flats in terms of circuits.

6.9.9 Theorem. *The following statements are equivalent for a non-empty set X of elements in a simple matroid M:*

(i) *X is a modular flat of M.*

(ii) *For every circuit C such that $C - X$ is non-empty, there is an element x of X such that $(C - X) \cup x$ is dependent.*

(iii) *For every circuit C and every element e of $C - X$, there is an element f of X and a circuit C' such that $e \in C' \subseteq f \cup (C - X)$.*

Proof. Assume that X satisfies (iii). We shall show that X satisfies (i). As M is simple and $\text{cl}(X) = X \cup \{x\colon\ M$ has a circuit D with $x \in D \subseteq X \cup x\}$, it follows easily that $\text{cl}(X) = X$. Now assume that X is not modular. Then, by Proposition 6.9.2(ii) and semimodularity, M has a flat Y disjoint from X such that $r(X)+r(Y) > r(X \cup Y)$. Let B_X and B_Y be bases for X and Y, respectively. Then, as $r(B_X \cup B_Y) < |B_X \cup B_Y|$, the set $B_X \cup B_Y$ must contain a circuit C, and C must meet both X and Y. Choose e in $C - X$. Then there is an element f in X and a circuit C' such that $e \in C' \subseteq f \cup (C - X)$. Hence $f \in X \cap \text{cl}(Y) = X \cap Y = \emptyset$; a contradiction. Thus (iii) implies (i).

Next we show that (i) implies (ii). Suppose that X is a modular flat and let C be a circuit such that $C - X$ is non-empty. If $C \cap X$ is empty, then (ii) certainly holds. Thus we may assume that $C \cap X$ is non-empty. Then both $C \cap X$ and $C - X$ are independent. Now (ii) will hold if we can show that $X \cap \text{cl}(C - X)$ is non-empty. Assume the contrary. Then, as X is modular,

$$\begin{aligned} r(X \cup \text{cl}(C - X)) &= r(X) + r(\text{cl}(C - X)) \\ &= r(X) + r(C - X) \\ &= r(X) + |C - X|. \end{aligned}$$

Therefore, as $r(X \cup \text{cl}(C - X)) = r(X \cup \text{cl}(C))$,

$$r(X \cup \text{cl}(C)) = r(X) + |C - X|. \tag{3}$$

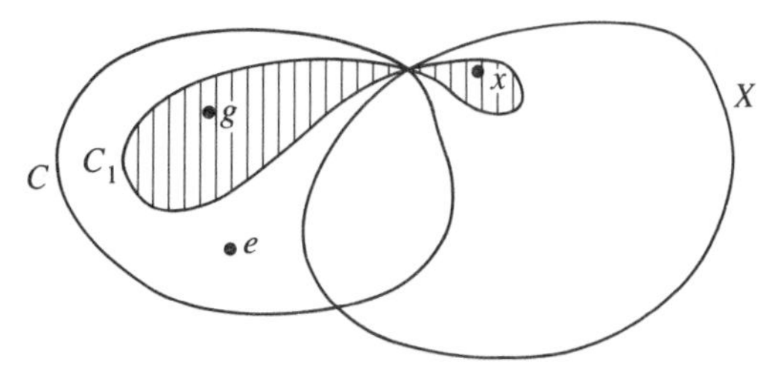

Fig. 6.35. Applying strong circuit elimination.

Again, as X is modular,

$$\begin{aligned} r(X \cup \mathrm{cl}(C)) &= r(X) + r(\mathrm{cl}(C)) - r(X \cap \mathrm{cl}(C)) \\ &\leq r(X) + r(C) - r(X \cap C). \end{aligned} \tag{4}$$

Therefore, as $r(X \cap C) = |X \cap C|$, we get, on combining (3) and (4), that $|C| \leq r(C)$, a contradiction to the fact that C is a circuit. Hence $X \cap \mathrm{cl}(C - X)$ is non-empty, and (i) implies (ii).

Now suppose that (ii) holds. We shall show that (iii) must hold. Let C be a circuit of M and e be an element of $C - X$. By (ii), there is an element x of X such that $(C - X) \cup x$ is dependent. We shall argue by induction on $|C - X|$ to show that the required circuit C' exists. As $x \in X \cap \mathrm{cl}(C - X)$, there is a circuit C_1 such that $x \in C_1 \subseteq (C - X) \cup x$. Since M is simple, $|C_1| \geq 3$. Thus $|C - X| \geq 2$, and if $|C - X| = 2$, then $e \in C_1$ and we can take $C' = C_1$ and $f = x$. Thus (iii) holds when $|C - X| = 2$. Assume it holds for $|C - X| < k$ and let $|C - X| = k$. If $e \in C_1$, we again take $C' = C_1$ and $f = x$. Thus assume $e \notin C_1$. Now take g in $C_1 \cap C$ (see Figure 6.35). Then, by the strong circuit elimination axiom, there is a circuit C_2 such that $x \in C_2 \subseteq (C \cup C_1) - g$. Since $(C_1 \cup C_2) - x \subseteq C$ and $(C_1 \cup C_2) - x$ contains a circuit, we must have that $(C_1 \cup C_2) - x = C$. Hence $C_2 \supseteq C - C_1$, so $e \in C_2$. As $C_2 - X$ is a subset of $(C - g) - X$, it follows that $|C_2 - X| < |C - X|$. Thus, by the induction assumption, as $e \in C_2$, there is a circuit C_3 and an element h of X such that $e \in C_3 \subseteq h \cup (C_2 - X)$. But $h \cup (C_2 - X) \subseteq h \cup (C - X)$ so, letting $C' = C$ and $f = h$, we have a circuit and an element with the required properties, and (iii) follows by induction. □

We close this section by noting two consequences of the last theorem, neither of which has a difficult proof. Some further consequences of the theorem will be considered in the exercises. The first application of modularity we shall meet will be in Section 7.2 where we consider the reverse of the deletion operation.

6.9.10 Corollary. *Let X be a flat of a simple matroid M and suppose that $X_1, X_2, \ldots, X_k$ are the components of $M|X$. Then X is modular if and*

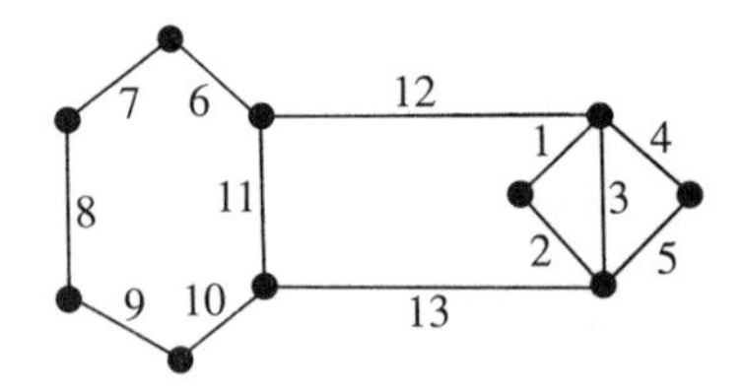

Fig. 6.36. Not every modular flat comes from a complete subgraph.

only if M has distinct components $M_1, M_2, \ldots, M_k$ such that X_i is a modular flat of M_i for all i in $\{1, 2, \ldots, k\}$. In particular, modular flats of simple connected matroids are connected.

Proof. This is left to the reader (see Exercise 7(ii)). □

One can determine when a set is a modular flat in a graphic matroid by combining the last corollary with the following result.

6.9.11 Proposition. *Let X be a set of edges of a simple graph G and suppose that $G[X]$ is connected. Then X is a modular flat in $M(G)$ if and only if, whenever a pair $\{u, v\}$ of distinct vertices of $G[X]$ is joined both by a path in X and a path that meets $G[X]$ in $\{u, v\}$, there is an edge of X joining u and v.* □

Note that an immediate consequence of this proposition is that if G is a simple graph and $G[X] \cong K_n$ for some $n \geq 2$, then X is a modular flat of $M(G)$. However, not every modular flat in a graphic matroid comes from a complete subgraph.

6.9.12 Example. Consider the graph G in Figure 6.36. By Proposition 6.9.11, $M(G)$ has several modular flats including $\{1, 2, 3, 4, 5\}$, $\{6, 7, 8, 9, 10, 11\}$, $\{1, 2, 3, 4, 5, 11, 12, 13\}$, and $\{3, 6, 7, 8, 9, 10, 11, 12, 13\}$. □

Exercises

1. Prove that the direct sum of two modular matroids is modular.

2. Show that a matroid is modular if and only if every hyperplane is a modular flat.

3. Prove that (ii) implies (i) in 6.9.2.

4. Find all simple $GF(q)$-representable matroids N with the property that if M is a simple $GF(q)$-representable matroid of which N is a restriction, and $E(N)$ is a flat of M, then $E(N)$ is a modular flat of M.

5. Prove that a line X in a connected, simple, binary matroid is a modular flat if and only if $|X| = 3$.

6. Prove Proposition 6.9.5.

7. (Brylawski 1975b)

(i) Let x and y be distinct elements in a simple matroid M. Prove that $\{x, y\}$ is a modular flat of M if and only if x and y are in different components of M.

(ii) Use (i) and Theorem 6.9.9 to prove Corollary 6.9.10.

8. (Brylawski 1975b) Suppose that X is a flat of the matroid M and assume $M/X = M_1 \oplus M_2$. Prove that $E(M_1) \cup X$ is a modular flat of M if and only if X is a modular flat of $M|(E(M_2) \cup X)$.

9. (Brylawski (1975b) Prove that the following statements are equivalent for a set X in a simple matroid M:

(a) X is a modular flat.

(b) If I is an independent subset of $E(M) - X$ such that $I \cup e$ is in $\mathcal{I}(M)$ for all e in X, then $I \cup I' \in \mathcal{I}(M)$ for all independent subsets I' of X.

10. (White 1971) Let C_1 and C_2 be distinct circuits in a matroid M such that $(E(M)-C_1, E(M)-C_2)$ is a modular pair of flats of M^*. Show the following:

(i) $r(C_1) + r(C_2) = r(C_1 \cup C_2) + r(C_1 \cap C_2)$.

(ii) $r(C_1 \cup C_2) = |C_1 \cup C_2| - 2$.

(iii) If $e \in C_1 \cap C_2$, then there is a unique circuit C_3 such that $C_3 \subseteq (C_1 \cup C_2) - e$. Moreover, $C_3 \supseteq (C_1 - C_2) \cup (C_2 - C_1)$.

(iv) (Fournier 1987) If B is a basis and C is a circuit of M such that $|C - B| \geq 2$, then M has two distinct circuits D_1 and D_2 such that $(D_1 - D_2) \cup (D_2 - D_1) \subseteq C$ and $(E(M) - D_1, E(M) - D_2)$ is a modular pair of flats of M^*.

11.* Let M be a simple rank-r matroid having W_1 points and W_{r-1} hyperplanes. Prove that:

(i) (Motzkin 1951; Basterfield and Kelly 1968; Greene 1970; Heron 1973) $W_1 \leq W_{r-1}$;

(ii) (Greene 1970) $W_1 = W_{r-1}$ if and only if M is modular.

(iii) (Mason 1973) $\mathcal{L}(M)$ has W_1 disjoint maximal chains each of which is from a point to a hyperplane.

7

Constructions

This chapter discusses several different ways of using some given set of matroids to form a new matroid. The five most basic examples of such constructions are deletion, contraction and the formation of duals, minors, and direct sums. Each of these operations has been looked at in detail in an earlier chapter. In this chapter and in Chapter 12, we shall consider some other matroid operations. The discussion here of matroid constructions will be far from complete. The reader seeking an encyclopaedic treatment of this subject is referred to the paper of Brylawski (1986a), which has been used as a source for much of this chapter. Another strong influence on the present treatment of constructions is Mason's (1977) expository paper.

The operation of direct sum provides a way to join matroids on disjoint sets. In Section 7.1, we look at three closely related ways of joining two matroids with exactly one common element. In Chapter 12, we shall discuss some ways of joining two matroids with more than one common element. Section 7.2 investigates the operations of extension and coextension, which reverse the operations of deletion and contraction, respectively. Finally, Section 7.3 uses the techniques of Section 7.2 to examine the relationship between two matroids such that every flat of one is a flat of the other.

While the results in Section 7.1 will be used quite frequently in later chapters, the reader could delay reading Sections 7.2 and 7.3 until Chapter 12 when various other constructions will be considered.

7.1 Series and parallel connection

In the last section of Chapter 5, we looked briefly at the operations of series and parallel connection of graphs. We now formally define these operations and show how they extend naturally to matroids. For $i = 1, 2$, let p_i be an edge of a graph G_i. Arbitrarily assign a direction to p_i and label its tail by u_i and its head by v_i. To form the *series* and *parallel*

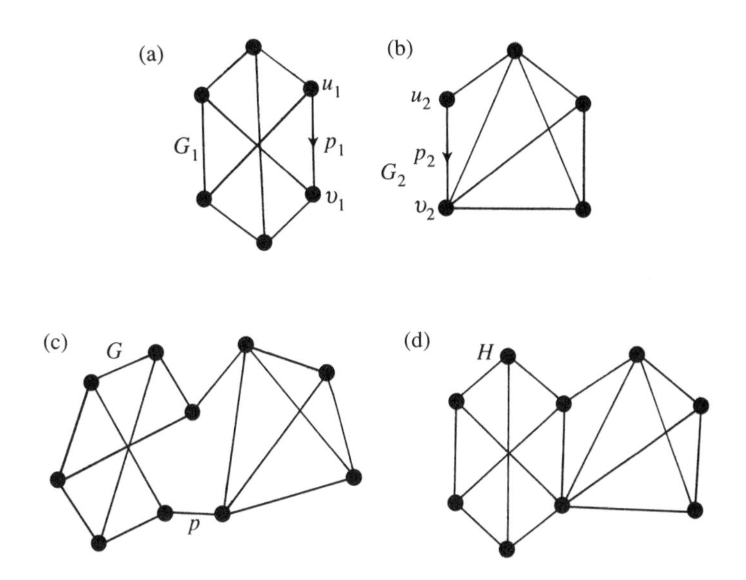

Fig. 7.1. Series and parallel connections of two graphs.

connections of G_1 and G_2 with respect to the directed edges p_1 and p_2, we begin by deleting p_1 from G_1 and p_2 from G_2; we then identify u_1 and u_2 as the vertex u. To complete the series connection, we add a new edge p joining v_1 and v_2. The parallel connection is completed by identifying v_1 and v_2 as the vertex v and then adding a new edge p joining u and v. Thus, unless exactly one of p_1 and p_2 is a loop, the parallel connection is obtained by simply identifying p_1 and p_2 so that their directions agree.

7.1.1 Example. The graphs G and H in Figure 7.1(c) and (d) are, respectively, the series and parallel connections of the graphs G_1 and G_2 in (a) and (b) with respect to the directed edges p_1 and p_2. □

Let $\mathcal{C}_S$ and $\mathcal{C}_P$ denote the collections of circuits of the cycle matroids of the series and parallel connections of the graphs G_1 and G_2. Then, in the last example, and indeed in general, it is not difficult to specify $\mathcal{C}_S$ and $\mathcal{C}_P$ in terms of $\mathcal{C}(M(G_1))$ and $\mathcal{C}(M(G_2))$. Writing M_1 for $M(G_1)$ and M_2 for $M(G_2)$ and assuming neither p_1 nor p_2 is a loop or a cut edge, we have

7.1.2
$$\mathcal{C}_S = \mathcal{C}(M_1\backslash p_1) \cup \mathcal{C}(M_2\backslash p_2)$$
$$\cup \{(C_1 - p_1) \cup (C_2 - p_2) \cup p : p_i \in C_i \in \mathcal{C}(M_i) \quad for \quad i = 1, 2\}$$

and

7.1.3
$$\begin{aligned}\mathcal{C}_P = \mathcal{C}(M_1\backslash p_1) &\cup \{(C_1 - p_1) \cup p : p_1 \in C_1 \in \mathcal{C}(M_1)\}\\ &\cup \mathcal{C}(M_2\backslash p_2) \cup \{(C_2 - p_2) \cup p : p_2 \in C_2 \in \mathcal{C}(M_2)\}\\ &\cup \{(C_1 - p_1) \cup (C_2 - p_2) : p_i \in C_i \in \mathcal{C}(M_i) \;\; for \;\; i = 1,2\}.\end{aligned}$$

Now suppose that M_1 and M_2 are arbitrary matroids on disjoint sets. Let p_1 and p_2 be elements of M_1 and M_2, respectively, such that neither p_1 nor p_2 is a loop or a coloop. Take p to be an element that is not in $E(M_1)$ or $E(M_2)$ and let $E = E(M_1\backslash p_1) \cup E(M_2\backslash p_2) \cup p$. Then, with $\mathcal{C}_S$ and $\mathcal{C}_P$ as specified in 7.1.2 and 7.1.3, respectively, we have the following result.

7.1.4 **Proposition.** *Each of $\mathcal{C}_S$ and $\mathcal{C}_P$ is the collection of circuits of a matroid on E.*

Proof. It is straightforward to check that both $\mathcal{C}_S$ and $\mathcal{C}_P$ satisfy (C1) and (C2). We shall show that $\mathcal{C}_S$ satisfies (C3), leaving the proof that $\mathcal{C}_P$ satisfies this condition to the reader.

Let C and D be distinct members of $\mathcal{C}_S$ having a common element e. If $\mathcal{C}(M_1\backslash p_1)$ or $\mathcal{C}(M_2\backslash p_2)$ contains both C and D, then $\mathcal{C}_S$ certainly has a member contained in $(C \cup D) - e$. Thus we may assume that $p \in C$ or $p \in D$.

Suppose that $p \in C \cap D$. Then, for $i = 1, 2,$ M_i has circuits C_i and D_i both containing p_i such that $C = (C_1 - p_1) \cup (C_2 - p_2) \cup p$ and $D = (D_1 - p_1) \cup (D_2 - p_2) \cup p$. As $C \neq D$, we may assume, without loss of generality, that $C_1 \neq D_1$. If $e = p$, or $e \in (C_2 - p_2) \cap (D_2 - p_2)$, then, on applying (C3) to the circuits C_1 and D_1 of M_1, eliminating the common element p_1, we deduce that M_1 has a circuit C_1' contained in $(C_1 \cup D_1) - p_1$. Evidently $C_1' \in \mathcal{C}_S$ and $C_1' \subseteq (C \cup D) - e$ so, in this case, (C3) holds for $\mathcal{C}_S$. We may now assume that $e \in (C_1 - p_1) \cap (D_1 - p_1)$. Then, applying (C3) in M_1 again, we get that M_1 has a circuit C_1'' contained in $(C_1 \cup D_1) - e$. If $p_1 \notin C_1''$, then C_1'' is a member of $\mathcal{C}_S$ contained in $(C \cup D) - e$. If $p \in C_1''$, then $(C_1'' - p_1) \cup (C_2 - p_2) \cup p$ is a member of $\mathcal{C}_S$ contained in $(C \cup D) - e$. We conclude that (C3) holds when $p \in C \cap D$. By symmetry, the only case that remains to be checked is when $p \in C$ and $p \notin D$. We leave the straightforward details of this to the reader. □

The matroids on E that have $\mathcal{C}_S$ and $\mathcal{C}_P$ as their sets of circuits will be denoted by $S((M_1; p_1), (M_2; p_2))$ and $P((M_1; p_1), (M_2; p_2))$, or briefly, $S(M_1, M_2)$ and $P(M_1, M_2)$. These matroids are called the *series* and *parallel connections* of M_1 and M_2 with respect to the *basepoints* p_1 and p_2 (Brylawski 1971).

It is often convenient to view $S(M_1, M_2)$ and $P(M_1, M_2)$ as being formed from two matroids M_1 and M_2 whose ground sets meet in a single element p. In this context, p is called the *basepoint* of the connection and we take $E = E(M_1) \cup E(M_2)$. Moreover, with $p_1 = p_2 = p$, the sets $\mathcal{C}_S$ and $\mathcal{C}_P$ are defined as in 7.1.2 and 7.1.3 provided neither M_1 nor M_2 has p as a loop or a coloop. If p is a loop of M_1, we define $P(M_1, M_2)$, $P(M_2, M_1)$, $S(M_1, M_2)$, and $S(M_2, M_1)$ by

7.1.5 $$P(M_1, M_2) = P(M_2, M_1) = M_1 \oplus (M_2/p) \quad \text{and}$$

7.1.6 $$S(M_1, M_2) = S(M_2, M_1) = (M_1/p) \oplus M_2.$$

If p is a coloop of M_1, we define

7.1.7 $$P(M_1, M_2) = P(M_2, M_1) = (M_1 \backslash p) \oplus M_2 \text{ and}$$

7.1.8 $$S(M_1, M_2) = S(M_2, M_1) = M_1 \oplus (M_2 \backslash p).$$

Note that if $\{p\}$ is a separator of both M_1 and M_2, then we have given two definitions of each of $P(M_1, M_2)$ and $S(M_1, M_2)$. To show that these definitions are consistent, one uses the fact that if $\{e\}$ is a separator of a matroid N, then $N \backslash e = N/e$ (Corollary 3.1.25). Furthermore, on combining 7.1.2, 7.1.3, and 7.1.5–7.1.8, we observe that, in all cases, $P(M_1, M_2) = P(M_2, M_1)$ and $S(M_1, M_2) = S(M_2, M_1)$.

Now let $\{p\}$ be a separator of M_1 or M_2 and suppose that $M_1 \cong M(G_1)$ and $M_2 \cong M(G_2)$ for some graphs G_1 and G_2. Then it is routine to verify that the matroids $S(M_1, M_2)$ and $P(M_1, M_2)$, defined as in 7.1.5–7.1.8, are isomorphic to the cycle matroids of the graphs that are the series and parallel connections of G_1 and G_2.

For matroids, just as for graphs, the operations of series and parallel connection generalize the operations of series and parallel extension. Indeed, if N is a series or parallel extension of M, then N is isomorphic to a series or parallel connection of M and $U_{1,2}$.

The next two examples illustrate the fact that, when M_1 and M_2 have small rank, it is easy to find a geometric representation for $P(M_1, M_2)$ from geometric representations for M_1 and M_2.

7.1.9 **Example.** Suppose that both M_1 and M_2 are isomorphic to $U_{2,4}$. Then geometric representations for $S(M_1, M_2)$ and $P(M_1, M_2)$ are given in Figure 7.2. In the former, the basepoint p is free in space, that is, p is in no circuits of size less than five. □

In the last example, the symmetry of M_1 and M_2 meant that, up to isomorphism, $S(M_1, M_2)$ and $P(M_1, M_2)$ did not depend on the choice of basepoints. In the next example, changing the basepoint in M_2 changes the isomorphism class of the parallel connection.

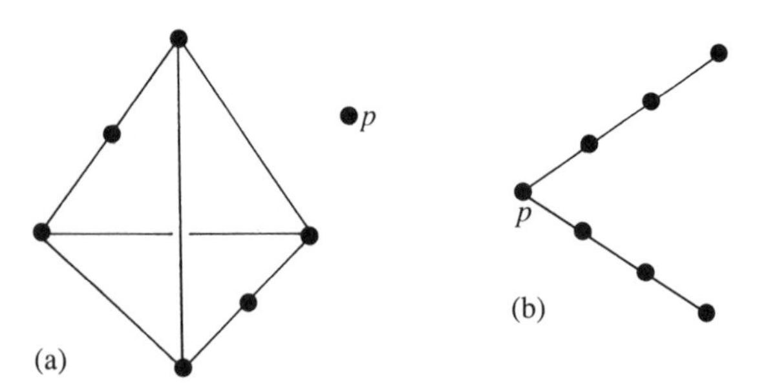

Fig. 7.2. (a) $S(U_{2,4}, U_{2,4})$. (b) $P(U_{2,4}, U_{2,4})$.

7.1.10 **Example.** Let M_1 and M_2 be the matroids for which geometric representations are shown in Figure 7.3 (a) and (b). Then geometric representations for $P((M_1; p_1), (M_2; x))$ and $P((M_1; p_1), (M_2; y))$ are as shown in Figure 7.3 (c) and (d). □

The operations of series and parallel connection are actually duals of each other. To show this, we shall determine the sets of bases of the series and parallel connections of two matroids. We first note that:

7.1.11 *The basepoint p is a coloop of $S(M_1, M_2)$ if and only if it is a coloop of M_1 or M_2.*

7.1.12 *The basepoint p is a loop of $P(M_1, M_2)$ if and only if it is a loop of M_1 or M_2.*

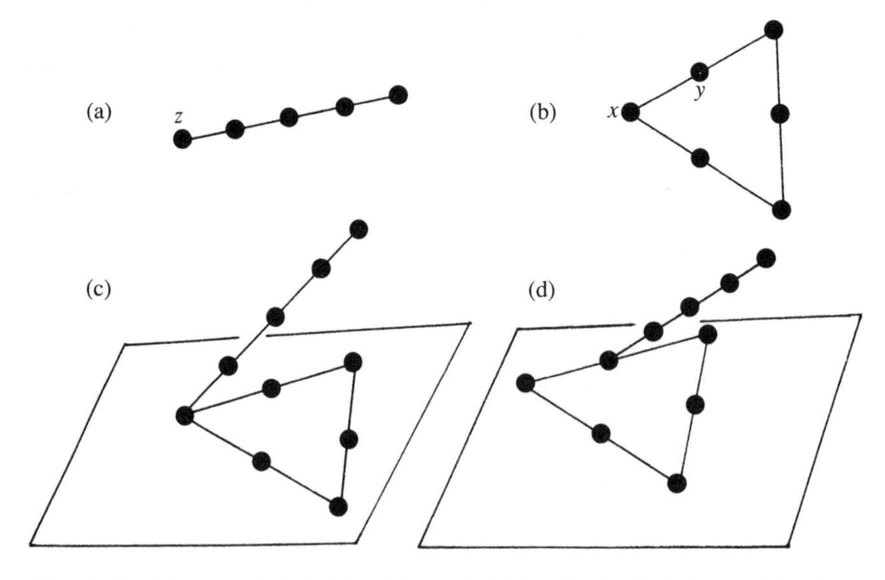

Fig. 7.3. (a) M_1. (b) M_2. (c) $P((M_1; z), (M_2; x))$. (d) $P((M_1; z), (M_2; y))$.

7.1.13 Proposition. *Let M_1 and M_2 be matroids with $E(M_1) \cap E(M_2) = \{p\}$. Let $E = E(M_1) \cup E(M_2)$ and B be a subset of E.*

(i) *Assume that, in at least one of M_1 and M_2, p is not a coloop. Then B is a basis of $S(M_1, M_2)$ if and only if B can be written as a disjoint union of bases of M_1 and M_2.*

(ii) *Assume that, in at least one of M_1 and M_2, p is not a loop. Then B is a basis of $P(M_1, M_2)$ containing p if and only if $B \cap E(M_i)$ is a basis of M_i containing p for $i = 1, 2$. Moreover, B is a basis of $P(M_1, M_2)$ not containing p if and only if $p \notin B$ and, for some distinct i and j in $\{1, 2\}$, the sets $[B \cap E(M_i)] \cup p$ and $B \cap E(M_j)$ are bases of M_i and M_j, respectively.*

Proof. If p is a loop or a coloop of M_1 or M_2, the result is not difficult to check using 7.1.5–7.1.8. Thus assume that neither M_1 nor M_2 has p as a loop or a coloop. We shall prove (i) and leave the proof of (ii) to the reader.

Let B be a basis of $S(M_1, M_2)$. For $i = 1, 2$, let $B_i = B \cap E(M_i)$. We shall first suppose that $p \notin B$. Then B_1 does not contain a member of $\mathcal{C}(M_1 \backslash p)$. However, if $e \in E(M_1) - (B_1 \cup p)$, then $B \cup e$ contains a circuit of $S(M_1, M_2)$. Hence $B_1 \cup e$ contains a member of $\mathcal{C}(M_1 \backslash p)$. Thus B_1 or $B_1 \cup p$ is a basis of M_1. But $B \cup p$ contains a circuit of $S(M_1, M_2)$ containing p. Hence $B_1 \cup p$ contains a circuit of M_1. We conclude that B_1 is a basis of M_1. Similarly, B_2 is a basis of M_2 and so B is a disjoint union of a basis of M_1 and a basis of M_2.

Now suppose that $p \in B$. Again B_i does not contain a member of $\mathcal{C}(M_i \backslash p)$ for $i = 1, 2$. Moreover, either B_1 does not contain a circuit of M_1 containing p, or B_2 does not contain a circuit of M_2 containing p, or both. Suppose that the last possibility occurs. Then, for all e in $E(M_1) - B_1$, the set $B \cup e$ contains a member $C(e)$ of $\mathcal{C}_S$. Evidently $C(e)$ cannot be of the form $C_1 \cup C_2$ where, for $i = 1, 2$, the set C_i is a circuit of M_i containing p. Thus $C(e) \in \mathcal{C}(M_1 \backslash p)$ for all e in $E(M_1) - B_1$. Hence $B_1 - p$ spans $M_1 \backslash p$. As $B_1 - p$ does not span M_1, it follows that p is a coloop of M_1; a contradiction. We conclude that exactly one of B_1 and B_2, say B_1, contains a circuit containing p. It is now straightforward to show that $B_1 - p$ is a basis of M_1 and B_2 is a basis of M_2.

To complete the proof of (i), we need to show that if B is the union of disjoint bases A_1 of M_1 and A_2 of M_2, then B is a basis of $S(M_1, M_2)$. Without loss of generality, we may assume that $p \notin A_2$. Evidently B is independent in $S(M_1, M_2)$. We show next that it is a maximal independent set. If $e \in E(M_2) - (A_2 \cup p)$, then $C_{M_2}(e, A_2)$ is a circuit of $S(M_1, M_2)$ contained in $B \cup e$, so $B \cup e$ is dependent in $S(M_1, M_2)$. If $e \in E(M_1) - A_1$, then either $C_{M_1}(e, A_1)$ contains p or not. In the first case, $B \cup e$ contains $C_{M_1}(e, A_1) \cup C_{M_2}(p, A_2)$, a circuit of $S(M_1, M_2)$. In the second case, $B \cup e$ contains $C_{M_1}(e, A_1)$, a circuit of $S(M_1, M_2)$. We conclude that B is indeed a maximal independent set of $S(M_1, M_2)$. □

We can now show that series and parallel connection are dual operations.

7.1.14 Proposition. *Let M_1 and M_2 be matroids such that $E(M_1) \cap E(M_2) = \{p\}$. Then*

$$S(M_1, M_2) = [P(M_1^*, M_2^*)]^*$$

and

$$P(M_1, M_2) = [S(M_1^*, M_2^*)]^*.$$

Proof. Since the first equation implies the second by duality, we need only prove the former. Assume initially that p is a loop of M_1. Then p is a coloop of M_1^*. Thus

$$\begin{aligned} S(M_1, M_2) &= (M_1/p) \oplus M_2 \\ &= (M_1^* \backslash p)^* \oplus (M_2^*)^* \\ &= [(M_1^* \backslash p) \oplus M_2^*]^* = [P(M_1^*, M_2^*)]^*. \end{aligned}$$

A similar argument establishes the result when p is a coloop of M_1.

We may now assume that neither M_1 nor M_2 has $\{p\}$ as a separator. Suppose that $E - B$ is a basis of $[P(M_1^*, M_2^*)]^*$ where $E = E(M_1) \cup E(M_2)$. We shall show that $E - B$ is a basis of $S(M_1, M_2)$. Evidently B is a basis of $P(M_1^*, M_2^*)$. We now distinguish two cases: (a) $p \in B$, and (b) $p \notin B$. In case (a), by Proposition 7.1.13(ii), $B \cap E(M_1^*)$ and $B \cap E(M_2^*)$ contain p and are bases of M_1^* and M_2^*, respectively. Thus $(E - B) \cap E(M_1)$ and $(E - B) \cap E(M_2)$ avoid p and are bases of M_1 and M_2, respectively. Hence $E - B$ can be written as a disjoint union of bases of M_1 and M_2. Therefore, by Proposition 7.1.13(i), $E - B$ is a basis of $S(M_1, M_2)$.

In case (b), Proposition 7.1.13(ii) implies that, for distinct i and j in $\{1, 2\}$, $[B \cap E(M_i^*)] \cup p$ and $B \cap E(M_j^*)$ are bases of M_i^* and M_j^*. Thus $[(E - B) \cap E(M_i)] - p$ and $(E - B) \cap E(M_j)$ are bases of M_i and M_j, respectively, the first avoiding p and the second containing p. Hence $E - B$ can be written as a disjoint union of bases of M_1 and M_2 and so is a basis of $S(M_1, M_2)$.

We conclude that $\mathcal{B}([P(M_1^*, M_2^*)]^*) \subseteq \mathcal{B}(S(M_1, M_2))$. A similar argument gives the reverse inclusion thereby finishing the proof. □

Brylawski (1971) noted numerous attractive properties of the operations of series and parallel connection. Some of these are summarized below, while others appear in the exercises. Several of these results are stated in terms of M_1, but, since $S(M_2, M_1) = S(M_1, M_2)$, and $P(M_2, M_1) = P(M_1, M_2)$, the corresponding results for M_2 also hold.

7.1.15 Proposition. *Let M_1 and M_2 be matroids with $E(M_1) \cap E(M_2) = \{p\}$. Then*

(i)
$$r(S(M_1, M_2)) = \begin{cases} r(M_1) + r(M_2) - 1, & \text{if } p \text{ is a coloop of both } M_1 \text{ and } M_2; \\ r(M_1) + r(M_2), & \text{otherwise.} \end{cases}$$

$$r(P(M_1, M_2)) = \begin{cases} r(M_1) + r(M_2), & \text{if } p \text{ is a loop of both } M_1 \text{ and } M_2; \\ r(M_1) + r(M_2) - 1, & \text{otherwise.} \end{cases}$$

(ii) $M_1 = S(M_1, M_2)/(E(M_2) - p)$ *unless p is a loop of M_1 and a coloop of M_2; and*

$M_1 = P(M_1, M_2)\backslash(E(M_2) - p)$ *unless p is a coloop of M_1 and a loop of M_2.*

(iii) $S(M_1, M_2)\backslash p = (M_1\backslash p) \oplus (M_2\backslash p)$ *and*

$P(M_1, M_2)/p = (M_1/p) \oplus (M_2/p)$.

(iv) $S(M_1, M_2)/p = P(M_1, M_2)\backslash p$.

(v) *If $e \in E(M_1) - p$, then*

$S(M_1, M_2)\backslash e = S(M_1\backslash e, M_2)$; $S(M_1, M_2)/e = S(M_1/e, M_2)$;

$P(M_1, M_2)\backslash e = P(M_1\backslash e, M_2)$; *and* $P(M_1, M_2)/e = P(M_1/e, M_2)$.

Proof. The proofs of these results use Propositions 7.1.13 and 7.1.14 along with 7.1.2, 7.1.3, and 7.1.5–7.1.8. The straightforward details are left to the reader. □

The operations of series and parallel connection arise naturally in the consideration of matroid structure because of the following fundamental link with matroid connectivity.

7.1.16 Theorem. *Let p be an element of a connected matroid M.*

(i) *If $M\backslash p = M_1 \oplus M_2$ where both M_1 and M_2 are non-empty, then $M = S(M/E(M_1), M/E(M_2))$.*

Dually,

(ii) *if $M/p = M_1 \oplus M_2$ where both M_1 and M_2 are non-empty, then $M = P(M\backslash E(M_1), M\backslash E(M_2))$.*

Proof. Evidently we need only prove (i). We shall show that $\mathcal{B}(M) = \mathcal{B}(N)$ where $N = S(M/E(M_1), M/E(M_2))$. First, we observe that, as $|E(M)| \geq 2$ and M is connected, $r(M) = r(M\backslash p)$, so

$$r(M) = r(M_1) + r(M_2). \tag{1}$$

If p is a coloop of $M/E(M_1)$, then

$$r(M\backslash p/E(M_1)) = r(M/E(M_1)) - 1 = r(M) - r(M_1) - 1.$$

But $M\backslash p/E(M_1) = M_2$, so $r(M_2) = r(M) - r(M_1) - 1$; a contradiction to (1). Thus p is not a coloop of $M/E(M_1)$ and, similarly, it is not a coloop of $M/E(M_2)$.

Now, by Proposition 7.1.15(iii),

$$\begin{aligned} N\backslash p &= (M/E(M_1)\backslash p) \oplus (M/E(M_2)\backslash p) \\ &= (M\backslash p/E(M_1)) \oplus (M\backslash p/E(M_2)) \\ &= M_2 \oplus M_1 = M\backslash p. \end{aligned}$$

Hence

$$\mathcal{B}(N\backslash p) = \mathcal{B}(M\backslash p).$$

Suppose next that B is a basis of N containing p. Then, without loss of generality, we may assume that $B = B_1 \dot{\cup} p \dot{\cup} B_2$ where $B_1 \cup p$ is a basis of $M/E(M_1)$ and B_2 is a basis of $M/E(M_2)$. As $p \notin B_2$, it follows that B_2 is a basis of $M/E(M_2)\backslash p$. But the last matroid is $M\backslash p/E(M_2) = M_1 = M|E(M_1)$. Therefore, as $B_1 \cup p$ is a basis of $M/E(M_1)$, it follows that $B_1 \cup p \cup B_2$ is a basis of M. Hence every basis of N containing p is a basis of M.

Finally, suppose that B is a basis of M containing p. Then $B - p$ is independent in $M\backslash p$. Hence, as $M\backslash p = M_1 \oplus M_2$,

$$|B - p| = r(M) - 1 = r(M_1) + r(M_2) - 1.$$

Therefore $B - p$ contains a basis of M_1 or of M_2. Without loss of generality, assume the latter, letting B_2 be such a basis. Then B_2 is a basis of $M|E(M_2)$. As B is a basis of M, it follows that $B - B_2$ is a basis of $M/E(M_2)$. Since M is connected, p is not a coloop of $M/E(M_1)$. Thus B_2, which is a basis of $M\backslash p/E(M_1)$, is also a basis of $M/E(M_1)$. Hence, by Proposition 7.1.13(i), B is a basis of N. We conclude that $\mathcal{B}(N) = \mathcal{B}(M)$, so $N = M$. □

By the last result, a connected matroid is a series or parallel connection if and only if it can be disconnected by deleting or contracting a single element. Next we show that connectivity is preserved under both of the operations of series and parallel connection.

7.1.17 Proposition. *If the matroids M_1 and M_2 each have at least two elements and $E(M_1) \cap E(M_2) = \{p\}$, then the following statements are equivalent:*

(i) *Both M_1 and M_2 are connected.*

(ii) *$S(M_1, M_2)$ is connected.*

(iii) *$P(M_1, M_2)$ is connected.*

Proof. Since the dual of a matroid is connected if and only if the matroid itself is connected, it suffices to establish the equivalence of (i) and (iii). Assume that (i) holds. Then it follows by Corollary 4.1.3 that $P(M_1, M_2)$ is connected since $M_1 = P(M_1, M_2)|E(M_1)$, $M_2 = P(M_1, M_2)|E(M_2)$, and $E(M_1) \cap E(M_2) = \{p\}$.

Now assume that $P(M_1, M_2)$ is connected but M_1 is not. Then, by 7.1.5 and 7.1.7, neither M_1 nor M_2 has $\{p\}$ as a separator. As M_1 is disconnected, there is a pair of distinct elements e and f of $E(M_1) - p$ such that no circuit of M_1 contains $\{e, f\}$. It follows from 7.1.3 that no circuit of $P(M_1, M_2)$ contains $\{e, f\}$; a contradiction. □

The assertion in 7.1.15(iv) that $P(M_1, M_2)\backslash p = S(M_1, M_2)/p$ is an important one. The next example illustrates this for graphic matroids.

7.1.18 Example. Consider Figure 7.4. Clearly the matroids $P(M(G_1), M(G_2))$ and $S(M(G_1), M(G_2))$ are isomorphic to the cycle matroids of the graphs in (c) and (d). The graph in (e) can be obtained both from the graph in (c) by deleting p and from the graph in (d) by contracting p. The cycle matroid of this graph is isomorphic to both $P(M(G_1), M(G_2))\backslash p$ and $S(M(G_1), M(G_2))/p$.

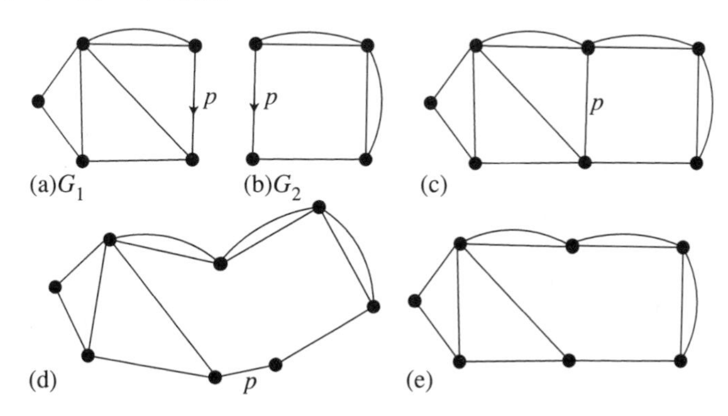

Fig. 7.4. The parallel and series connections and 2-sum of two graphs.

The graph in Figure 7.4(e) can be obtained directly from G_1 and G_2 by identifying the edges labelled p and then deleting the identified edge. Following Robertson and Seymour (1984), we call a graph obtained in this way a *2-sum* of G_1 and G_2. To ensure that this operation is well-defined, we insist that if the edge p is a loop in one of G_1 and G_2, then it is a loop in the other. □

By analogy with the above situation for graphs, we have the following definition for matroids. Let M and N be matroids, each with at least three elements. Let $E(M) \cap E(N) = \{p\}$ and suppose that neither M nor N has $\{p\}$ as a separator. Then the *2-sum* $M \oplus_2 N$ of M and N is $S(M,N)/p$ or, equivalently, $P(M,N)\backslash p$. Clearly $N \oplus_2 M = M \oplus_2 N$. The element p is called the ***basepoint*** of the 2-sum, and M and N are called the ***parts*** of the 2-sum.

This operation of 2-sum is of fundamental importance. We noted in Chapter 4 that a matroid is not 2-connected if and only if it can be written as a direct sum. In Chapter 8, where higher connectivity for matroids is discussed, we shall see that a 2-connected matroid is not 3-connected if and only if it can be written as a 2-sum. The next result contains a basic and frequently used property of 2-sums.

7.1.19 Proposition. *Both M and N are isomorphic to proper minors of $M \oplus_2 N$.*

Proof. It suffices to show that M is a minor of $M \oplus_2 N$. As p is neither a loop nor a coloop of N, there is a circuit C of N that contains p and has at least two elements. Choose q in $C - p$. Now consider the matroid obtained from $P(M,N)$ by deleting $E(N) - C$ and then contracting $C - \{p,q\}$. By Proposition 7.1.15(v), this matroid is the parallel connection of M and $N\backslash(E(N)-C)/(C-\{p,q\})$. But $N\backslash(E(N)-C)/(C-\{p,q\})$ is a circuit with elements p and q. It follows that the matroid $(M \oplus_2 N)\backslash(E(N)-C)/(C-\{p,q\})$ is isomorphic to M. □

Two of the many other attractive properties of the 2-sum operation are contained in the following proposition, the proof of which is left to the reader (Exercise 7).

7.1.20 Proposition.

(i) $(M \oplus_2 N)^* = M^* \oplus_2 N^*$.

(ii) *If $|E(N_i)| \geq 2$ for $i = 1, 2$, then $P(N_1, N_2)\backslash p$ is connected if and only if both N_1 and N_2 are connected. In particular, $M \oplus_2 N$ is connected if and only if both M and N are connected.* □

We have seen that both the series and parallel connections of two graphic matroids are graphic. Moreover, Example 7.1.9 shows that, not

$$
\begin{array}{cc}
E(M_1)-p & p \\
\left[\begin{array}{c|c}
 & 0 \\
 & 0 \\
A_1 & \vdots \\
 & 0 \\
 & 1
\end{array}\right]
\end{array}
\qquad\qquad
\begin{array}{cc}
p & E(M_2)-p \\
\left[\begin{array}{c|c}
1 & \\
0 & \\
\vdots & A_2 \\
0 & \\
0 &
\end{array}\right]
\end{array}
$$

(a) (b)

Fig. 7.5. Representations for M_1 and M_2.

surprisingly, neither the series nor the parallel connection of two uniform matroids need be uniform. We now consider the effect of the operations of series and parallel connection on various other classes of matroids.

7.1.21 Proposition. *Let F be a field. If M_1 and M_2 are F-representable matroids such that $E(M_1) \cap E(M_2) = \{p\}$, then both $P(M_1, M_2)$ and $S(M_1, M_2)$ are F-representable.*

Proof. If p is a loop or a coloop of M_1 or M_2, then, since $S(M_1, M_2)$ and $P(M_1, M_2)$ are direct sums of minors of M_1 and M_2, each is F-representable. Now assume that neither M_1 nor M_2 has $\{p\}$ as a separator. Then we may assume that M_1 and M_2 are represented over F by the matrices in Figure 7.5(a) and (b), respectively. It is now not difficult to check that $P(M_1, M_2)$ and $S(M_1, M_2)$ are represented over F by the matrices in Figure 7.6(a) and (b). The details of this argument are left to the reader. □

We remark that once we had shown $P(M_1, M_2)$ to be F-representable in the last proof, duality could have been used to deduce that $S(M_1, M_2)$ is F-representable. However, this approach does not give the F-representation in Figure 7.6(b).

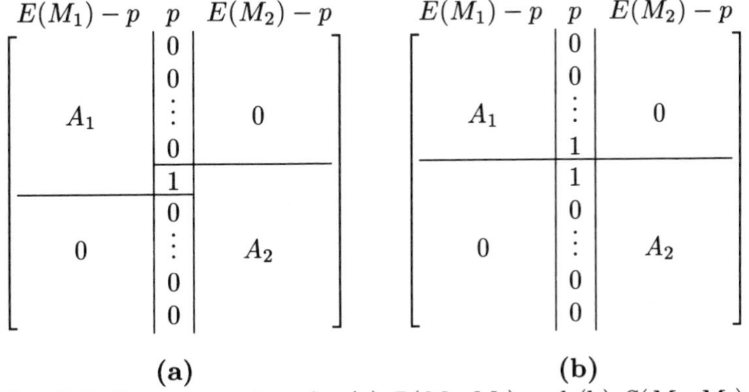

Fig. 7.6. Representations for (a) $P(M_1, M_2)$ and (b) $S(M_1, M_2)$.

7.1.22 Corollary. *If M_1 and M_2 are regular matroids with $E(M_1) \cap E(M_2) = \{p\}$, then both $P(M_1, M_2)$ and $S(M_1, M_2)$ are regular.*

Proof. By Theorem 6.6.3, a matroid is regular if and only if it is both binary and ternary. The corollary follows from this and the last proposition. □

We leave it to the reader to show that the converses of the last two results hold. Now suppose, in Corollary 7.1.22, that p is a loop of neither M_1 nor M_2. Then M_1 and M_2 are represented by totally unimodular matrices of the form shown in Figure 7.5(a) and (b). Moreover, it is not difficult to check that the matrices in Figure 7.6(a) and (b) are totally unimodular representations for $P(M_1, M_2)$ and $S(M_1, M_2)$.

Since the 2-sum of M_1 and M_2 is a minor of their parallel connection, the following corollary is a straightforward consequence of earlier results. A class of matroids is *closed under 2-sums* if the 2-sum of any two members of the class is also in the class.

7.1.23 Corollary. *The classes of graphic, cographic, F-representable, and regular matroids are all closed under 2-sums.* □

We saw in the proof of Proposition 7.1.21 how to obtain an F-representation for $P(M_1, M_2)$ from F-representations for M_1 and M_2. Because $M_1 \oplus_2 M_2 = P(M_1, M_2) \backslash p$, we get an F-representation for $M_1 \oplus_2 M_2$ simply by deleting the column in Figure 7.6(a) labelled by p.

In contrast to the last corollary, the next result shows that the class of transversal matroids is not closed under 2-sums or parallel connections. However, it is closed under series connections. A proof of the last fact will be given in Chapter 12, while Exercise 9 characterizes precisely when the parallel connection of two transversal matroids is transversal.

7.1.24 Example. We noted in Example 1.6.3 that, for the graph G in Figure 7.7, $M(G)$ is not transversal. But $M(G)$ is isomorphic to both $P(M(G_2), M(G_3))$ and $M(G_1) \oplus_2 M(G_3)$, and each of $M(G_1)$, $M(G_2)$, and $M(G_3)$ is easily shown to be transversal. □

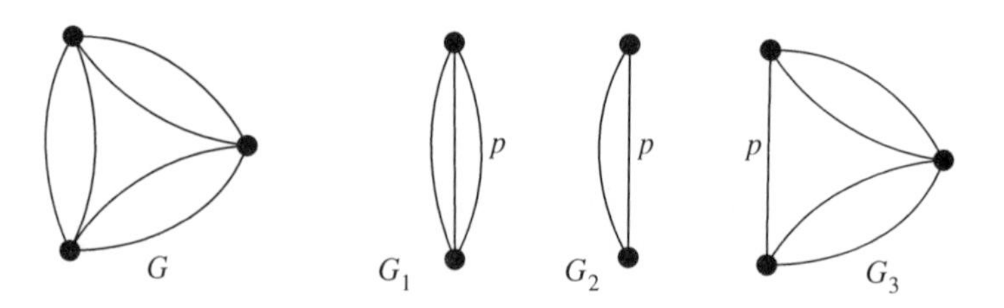

Fig. 7.7.

Exercises

1. Let M_1 and M_2 be matroids such that $E(M_1) \cap E(M_2) = \{p\}$.

(i) (Jaeger, Vertigan, and Welsh 1990) Let $M = P(M_1, M_2)\backslash p$. Show that if $A_1 \subseteq E(M_1)$ and $A_2 \subseteq E(M_2)$, then

$$r_M(A_1 \cup A_2) = r_1(A_1) + r_2(A_2) - \theta(A_1, A_2) + \theta(\emptyset, \emptyset)$$

where $\theta(X, Y)$ is 1 if $r_1(X \cup p) = r_1(X)$ and $r_2(Y \cup p) = r_2(Y)$, and $\theta(X, Y)$ is 0 otherwise.

(ii) For $N = P(M_1, M_2)$, determine the rank function, closure operator, hyperplanes, cocircuits, flats, and independent sets of N.

(iii) Repeat (ii) with $N = S(M_1, M_2)$.

2. Let M_1, M_2, and M_3 be matroids such that $E(M_1) \cap E(M_2) = \{p\}$ and $E(M_3) \cap (E(M_1) \cup E(M_2)) = \emptyset$. Show that $S(M_1, M_2) \oplus M_3 = S(M_1, M_2 \oplus M_3)$, and $P(M_1, M_2) \oplus M_3 = P(M_1, M_2 \oplus M_3)$.

3. Prove Proposition 7.1.15.

4. Show that the 2-sum of two identically self-dual matroids is identically self-dual.

5. Find all fields F for which $P(AG(2,3), AG(3,2))$ is F-representable (see Exercises 3 and 9 of Section 6.4).

6. Does Theorem 7.1.16 hold for all matroids M?

7. Prove Proposition 7.1.20.

8. Let $M = M_1 \oplus_2 M_2$ and suppose that $E(M_2) - p$ has no element e such that M_1 is isomorphic to a minor of both $M\backslash e$ and M/e. Prove that M_2 is a uniform matroid of rank or corank one.

9.* Suppose that N_1 and N_2 are connected matroids such that $|E(N_1)|$, $|E(N_2)| \geq 2$. Prove that $P((N_1; p), (N_2; p))$ is transversal if and only if N_1 and N_2 are both transversal and each has a presentation in which p occurs in exactly one of the sets of the presentation.

7.2 Single-element extensions

If a matroid M is obtained from a matroid N by deleting a non-empty subset T of $E(N)$, then N is called an *extension* of M. In particular, if $|T| = 1$, then N is a *single-element extension* of M. In this section, following Crapo (1965), we shall investigate how to find all of the single-element extensions of a given matroid. The term 'single-element extension' is often abbreviated in the literature to 'extension', and we shall follow this practice here when it should not cause confusion. Another term that is sometimes used instead of 'single-element extension' is 'addition' (see, for example, Truemper 1985a).

Two obvious ways to extend a matroid M are to adjoin a loop or to adjoin a coloop. In these cases, the resulting matroids are isomorphic to $M \oplus U_{0,1}$ and $M \oplus U_{1,1}$, respectively. Another type of extension that we have already met is a parallel extension. Here we add a new element in parallel to some existing element of M.

If N^* is an extension of M^*, then N is called a *coextension* of M. In this case, $M = N/T$ for some subset T of $E(N)$. A single-element coextension of M has also been called a 'lift' of M (see, for example, Mason 1977), but we shall reserve this term for a slightly different construction to be considered towards the end of Section 7.3. We remark that if N is a series extension of M, then N is actually a coextension of M rather than an extension. Nevertheless, we shall continue to use the well-established term 'series extension' instead of the more accurate 'series coextension'.

Now suppose that the matroid M is given and we wish to adjoin the element e to M to form an extension N. By Proposition 3.3.1, the flats of M are of the form $X - e$ where X is a flat of N. Thus to specify N in terms of M, we consider the effect on the flats of M of the addition of e. Clearly, there are three possibilities for each flat F of M:

(i) $F \cup e$ is a flat of N and $r(F \cup e) = r(F)$;

(ii) $F \cup e$ is a flat of N and $r(F \cup e) = r(F) + 1$; or

(iii) $F \cup e$ is not a flat of N.

Consider the set $\mathcal{M}$ of flats F of M satisfying (i).

7.2.1 Lemma. *$\mathcal{M}$ has the following properties:*

(i) *If $F \in \mathcal{M}$ and F' is a flat of M containing F, then $F' \in \mathcal{M}$.*

(ii) *If $F_1, F_2 \in \mathcal{M}$ and (F_1, F_2) is a modular pair, then $F_1 \cap F_2 \in \mathcal{M}$.*

Proof. (i) As $F \in \mathcal{M}$, $r(F \cup e) = r(F)$. Therefore $e \in \mathrm{cl}_N(F) \subseteq \mathrm{cl}_N(F')$. Thus $r(F' \cup e) = r(F')$ and so $F' \cup e$ is a flat of N. Hence $F' \in \mathcal{M}$.

(ii) Clearly $r(F_i \cup e) = r(F_i)$ for $i = 1, 2$. Thus, as (F_1, F_2) is a modular pair,

$$r(F_1 \cap F_2) + r(F_1 \cup F_2) = r(F_1) + r(F_2) = r(F_1 \cup e) + r(F_2 \cup e).$$

Applying the semimodularity of r to the last line, we get that

$$r(F_1 \cap F_2) + r(F_1 \cup F_2) \geq r((F_1 \cap F_2) \cup e) + r(F_1 \cup F_2 \cup e). \quad (1)$$

As $r(F_1 \cup e) = r(F_1)$, it follows that $r(F_1 \cup F_2 \cup e) = r(F_1 \cup F_2)$. Thus (1) implies that $r(F_1 \cap F_2) \geq r((F_1 \cap F_2) \cup e)$. Hence $r(F_1 \cap F_2) = r((F_1 \cap F_2) \cup e)$ and so $F_1 \cap F_2 \in \mathcal{M}$. □

An arbitrary set $\mathcal{M}$ of flats of a matroid M is called a *modular cut* if it satisfies (i) and (ii) of Lemma 7.2.1. By that lemma, every single-element extension of a matroid gives rise to a modular cut. So, for example, the modular cut corresponding to the extension $M \oplus U_{0,1}$ of M consists of the set of all flats of M, while the modular cut corresponding to $M \oplus U_{1,1}$ is empty. The next result establishes that every modular cut gives rise to a unique extension.

7.2.2 Theorem. *Let $\mathcal{M}$ be a modular cut of a matroid M on a set E. Then there is a unique extension N of M on $E \dot\cup e$ such that $\mathcal{M}$ consists of those flats F of M for which $F \cup e$ is a flat of N having the same rank as F. Moreover, for all subsets X of E,*

$$r_N(X) = r_M(X) \quad \textit{and}$$

$$r_N(X \cup e) = \begin{cases} r_M(X), & \textit{if} \quad \mathrm{cl}_M(X) \in \mathcal{M}, \\ r_M(X) + 1, & \textit{if} \quad \mathrm{cl}_M(X) \notin \mathcal{M}. \end{cases}$$

Proof. It is straightforward to check that if N is an extension of M satisfying the specified condition on flats, then N must have rank function r_N. Hence if such an extension exists, it is unique.

To complete the proof of the theorem, we must show that r_N is the rank function of a matroid on $E \cup e$. One can easily check that r_N satisfies (R1) and (R2). Since r_M satisfies (R3), to show that r_N satisfies (R3), we need only check that the semimodular inequality holds for all pairs of sets having one of the forms $(X \cup e, Y)$ and $(X \cup e, Y \cup e)$ where $X, Y \subseteq E$.

In the first case,

$$r_N(X \cup e) + r_N(Y) = r_M(X) + r_M(Y) + \delta_X$$

where δ_X is 0 or 1 according to whether or not $\mathrm{cl}_M(X)$ is in $\mathcal{M}$. Thus

$$\begin{aligned} r_N(X \cup e) + r_N(Y) &\geq r_M(X \cup Y) + r_M(X \cap Y) + \delta_X \\ &= r_M(X \cup Y) + r_N((X \cup e) \cap Y) + \delta_X. \quad (2) \end{aligned}$$

But if $\mathrm{cl}_M(X) \in \mathcal{M}$, then $\mathrm{cl}_M(X \cup Y) \in \mathcal{M}$, so

$$r_M(X \cup Y) + \delta_X \geq r_N(X \cup Y \cup e). \quad (3)$$

If $\mathrm{cl}_M(X) \notin \mathcal{M}$, then $\delta_X = 1$ and again (3) holds. On substituting from (3) into (2), we get that (R3) holds for the pair $(X \cup e, Y)$.

Finally, we show that (R3) holds for the pair $(X \cup e, Y \cup e)$. Clearly

$$\begin{aligned} r_N(X \cup e) + r_N(Y \cup e) &= r_M(X) + \delta_X + r_M(Y) + \delta_Y \\ &= r_M(\mathrm{cl}_M(X)) + \delta_X + r_M(\mathrm{cl}_M(Y)) + \delta_Y \\ &\geq r_M(\mathrm{cl}_M(X) \cup \mathrm{cl}_M(Y)) + \delta_X \\ &\quad + r_M(\mathrm{cl}_M(X) \cap \mathrm{cl}_M(Y)) + \delta_Y. \end{aligned} \tag{4}$$

If $\delta_X = \delta_Y = 1$, then

$$r_N(X \cup e) + r_N(Y \cup e) \geq r_M(X \cup Y) + 1 + r_M(X \cap Y) + 1,$$

so

$$\begin{aligned} r_N(X \cup e) + r_N(Y \cup e) &\geq r_N((X \cup e) \cup (Y \cup e)) \\ &\quad + r_N((X \cup e) \cap (Y \cup e)). \end{aligned} \tag{5}$$

If one of δ_X and δ_Y is 0 and the other is 1, then exactly one of $\mathrm{cl}_M(X)$ and $\mathrm{cl}_M(Y)$ is in $\mathcal{M}$. Hence $\mathrm{cl}_M(\mathrm{cl}_M(X) \cup \mathrm{cl}_M(Y))$ is in $\mathcal{M}$ but $\mathrm{cl}_M(X) \cap \mathrm{cl}_M(Y)$ is not. Thus, from (4),

$$\begin{aligned} r_N(X \cup e) + r_N(Y \cup e) &\geq r_N(\mathrm{cl}_M(X) \cup \mathrm{cl}_M(Y) \cup e) \\ &\quad + r_N((\mathrm{cl}_M(X) \cup e) \cap (\mathrm{cl}_M(Y) \cup e)) \end{aligned} \tag{6}$$

and again (5) holds. Similarly, (5) holds if $\delta_X = \delta_Y = 0$ unless equality holds in (4). But, in the exceptional case, $(\mathrm{cl}_M(X), \mathrm{cl}_M(Y))$ is a modular pair of flats of M, so $\mathrm{cl}_M(X) \cap \mathrm{cl}_M(Y) \in \mathcal{M}$. As $\mathrm{cl}_M(\mathrm{cl}_M(X) \cup \mathrm{cl}_M(Y))$ is also in $\mathcal{M}$, (6) follows and hence so does (5). This completes the proof that r_N satisfies (R3). □

On combining Theorem 7.2.2 and Lemma 7.2.1, we get that there is a one-to-one correspondence between single-element extensions of a matroid M and modular cuts of M. In view of this, we shall often refer to a particular extension as being *determined* by the corresponding modular cut. If $M = N \backslash e$ and $\mathcal{M}$ is the modular cut corresponding to the extension N, we shall often write N as $M +_{\mathcal{M}} e$.

By finding all modular cuts of a matroid M, we can determine all possible single-element extensions of M. This gives a systematic technique by which, theoretically, one can determine all matroids on some fixed number of elements. Although this technique quickly becomes computationally impractical, it was used, for example, by Blackburn, Crapo, and Higgs (1973) to find all simple matroids on at most eight elements.

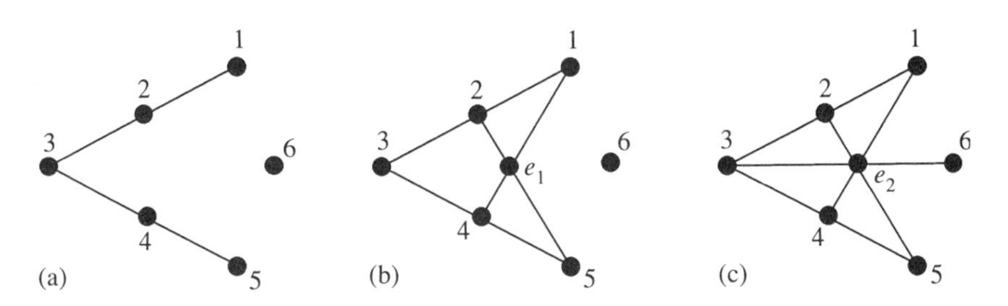

Fig. 7.8. (a) Q_6. (b) N_1 and (c) N_2 are incompatible extensions.

7.2.3 Example. In the matroid Q_6, for which a geometric representation is shown in Figure 7.8(a), let $\mathcal{M}_1 = \{\{1,4\},\{2,5\},E(Q_6)\}$ and $\mathcal{M}_2 = \{\{1,4\},\{2,5\},\{3,6\},E(Q_6)\}$. Each of $\mathcal{M}_1$ and $\mathcal{M}_2$ is easily shown to be a modular cut of Q_6, so each determines an extension. Geometric representations for these extensions are shown in Figure 7.8(b) and (c), the added elements being e_1 and e_2, respectively. The extensions N_1 and N_2 are examples of *incompatible single-element extensions* of Q_6, for both are extensions of Q_6, yet there is no matroid N such that $N\backslash e_1 = N_2$ and $N\backslash e_2 = N_1$. We defer to the exercises consideration of the problem of determining precisely when two single-element extensions are compatible.□

The rank function of $M +_{\mathcal{M}} e$ is given in Theorem 7.2.2. Using this, it follows that $r(M +_{\mathcal{M}} e) = r(M) + 1$ if and only if $\mathcal{M} = \emptyset$.

7.2.4 Corollary. *The flats of $M +_{\mathcal{M}} e$ fall into three disjoint classes:*

(i) *flats F of M that are not in $\mathcal{M}$;*

(ii) *sets $F \cup e$ where F is a flat of M that is in $\mathcal{M}$; and*

(iii) *sets $F \cup e$ where F is a flat of M that is not in $\mathcal{M}$ and F is not covered in $\mathcal{L}(M)$ by any member of $\mathcal{M}$.* □

This corollary is illustrated in Figure 7.9 where a schematic representation of an arbitrary modular cut in $\mathcal{L}(M)$ is shown along with the lattice $\mathcal{L}(M +_{\mathcal{M}} e)$ of flats of the corresponding extension (see Mason 1977).

If F is a flat of a matroid M, then the set $\mathcal{M}_F$ of flats containing F is easily shown to be a modular cut. We call $\mathcal{M}_F$ a *principal modular cut* and the extension determined by $\mathcal{M}_F$ a *principal extension*. We shall denote this principal extension by $M +_F e$ and we shall also say, in this case, that the element e has been *freely added* to the flat F or, if $F = E(M)$, that e has been freely added to M. In the last case, e is said to be *free* in $M +_{E(M)} e$ and this matroid is called the *free extension* of M. When adding the element e freely to the flat F, we add e to F so that the only

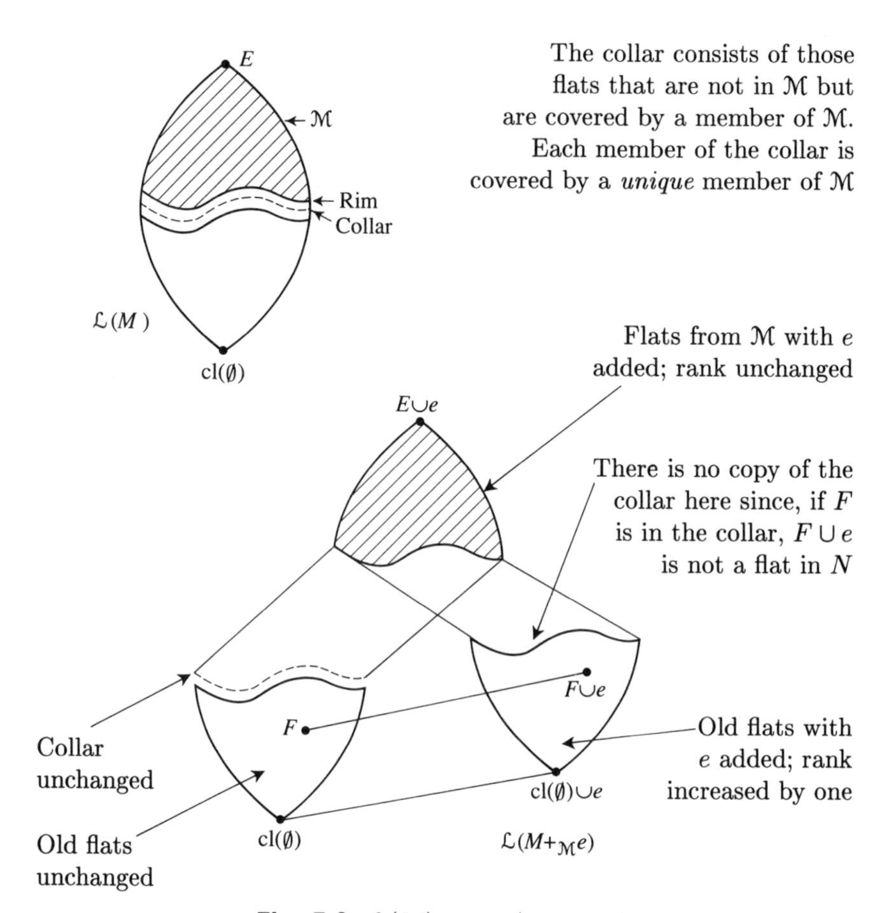

Fig. 7.9. $\mathcal{L}(M)$ and $\mathcal{L}(M +_{\mathcal{M}} e)$

new circuits created are those which are forced by the fact that e has been placed on F.

7.2.5 Proposition. *Let F be a flat of a matroid M and $N = M +_F e$. Then*

$$\mathcal{I}(N) = \mathcal{I}(M) \cup \{I \cup e : I \in \mathcal{I}(M) \text{ and } \mathrm{cl}_M(I) \not\supseteq F\}.$$

Proof. This follows without difficulty from Corollary 7.2.4 and is left as an exercise. □

Descriptions of principal extensions in terms of circuits, bases and so on will be considered in the exercises.

One attractive feature of modular cuts is that the intersection of two modular cuts is a modular cut. Suppose that we wish to add the element

e to each of the flats $F_1, F_2, \ldots, F_m$. Then the modular cut corresponding to such an extension must contain $\{F_1, F_2, \ldots, F_m\}$. The smallest modular cut containing this set, which is the intersection of all modular cuts containing $\{F_1, F_2, \ldots, F_m\}$, is called the modular cut *generated* by $\{F_1, F_2, \ldots, F_m\}$. Frequently this modular cut is equal to the set of all flats of the matroid, in which case the only way that e can be added to all of $F_1, F_2, \ldots, F_m$ is as a loop.

7.2.6 Example. Consider the matroid M for which a geometric representation is shown in Figure 7.10(a). We wish to adjoin the element 1 to M so that it is on both the line through 2 and 3 and the line through 4 and 5. Thus we take the modular cut $\mathcal{M}$ of M that is generated by $\{\{2,3\},\{4,5\}\}$. Evidently $\{\{2,3,4,5\},\{2,3,6,7\},\{4,5,6,7\}\} \subseteq \mathcal{M}$. Hence, as $(\{2,3,6,7\},\{4,5,6,7\})$ is a modular pair, $\{6,7\} \in \mathcal{M}$. Moreover, it is not difficult to check that $\mathcal{M} = \{\{2,3\},\{4,5\},\{6,7\},\{2,3,4,5\},\{2,3,6,7\},\{4,5,6,7\},E(M)\}$. Thus the extension $M +_{\mathcal{M}} 1$ has the geometric representation shown in Figure 7.10(b). It should be noted that this is consistent with Example 1.5.5 where it was shown that if, in Figure 7.10(b), the points 1, 6, and 7 are not collinear, then the resulting diagram is not a geometric representation of a matroid. We conclude that, in order to be able to add 1 on the lines through 2 and 3 and through 4 and 5, we must also place 1 on the line through 6 and 7. □

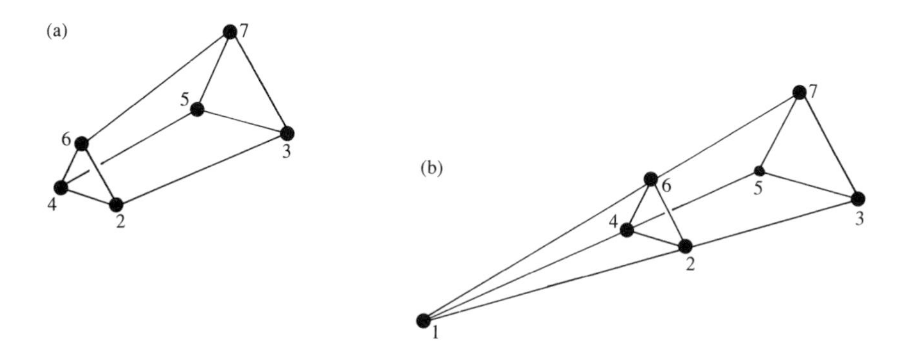

Fig. 7.10. (a) A matroid. (b) The extension generated by $\{\{2,3\},\{4,5\}\}$.

Exercises

1. Find all non-isomorphic single-element extensions of Q_6 and specify the corresponding modular cuts.

2. Suppose that $N \backslash e = M$. Show that e is free in N if and only if $r(N) = r(M)$ and every circuit of N containing e has $r(N)+1$ elements.

3. Prove Proposition 7.2.5.

4. Let $\mathcal{M}$ be a modular cut of a matroid M, let N be $M +_{\mathcal{M}} e$, and let $\mathcal{N}$ be $\{X \subseteq E(M) : \mathrm{cl}_M(X) \in \mathcal{M}\}$. Show that:

(i) If $X \in \mathcal{N}$ and $X \subseteq Y \subseteq E(M)$, then $Y \in \mathcal{N}$.

(ii) A subset X of $E(M)$ is in $\mathcal{N}$ if and only if $r_N(X \cup e) = r_M(X)$.

(iii) If $X_1, X_2 \in \mathcal{N}$ and $r(X_1) + r(X_2) = r(X_1 \cup X_2) + r(X_1 \cap X_2)$, then $X_1 \cap X_2 \in \mathcal{N}$.

5. Suppose that F is a flat of a matroid M. Let $E(M) = E$ and let $N = M +_F e$. Show the following:

(i) $\mathcal{B}(N) = \mathcal{B}(M) \cup \{(B - f) \cup e : B \in \mathcal{B}(M)$ and $f \in B \cap F\}$.

(ii) $\mathcal{C}(N) = \mathcal{C}(M) \cup \{X \cup e : \mathrm{cl}_M(X) \supseteq F$ and $\mathrm{cl}_M(X - x) \not\supseteq F$ for all x in $X\}$.

(iii) For all $X \subseteq E$, $r_N(X) = r_M(X)$, and

$$r_N(X \cup e) = \begin{cases} r_M(X) & \text{if } r_M(X \cup F) = r_M(X), \\ r_M(X) + 1 & \text{otherwise.} \end{cases}$$

(iv) For all $X \subseteq E$,

$$\mathrm{cl}_N(X) = \begin{cases} \mathrm{cl}_M(X) \cup e & \text{if } \mathrm{cl}_M(X) \supseteq F, \\ \mathrm{cl}_M(X) & \text{otherwise,} \end{cases}$$

and

$$\mathrm{cl}_N(X \cup e) = \begin{cases} \mathrm{cl}_M(X \cup F) \cup e & \text{if } r_M(X \cup F) = r_M(X) + 1, \\ \mathrm{cl}_M(X) \cup e & \text{otherwise.} \end{cases}$$

6. (Crapo 1965) A *linear subclass* of a matroid M is a subset $\mathcal{H}'$ of the set of hyperplanes of M such that if H_1 and H_2 are members of $\mathcal{H}'$ for which $r(H_1 \cap H_2) = r(M) - 2$, and H_3 is a hyperplane containing $H_1 \cap H_2$, then $H_3 \in \mathcal{H}'$. Show the following:

(i) If $\mathcal{M}$ is a modular cut of M, then the hyperplanes of M in $\mathcal{M}$ form a linear subclass.

(ii) If $\mathcal{H}'$ is a linear subclass of M and $\mathcal{M}$ consists of all flats X of M for which every hyperplane containing X is in $\mathcal{H}'$, then $\mathcal{M}$ is a modular cut of M.

(iii) If $\mathcal{H}'$ is a linear subclass of M, then there is a unique extension N of M by e such that $\mathcal{H}'$ is the set of hyperplanes H of M for which $H \cup e$ is a hyperplane of N.

7. (Oxley 1984b) Verify the following: Let $\{e_1, e_2, \ldots, e_k\}$ be a circuit in a matroid M where $k \geq 3$ and suppose that e_1 is in every dependent flat of M. Then a flat F of M is in the modular cut generated by $\text{cl}(\{e_1, e_2\})$ and $\text{cl}(\{e_3, e_4, \ldots, e_k\})$ if and only if F contains one of the two generating flats. Moreover, the generating flats are disjoint.

8.* (Cheung 1974) Let M, N_1, and N_2 be matroids such that $N_1\backslash e_1 = M = N_2\backslash e_2$ where $e_1 \neq e_2$. Let $\mathcal{M}_1$ and $\mathcal{M}_2$ be the modular cuts of M corresponding to N_1 and N_2, respectively. We call N_1 and N_2 *compatible extensions* of M if there is a matroid N on the set $E(M) \cup \{e_1, e_2\}$ such that $N\backslash e_2 = N_1$ and $N\backslash e_1 = N_2$.

   (i) Prove that N_1 and N_2 are compatible extensions of M if and only if $N_{12} = N_{21}$, where N_{12} is the extension of N_1 by e_2 that is determined by the modular cut generated by $\{\text{cl}_{N_1}(X) : X \in \mathcal{M}_2\}$, and N_{21} is defined analogously.

   (ii) Show that if N_1 and N_2 are both principal extensions, then N_1 and N_2 are compatible extensions of M.

7.3 Quotients and related operations

In this section, we shall consider various matroid operations that are closely related to the operation of extension. As in the last section, modular cuts will play an important role here.

A modular cut of a matroid M will be called *proper* if it is not equal to the set of flats of M. Let $\mathcal{M}$ be such a modular cut and assume also that $\mathcal{M}$ is non-empty. The *elementary quotient* of M with respect to $\mathcal{M}$ is the matroid $(M +_{\mathcal{M}} e)/e$. Evidently its rank is one less than that of M and its ground set is the same as that of M. The reason for not allowing $\mathcal{M}$ to be empty or non-proper is that otherwise e is a coloop or a loop of $M +_{\mathcal{M}} e$ and so $(M +_{\mathcal{M}} e)/e = M$. Hence the effect of these constraints is only to exclude two trivial cases. In general, a matroid Q is a *quotient* of the matroid M if there is a matroid N such that, for some subset X of $E(N)$, $M = N\backslash X$ and $Q = N/X$. Thus if Q is a quotient of M, then $E(Q) = E(M)$. Another easy consequence of the definition is the following.

7.3.1 Proposition. *Let M_1 and M_2 be matroids. Then M_1 is a quotient of M_2 if and only if M_2^* is a quotient of M_1^*.* □

7.3.2 Example. In Figure 7.11, M_1 and M_2 are two of the many elementary quotients of $M(K_4)$. □

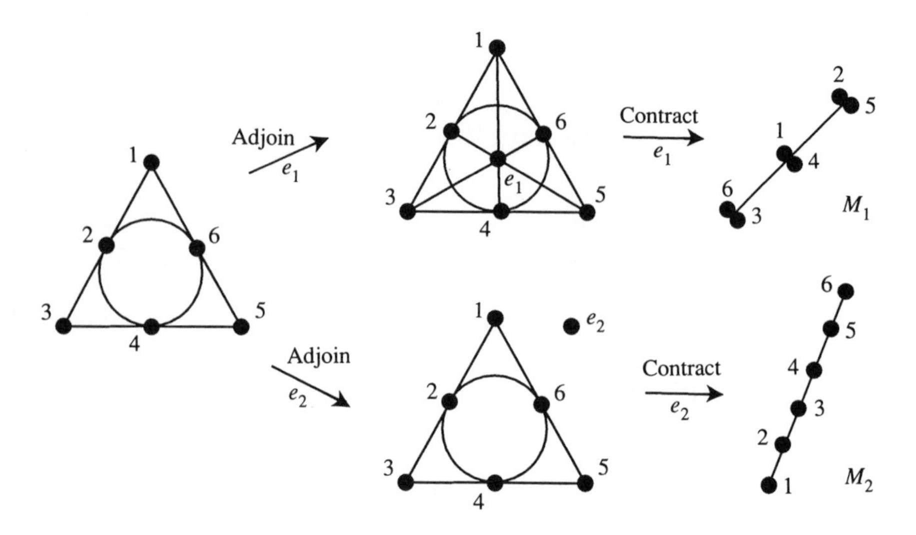

Fig. 7.11. M_1 and M_2 are two of the quotients of $M(K_4)$.

The next proposition shows that an arbitrary quotient can be formed by taking a (possibly empty) sequence of elementary quotients. The proof will use the following result.

7.3.3 Lemma. *Q is a quotient of M if and only if there is a matroid N having an independent set I such that $r(N) = r(M)$, $M = N\backslash I$, and $Q = N/I$.*

Proof. If N and I exist satisfying the specified conditions, then Q is certainly a quotient of M. Conversely, suppose that Q is a quotient of M. Then there is a matroid N_1 and a subset X of $E(N_1)$ such that $M = N_1\backslash X$ and $Q = N_1/X$. Choose such a matroid N_1 to have minimum rank satisfying these conditions. Let I be a basis of X in N_1 and $N = N_1\backslash(X - I)$. Then $N\backslash I = M$ and $N/I = N_1/I\backslash(X - I)$. But the last matroid is Q since every element of $X - I$ is a loop of N_1/I. Moreover, by the choice of N_1, $r(N) = r(N_1)$. Assume that $r(N) \neq r(M)$. Then, for some x in I, $r(E(M) \cup x) > r(M)$. Thus $N\backslash(I - x)/x = M$ and $N/x/(I - x) = Q$. Since N/x contradicts the choice of N_1, we conclude that $r(N) = r(M)$ and this completes the proof of the lemma. □

An immediate consequence of this lemma is the following.

7.3.4 Corollary. *If Q is a quotient of M and $r(Q) = r(M)$, then $Q = M$.* □

7.3.5 Proposition. *A matroid Q is a quotient of M with $r(M) - r(Q) = k$ if and only if there is a sequence $M_0, M_1, M_2, \ldots, M_k$ of matroids such that*

$M_0 = M$, $M_k = Q$, and, for all i in $\{1, 2, \ldots, k\}$, M_i is an elementary quotient of M_{i-1}.

Proof. Suppose Q is a quotient of M with $r(M) - r(Q) = k$. Then, by Lemma 7.3.3, there is a matroid N having an independent set I such that $r(M) = r(N)$, $M = N \backslash I$, and $Q = N/I$. Evidently $|I| = k$, so let $I = \{e_1, e_2, \ldots, e_k\}$. For each i in $\{0, 1, \ldots, k\}$, let M_i be the matroid $N\backslash(I - \{e_1, e_2, \ldots, e_i\})/\{e_1, e_2, \ldots, e_i\}$. Then it is straightforward to check that $M_0 = M$, $M_k = Q$, and, for each i in $\{1, 2, \ldots, k\}$, M_i is an elementary quotient of M_{i-1}.

Conversely, assume that a sequence $M_0, M_1, \ldots, M_k$ of matroids exists with the specified properties. Then, for each i in $\{1, 2, \ldots, k\}$, there is an extension N_i of M_{i-1} on $E(M_{i-1}) \cup e_i$ such that $N_i/e_i = M_i$, $r(N_i) = r(M_{i-1})$, and $r(M_i) = r(N_i) - 1$. To complete the proof we show, by induction on i, that there is an extension N_i' of M on $E(M) \cup \{e_1, e_2, \ldots, e_i\}$ such that $\{e_1, e_2, \ldots, e_i\}$ is independent in N_i' and $N_i'/\{e_1, e_2, \ldots, e_i\} = M_i$. This is certainly true if $i = 1$. Assume it true for $i = j$ and let $i = j + 1$. Then, by the induction assumption, M has an extension N_j' such that $M_j = N_j'/\{e_1, e_2, \ldots, e_j\}$ and $\{e_1, e_2, \ldots, e_j\}$ is independent in N_j'. Now $N_{j+1}\backslash e_{j+1} = M_j$, $r(N_{j+1}) = r(M_j)$, $M_{j+1} = N_{j+1}/e_{j+1}$, and $r(M_{j+1}) = r(N_{j+1}) - 1$. Thus $N_j'/\{e_1, e_2, \ldots, e_j\}$ has a proper non-empty modular cut corresponding to the extension N_{j+1}. Let $\mathcal{M}' = \{F \cup \{e_1, e_2, \ldots, e_j\} : F \in \mathcal{M}\}$. Then it is routine to check that $\mathcal{M}'$ is a proper non-empty modular cut of N_j'. Let the corresponding extension of N_j' have ground set $E(N_j') \cup e_{j+1}$ and call this extension N_{j+1}'. By comparing their rank functions, one can show that N_{j+1} and $N_{j+1}'/\{e_1, e_2, \ldots, e_j\}$ are equal. Therefore, as $N_{j+1}/e_{j+1} = M_{j+1}$, we get that $M_{j+1} = N_{j+1}'/\{e_1, e_2, \ldots, e_{j+1}\}$. Finally, as M_{j+1} is an elementary quotient of M_j, the set $\{e_{j+1}\}$ is independent in $N_{j+1}'/\{e_1, e_2, \ldots, e_j\}$. Thus $\{e_1, e_2, \ldots, e_{j+1}\}$ is independent in N_{j+1}', so N_{j+1}' is the required extension of M. The proposition now follows by induction. □

The next result contains a number of alternative characterizations of quotients including an attractive characterization in terms of flats. Several other such characterizations will be considered in the exercises.

7.3.6 Proposition. *Let M_1 and M_2 be matroids having rank functions r_1 and r_2, closure operators cl_1 and cl_2, and a common ground set E. The following statements are equivalent:*

(i) *M_2 is a quotient of M_1.*

(ii) *Every flat of M_2 is a flat of M_1.*

(iii) *If $X \subseteq Y \subseteq E$, then $r_1(Y) - r_1(X) \geq r_2(Y) - r_2(X)$.*

(iv) *Every circuit of M_1 is a union of circuits of M_2.*

(v) *For all $X \subseteq E$, $\mathrm{cl}_1(X) \subseteq \mathrm{cl}_2(X)$.*

Proof. It is straightforward to show using Proposition 3.3.1 and the definition of a quotient that (i) implies (ii). Next we establish the equivalence of (ii)–(v) by showing that each of (ii)–(iv) implies its successor and that (v) implies (ii).

Assume (ii) holds. We argue by induction on $|Y - X|$ to show that (iii) holds. This is certainly true if $|Y - X| = 0$. Now assume (iii) holds for $|Y - X| < k$ and let $|Y - X| = k$. Choose y in $Y - X$. If $r_1(Y) - r_1(X) = r_1(Y-y) - r_1(X) + 1$, then, by the induction assumption, $r_1(Y) - r_1(X) \geq r_2(Y - y) - r_2(X) + 1 \geq r_2(Y) - r_2(X)$. Thus we may suppose that $r_1(Y) - r_1(X) = r_1(Y-y) - r_1(X)$. Hence $r_1(Y) = r_1(Y-y)$, so y is in $\mathrm{cl}_1(Y-y)$. If $r_2(Y) = r_2(Y-y)$, then the required result follows. Thus we may assume that $r_2(Y) \neq r_2(Y-y)$, that is, $y \notin \mathrm{cl}_2(Y-y)$. Then $\mathrm{cl}_2(Y-y)$ and $\mathrm{cl}_2(Y)$ are distinct flats of M_2 and hence of M_1. Therefore $\mathrm{cl}_2(Y-y) = \mathrm{cl}_1(\mathrm{cl}_2(Y-y)) \supseteq \mathrm{cl}_1(Y-y) \supseteq Y$, so $\mathrm{cl}_2(Y-y) = \mathrm{cl}_2(Y)$; a contradiction. Hence (ii) implies (iii).

Now assume (iii) holds and let C be a circuit of M_1. If C is not a union of circuits of M_2, then C has an element x that is not in a circuit of $M_2|C$. Thus $r_2(C) - r_2(C-x) = 1$. Hence, by (iii), $r_1(C) - r_1(C-x) \geq 1$, contradicting the fact that C is a circuit. Therefore (iii) implies (iv).

Next assume that (iv) holds and suppose $X \subseteq E$. Then

$$\begin{aligned}\mathrm{cl}_1(X) &= X \cup \{x : M_1 \text{ has a circuit } C \text{ such that } x \in C \subseteq X \cup x\} \\ &\subseteq X \cup \{x : M_2 \text{ has a circuit } D \text{ such that } x \in D \subseteq X \cup x\} \\ &= \mathrm{cl}_2(X),\end{aligned}$$

that is, (iv) implies (v).

To complete the proof of the equivalence of (ii)–(v), we now show that (v) implies (ii). Hence suppose that (v) holds. Let F be a flat of M_2. Then $F = \mathrm{cl}_2(F) \supseteq \mathrm{cl}_1(F) \supseteq F$, so F is a flat of M_1, and (ii) holds. Thus (ii)–(v) are indeed equivalent.

Now, to finish the proof of Proposition 7.3.6, we shall show that (ii) implies (i), the proof of this being quite long. Thus suppose that every flat of M_2 is a flat of M_1. Then, from the above, each of (iii) – (v) holds. By (iii), $r_1(M_1) \geq r_2(M_2)$. We shall show, by induction on $r_1(M_1) - r_2(M_2)$, that M_2 is a quotient of M_1. The following lemma will be useful in this proof.

7.3.7 Lemma. *Let X be a flat of M_2 for which $r_1(X) - r_2(X) = r_1(M_1) - r_2(M_2)$, and let Y be a flat of M_1 that covers X in $\mathcal{L}(M_1)$. Then Y is a flat of M_2 covering X in $\mathcal{L}(M_2)$, and $r_1(Y) - r_2(Y) = r_1(M_1) - r_2(M_2)$.*

Proof. Choose y in $Y - X$. Then, as X is a flat of M_2, $\mathrm{cl}_2(X \cup y)$ is a flat of M_2 covering X in $\mathcal{L}(M_2)$. Thus $\mathrm{cl}_2(X \cup y)$ is a flat of M_1. By (iii),

$$r_1(E) - r_1(\mathrm{cl}_2(X \cup y)) \geq r_2(E) - r_2(\mathrm{cl}_2(X \cup y)).$$

Hence

$$[r_1(M_1) - r_2(M_2)] + r_2(\mathrm{cl}_2(X \cup y)) \geq r_1(\mathrm{cl}_2(X \cup y)),$$

so

$$[r_1(X) - r_2(X)] + r_2(\mathrm{cl}_2(X \cup y)) \geq r_1(\mathrm{cl}_2(X \cup y)).$$

Therefore

$$r_1(X) + [r_2(\mathrm{cl}_2(X \cup y)) - r_2(X)] \geq r_1(\mathrm{cl}_2(X \cup y)),$$

that is,

$$r_1(X) + 1 \geq r_1(\mathrm{cl}_2(X \cup y)). \tag{1}$$

As $\mathrm{cl}_2(X \cup y) \neq X$, equality must hold in (1). Thus both $\mathrm{cl}_2(X \cup y)$ and Y cover X in $\mathcal{L}(M_1)$. Since the intersection of $\mathrm{cl}_2(X \cup y)$ and Y properly contains X, we conclude that $\mathrm{cl}_2(X \cup y) = Y$ and the lemma follows. □

Now to show that M_2 is a quotient of M_1, suppose first that $r_1(M_1) - r_2(M_2) = 0$. We shall show that the sets of flats of M_1 and M_2 coincide. By assumption, every flat of M_2 is a flat of M_1. If M_1 has a flat that is not a flat of M_2, take Y to be such a flat of smallest rank. By (iii), $r_1(M_1) - r_1(\mathrm{cl}_2(\emptyset)) \geq r_2(M_2) - r_2(\mathrm{cl}_2(\emptyset))$. Thus $r_1(\mathrm{cl}_2(\emptyset)) = 0$, so, by (ii), $\mathrm{cl}_2(\emptyset) = \mathrm{cl}_1(\emptyset)$. Hence $Y \neq \mathrm{cl}_1(\emptyset)$ so $r_1(Y) > 0$. By the choice of Y, there is a flat X of M_2 that is covered by Y in $\mathcal{L}(M_1)$. But then Lemma 7.3.7 implies the contradiction that Y is a flat of M_2. We conclude that if $r_1(M_1) - r_2(M_2) = 0$, then $M_1 = M_2$, so M_2 is certainly a quotient of M_1.

Now assume that if $r_1(M_1) - r_2(M_2) < k$, then M_2 is a quotient of M_1 and let

$$r_1(M_1) - r_2(M_2) = k.$$

Let $\mathcal{M}$ be the set of flats F of M_1 such that F is a flat of M_2 with

$$r_2(F) = r_1(F) - k.$$

Notice that $\mathcal{M}$ consists of those flats X for which equality holds in (iii) when $Y = E$.

7.3.8 Lemma. *$\mathcal{M}$ is a modular cut of M_1.*

Proof. Let F be a member of $\mathcal{M}$. By Lemma 7.3.7, every flat covering F in $\mathcal{L}(M_1)$ is in $\mathcal{M}$. It follows easily that every flat in M_1 containing F is in $\mathcal{M}$. Using this and (iii), it is not difficult to complete the proof that $\mathcal{M}$ is a modular cut. □

As $\mathcal{M}$ is neither empty nor equal to the set of all flats of M_1, we can form the elementary quotient $(M_1 +_{\mathcal{M}} e)/e$ of M_1. Call this elementary quotient N. We show next that every flat of M_2 is a flat of N so that we can apply the induction assumption to M_2 and N.

Let F be a flat of M_2. Certainly F is a flat of M_1. If $F \in \mathcal{M}$, then $F \cup e$ is a flat of $M_1 +_{\mathcal{M}} e$, so F is a flat of $(M_1 +_{\mathcal{M}} e)/e$, that is, F is a flat of N. Now suppose that $F \notin \mathcal{M}$ and assume that F is covered in $\mathcal{L}(M_1)$ by a flat F' in $\mathcal{M}$. Then, by (iii),

$$1 = r_1(F') - r_1(F) \geq r_2(F') - r_2(F). \tag{2}$$

Since $r_2(F') - r_2(F) \geq 1$, equality must hold in (2), so $r_1(F) - r_2(F) = r_1(F') - r_2(F') = k$. Thus $F \in \mathcal{M}$; a contradiction. We conclude that F is not covered in $\mathcal{L}(M_1)$ by a flat in $\mathcal{M}$. Hence $F \cup e$ is a flat of $M_1 +_{\mathcal{M}} e$, so F is a flat of N.

We have now shown that every flat of M_2 is a flat of N. As $r(N) - r(M_2) < r(M_1) - r(M_2)$, we can apply the induction assumption to get that M_2 is a quotient of N. But N is a quotient of M_1, and so, by Proposition 7.3.5, M_2 is a quotient of M_1. This completes the proof of Proposition 7.3.6. □

Next we shall look briefly at the reverse of the quotient operation. A matroid M_1 is a *lift* of a matroid M_2 if M_2 is a quotient of M_1. If M_2 is an elementary quotient of M_1, then M_1 is an *elementary lift* of M_2. Thus M_1 is a lift of M_2 if there is a matroid N such that, for some subset Y of $E(N)$, $N/Y = M_1$ and $N\backslash Y = M_2$. Hence M_1 is a lift of M_2 if and only if M_1^* is a quotient of M_2^*.

To form an elementary lift of the matroid M, we first coextend M by an element e so that e is neither a loop nor a coloop in the resulting matroid N. The elementary lift is then obtained from N by deleting e.

Now let N be a coextension of M by e. We consider next how to represent N geometrically when e is not a loop of it. Since contraction of e corresponds geometrically to projection from e, to coextend M by e, first one adds a new point to correspond to e, this point being added in a new dimension. Then, each non-loop element f of M is slid up along the line joining e and f. More formally, the point in the coextension corresponding to f is some point that is on the line joining e and f but is distinct from e. Three examples of this construction are shown in Figures 7.12–7.14. The corresponding elementary lifts of M are obtained

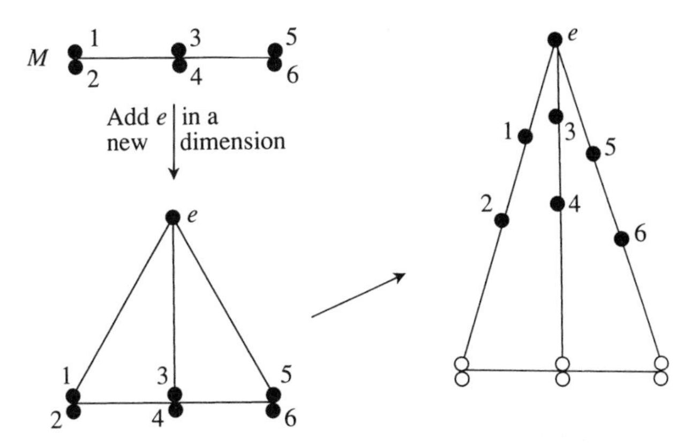

Fig. 7.12. Constructing the free coextension of M.

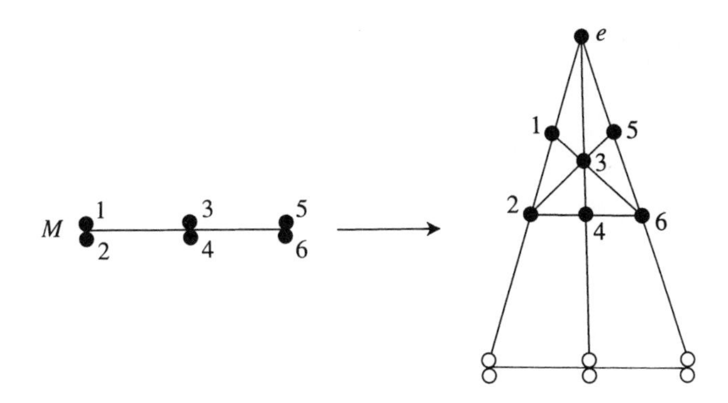

Fig. 7.13. A non-free coextension of M.

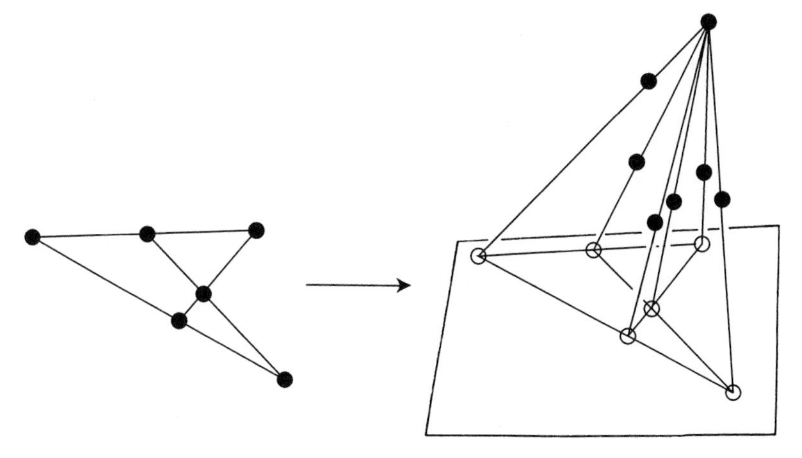

Fig. 7.14. Constructing the free coextension of $M(K_4)$.

in each case by deleting e. Clearly these elementary lifts are isomorphic to $U_{3,6}$, $\mathcal{W}^3$, and $U_{4,6}$, respectively.

It is clear that, in general, a matroid will have many different single-element extensions and many different single-element coextensions. The particular coextensions shown in Figures 7.12 and 7.14 are special in the sense that, in each case, every dependent flat of the coextension contains e, where a *dependent flat* is a flat that is also a dependent set. In general, if N is the coextension of M by e in which e is a non-loop that is in every dependent flat, then N is called the *free coextension* of M. It is an immediate consequence of the next result that N is the free coextension of M if and only if N^* is the free extension of M^*.

7.3.9 Proposition. *Let e be an element of a matroid N and suppose that e is not a loop. Then e is in every dependent flat of N if and only if e is free in N^*.*

Proof. Since e is not a loop of N, each of statements (ii)–(vi) in the following list is equivalent to its predecessor. The fact that (ii) implies (iii) follows because a circuit and a cocircuit cannot have exactly one common element (Proposition 2.1.11).

(i) The element e is in every dependent flat of N.

(ii) The element e is in the closure of every circuit of N.

(iii) N does not have a circuit D and a cocircuit C such that $e \in C$ and $C \cap D$ is empty.

(iv) N^* does not have a cocircuit D and a circuit C such that $e \in C$ and $C \cap D$ is empty.

(v) Every circuit of N^* containing e is spanning.

(vi) The element e is free in N^*. □

An important special case of the quotient operation is the operation of truncation. The *truncation* $T(M)$ of M is the elementary quotient corresponding to the free extension, that is, $T(M) = (M +_{E(M)} e)/e$. Evidently $r(T(M)) = r(M) - 1$. More generally, for a flat F of M of positive rank, the *principal truncation* $T_F(M)$ (Brown 1971) is $(M +_F e)/e$. Geometrically, $T_F(M)$ is obtained from M by freely adding e to the flat F and then contracting e. This operation can be iterated up to $r(F)$ times by using the following inductive definition: $T_F^i(M) = T_{T^{i-1}(M|F)}(T_F^{i-1}(M))$. Similarly, if $i \leq r(M)$, then $T^i(M) = T(T^{i-1}(M))$. Thus, for example, $T^k(U_{n,n}) = U_{n-k,n}$. The *complete principal truncation* is $T_F^i(M)$ where $i = r(F) - 1$. Geometrically, it is obtained by freely adding an $(r(F)-1)$-element independent set I to F and then contracting I.

It is particularly simple to describe the lattice of flats and the set of independent sets of the ith truncation $T^i(M)$. We defer the derivation of the corresponding result for the ith principal truncation to the exercises.

7.3.10 Proposition. *Let M be a matroid and i be a positive integer not exceeding $r(M)$. Then*

$$\mathcal{I}(T^i(M)) = \{X \in \mathcal{I}(M) : |X| \leq r(M) - i\}.$$

Moreover, $\mathcal{L}(T^i(M))$ is obtained from $\mathcal{L}(M)$ by removing all flats of the latter of rank exceeding $r(M) - i - 1$, and making $E(M)$ the unique flat of rank $r(M) - i$.

Proof. This is straightforward by induction and is omitted. □

The theory of quotients of matroids is part of an extensive theory of maps between matroids developed by Crapo (1965, 1967) and Higgs (1966a,b, 1968). There are two types of such maps: strong and weak, and, in general, these maps are defined from a matroid M_1 on a set E_1 to a matroid M_2 on a set E_2. We now look briefly at such maps in the case when $E_1 = E_2$ and the map from E_1 to E_2 is the identity map ι. In that case, ι is a *strong map* if every flat of M_2 is a flat of M_1. Hence, by Proposition 7.3.6, ι is a strong map from M_1 to M_2 if and only if M_2 is a quotient of M_1.

We say that the identity map is a *weak map* from M_1 to M_2 if every independent set in M_2 is also independent in M_1. In this case, M_2 is called a *weak-map image* of M_1. The following result contains a number of alternative characterizations of weak maps. Its straightforward proof is left as an exercise.

7.3.11 Proposition. *Let M_1 and M_2 be matroids having rank functions r_1 and r_2 and a common ground set E. The following are equivalent:*

(i) *The identity map on E is a weak map from M_1 to M_2.*

(ii) *Every dependent set in M_1 is dependent in M_2.*

(iii) *Every circuit of M_1 contains a circuit of M_2.*

(iv) *For every subset X of E, $r_1(X) \geq r_2(X)$.* □

Two easy consequences of this result are:

7.3.12 Corollary. *If the identity map ι on E is a strong map from the matroid M_1 to the matroid M_2, then ι is a weak map from M_1 to M_2.* □

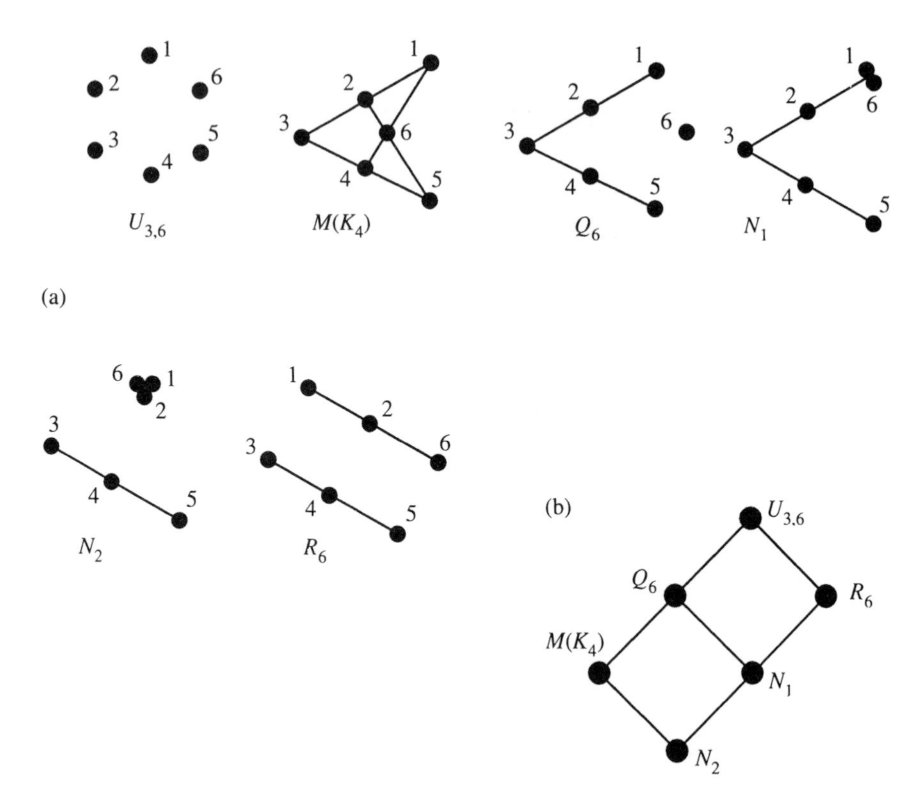

Fig. 7.15. (a) Six matroids and (b) the weak order on them.

7.3.13 Corollary. *If M_1 and M_2 are matroids of the same rank on E and the identity map ι on E is a weak map from M_1 to M_2, then ι is a weak map from M_1^* to M_2^*.* □

The collection $\mathcal{E}$ of matroids on a fixed set E can be partially ordered by taking $M_1 \geq M_2$ if the identity map on E is a weak map from M_1 to M_2. Under this partial order, the *weak order* on $\mathcal{E}$, the set $\mathcal{E}$ has both a one and a zero, namely the free matroid on E and the rank-0 matroid on E. We say that M_1 is *freer* than M_2 if $M_1 \geq M_2$. Making a matroid freer corresponds to destroying small circuits, that is, circuits of size less than the rank. Clearly a relaxation of a matroid M is freer than M itself. In Figure 7.15, six matroids on six elements are shown along with the poset induced on them by the weak order.

We have taken a brief glimpse at the theory of strong and weak maps. For a much more detailed exposition of this theory, the reader is referred to White (1986) where there are chapters devoted to both types of maps.

Exercises

1. Show that $T(AG(3,2)) \cong U_{3,8}$.

2. Find all non-isomorphic elementary quotients of $M(K_4)$.

3. Let M be the free coextension of F_7 by the element e. If f is in $E(F_7) - e$, give a geometric representation for M/f.

4. Is a matroid of rank at least one uniquely determined by its collection of elementary quotients?

5. Prove that the following are equivalent for matroids M_1 and M_2 on a set E:

   (a) M_2 is a quotient of M_1.

   (b) M_1^* is a quotient of M_2^*.

   (c) Every cocircuit of M_2 is a union of cocircuits of M_1.

   (d) Every hyperplane of M_2 is the intersection of a set of hyperplanes of M_1.

6. The matroids P_6 and $\mathcal{W}^3$ can be formed by relaxing $\{1,2,3\}$ in the matroids Q_6 and $M(K_4)$, respectively, in Figure 7.15(a). Find how these matroids fit into the weak order in Figure 7.15(b).

7. Suppose that M_1 is a graphic matroid and M_2 is an elementary quotient of M_1. Prove that M_2 is graphic if and only if there are graphs G_1 and G_2 such that $M_1 = M(G_1)$, $M_2 = M(G_2)$, and G_2 is obtained from G_1 by identifying two distinct vertices of the latter.

8. Let M_1 and M_2 be matroids having rank functions r_1 and r_2 and a common ground set E. Assume that every flat of M_2 is a flat of M_1. Prove that if X is a flat of M_2 such that $r_1(X) - r_2(X) = r_1(M_1) - r_2(M_2)$, then $M_1/X = M_2/X$.

9. Let $\mathcal{M}$ be a proper non-empty modular cut of a matroid M. If $A \subseteq E(M)$, let

$$r'(A) = \begin{cases} r_M(A) & \text{if } \mathrm{cl}_M(A) \notin \mathcal{M}; \\ r_M(A) - 1 & \text{otherwise.} \end{cases}$$

   Show that r' is the rank function of an elementary quotient of M and every elementary quotient of M arises in this way.

10. (Brylawski 1986a) If M_1 and M_2 are matroids on the same set E, prove that M_2 is a quotient of M_1 if and only if, for all subsets X of E, the matroid M_2/X is a weak-map image of M_1/X.

11. (Lucas 1975) Suppose that M_2 is a weak-map image of M_1 and $r(M_2) = r(M_1)$. Prove that every separator of M_1 is a separator of M_2.

12. Let F be a flat of a matroid M and N be $T_F(M)$.

(i) Specify $\mathcal{L}(N)$ in terms of $\mathcal{L}(M)$.

(ii) Show that

$$r_N(X) = \begin{cases} r_M(X) - 1 & \text{if } r_M(X) = r_M(X \cup F), \\ r_M(X) & \text{otherwise.} \end{cases}$$

(iii) Find $\mathcal{I}(N)$.

(iv) Show that

$$\text{cl}_N(X) = \begin{cases} \text{cl}_M(X \cup F) & \text{if } r_M(X \cup F) = r_M(X) + 1, \\ \text{cl}_M(X) & \text{otherwise.} \end{cases}$$

13. (Oxley and Whittle 1991) Let $\mathcal{C}'$ be the set of non-spanning circuits of a rank-r matroid M on a set E.

(i) Specify $\mathcal{C}(M)$ in terms of $\mathcal{C}'$ and r.

(ii) Show that there is a matroid N and a non-negative integer k such that the set of matroids on E having $\mathcal{C}'$ as its set of non-spanning circuits is $\{N, T(N), T^2(N), \dots, T^k(N)\}$.

14.* (Kung 1977) Let M_1 and M_2 be matroids on a set E and cl denote the closure operator of M_1. Prove that the following are equivalent:

(a) M_2 is a quotient of M_1.

(b) For all pairs $\{I_1, I_2\}$ of independent sets in M_1 such that $\text{cl}(I_1) = \text{cl}(I_2)$, if I_1 is dependent in M_2, then I_2 is dependent in M_2.

15.* (Lucas 1975) Let M_2 be a weak-map image of a binary matroid M_1 and suppose that $r(M_2) = r(M_1)$. Prove that:

(i) M_2 is binary;

(ii) if $M_2 \neq M_1$, then M_2 is disconnected.

8
Higher connectivity

The property of 2-connectedness for matroids has already proved to be of basic structural importance. When we introduced this property in Chapter 4, we noted that it is closely related to the property of 2-connectedness for graphs. In view of this, it is natural to ask whether, for arbitrary n, the property of n-connectedness for graphs can be extended to matroids. This chapter addresses this and several related questions focusing particularly on the case when $n = 3$. Some further matroid connectivity results will be presented in Section 10.2.

8.1 Tutte's definition

The concept of n-connectedness for matroids was introduced by W. T. Tutte (1966b). The motivation for Tutte's definition appears to derive from two sources: a desire to generalize the corresponding concept for graphs, and a wish to incorporate duality into the theory. These two aims are not totally compatible.

8.1.1 Example. It follows from the definition of graph connectivity in Chapter 4 that the planar graph G in Figure 8.1 is 3-connected. However, the graph G^*, a geometric dual of G, has a degree-2 vertex and so is not 3-connected. Thus if we make the matroid connectivities of $M(G)$ and its dual $M(G^*)$ equal, then, for at least one of G and G^*, the graph connectivity and the connectivity of the corresponding cycle matroid will be different. □

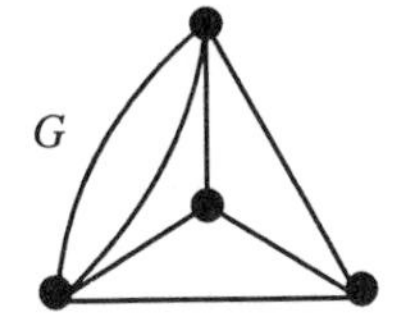

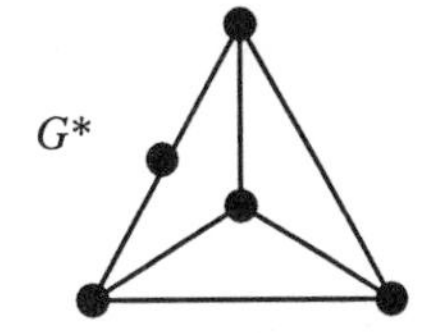

Fig. 8.1. The connectivities of G and G^* are different.

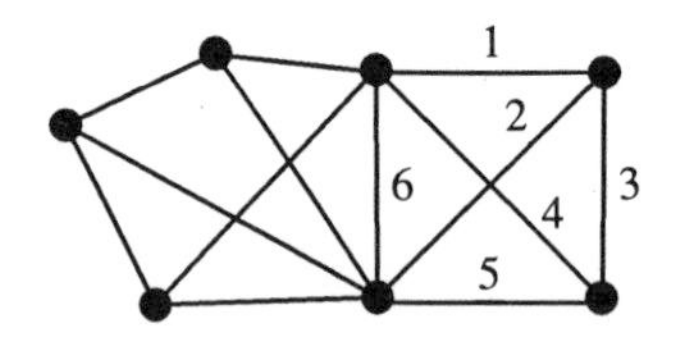

Fig. 8.2. A graph with connectivity 2.

The problems that occur in the last example are predictable and, indeed, in the most frequently studied case, when $n = 3$, Tutte's matroid definition of n-connectedness essentially succeeds both in generalizing graphical n-connectedness and in being duality invariant.

In order to extend the notion of 2-connectedness from graphs to matroids, we needed to find a reformulation of the definition of 2-connectedness that did not mention vertices, for in an arbitrary matroid there is no direct generalization of the concept of a vertex. This approach led to a matroid being defined to be 2-connected if, for every pair of distinct elements, there is a circuit containing both. For $n \geq 3$, the matroid definition of n-connectedness is not quite as straightforward. It generalizes the rank formulation of 2-connectedness for matroids. Recall from 4.2.3 that if X is a subset of the ground set E of a matroid M, then

$$r(X) + r(E - X) - r(M) \geq 0. \tag{1}$$

Moreover, by Proposition 4.2.1, M is not 2-connected if and only if, for some proper non-empty subset X of E, equality holds in (1). Before formally defining n-connectedness for matroids, we consider the following.

8.1.2 Example. Let G be the graph shown in Figure 8.2. Evidently $\kappa(G) = 2$. As G is loopless and has at least three vertices, it follows by Proposition 4.1.1 that $M(G)$ is 2-connected. Thus, by the remarks preceding this example, for all partitions (X, Y) of $E(G)$ such that

$$\min\{|X|, |Y|\} \geq 1, \tag{2}$$

we have

$$r(X) + r(Y) - r(M(G)) > 0. \tag{3}$$

Now let $X = \{1, 2, \ldots, 6\}$ and $Y = E(G) - X$. Then

$$r(X) + r(Y) - r(M(G)) = 1,$$

that is, with this choice of X and Y, the value of $r(X) + r(Y) - r(M(G))$ is minimized over all partitions of $E(G)$ satisfying (2) and (3). □

Let k be a positive integer. Then, for a matroid M, a partition (X, Y) of $E(M)$ is a *k-separation* if

8.1.3 $$\min\{|X|, |Y|\} \geq k$$

and

8.1.4 $$r(X) + r(Y) - r(M) \leq k - 1.$$

When equality occurs in 8.1.4, we call (X, Y) an *exact k-separation* of M. When equality occurs in 8.1.3, (X, Y) is a *minimal k-separation.* If M has a k-separation, then M is called *k-separated* or *k-separable.* Evidently a matroid is 1-separated if and only if it is disconnected. The matroid $M(G)$ in Example 8.1.2 is 2-separated but is not 1-separated.

If M is k-separated for some k, then the *connectivity* $\lambda(M)$ of M is $\min\{j : M \text{ is } j\text{-separated}\}$; otherwise we take $\lambda(M)$ to be ∞. Sometimes $\lambda(M)$ will be called the *Tutte connectivity* of M to distinguish it from two other types of matroid connectivity which will be discussed in the next section. Evidently M is 2-connected if and only if $\lambda(M) \geq 2$. In general, if n is an integer exceeding one, we shall say that M is *n-connected* if $\lambda(M) \geq n$. Thus, for the graphic matroid $M(G)$ in Example 8.1.2, $\lambda(M(G)) = 2$ so, like the graph G, the matroid $M(G)$ is 2-connected but not 3-connected. The exact relationship between matroid connectivity and graph connectivity will be examined in the next section. The rest of this section will concentrate on presenting some elementary properties of n-connected matroids. First, we note that a matroid is n-connected if and only if its dual is n-connected.

8.1.5 Proposition. *If M is a matroid, then $\lambda(M) = \lambda(M^*)$.*

Proof. By Lemma 4.2.5 and its dual, if $T \subseteq E(M)$, then

$$r(T)+r(E-T)-r(E) = r(T)+r^*(T)-|T| = r^*(T)+r^*(E-T)-r^*(E).$$

Thus $(T, E-T)$ is a k-separation of M if and only if it is a k-separation of M^*. □

It is easy to see that an n-connected graph G has no vertices of degree less than n and, more generally, that such a graph has no bonds of size less than n. This prompts one to ask what can be said about the size of cocircuits in an n-connected matroid. The next result answers this question. Note that, since n-connectedness is duality invariant, the result also gives information about circuit sizes.

8.1.6 Proposition. *If M is an n-connected matroid and $|E(M)| \geq 2(n-1)$, then all circuits and all cocircuits of M have at least n elements.*

Proof. If, for some $j < n$, M has a j-element subset X that is a circuit or a cocircuit, then it is easy to check that $(X, E(M) - X)$ is a j-separation of M, contradicting the fact that M is n-connected. □

The proof of the next result is similar to the last proof; the details are left as an exercise.

8.1.7 Proposition. *Let M be an n-connected matroid having at least $2n-1$ elements. Then M has no n-element subset that is both a circuit and a cocircuit.* □

It is easy to see that an n-connected matroid having fewer than $2n$ elements must have infinite connectivity. Using this fact in conjunction with Proposition 8.1.6, it is not difficult to determine all the matroids that have infinite connectivity, a result of Richardson (1973) and Inukai and Weinberg (1978). We leave it to the reader to verify their result, which is as follows.

8.1.8 Corollary. *If a matroid has infinite connectivity, then it is uniform. Moreover,*

$$\lambda(U_{r,n}) = \begin{cases} r+1, & \text{if } n \geq 2r+2; \\ n-r+1, & \text{if } n \leq 2r-2; \\ \infty, & \text{otherwise.} \end{cases}$$

□

8.1.9 Example. From the last result, we deduce that $U_{2,4}$ has infinite connectivity, so, in particular, $U_{2,4}$ is 4-connected. Yet $U_{2,4}$ has both 3-circuits and 3-cocircuits. Clearly, then, the assumption that $|E(M)| \geq 2(n-1)$ is needed in Proposition 8.1.6. If we now add an element to $U_{2,4}$ in parallel to an existing element, then we obtain a matroid M that is certainly 2-connected. However, because M has at least four elements and a 2-circuit, Proposition 8.1.6 implies that it is not 3-connected. Thus adding a non-loop element to a matroid without altering the rank can actually decrease the connectivity. This is one instance of where matroid connectivity behaves differently from graph connectivity. In the next section, we shall see that the definition of matroid connectivity can be modified so that it directly generalizes its graphical namesake. When this is done, however, one loses the invariance of connectivity under duality. □

It was noted above that adding an element to a 3-connected matroid so as to create a small circuit can result in the connectivity decreasing. Of course, adding a coloop will also decrease the connectivity. We now show that these are the only ways that extension by a single element can decrease the connectivity.

8.1.10 Proposition. *Let M and N be matroids such that $M\backslash e = N$. Suppose that N is n-connected but M is not. Then either e is a coloop of M, or M has a circuit that contains e and has fewer than n elements.*

Proof. As M is not n-connected, M is $(n-j)$-separated for some positive integer j, that is, there is a partition (X, Y) of $E(M)$ with

$$\min\{|X|, |Y|\} \geq n - j$$

and

$$r(X) + r(Y) - r(M) \leq n - j - 1. \tag{4}$$

Suppose, without loss of generality, that $e \in X$. Moreover, assume that e is not a coloop of M. Then, as $M\backslash e = N$, it follows from (4) that

$$r(X - e) + r(Y) - r(N) \leq n - j - 1 - (r(X) - r(X - e)). \tag{5}$$

But N is n-connected, so $(X - e, Y)$ is not an $(n - j)$-separation of N. Hence, $|X - e| < n - j$. As $|X| \geq n - j$, it follows that $|X| = n - j$. Moreover, $r(X) = r(X - e)$, otherwise $(X - e, Y)$ is an $(n - j - 1)$-separation of N. Thus X contains a circuit containing e, and this circuit has at most $n - j$ elements. □

Next we note another consequence of Proposition 8.1.6. Its straightforward proof is left as an exercise.

8.1.11 Corollary. *Let (X, Y) be a k-separation of a k-connected matroid and suppose that $|X| = k$. Then X is either a coindependent circuit or an independent cocircuit.* □

The last corollary provides useful information about the structure of minimal k-separations. The next result provides equally important information for non-minimal k-separations. The elementary proof is left to the reader.

8.1.12 Proposition. *Let (X, Y) be a k-separation of a matroid M and suppose that $|Y| \geq k + 1$. Then either X is both a flat and a coflat of M or, for some element e of Y, $(X \cup e, Y - e)$ is a k-separation of M.* □

Another elementary property of matroid connectivity which is similar to a property of graph connectivity is that deletion or contraction of a single element drops the connectivity by at most one provided the number of elements is not too small. Formally:

8.1.13 Proposition. *If e is an element of an n-connected matroid M, then, provided $|E(M)| \geq 2(n - 1)$, both $M\backslash e$ and M/e are $(n - 1)$-connected.*

Proof. This is straightforward and is left as an exercise. □

To determine the connectivity of a matroid M, we need to evaluate quantities of the form $r(X) + r(E(M) - X) - r(M)$. We now define this quantity to be $k(X)$ and view k as a function defined on all subsets X of $E(M)$. This function will be called the *connectivity function* of M.

How much information about M does the connectivity function convey? In particular, when can two distinct matroids have the same connectivity function? The next lemma points to two situations when this occurs.

8.1.14 **Lemma.** *A matroid and its dual have the same connectivity function. Moreover, if M_1 and M_2 are arbitrary matroids on disjoint sets, then the connectivity functions of $M_1 \oplus M_2$ and $M_1 \oplus M_2^*$ coincide.*

Proof. The first part follows easily from Lemma 4.2.5. The straightforward proof of the second part is left to the reader. □

It was conjectured by Cunningham (unpublished) that the last lemma describes the only ways that two distinct matroids can have the same connectivity function. Specifically, he proposed the following.

8.1.15 **Conjecture.** *If M and N are connected matroids on the same set having the same connectivity function, then either $M = N$ or $M = N^*$.* □

Seymour (1988) proved that this conjecture holds when M and N are binary, and Lemos (1990) proved it when $r(M) \neq r(M^*)$. However, Seymour also showed that the conjecture fails in general:

8.1.16 **Example.** Let $E = \{1, 2, \ldots, 8\}$ and let M and N be the rank-4 matroids on E for which geometric representations are shown in Figure 8.3, where we note that the elements 7 and 8 are free in M, while 5 and 6 are free in N. Evidently M and N are connected, and M is not equal to N or N^*, although M is isomorphic to both these matroids. The reader can easily

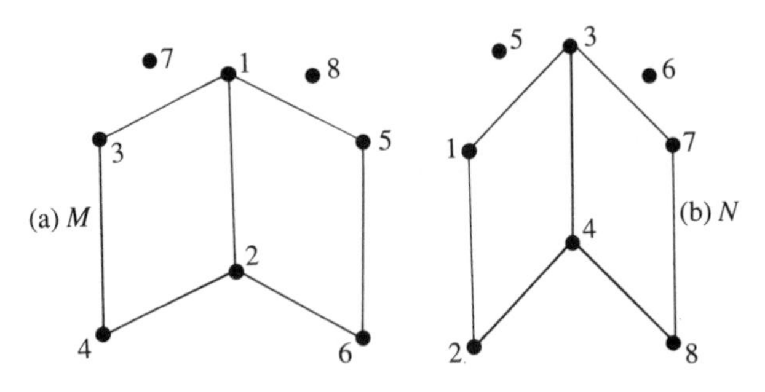

Fig. 8.3. Seymour's counterexample to Conjecture 8.1.15.

check that the connectivity functions of both M and N are equal to the function k defined by

$$k(X) = \begin{cases} |X|, & \text{if } |X| \leq 3; \\ |E - X|, & \text{if } |X| \geq 5; \\ 3, & \text{if } X \text{ or } E - X \text{ is } \{1,2,3,4\} \text{ or } \{1,2,5,6\}; \\ 4, & \text{otherwise.} \end{cases}$$

□

Exercises

1. Show that F_7 and F_7^- are 3-connected.

2. Find $\lambda(AG(3,2))$ and $\lambda(V_8)$.

3. Let M be a rank-3 matroid having at least four elements. Give necessary and sufficient conditions, in terms of its geometric representation, for M to be 3-connected.

4. Let M be a 3-connected matroid having at least four elements and let N be a single-element extension of M. Prove that N is 3-connected if and only if it is simple and has the same rank as M.

5. Let N be an elementary quotient of M. Give examples to show that $\lambda(M)$ can exceed $\lambda(N)$, that $\lambda(M)$ can equal $\lambda(N)$, and that $\lambda(N)$ can exceed $\lambda(M)$.

6. (Akkari 1988) Let M be an n-connected matroid having at least $2(n-1)$ elements and suppose that X is an independent subset of $E(M)$. Show that if $Y \subseteq X$ and M/X is n-connected, then M/Y is n-connected.

7. Prove Corollary 8.1.11.

8. (Oxley 1981b) Suppose that x and y are distinct elements of an n-connected matroid M where $n \geq 2$ and $|E(M)| \geq 2(n-1)$. Assume that $M\backslash x/y$ is n-connected but that $M\backslash x$ is not. Prove that M has a cocircuit of size n containing x and y.

9. Prove Proposition 8.1.12.

8.2 Matroid versus graph connectivity

In this section we indicate the precise relationship between matroid connectivity and graph connectivity. We also show how Tutte's definition of matroid connectivity can be modified to give a direct generalization of graph connectivity.

If G is a connected graph, then the addition of isolated vertices to G produces a disconnected graph whose cycle matroid equals that of G. For this reason, many of the results in this section will be stated for graphs that have no isolated vertices.

Graph connectivity was defined in terms of vertex deletions. An alternative and very useful characterization is contained in the following well-known theorem of Menger (1927). Let $\{P_1, P_2, \ldots, P_m\}$ be a set of paths in a graph G, each of which joins the distinct vertices u and v. These paths are said to be *internally disjoint* if no vertex in $V(G) - \{u, v\}$ is in more than one of the paths.

8.2.1 **Menger's Theorem.** *Let G be a graph having at least $n+1$ vertices. Then G is n-connected if and only if all pairs of distinct vertices of G are joined by at least n internally disjoint paths.*

Proof. This can be found in most basic graph theory texts (see, for example, Bondy and Murty 1976, Chapter 11). □

A link between 2-connected graphs and 2-connected matroids was noted in Proposition 4.1.8. The following slight strengthening of that result is easy to deduce from it.

8.2.2 **Corollary.** *Let G be a graph having no isolated vertices. If $|V(G)| \geq 3$, then $M(G)$ is 2-connected if and only if G is 2-connected and loopless.* □

This result points to the basic difference between n-connectedness in graphs and Tutte's definition of n-connectedness in matroids: an n-connected matroid cannot have small circuits whereas an n-connected graph can. This observation will be stated more precisely later in the section (Corollary 8.2.7). Before this, however, we indicate how Tutte's definition of matroid connectivity can be modified to yield a concept that directly generalizes the notion of connectivity for graphs. The disadvantage of this alternative concept is that one loses invariance under duality. We note here that Tutte's response to the difference between graph and matroid connectivity was to use a modified definition of graph connectivity (1961a), one that is directly generalized by his definition of matroid connectivity.

For a positive integer k, we say that a matroid M is *vertically k-separated* if there is a partition (X, Y) of $E(M)$ such that

8.2.3 $$\min\{r(X), r(Y)\} \geq k$$

and

8.2.4 $$r(X) + r(Y) - r(M) \leq k - 1.$$

Comparing these two inequalities with 8.1.3 and 8.1.4, we see that 8.1.4 and 8.2.4 are the same, while 8.2.3 strengthens 8.1.3. Hence if M is vertically k-separated, it is certainly k-separated. It is straightforward to check that M is vertically k-separated for some positive integer k if and only if M has a pair of disjoint cocircuits. Thus, provided the last condition holds, we define the *vertical connectivity* $\kappa(M)$ of M to be the least positive integer j such that M is vertically j-separated; otherwise we let $\kappa(M) = r(M)$. Hence, for the uniform matroid $U_{r,n}$, we have that

$$\kappa(U_{r,n}) = \begin{cases} n - r + 1, & \text{if } n \leq 2r - 2; \\ r, & \text{otherwise.} \end{cases}$$

In general, a matroid M is called *vertically n-connected* if n is an integer for which $2 \leq n \leq \kappa(M)$.

Vertical n-connectedness in matroids is a direct generalization of the notion of n-connectedness in graphs.

8.2.5 Theorem. *Let G be a connected graph. Then*

$$\kappa(M(G)) = \kappa(G).$$

Before proving this result, we shall discuss the relationship between vertical connectivity and Tutte connectivity, showing that the latter is obtained from the former by adding the restriction that small circuits be excluded. The precise statement of this relationship uses the notion of the *girth* $g(M)$ of a matroid M which is defined to be ∞ if M has no circuits and $\min\{|C| : C \text{ is a circuit of } M\}$ otherwise. The *girth* $g(G)$ of a graph G is the girth $g(M(G))$ of its cycle matroid.

8.2.6 Theorem. *Let M be a matroid and suppose that M is not isomorphic to any uniform matroid $U_{r,n}$ with $n \geq 2r - 1$. Then*

$$\lambda(M) = \min\{\kappa(M), g(M)\}.$$

Proof. By Corollary 8.1.8, $\lambda(M)$ is finite taking the value k, say. Thus M has a k-separation (X, Y). Hence $|E(M)| \geq 2k$ and so, by Proposition 8.1.6,

$$g(M) \geq k.$$

If $g(M) > k$, then, as $\min\{|X|, |Y|\} \geq k$, we have $\min\{r(X), r(Y)\} \geq k$. Hence (X, Y) is a vertical k-separation of M, so $\kappa(M) \leq k$. But

$$\begin{aligned} \kappa(M) &= \min\{j : M \text{ is vertically } j\text{-separated}\} \\ &\geq \min\{j : M \text{ is } j\text{-separated}\} = k = \lambda(M). \end{aligned} \tag{1}$$

We conclude that if $g(M) > k$, then the theorem holds.

We may now suppose that $g(M) = k$. If M has two disjoint cocircuits, then the theorem follows by (1). Thus we may assume that M does not have two disjoint cocircuits. Hence $\kappa(M) = r(M)$. Thus if $r(M) \geq g(M) = k$, then $\lambda(M) = k = \min\{\kappa(M), g(M)\}$. Hence we may suppose that $r(M) < g(M) = k$. Therefore $g(M) = r(M)+1$ and so $M \cong U_{r,n}$ for some non-negative integer n. As M does not have two disjoint cocircuits, $2(n - r + 1) > n$, that is, $n \geq 2r - 1$. This contradiction completes the proof of the theorem. □

For the uniform matroids excluded in the last theorem, the connectivity, λ, is either ∞ or $r + 1$, whereas the vertical connectivity, κ, is r. On combining the last result with Theorem 8.2.5, it is not difficult to deduce the corresponding result for graphic matroids.

8.2.7 Corollary. *Let G be a connected graph having at least three vertices and suppose that $G \not\cong K_3$. Then*

$$\lambda(M(G)) = \min\{\kappa(G), g(G)\}.$$

Proof. As G is connected and $|V(G)| \geq 3$, $r(M(G)) \geq 2$. Since $U_{2,4}$ is not graphic, $U_{r,n}$ is not graphic if both r and $n-r$ exceed one. It follows, since $G \not\cong K_3$, that $M(G)$ is not isomorphic to any $U_{r,n}$ with $n \geq 2r - 1$. The corollary now follows easily from the last two theorems. □

As an easy consequence of the last corollary, we have the following result which gives the relationship between 3-connectedness in matroids and 3-connectedness in graphs.

8.2.8 Corollary. *Let G be a graph without isolated vertices and suppose that $|V(G)| \geq 3$ and $G \not\cong K_3$. Then $M(G)$ is 3-connected if and only if G is 3-connected and simple.* □

From this result and Corollary 8.2.2, we see that, for n equal to two or three, n-connectedness for matroids is very closely linked to n-connectedness for graphs. Indeed, for simple graphs the concepts essentially coincide. For larger values of n, the matroid and graph concepts are not as closely related and one must choose whether to use Tutte n-connectedness, which is duality invariant, or vertical n-connectedness, a direct generalization of graphical n-connectedness. Most of the small amount of research that has been done in this case has used the former.

The following example illustrates the use of Corollary 8.2.8 on a class of graphic matroids which, as will be seen in Section 8.4, is of fundamental importance in any discussion of matroid connectivity.

8.2.9 Example. The r-spoked wheel graph $\mathcal{W}_r$ was introduced in Example 5.1.5. The first three wheels are shown in Figure 8.4. Clearly $\mathcal{W}_3 \cong K_4$

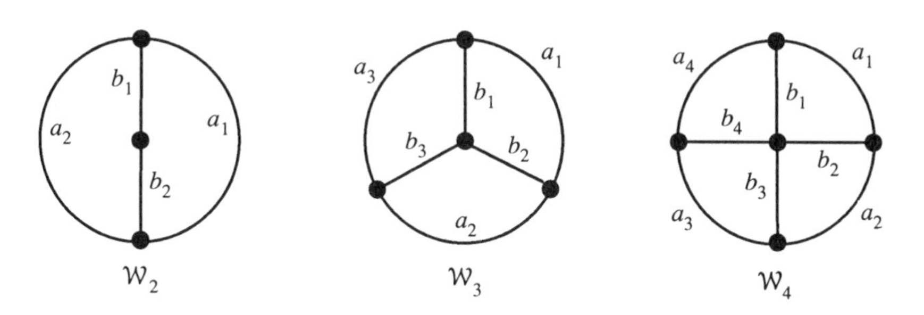

Fig. 8.4. The three smallest wheels.

(Exercise 10, Section 2.3) and $\widetilde{\mathcal{W}}_2 \cong K_3$. Thus $\kappa(\mathcal{W}_3) = 3$ and $\kappa(\mathcal{W}_2) = 2$. Moreover, for $r \geq 4$, it is easy to show using Menger's Theorem that $\kappa(\mathcal{W}_r) \geq 3$. But $\mathcal{W}_r$ has a degree-3 vertex, so $\kappa(\mathcal{W}_r) \leq 3$ and therefore $\kappa(\mathcal{W}_r) = 3$. It now follows from Corollary 8.2.8 that $\lambda(M(\mathcal{W}_r))$ is 2 if r is 2 and $\lambda(M(\mathcal{W}_r))$ is 3 otherwise.

Geometric representations for the matroids $M(\mathcal{W}_2)$, $M(\mathcal{W}_3)$, and $M(\mathcal{W}_4)$ are shown in Figure 8.5. □

We shall now prove Theorem 8.2.5. The proof will use the next four lemmas, the first two of which are elementary. We leave the proof of the first to the reader (Exercise 1).

8.2.10 Lemma. *Let X_1, X_2, Y_1, and Y_2 be subsets of the ground set of a matroid M. If $X_1 \supseteq Y_1$ and $X_2 \supseteq Y_2$, then*

$$r(X_1) + r(X_2) - r(X_1 \cup X_2) \geq r(Y_1) + r(Y_2) - r(Y_1 \cup Y_2). \quad \square$$

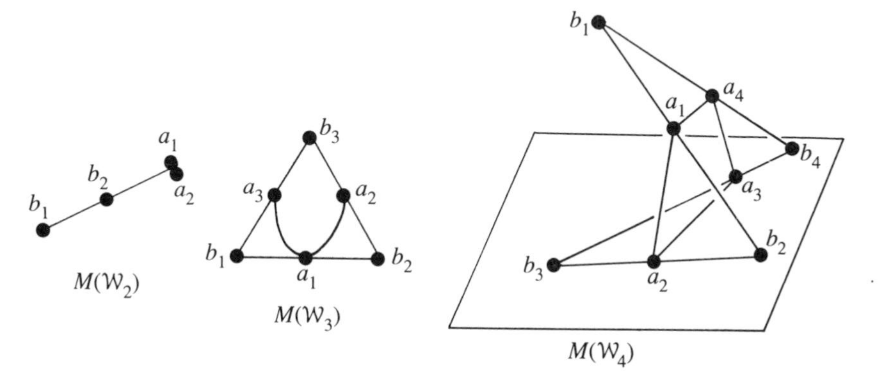

Fig. 8.5. The cycle matroids of the three smallest wheels.

The next lemma is a straightforward consequence of the fact noted in 1.3.8 that if X is a set of edges of a graph G and $G[X]$, the subgraph of G induced by X, has $\omega(G[X])$ components, then

$$r(X) = |V(G[X])| - \omega(G[X]).$$

The details of the proof of this lemma are left as an exercise.

8.2.11 Lemma. *Let (E_1, E_2) be a partition of the edge set of a connected graph G. Then*

$$r(E_1) + r(E_2) - r(M(G)) \leq |V(G[E_1]) \cap V(G[E_2])| - 1.$$

Moreover, if $G[E_1]$ and $G[E_2]$ are both connected, then equality holds here. □

The core of the proof of Theorem 8.2.5 is contained in the next two lemmas.

8.2.12 Lemma. *Let G be a connected graph and suppose that G has a k-element minimal vertex cut V'. Then $M(G)$ is vertically k-separated.*

Proof. Let H_1 and H_2 be distinct components of $G - V'$ and, for $i = 1, 2$, let E_i be the set of edges of $G[V' \cup V(H_i)]$. Since the last graph is clearly connected and has at least $k+1$ vertices, $r(E_i) \geq k$. Let $E'_2 = E(G) - E_1$. As $E'_2 \supseteq E_2$, $r(E'_2) \geq k$. Moreover, $V(G[E_1])$ and $V(G[E'_2])$ meet in V'. Therefore, by Lemma 8.2.11, $r(E_1) + r(E'_2) - r(M(G)) \leq |V'| - 1 = k - 1$. Thus $M(G)$ is vertically k-separated. □

In the next lemma, we shall refer to two vertices as being *connected* in a graph G if they lie in the same component of G.

8.2.13 Lemma. *If G is an n-connected graph, then $M(G)$ is not vertically k-separated for any $k < n$.*

Proof. Suppose that, for some $k < n$, there is a vertical k-separation (E_1, E_2) of $M(G)$. Let $G_1 = G[E_1]$ and $G_2 = G[E_2]$. Now assume that G has two vertices u and v that are not connected in either G_1 or G_2. Then u and v are not adjacent in G. As G is n-connected, $|V(G)| \geq n + 1$. Therefore, by Menger's Theorem (8.2.1), there are n internally disjoint paths $P_1, P_2, \ldots, P_n$ in G joining u to v. By assumption, if $j \in \{1, 2, \ldots, n\}$ and $i \in \{1, 2\}$, then $E(P_j) \not\subseteq E_i$. Now, for $i = 1, 2$, let $Y_i = E_i \cap \left(\bigcup_{j=1}^{n} E(P_j)\right)$. Then $r(Y_i) = |Y_i|$ and $r(Y_1 \cup Y_2) = |Y_1 \cup Y_2| - n + 1$. Thus $r(Y_1) + r(Y_2) - r(Y_1 \cup Y_2) = n - 1$. But $E_1 \supseteq Y_1$ and $E_2 \supseteq Y_2$, hence, by Lemma 8.2.10, $r(E_1) + r(E_2) - r(E(G)) \geq n - 1 \geq k$. This contradiction to the fact that (E_1, E_2) is a vertical k-separation of $M(G)$ establishes that every two vertices of G are connected in at least one of G_1 and G_2.

As $\min\{r(E_1), r(E_2)\} \geq k$ and $r(E_1) + r(E_2) - r(M(G)) \leq k - 1$, it follows that $\max\{r(E_1), r(E_2)\} < r(M(G))$. Hence neither G_1 nor G_2 is a connected graph whose edge set is spanning in $M(G)$. Thus G has two distinct vertices u and v that are in the same component H_1 of G_1 but are not connected in G_2, otherwise G_2 is connected and E_2 is spanning in $M(G)$; a contradiction. Choose a vertex w of G that is not in $V(H_1)$. Then v and w are not connected in G_1, so must be connected in G_2. Likewise, u and w must be connected in G_2. Hence u and w are connected in G_2; a contradiction. □

Using the above lemmas, it is now not difficult to prove Theorem 8.2.5.

Proof of Theorem 8.2.5. If $\widetilde{G}$ is complete, then $\kappa(G) = |V(G)| - 1$. Moreover, $M(G)$ does not have two disjoint cocircuits, so $\kappa(M(G)) = r(M(G)) = |V(G)| - 1$. Thus if $\widetilde{G}$ is complete, then $\kappa(G) = \kappa(M(G))$. We may now assume that $\widetilde{G}$ is not complete. Let $\kappa(G) = n$. Then G has an n-element minimal vertex cut. Hence, by Lemma 8.2.12, $M(G)$ is vertically n-separated and so $\kappa(M(G)) \leq n$, that is, $\kappa(M(G)) \leq \kappa(G)$. But, by Lemma 8.2.13, as G is n-connected, $\kappa(M(G)) \geq n$. Thus $\kappa(M(G)) = \kappa(G)$ and the theorem is proved. □

The notion of vertical connectivity for matroids was introduced independently by Inukai and Weinberg (1981), Cunningham (1981), and Oxley (1981d), and all three of these papers prove results equivalent to Theorem 8.2.5. The proof given here is Cunningham's, as is the terminology, which was borrowed from Tutte's terminology (1977) for the graph-theoretic notion of (vertex) connectivity. We remark that Inukai and Weinberg (1981) take $\kappa(M)$ to be ∞ instead of $r(M)$ when M does not have two disjoint cocircuits. The first explicit statement of Corollary 8.2.7 seems to be in Graver and Watkins' book (1977), although the statement there makes no reference to matroids.

In addition to examining both Tutte connectivity and vertical connectivity, Cunningham (1981) discussed the dual of vertical connectivity calling it cyclic connectivity. If k is a positive integer, a partition (X, Y) of the ground set $E(M)$ of a matroid M is a *cyclic k-separation* of M if $r(X) + r(Y) - r(M) \leq k - 1$ and both X and Y contain circuits of M. It is straightforward to check that M is cyclically k-separated for some positive integer k if and only if M has a pair of disjoint circuits. Thus, provided the last condition holds, we define the *cyclic connectivity* $\kappa^*(M)$ of M to be the least positive integer j such that M is cyclically

j-separated; otherwise, we let $\kappa^*(M) = r^*(M)$. It is straightforward to check that, for all matroids M,

8.2.14 $\kappa^*(M) = \kappa(M^*)$.

We call the matroid M *cyclically n-connected* if n is an integer such that $2 \leq n \leq \kappa^*(M)$.

It should, perhaps, be emphasized here that, although we now have three different types of n-connectedness for matroids, when we refer to an 'n-connected' matroid, we mean one that is n-connected in Tutte's sense. The next result shows that Tutte n-connectedness is basically the conjunction of vertical n-connectedness and cyclic n-connectedness. The proof is left to the reader.

8.2.15 Proposition. *Suppose that M is a matroid for which $\lambda(M)$ is finite and let n be a positive integer. Then M is n-connected if and only if M is both vertically and cyclically n-connected.* □

To close this section, we use the links established here between graph and matroid connectivity, together with some results from Section 5.3 to prove two of Whitney's results: a 3-connected simple planar graph has a unique planar embedding, and a 3-connected loopless planar graph has a unique abstract dual. Evidently to formulate the first of these results precisely, we shall need to define when two planar embeddings of a graph are the same. Let G be a planar graph. Two planar embeddings G_1 and G_2 of G are said to be *equivalent* if the set of edges on the boundary of a face in G_1 always corresponds to the set of edges on the boundary of a face in G_2. We say that G is *uniquely embeddable in the plane* if any two planar embeddings of G are equivalent. Since stereographic projection allows us to transform a plane graph into a graph embedded on the surface of a sphere and vice versa, we shall also say that G is *uniquely embeddable on the sphere* if it is uniquely embeddable in the plane. The plane graphs G_1 and G_2 in Figure 8.6 are equivalent. Each can be obtained from the same graph on the sphere by stereographic projection: for G_1, the point P of projection is taken inside a face bounded by four edges, while for G_2, we take P inside a face bounded by three edges. It is not difficult to see

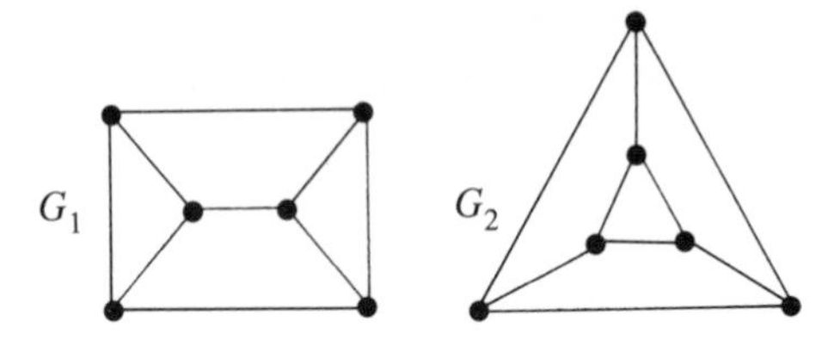

Fig. 8.6. Two equivalent plane graphs.

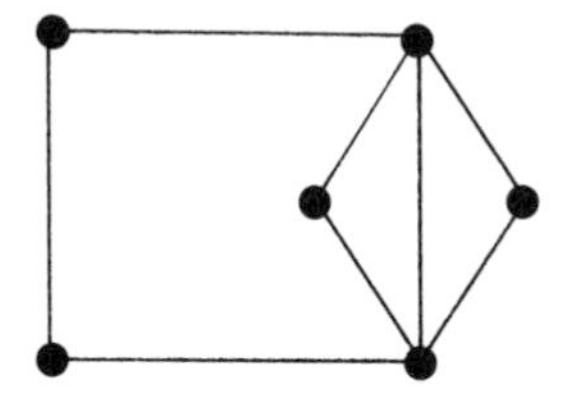
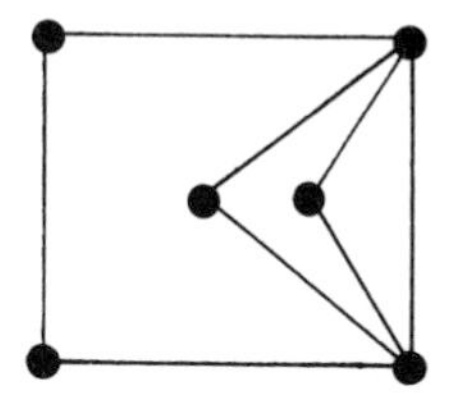

Fig. 8.7. Two isomorphic but inequivalent plane graphs.

that the isomorphic plane graphs in Figure 8.7 are not equivalent. The essential difference between the pairs of graphs in these two diagrams is their connectivities.

8.2.16 Theorem. *(Whitney 1932b) Let G be a simple 3-connected planar graph. Then G is uniquely embeddable on the sphere.*

The proof will use the following.

8.2.17 Lemma. *Let G' be an abstract dual of a 3-connected simple planar graph G and suppose G' has no isolated vertices. Then G' is 3-connected and simple.*

Proof. This is a straightforward combination of Corollary 8.2.8 and Proposition 8.1.5, and the details are left to the reader. □

Proof of Theorem 8.2.16. Suppose that G_1 and G_2 are inequivalent planar embeddings of G and let G_1^* and G_2^* be their geometric duals. Then, by the last lemma, both G_1^* and G_2^* are 3-connected. If we label the edges of G by the set E, then there is a natural labelling of the edge sets of all of G_1, G_2, G_1^*, and G_2^* by E. The identity map ι on E induces an isomorphism between $M(G_1)$ and $M(G_2)$ and hence between $M^*(G_1)$ and $M^*(G_2)$. Thus, by Lemmas 2.3.7 and 5.3.2, ι induces an isomorphism between $M(G_1^*)$ and $M(G_2^*)$ and hence between G_1^* and G_2^*.

As G_1 and G_2 are inequivalent, we may assume that there is a set X of edges that bounds a face in G_1 but not in G_2. Then, since G_1^* and G_2^* are 3-connected, X is a vertex bond in G_1^* but not in G_2^*; a contradiction. □

Now, as promised in Chapter 5, we give a proof of Proposition 5.3.7, first restating the proposition for convenient reference.

8.2.18 Proposition. *Let G be a 3-connected loopless planar graph. If G_1 and G_2 are abstract duals of G, each having no isolated vertices, then $G_1 \cong G_2$.*

Proof. We first observe that if G is simple, the result follows easily on combining Lemmas 8.2.17 and 5.3.2. Now suppose that G is non-simple and let X be a subset of $E(G)$ that is formed by choosing all but one edge from each parallel class of G. For $i = 1, 2$, there is a bijection ψ_i from $E(G)$ onto $E(G_i)$ that induces an isomorphism between $M^*(G)$ and $M(G_i)$. Evidently $G\backslash X$ is simple and has both $G_1/\psi_1(X)$ and $G_2/\psi_2(X)$ as abstract duals. Moreover, by Lemma 8.2.17, the last two graphs are 3-connected, and so, by Lemma 5.3.2, the restriction of $\psi_2 \circ \psi_1^{-1}$ to $E(G_1) - \psi_1(X)$ induces an isomorphism between these two graphs.

For $i = 1, 2$, every edge of $\psi_i(X)$ is in series in G_i with some edge of $E(G_i) - \psi_i(X)$. Furthermore, as $G_i/\psi_i(X)$ is 3-connected, G_i is a subdivision of $G_i/\psi_i(X)$. Indeed, G_i can be constructed from $G_i/\psi_i(X)$ by replacing each edge e by a path, the length of which equals the size of the parallel class in G that contains $\psi_i^{-1}(e)$. It now follows easily that $\psi_2 \circ \psi_1^{-1}$ induces an isomorphism between G_1 and G_2. □

Exercises

1. Prove Lemma 8.2.10. (Hint: For $i = 1, 2$, let B_i be a basis for $X_i - Y_i$ in M/Y_i.)

2. **(i)** Prove that $\kappa^*(M) = \kappa(M^*)$ for all matroids M.

(ii) Find $\kappa^*(U_{r,n})$.

3. Show that, for a matroid M, $\kappa(M) - \lambda(M)$ can be arbitrarily large, and $\kappa^*(M)$ and $\kappa(M)$ need not be equal.

4. Show that both $M^*(K_{3,3})$ and $M^*(K_5)$ are 3-connected.

5. (Seymour 1980b) Prove that the following are equivalent for a connected matroid M:

(a) M is cyclically 3-connected.

(b) M can be obtained from a 3-connected matroid by a sequence of series extensions.

(c) For every partition (X, Y) of $E(M)$ with $r(X)+r(Y) \leq r(M)+1$, either X or Y is contained in a series class of M.

6. (Inukai and Weinberg 1981) Prove that the following four statements are equivalent for a matroid M:

(a) There is no positive integer k for which M is vertically k-separated.

(b) For every non-empty proper subset X of $E(M)$, at least one of X and $E(M) - X$ is spanning.

(c) M has no two disjoint cocircuits.

(d) For every non-empty proper subset X of $E(M)$, $r(M.X) = 0$ or $r(M/X) = 0$.

7. Prove Proposition 8.2.15.

8. (Hausmann and Korte 1981) If M is a matroid and $X \subseteq E$, the girth of $M|X$ will be denoted $g(X)$. This defines a function from 2^E into $\mathbb{Z}^+ \cup \{0, \infty\}$. Prove that an arbitrary such function is the girth function of a matroid on E if and only if g satisfies the following conditions:

(G1) *If $X \subseteq E$ and $g(X) < \infty$, then X has a subset Y such that $g(X) = g(Y) = |Y|$.*

(G2) *If $X \subseteq Y \subseteq E$, then $g(X) \geq g(Y)$.*

(G3) *If X and Y are distinct subsets of E with $g(X) = |X|$, $g(Y) = |Y|$, and e in $X \cap Y$, then $g((X \cup Y) - e) < \infty$.*

9. (Cunningham 1981) Prove that if $M \backslash e$ is vertically n-connected and e is not a coloop of M, then M is vertically n-connected.

10. In the following classes of matroids, find all members that have no two disjoint cocircuits:

(i) graphic matroids;

(ii) cographic matroids;

(iii)* (Li Weixuan 1983) simple binary matroids.

8.3 3-connected matroids and 2-sums

In this section we shall establish a basic link between the operation of 2-sum and the property of 3-connectedness. To begin, recall that a matroid M fails to be 2-connected if and only if it can be written as a direct sum of two non-empty matroids, each of which is isomorphic to a minor of M. The following analogue of this result for 3-connectedness was proved independently by Bixby (1972), Cunningham (1973), and Seymour (1980b). The proof given here is Seymour's.

8.3.1 Theorem. *A 2-connected matroid M is not 3-connected if and only if $M = M_1 \oplus_2 M_2$ for some matroids M_1 and M_2, each of which is isomorphic to a proper minor of M.*

Proof. It is an easy consequence of Proposition 7.1.15(i) that if $M = M_1 \oplus_2 M_2$, then M is 2-separated and hence is not 3-connected. The proof of the converse is not quite as straightforward. Suppose that M is not 3-connected. Then M has an exact k-separation for some $k < 3$. As M is 2-connected, $k = 2$. Thus there is a partition (X_1, X_2) of $E(M)$ such that $\min\{|X_1|, |X_2|\} \geq 2$ and

$$r(X_1) + r(X_2) - r(M) = 1. \tag{1}$$

The next two lemmas are fundamental to the construction of the matroids M_1 and M_2 of which M is the 2-sum.

8.3.2 Lemma. *Let C_1 and C_2 be circuits of M that meet both X_1 and X_2. Then $C_1 \cap X_1$ is not a proper subset of $C_2 \cap X_1$.*

Proof. Assume the contrary and choose x_1 in $C_1 \cap X_1$ and x_2 in $(C_2 - C_1) \cap X_1$. As $C_2 \cap X_1$ is independent in M, it is contained in a basis B_1 of X_1. Let B_2 be a basis of X_2. Then $\mathrm{cl}((B_1 \cup B_2) - \{x_1, x_2\})$ contains X_2. Therefore it also contains x_1 and hence x_2; that is, $(B_1 \cup B_2) - \{x_1, x_2\}$ spans M. Hence $r(M) \leq |B_1| + |B_2| - 2 = r(X_1) + r(X_2) - 2$. This contradiction to (1) completes the proof of the lemma. □

8.3.3 Lemma. *Let Y_1 and Y_2 be non-empty subsets of X_1 and X_2 respectively. Suppose that M has circuits C_1 and C_2 such that $C_1 \cap X_1 = Y_1$, $C_2 \cap X_2 = Y_2$, and both $C_1 \cap X_2$ and $C_2 \cap X_1$ are non-empty. Then $Y_1 \cup Y_2$ is a circuit of M.*

Proof. Choose circuits C_1 and C_2 with the properties described such that $C_1 \cup C_2$ is minimal (see Figure 8.8). If $C_1 = C_2$, then the lemma holds so assume that $C_1 \neq C_2$. Suppose that $(C_1 \cup C_2) \cap X_1$ is dependent and let C be a circuit contained in it. Now $C \neq C_1$ since C_1 meets X_2 but C does not. Thus there is an element x in $C - C_1$. As $C \subseteq C_1 \cup C_2$, $x \in C_2$. Choose y in $C_2 \cap X_2$. Then $y \in C_2 - C$ and $x \in C_2 \cap C$. Thus, by the strong circuit elimination axiom, there is a circuit C_2' that contains y such that $C_2' \subseteq (C_2 \cup C) - x$. Clearly $C_2' \cap X_2 \subseteq (C_2 \cup C) \cap X_2 = C_2 \cap X_2$. It

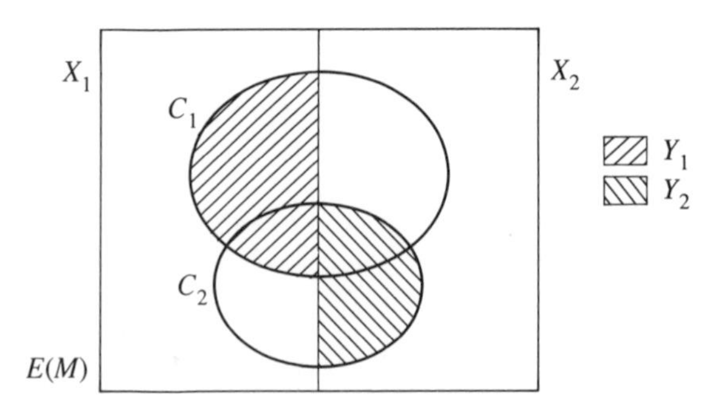

Fig. 8.8.

follows, since $C'_2 \neq C_2$, that $C'_2 \cap X_1$ is non-empty. Moreover, $C'_2 \cap X_2$ is non-empty since it contains y. Applying Lemma 8.3.2 to C_2 and C'_2, we deduce, since $C'_2 \cap X_2 \subseteq C_2 \cap X_2$, that $C'_2 \cap X_2 = C_2 \cap X_2 = Y_2$. But $C_1 \cup C'_2 \subseteq (C_1 \cup C_2) - x$. This contradicts the minimality of $C_1 \cup C_2$. Thus $(C_1 \cup C_2) \cap X_1$ is independent. Similarly, $(C_1 \cup C_2) \cap X_2$ is independent.

Now, for $i = 1, 2$, let B_i be a basis of X_i containing $(C_1 \cup C_2) \cap X_i$. If $C_1 \cap X_2 = Y_2$, then $Y_1 \cup Y_2 = C_1$ and the lemma holds. We may therefore assume that $C_1 \cap X_2 \neq Y_2$. Thus, by Lemma 8.3.2, $C_1 \cap X_2 \not\subseteq Y_2$ and we can choose an element x_2 from $(C_1 \cap X_2) - Y_2$. Similarly, we can choose an element x_1 from $(C_2 \cap X_1) - Y_1$. Evidently $x_1 \in B_1$ and $x_2 \in B_2$. Moreover, $(B_1 \cup B_2) - \{x_1, x_2\}$ contains both $C_1 - x_2$ and $C_2 - x_1$. Therefore $(B_1 \cup B_2) - \{x_1, x_2\}$ spans M. Hence

$$r(M) \leq |B_1| + |B_2| - 2 = r(X_1) + r(X_2) - 2.$$

This contradiction to (1) completes the proof of the lemma. □

To finish the proof of Theorem 8.3.1, we now construct matroids M_1 and M_2 whose 2-sum is M. Let p be an element that is not in $E(M)$ and let $\mathcal{C}_1 = \mathcal{C}(M|X_1) \cup \{(C \cap X_1) \cup p\colon\ C$ is a circuit of M that meets both X_1 and $X_2\}$. It is not difficult to check using the last lemma that $\mathcal{C}_1$ is the set of circuits of a matroid on $X_1 \cup p$. We call this matroid M_1. Define M_2 similarly as the matroid on $X_2 \cup p$ whose set of circuits is $\mathcal{C}(M|X_2) \cup \{(C \cap X_2) \cup p\colon\ C$ is a circuit of M that meets both X_1 and $X_2\}$. It now follows easily that $M = M_1 \oplus_2 M_2$. Finally, since Proposition 7.1.19 showed that the parts of a 2-sum are isomorphic to proper minors of the 2-sum itself, we deduce that M_1 and M_2 are isomorphic to proper minors of M. □

The next result, a straightforward combination of Theorem 8.3.1 and Proposition 4.2.11, emphasizes the structural importance of 3-connected matroids within the class of all matroids.

8.3.4 **Corollary.** *Every matroid that is not 3-connected can be constructed from 3-connected proper minors of it by a sequence of the operations of direct sum and 2-sum.* □

This result will be used frequently in Chapter 11 where we shall specify the structure of many classes of matroids. We already know, by virtue of a host of results in Chapter 4, that in many matroid arguments the general result can be obtained by restricting attention to the 2-connected case. It follows, on combining Corollary 8.3.4 with the results on 2-sums at the end of Section 7.1, that one can frequently be even more selective and concentrate just on 3-connected matroids.

The following result (Cunningham and Edmonds 1980, Seymour 1981b) gives an explicit description of the structure of those connected matroids that are not 3-connected. The proof, which is not difficult, is omitted.

8.3.5 Proposition. *Let M be a connected matroid. Then, for some positive integer k, there is a collection $M_1, M_2, \ldots, M_k$ of 3-connected matroids and a k-vertex tree T with edges labelled $e_1, e_2, \ldots, e_{k-1}$ and vertices labelled $M_1, M_2, \ldots, M_k$ such that*

(i) $E(M_1) \cup E(M_2) \cup \cdots \cup E(M_k) = E(M) \cup \{e_1, e_2, \ldots, e_k\}$;

(ii) *if the edge e_i joins the vertices M_{j_1} and M_{j_2}, then $E(M_{j_1}) \cap E(M_{j_2})$ is $\{e_i\}$; and*

(iii) *if no edge joins the vertices M_{j_1} and M_{j_2}, then $E(M_{j_1}) \cap E(M_{j_i})$ is empty.*

Moreover, M is the matroid that labels the single vertex of the tree $T/e_1, e_2, \ldots, e_{k-1}$ at the conclusion of the following process: contract the edges $e_1, e_2, \ldots, e_{k-1}$ of T one by one in order; when e_i is contracted, its ends are identified and the vertex formed by this identification is labelled by the 2-sum of the matroids that previously labelled the ends of e_i. □

8.3.6 Example. Let M be the cycle matroid of the graph G shown in Figure 8.9(a). Then the tree T in Figure 8.9(b) satisfies the conclusion of the last proposition where $G_1, G_2, \ldots, G_5$ are as shown in Figure 8.9(c). However, T is not the only such tree. Indeed, the tree T' in Figure 8.9(e) also satisfies the conclusion of the proposition if G_3, G_4, and G_5 are relabelled as in Figure 8.9(d). □

The last example showed that the labelled tree in Proposition 8.3.5 need not be unique. However, Cunningham and Edmonds (1980) showed that by making some slight modifications to the proposition one can get a unique decomposition result. The non-uniqueness arose in the last example because we were able to decompose a cocircuit in more than one way. Dually, a circuit can be decomposed in more than one way. By restricting the ways in which circuits and cocircuits can occur in the decomposition, Cunningham and Edmonds were able to guarantee uniqueness of the decomposition. Specifically, they proved that the tree T in Proposition 8.3.5 is unique to within a relabelling of its edges provided that

(i) $M_1, M_2, \ldots, M_k$ are allowed to be circuits or cocircuits as well as 3-connected matroids; and

(ii) T does not have two adjacent vertices that are both labelled by circuits or that are both labelled by cocircuits.

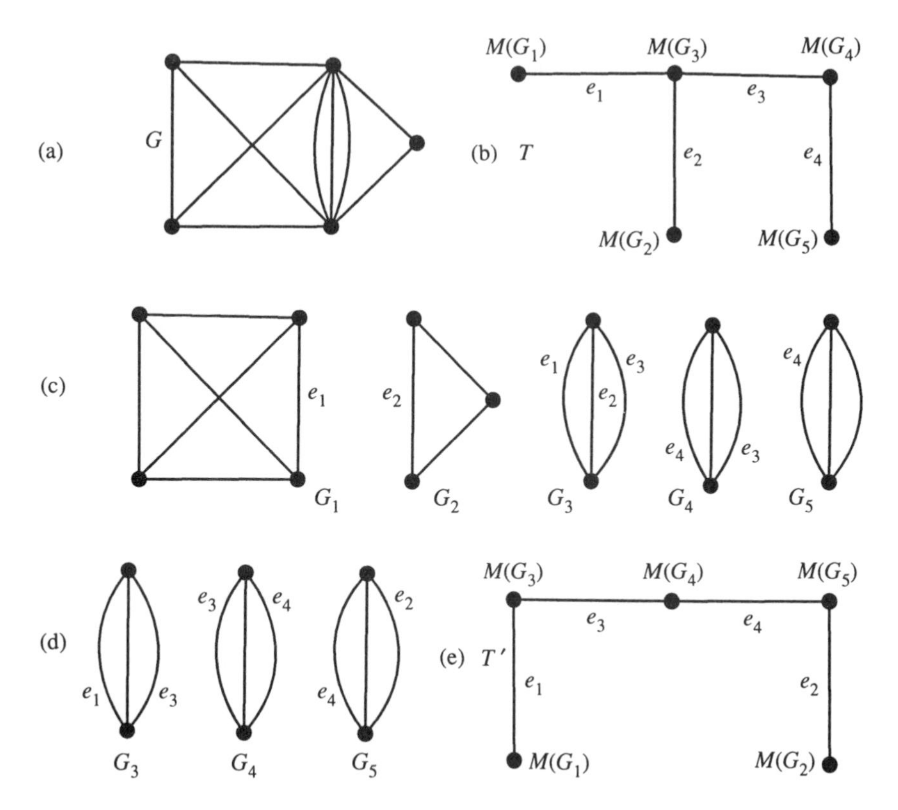

Fig. 8.9. A graph G and two decompositions of $M(G)$.

Exercises

1. Show that the matroid R_6, for which a geometric representation is shown in Figure 8.10, is not 3-connected.

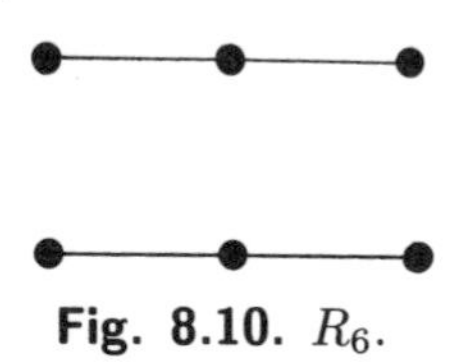

Fig. 8.10. R_6.

2. Let F be an arbitrary field and M be a single-element extension of a 3-connected F-representable matroid N. Prove that if M is not F-representable, then M is 3-connected.

3. Prove that M_1 and M_2 are series minors and parallel minors of $M_1 \oplus_2 M_2$.

4. Prove Proposition 8.3.5.

5. (Seymour 1980b) If X and Y are disjoint subsets of the ground set of a matroid M, let $k_M(X,Y) = \min\{r(X') + r(Y') - r(M) : (X', Y')$ is a partition of $E(M)$ with $X \subseteq X'$ and $Y \subseteq Y'\}$. Prove the following:

(i) $k_M(X,Y) = k_{M^*}(X,Y)$.

(ii) If N is a minor of M and X and Y are disjoint subsets of $E(N)$, then $k_N(X,Y) \leq k_M(X,Y)$.

(iii) If N is a j-connected minor of M and (X_1, Y_1) is an m-separation of M for some m with $1 \leq m < j$, then $\min\{|X_1 \cap E(N)|, |Y_1 \cap E(N)|\} \leq m - 1$.

(iv) If $e \in E(M) - (X \cup Y)$, then $k_M(X,Y)$ equals $k_{M \backslash e}(X,Y)$ or $k_{M/e}(X,Y)$.

8.4 More properties of 3-connected matroids

In this section we note some more examples of 3-connected matroids. In addition, we show that such matroids have numerous attractive properties, many of which are analogues of properties of 2-connected matroids that were noted in Chapter 4.

Table 4.1 (p. 126) listed all 2-connected matroids on at most four elements. All of those matroids with three or fewer elements have infinite connectivity and so are 3-connected. Table 8.1 lists all 3-connected matroids on at most five elements. The reader is urged to check the completeness of this table. One false impression that the table may convey is that all 3-connected matroids are uniform. But we recall that the wheels $M(\mathcal{W}_r)$ for $r \geq 3$ are 3-connected and none of these is uniform. In fact,

Table 8.1. All 3-connected matroids on at most five elements.

Number of elements	3-connected matroids
0	$U_{0,0}$
1	$U_{0,1}, U_{1,1}$
2	$U_{1,2}$
3	$U_{1,3}, U_{2,3}$
4	$U_{2,4}$
5	$U_{2,5}, U_{3,5}$

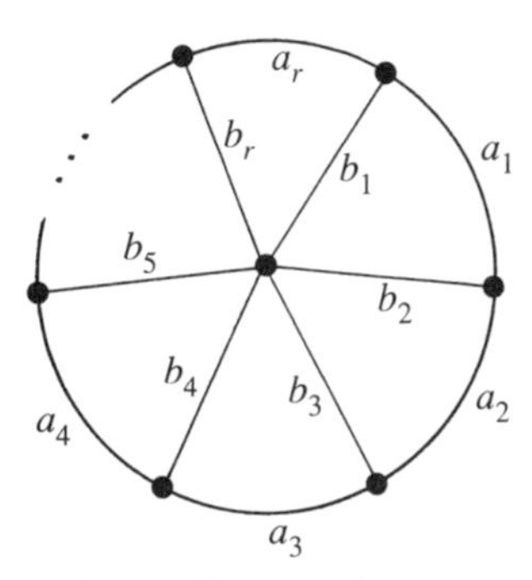

Fig. 8.11. The whirl $\mathcal{W}^r$ has $E(\mathcal{W}_r)$ as its ground set.

the table reflects the fact that all 3-connected matroids having rank or corank at most two are uniform (Exercise 1). We shall see many examples in this section of 3-connected matroids with rank and corank exceeding two that are not uniform.

The operation of relaxation of a circuit–hyperplane was introduced in Section 1.5 and has been referred to many times since. This operation reappears in the construction of one of the two fundamental families of 3-connected matroids. We shall use the following result (Kahn 1985), the proof of which is straightforward.

8.4.1 Proposition. *Let M' be the matroid that is obtained by relaxing a circuit–hyperplane of a matroid M. Then $\lambda(M') \geq \lambda(M)$.* □

It was remarked in Section 8.2 that the set of matroids $M(\mathcal{W}_r)$ for $r \geq 3$ is an important family of 3-connected matroids. By relaxing these matroids, we get another basic family of 3-connected matroids.

8.4.2 Example. For $r \geq 2$, let the edge set of $\mathcal{W}_r$ be labelled as in Figure 8.11. Then the rim, $\{a_1, a_2, \ldots, a_r\}$, is the unique circuit–hyperplane of $M(\mathcal{W}_r)$. We define $\mathcal{W}^r$, the *rank-r whirl*, to be the matroid obtained from $M(\mathcal{W}_r)$ by relaxing this circuit–hyperplane. Thus $E(\mathcal{W}^r) = E(\mathcal{W}_r)$, while the set of bases of $\mathcal{W}^r$ consists of the rim together with all edge sets of spanning trees of $\mathcal{W}_r$. By Proposition 1.5.13, the set of circuits of $\mathcal{W}^r$ consists of $\{C : C$ is the edge set of a cycle of $\mathcal{W}_r$ other than the rim$\} \cup \{\{a_1, a_2, \ldots, a_r\} \cup \{b_i\} : 1 \leq i \leq r\}$. Geometric representations for the matroids $M(\mathcal{W}_2)$, $M(\mathcal{W}_3)$, and $M(\mathcal{W}_4)$ were shown in Figure 8.5. In Figure 8.12, we give geometric representations for $\mathcal{W}^2$, $\mathcal{W}^3$, and $\mathcal{W}^4$. All three of these matroids have been considered earlier in the book (see, for example, 6.5.4 and Figures 2.1 and 6.12). Evidently $\mathcal{W}^2 \cong U_{2,4}$ so $\lambda(\mathcal{W}^2) = \infty$. Moreover, for $r \geq 3$, by Proposition 8.4.1, $\lambda(\mathcal{W}^r) \geq \lambda(M(\mathcal{W}_r)) = 3$. Since $\mathcal{W}^r$ has a 3-circuit, it follows that $\lambda(\mathcal{W}^r) = 3$. Hence, for all $r \geq 2$, the matroid $\mathcal{W}^r$ is 3-connected. □

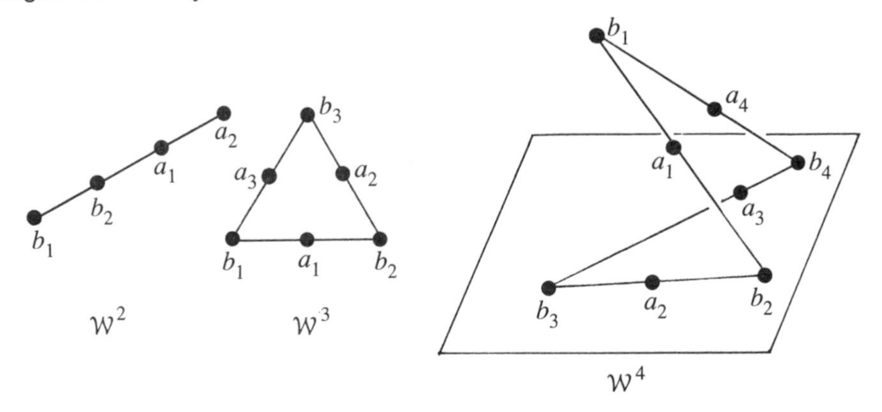

Fig. 8.12. The three smallest whirls.

The families $\{M(\mathcal{W}_r) : r \geq 3\}$ and $\{\mathcal{W}^r : r \geq 2\}$ are the two fundamental families of 3-connected matroids referred to earlier. The result that distinguishes the fundamental role of these two families is Tutte's Wheels and Whirls Theorem, which will be stated shortly. The matroids $M(\mathcal{W}_r)$ will frequently be referred to simply as *wheels* rather than as cycle matroids of wheels. These wheels and the closely related whirls have numerous attractive properties, some of which will now be discussed.

8.4.3 Proposition. *For all $r \geq 2$, both $M(\mathcal{W}_r)$ and $\mathcal{W}^r$ are self-dual.*

Proof. The geometric dual of the planar embedding of $\mathcal{W}_r$ in Figure 8.11 is clearly isomorphic to $\mathcal{W}_r$ (Exercise 10 of Section 2.3). Hence $M^*(\mathcal{W}_r) \cong M(\mathcal{W}_r)$. Furthermore, one easily checks that $\mathcal{W}^r$ is self-dual since the map that interchanges a_i and b_i for all i is an isomorphism between $\mathcal{W}^r$ and its dual. □

The terms 'rim' and 'spoke' will be used in the obvious way in whirls of rank at least three. Hence the spokes of a whirl are the elements that are in more than one triangle, while the rim consists of all elements that are not spokes. Although $\mathcal{W}^2$ has been specifically excluded here because of its symmetry, it will sometimes be convenient to view two of its elements as spokes and the other two as rim elements.

8.4.4 Example. Each of the last four matroids in Figure 8.13 can be obtained from its predecessor by relaxing a circuit–hyperplane. Since the first matroid in the list is certainly 3-connected, all five of these matroids are 3-connected. Moreover, as noted in Section 2.1, all of these matroids are self-dual. We shall see in Chapter 11 that these five matroids are the only rank-3, 6-element, 3-connected matroids. The only other 6-element 3-connected matroids are $U_{2,6}$ and $U_{4,6}$. □

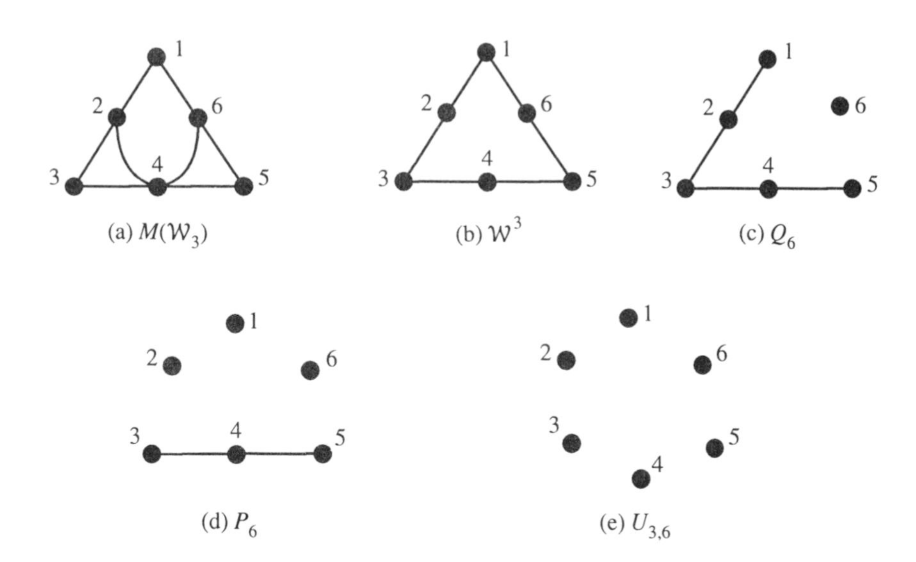

Fig. 8.13. $M(\mathcal{W}_3)$ and four successive relaxations of it.

What makes the wheels and whirls special within the class of 3-connected matroids? One obvious property of these matroids is that every element occurs in both a *triangle*, a 3-element circuit, and a *triad*, a 3-element cocircuit. Thus if M is a wheel or whirl with at least six elements and e is an element of M, then $M\backslash e$ and M/e have a 2-element cocircuit and a 2-element circuit, respectively. Thus, by Proposition 8.1.6, neither $M\backslash e$ nor M/e is 3-connected. The next result, Tutte's Wheels and Whirls Theorem (1966b), distinguishes the wheels and whirls among all 3-connected matroids. The full significance of this fundamental result will become clearer in Chapter 11.

8.4.5 Theorem. *The following statements are equivalent for a 3-connected matroid M having at least one element:*

(i) *For every element e of M, neither $M\backslash e$ nor M/e is 3-connected.*

(ii) *M has rank at least three and is isomorphic to a wheel or a whirl.*

Proof. We noted above that (ii) implies (i). The proof of the converse will be delayed until Chapter 11 where it will be deduced as a consequence of a result of Seymour (1980b) called the 'Splitter Theorem'. □

One of the most useful properties of 2-connectedness is that if M is a 2-connected matroid and e is an element of M, then at least one of $M\backslash e$ and M/e is 2-connected (Proposition 4.3.1). The last theorem shows that this proposition no longer holds if we replace '2-connected' throughout by '3-connected'. However, there is a generalization of the proposition to 3-connected matroids that does hold. The statement of this result involves $\widetilde{M/e}$, the simple matroid associated with M/e, and $\underset{\sim}{M\backslash e}$, the cosimple matroid associated with $M\backslash e$ where, for an arbitrary matroid N, the matroid $\underset{\sim}{N}$ is defined to be $(\widetilde{N^*})^*$. Thus $\underset{\sim}{N}$ is obtained from N by contracting all coloops and contracting all but one element from each non-trivial series class. Bixby (1982) proved that if e is an element of a 3-connected matroid M, then $\underset{\sim}{M\backslash e}$ or $\widetilde{M/e}$ is 3-connected.

8.4.6 Proposition. *Let M be a 3-connected matroid and e be an element of M. Then either $M\backslash e$ or M/e has no non-minimal 2-separations. Moreover, in the first case, $\underset{\sim}{M\backslash e}$ is 3-connected, while in the second case, $\widetilde{M/e}$ is 3-connected.*

The proof of this result will use the following lemma. If X_1, X_2, Y_1, and Y_2 are sets, the pairs (X_1, X_2) and (Y_1, Y_2) will be said to *cross* if all four of the sets $X_1 \cap Y_1$, $X_1 \cap Y_2$, $X_2 \cap Y_1$, and $X_2 \cap Y_2$ are non-empty.

8.4.7 Lemma. *Let M be a $(k+1)$-connected matroid for some $k \geq 1$ and suppose $e \in E(M)$. Assume that (X_1, X_2) and (Y_1, Y_2) are k-separations of $M\backslash e$ and M/e, respectively. Then (X_1, X_2) and (Y_1, Y_2) cross. Moreover,*

(i) $|X_1 \cap Y_1| < k$ *or* $|X_2 \cap Y_2| < k$; *and*

(ii) $|X_1 \cap Y_2| < k$ *or* $|X_2 \cap Y_1| < k$.

Proof. As (X_1, X_2) and (Y_1, Y_2) are k-separations of $M\backslash e$ and M/e, respectively,

$$r(X_1) + r(X_2) \leq r(M\backslash e) + k - 1 \tag{1}$$

and

$$r_{M/e}(Y_1) + r_{M/e}(Y_2) \leq r(M/e) + k - 1. \tag{2}$$

Since $|E(M)| = |X_1| + |X_2| + 1 \geq 2k + 1$, Proposition 8.1.6 implies that M has no loops or coloops. Thus, from (1) and (2), we deduce using 3.1.7 that

$$r(X_1) + r(X_2) \leq r(M) + k - 1$$

and

$$r(Y_1 \cup e) + r(Y_2 \cup e) \leq r(M) + k.$$

Adding these inequalities gives

$$[r(X_1) + r(Y_1 \cup e)] + [r(X_2) + r(Y_2 \cup e)] \leq 2r(M) + 2k - 1.$$

Thus, using semimodularity and the fact that e is not in X_1 or X_2, we find that

$$[r(X_1 \cup Y_1 \cup e) + r(X_1 \cap Y_1)] + [r(X_2 \cup Y_2 \cup e) + r(X_2 \cap Y_2)] \leq 2(r(M) + k - 1) + 1,$$

that is,

$$[r(X_1 \cup Y_1 \cup e) + r(X_2 \cap Y_2)] + [r(X_2 \cup Y_2 \cup e) + r(X_1 \cap Y_1)] \leq 2(r(M) + k - 1) + 1.$$

Hence either

$$r(X_1 \cup Y_1 \cup e) + r(X_2 \cap Y_2) \leq r(M) + k - 1$$

or

$$r(X_2 \cup Y_2 \cup e) + r(X_1 \cap Y_1) \leq r(M) + k - 1.$$

But both $(X_1 \cup Y_1 \cup e, X_2 \cap Y_2)$ and $(X_2 \cup Y_2 \cup e, X_1 \cap Y_1)$ partition $E(M)$ and M is not k-separated. Hence either $|X_1 \cap Y_1| < k$ or $|X_2 \cap Y_2| < k$, that is, (i) holds. Moreover, as the roles of Y_1 and Y_2 are indistinguishable above, we may interchange Y_1 and Y_2 to obtain (ii).

To deduce that (X_1, X_2) and (Y_1, Y_2) cross, we note that if not, then one of the following holds: $X_1 \subseteq Y_1$, $X_2 \subseteq Y_2$, $X_1 \subseteq Y_2$, or $X_2 \subseteq Y_1$. In the first two cases, (i) fails; in the last two cases, (ii) fails. □

The last lemma was proved in the case $k = 2$ by Bixby (1982) on his way to proving Proposition 8.4.6. Coullard (1985) noted that Bixby's argument also holds for arbitrary k. The type of argument used in the last proof originated in Tutte's (1966b) work on matroid connectivity and is used frequently in this area. We remark here that if $k = 1$ in Lemma 8.4.7, then, by (i), $X_1 \cap Y_1$ or $X_2 \cap Y_2$ is empty, that is, (X_1, X_2) and (Y_1, Y_2) do not cross. Hence, in this case, the lemma implies that at least one of $M\backslash e$ and M/e must be 2-connected. Thus the above argument gives an alternative proof of Theorem 4.3.1.

Proof of Proposition 8.4.6. First suppose that $|E(M)| < 4$. Then, since M is 3-connected, $\lambda(M) = \infty$. Hence, by Corollary 8.1.8, as $|E(M)| \geq 1$, M is isomorphic to one of $U_{0,1}$, $U_{1,1}$, $U_{1,2}$, $U_{1,3}$, or $U_{2,3}$. It is straightforward to check in each of these cases that the conclusions of the proposition hold. We may now assume that $|E(M)| \geq 4$. Hence, by Proposition 8.1.6, all the circuits and cocircuits of M have at least three elements. Moreover, by Proposition 8.1.13, both $M\backslash e$ and M/e are 2-connected.

Assume that (X_1, X_2) and (Y_1, Y_2) are non-minimal 2-separations of $M\backslash e$ and M/e, respectively. Then, by Lemma 8.4.7, (X_1, X_2) and (Y_1, Y_2)

cross and we may assume that $|X_1 \cap Y_1| = 1$. Since $|X_1| \geq 3$ and $|Y_1| \geq 3$, it follows that $|X_1 \cap Y_2| \geq 2$ and $|X_2 \cap Y_1| \geq 2$. But this contradicts Lemma 8.4.7(ii). Hence one of $M \backslash e$ and M/e has no non-minimal 2-separations. We shall assume the latter and complete the proof in that case. If the former holds, then the proof can be completed by dualizing the argument that follows.

Consider $\widetilde{M/e}$. If it is 3-connected, then the result holds so assume the contrary. Then, as M/e is 2-connected, so too is $\widetilde{M/e}$. Thus $\widetilde{M/e}$ has an exact 2-separation (Z_1, Z_2). Each element of M/e that is not in $\widetilde{M/e}$ is parallel to some element of Z_1 or Z_2. Thus M/e has a 2-separation (Z_1', Z_2') where $Z_1 \subseteq Z_1'$ and $Z_2 \subseteq Z_2'$. By assumption, (Z_1', Z_2') must be a minimal 2-separation of M/e so we may suppose that $|Z_1'| = 2$. Since $|Z_1| \geq 2$ and $Z_1' \supseteq Z_1$, it follows that $Z_1' = Z_1$. By Corollary 8.1.11, Z_1' is either a circuit or a cocircuit of M/e. The former cannot occur since $Z_1' \subseteq E(\widetilde{M/e})$ and $\widetilde{M/e}$ has no 2-circuits. Thus Z_1' is a 2-element cocircuit of M/e. But this implies that Z_1' is a 2-element cocircuit of M; a contradiction. □

A result that is closely related to the last proposition can be found in Lemma 13.3.6. In all the examples of 3-connected matroids M considered so far, if e is an element such that neither $M \backslash e$ nor M/e is 3-connected, then e is in a triangle or a triad of M. Tutte (1966b) proved that this is true in general and we leave the reader to deduce his result from Proposition 8.4.6.

8.4.8 Corollary. *Let M be a 3-connected matroid and e be an element of M such that neither $M \backslash e$ nor M/e is 3-connected. Then e is in a triangle or a triad of M.* □

Triangles and triads feature prominently in any discussion of 3-connected matroids. In the exercises we shall consider the properties of *minimally 3-connected matroids*, those 3-connected matroids M such that, for all elements e, $M \backslash e$ is not 3-connected. It will be noted there that such matroids have a relatively large number of triads. In Chapter 11, we shall prove Seymour's Splitter Theorem and various extensions of it. Again, triangles and triads will be of basic importance in those arguments. The final result of this section is due to Tutte (1966b). It, too, concerns triangles and triads in 3-connected matroids. It featured prominently in Tutte's proof of the Wheels and Whirls Theorem. We shall use it as a lemma for the proof of the Splitter Theorem. The proof given here, a modification of Tutte's original proof, is due to Coullard (private communication). Although it is quite long, the basic technique used is similar to that used in the proof of Lemma 8.4.7.

8.4.9 Lemma. *Let M be a 3-connected matroid having at least four elements and suppose that $\{e, f, g\}$ is a triangle of M such that neither $M\backslash e$ nor $M\backslash f$ is 3-connected. Then M has a triad that contains e and exactly one of f and g.*

Proof. If $|E(M)| = 4$, then $M \cong U_{2,4}$, and every single-element deletion of this matroid is 3-connected. Thus we may assume that $|E(M)| \geq 5$. Now if we can show that M has a triad T containing e, then, by Proposition 2.1.11, we can conclude that T meets $\{f, g\}$. But, as $|E(M)| \geq 5$, Proposition 8.1.7 implies that $T \not\supseteq \{e, f, g\}$. Hence to prove the lemma, it suffices to show that M has a triad containing e.

Suppose first that $M\backslash e, f$ is disconnected letting (X_1, X_2) be a 1-separation of this matroid with g in X_1. Then

$$r(X_1) + r(X_2) - r(M\backslash e, f) = 0. \tag{3}$$

By Proposition 8.1.6, M has no 1- or 2-cocircuits, hence $r(M\backslash e, f) = r(M)$. Thus $r(X_1)+r(X_2)-r(M) = 0$ and so, since $\{e, f, g\}$ is a triangle of M and $g \in X_1$, $r(X_1 \cup \{e, f\}) + r(X_2) - r(M) \leq 1$. Since M is not 2-separated but $(X_1 \cup \{e, f\}, X_2)$ is a partition of $E(M)$, we must have that $|X_2| = 1$, so let $X_2 = \{x\}$. By (3), x is a loop or a coloop of $M\backslash e, f$. In the first case, x is a loop of M; a contradiction. Hence x is a coloop of $M\backslash e, f$ and we get, by Proposition 8.1.6 again, that $X_2 \cup \{e, f\}$is a cocircuit of M; that is, e is in a triad of M.

We may now assume that $M\backslash e, f$ is 2-connected. Since neither $M\backslash e$ nor $M\backslash f$ is 3-connected, it follows by Proposition 8.1.13 that each is 2-separated. Let (E_1, E_2) and (F_1, F_2) be 2-separations of $M\backslash e$ and $M\backslash f$, respectively, and suppose that $f \in E_1$ and $e \in F_1$ (see Figure 8.14(a) and (b)). If $g \in E_1$, then $(E_1 \cup e, E_2)$ is a 2-separation of M; a contradiction. Hence $g \in E_2$ and, similarly, $g \in F_2$ (Figure 8.14(c)). Now $r(M\backslash e, f) = r(M\backslash e) = r(M\backslash f) = r(M)$. Thus

$$r(E_1) + r(E_2) \leq r(M) + 1, \tag{4}$$

and

$$r(F_1) + r(F_2) \leq r(M) + 1. \tag{5}$$

Moreover, $|E_1| \geq 3$ and $|F_1| \geq 3$ otherwise $(E_1 - f, E_2)$ or $(F_1 - e, F_2)$ is a 1-separation of the 2-connected matroid $M\backslash e, f$. Adding (4) and (5) and applying semimodularity, we obtain

$$r(E_1 \cup F_1) + r(E_1 \cap F_1) + r(E_2 \cup F_2) + r(E_2 \cap F_2) \leq 2r(M) + 2 \tag{6}$$

and

$$r(E_1 \cup F_2) + r(E_1 \cap F_2) + r(E_2 \cup F_1) + r(E_2 \cap F_1) \leq 2r(M) + 2. \tag{7}$$

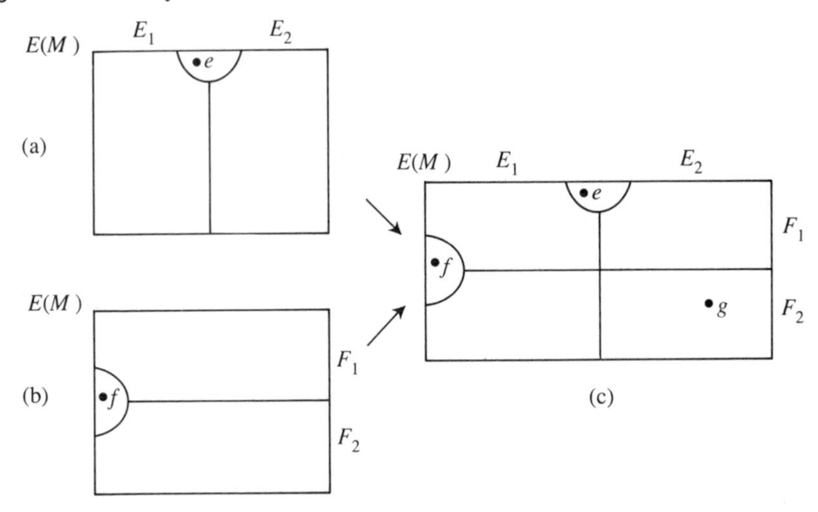

Fig. 8.14. Combining the 2-separations of $M\backslash e$ and $M\backslash f$.

From (7), we have that either

$$r(E_1 \cup F_2) + r(E_2 \cap F_1) \leq r(M) + 1 \tag{8}$$

or

$$r(E_2 \cup F_1) + r(E_1 \cap F_2) \leq r(M) + 1. \tag{9}$$

Now $E_1 \cup F_2$ contains $\{f, g\}$, and $E_2 \cup F_1$ contains $\{e, g\}$. Therefore $e \in \text{cl}(E_1 \cup F_2)$ and $f \in \text{cl}(E_2 \cup F_1)$. Thus, by (8) and (9), either

$$r(E_1 \cup F_2 \cup e) + r(E_2 \cap F_1) \leq r(M) + 1 \tag{10}$$

or

$$r(E_2 \cup F_1 \cup f) + r(E_1 \cap F_2) \leq r(M) + 1. \tag{11}$$

But both $(E_1 \cup F_2 \cup e,\ E_2 \cap F_1)$ and $(E_2 \cup F_1 \cup f,\ E_1 \cap F_2)$ partition $E(M)$, and M is 3-connected. Therefore either $|E_2 \cap F_1| \leq 1$ or $|E_1 \cap F_2| \leq 1$. We conclude that $E_1 \cap F_1$ is non-empty since this follows in each of these two cases because both $|E_1|$ and $|F_1|$ exceed two.

We now know that $(E_1 \cap F_1,\ E_2 \cup F_2)$ is a partition of $E(M\backslash e, f)$ with $\min\{|E_1 \cap F_1|, |E_2 \cup F_2|\} \geq 1$. But $M\backslash e, f$ is 2-connected and therefore

$$r(E_1 \cap F_1) + r(E_2 \cup F_2) \geq r(M) + 1.$$

On combining this with (6), we deduce that

$$r(E_1 \cup F_1) + r(E_2 \cap F_2) \leq r(M) + 1.$$

But $(E_1 \cup F_1,\ E_2 \cap F_2)$ is a partition of $E(M)$ and so, as M is 3-connected, $|E_2 \cap F_2| \leq 1$. Since $g \in E_2 \cap F_2$, it follows that

$$|E_2 \cap F_2| = 1 \tag{12}$$

and therefore

$$|E_2 \cap F_1| \geq 1 \tag{13}$$

and

$$|E_1 \cap F_2| \geq 1. \tag{14}$$

As $(E_1 \cup F_2,\ E_2 \cap F_1)$ and $(E_2 \cup F_1,\ E_1 \cap F_2)$ partition $E(M\backslash e)$ and $E(M\backslash f)$, respectively,

$$r(E_1 \cup F_2) + r(E_2 \cap F_1) \geq r(M) + 1 \tag{15}$$

and

$$r(E_2 \cup F_1) + r(E_1 \cap F_2) \geq r(M) + 1. \tag{16}$$

By (7), equality must hold in each of (15) and (16). Thus equality holds in each of (8) and (9) and hence in each of (10) and (11). From (10) we deduce that $|E_2 \cap F_1| \leq 1$. Therefore, by (13), $|E_2 \cap F_1| = 1$ and so, by (12), $|E_2| = 2$. Thus (E_1, E_2) is a minimal 2-separation of $M\backslash e$. Hence, by Corollary 8.1.11, E_2 is a circuit or a cocircuit of $M\backslash e$. As M has no 2-element circuits, the first possibility cannot occur. Hence E_2 is a cocircuit of $M\backslash e$ and so $E_2 \cup e$ is a triad of M. □

As an indication of the significance of the last lemma, we remark that, apart from being used in the proof of Tutte's Wheels and Whirls Theorem, it has also been called upon frequently in subsequent work on 3-connected matroids (see, for example, Bixby and Coullard 1986, 1987, Lemos 1989, and Oxley 1981b,c).

To conclude this chapter, we mention briefly a class of matroids which lies properly between the classes of 3- and 4-connected matroids and which has appeared several times in the literature (see, for example, Seymour 1985b, 1986a). A 3-connected *internally 4-connected matroid* is a 3-connected matroid whose only 3-separations are minimal. Such matroids are particularly natural objects to study for, while triangles and triads prevent a matroid M with six or more elements from being 4-connected, they do not stop M from being internally 4-connected.

Exercises

1. Prove that all 3-connected matroids having rank or corank at most two are uniform.

2. Prove Proposition 8.4.1.

3. (Coullard 1985) Let e be an element of a 4-connected matroid M. Prove that, for some N in $\{M\backslash e, M/e\}$, every 3-separation (X, Y) of N has the property that $\min\{|X|, |Y|\} \leq 4$.

4. (Wong 1978) Let M be a 4-connected matroid having at least eight elements. Prove that if M has an element e such that neither $M\backslash e$ nor M/e is 4-connected, then M has a circuit or a cocircuit having exactly four elements.

5. **(i)** Let M be a self-dual matroid on a set E and M' be the matroid that is obtained by relaxing a circuit–hyperplane X of M. Prove that M' is self-dual if and only if there is an isomorphism ψ from M to M^* such that $\psi(X) = E - X$.

(ii) Give an example of a relaxation of a self-dual matroid that is not self-dual.

6. (Oxley 1981b) Let M be a *minimally n-connected matroid*, that is, an n-connected matroid for which no single-element deletion is n-connected. Let $r(M) = r \geq 2$. Prove the following:

(i) If $n \leq r$, then $|E(M)| \geq r + n - 1$ with equality being attained if and only if $M \cong U_{r,r+n-1}$.

(ii) If $n > r$, then $|E(M)| \geq 2r - 1$ with equality being attained if and only if $M \cong U_{r,2r-1}$.

7. (Oxley 1981b) Let M be a minimally 3-connected matroid. Prove the following:

(i) If e is an element of M such that M/e is not 3-connected, then e is in a triad.

(ii) If $|E(M)| \geq 4$ and f is an element of M that is not in a triad, then M/e is minimally 3-connected.

8. (Lemos 1989) Let e and f be distinct elements of a matroid M such that $M\backslash e$ is not 3-connected. Suppose that M has triangles T and T' containing e and f, respectively, such that $|T \cap T'| = 1$ and $T \cup f$ is a cocircuit of M. Prove that e is in a triad of M.

9. (Oxley 1981a,b)

(i) Prove that if M is a minimally 3-connected matroid and $|E(M)| \geq 4$, then every circuit of M meets at least two distinct triads.

(ii) Deduce that such a matroid M has at least $\frac{1}{2}r^*(M)+1$ triads.

9
Binary matroids

In earlier chapters we have seen several examples of graph-theoretic results that have matroid analogues or generalizations. Numerous other examples will appear later in the book. When a result for graphs does not generalize to all matroids, there are two natural classes of matroids for which the result may still hold: the class of binary matroids and the smaller class of regular matroids. Regular matroids will be looked at in detail in Chapter 13. In this chapter we shall examine binary matroids and their properties.

Numerous characterizations of binary matroids will be proved here including Tutte's excluded-minor result stated in Chapter 6. In addition, we shall look at a number of special properties of binary matroids that are not possessed by matroids in general.

9.1 Characterizations

Binary matroids have been characterized in many ways. Seven such characterizations are stated in the first theorem of this section. Several more will be proved later in the section including the excluded-minor characterization stated in Chapter 6. Apart from this excluded-minor characterization, the descriptions of binary matroids in the next two theorems are of two basic types: those that concern the cardinality of circuit–cocircuit intersections and those that concern symmetric differences of circuits. If X and Y are sets, then their *symmetric difference*, $X \triangle Y$, is the set $(X - Y) \cup (Y - X)$. One easily checks that the operation of symmetric difference is both commutative and associative. Indeed, for sets $X_1, X_2, \ldots, X_n$, their symmetric difference $X_1 \triangle X_2 \triangle \cdots \triangle X_n$ consists of those elements that are in an odd number of the sets $X_1, X_2, \ldots, X_n$.

The proof of the first theorem will use the following easy lemma.

9.1.1 Lemma. *If e is an element of an independent set I of a matroid M, then M has a cocircuit C^* such that $C^* \cap I = \{e\}$.* □

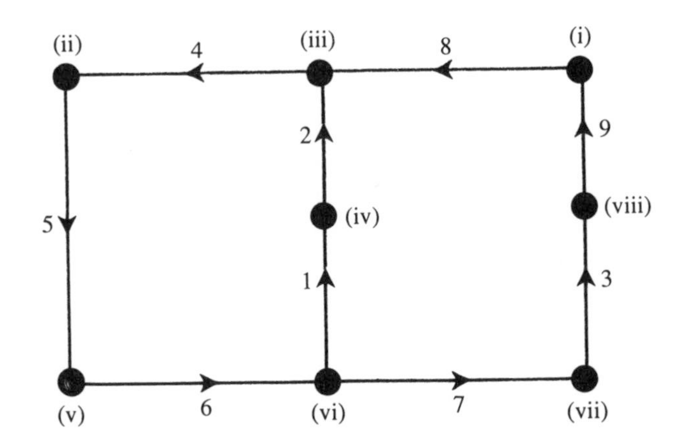

Fig. 9.1. Each arc corresponds to an implication that will be proved.

In the statement of the following theorem, it should be noted that a disjoint union of sets may consist of the disjoint union of the empty collection of sets and hence may be empty.

9.1.2 Theorem. *The following statements are equivalent for a matroid M:*

(i) *M is binary.*

(ii) *If C is a circuit and C^* is a cocircuit, then $|C \cap C^*|$ is even.*

(iii) *If C_1 and C_2 are distinct circuits, then $C_1 \triangle C_2$ contains a circuit.*

(iv) *If C_1 and C_2 are distinct circuits, then $C_1 \triangle C_2$ is a disjoint union of circuits.*

(v) *The symmetric difference of any set of circuits is either empty or contains a circuit.*

(vi) *The symmetric difference of any set of circuits is a disjoint union of circuits.*

(vii) *If B is a basis and C is a circuit, then $C = \triangle_{e \in C-B} C(e, B)$.*

(viii) *There is a basis B of M such that, if C is a circuit, then $C = \triangle_{e \in C-B} C(e, B)$.*

Proof. The digraph in Figure 9.1 summarizes the proof, with each arc corresponding to an implication that will be proved. The numbers on the arcs correspond to the order in which the implications are to be proved. The first three are immediate.

Now suppose that M satisfies (iii) but not (ii). Let C be a circuit and C^* be a cocircuit for which $|C \cap C^*|$ is odd so that, among such pairs (C, C^*), $|C \cap C^*|$ is a minimum. Then, by Proposition 2.1.11, C and C^* do not have exactly one common element, so we can find three distinct elements x, y, and z in their intersection. By the dual of Proposition 4.2.9, M has a circuit that meets C^* in some subset of $(C \cap C^*) - y$ containing x. Choose such a circuit C_1 so that $C \cup C_1$ is minimal.

Now $x \in C \cap C_1$ and $y \in C - C_1$, so there is a circuit C_2 such that $y \in C_2 \subseteq (C \cup C_1) - x$. Since $y \in C \cap C_2$ and $x \in C - C_2$, there is a circuit C_3 such that $x \in C_3 \subseteq (C \cup C_2) - y$. Clearly

$$C \cup C_3 \subseteq C \cup C_2 \subseteq C \cup C_1. \tag{1}$$

As x is in C_3 but y is not, it follows by (1) and the choice of C_1 that equality must hold throughout (1). Thus $C_2 - C = C_3 - C$, so $C_2 \triangle C_3 \subseteq C$. Since $x \in C_3 - C_2$, the circuits C_2 and C_3 are distinct. Thus, by (iii), $C_2 \triangle C_3$ contains a circuit. Hence

$$C_2 \triangle C_3 = C. \tag{2}$$

Therefore $C_2 \cap C^*$ and $C_3 \cap C^*$ partition $C \cap C^*$. Moreover, as $y \in C_2 \cap C^*$ and $x \in C_3 \cap C^*$, both $|C_2 \cap C^*|$ and $|C_3 \cap C^*|$ are less than $|C \cap C^*|$. It follows by the choice of (C, C^*) that both $|C_2 \cap C^*|$ and $|C_3 \cap C^*|$ are even. Hence $|C \cap C^*|$ is even; a contradiction. We conclude that (iii) implies (ii).

Now assume that (ii) holds but (v) does not. Then M has a set of distinct circuits $C_1, C_2, \ldots C_k$ such that $C_1 \triangle C_2 \triangle \cdots \triangle C_k$ is non-empty and independent. Therefore, by Lemma 9.1.1, there is a cocircuit C^* that meets $C_1 \triangle C_2 \triangle \cdots \triangle C_k$ in a single element. Thus $1 = |C^* \cap (C_1 \triangle C_2 \triangle \cdots \triangle C_k)| = |(C^* \cap C_1) \triangle (C^* \cap C_2) \triangle \cdots \triangle (C^* \cap C_k)|$. But, by (ii), $|C^* \cap C_i|$ is even for all i. Hence $|(C^* \cap C_1) \triangle (C^* \cap C_2) \triangle \cdots \triangle (C^* \cap C_k)|$ is even. It follows from this contradiction that (ii) implies (v).

Next assume that (v) holds but (vi) does not. Let $\{C_1, C_2, \ldots, C_k\}$ be a set of circuits for which $C_1 \triangle C_2 \triangle \cdots \triangle C_k$ is not a disjoint union of circuits so that, among such sets, $|C_1 \triangle C_2 \triangle \cdots \triangle C_k|$ is a minimum. By (v), $C_1 \triangle C_2 \triangle \cdots \triangle C_k$ contains a circuit D_1, and clearly $|C_1 \triangle C_2 \triangle \cdots \triangle C_k \triangle D_1| < |C_1 \triangle C_2 \triangle \cdots \triangle C_k|$. Thus $C_1 \triangle C_2 \triangle \cdots \triangle C_k \triangle D_1$ is a disjoint union of circuits; hence so too is $C_1 \triangle C_2 \triangle \cdots \triangle C_k$. This contradiction completes the proof that (v) implies (vi).

To show that (vi) implies (vii), assume that (vi) holds and let C be a circuit and B be a basis of M. Evidently $C - B = (\triangle_{e \in C-B}\, C(e, B)) - B$. If $C \neq \triangle_{e \in C-B}\, C(e, B)$, then $C \triangle (\triangle_{e \in C-B}\, C(e, B))$ is non-empty and so, by (vi), is dependent. But this is a contradiction since $C \triangle (\triangle_{e \in C-B}\, C(e, B)) \subseteq B$. Thus $C = \triangle_{e \in C-B}\, C(e, B)$ and so (vi) implies (vii).

Next we show that (i) implies (iii). Thus assume that (i) holds and let C_1 and C_2 be distinct circuits of M. Let $\psi : E(M) \to V(r, 2)$ be a coordinatization of M where $r = r(M)$. As C_1 and C_2 are both circuits, $\sum_{e \in C_1} \psi(e) = \underline{0} = \sum_{e \in C_2} \psi(e)$. Thus $\sum_{e \in C_1} \psi(e) + \sum_{e \in C_2} \psi(e) = \underline{0}$. If $e \in C_1 \cap C_2$, then $\psi(e)$ appears exactly twice on the left-hand side of the last equation. Thus $\sum_{e \in C_1 \triangle C_2} \psi(e) = \underline{0}$, so $C_1 \triangle C_2$ contains a circuit of M. Hence (i) implies (iii).

Finally, we shall show that (viii) implies (i). The proof will make use of the implications established above. In particular, we shall use the fact that, since a binary matroid satisfies (iii), it satisfies (ii)–(vii). Assume that M satisfies (viii), that is, M has a basis B such that if C is a circuit of M, then $C = \triangle_{e \in C - B} C(e, B)$. Let X be the B-fundamental-circuit incidence matrix of M and let $r(M) = r$. Then, viewed over $GF(2)$, the matrix $[I_r|X]$ represents a binary matroid. We shall show that this matroid is equal to M. First we note that B is a basis of $M[I_r|X]$ and that if $e \in E(M) - B$, then $C_M(e, B)$ is a circuit of $M[I_r|X]$. Hence we may drop the subscript on $C_M(e, B)$ without creating ambiguity.

Let C be a circuit of M. Then, by (viii), $C = \triangle_{e \in C - B} C(e, B)$. As $C(e, B)$ is a circuit of the binary matroid $M[I_r|X]$, it follows, by (vi), that $\triangle_{e \in C - B} C(e, B)$ is a disjoint union of circuits of $M[I_r|X]$. Thus, by Proposition 7.3.6, $M[I_r|X]$ is a quotient of M. But, since M and $M[I_r|X]$ both have rank r, it follows, by Corollary 7.3.4, that they are equal, that is, M is binary. We conclude that (viii) implies (i) and this completes the proof of the theorem. □

The proof just given that (viii) implies (i) is somewhat less elementary than the other parts of the proof, relying as it does on results on quotients. Since it is this implication that completes the proof that all of (ii)–(viii) are equivalent to (i), the reader may prefer a more elementary proof. We now present such a proof.

Alternative proof that (viii) implies (i). The proof proceeds as before to the point where it is shown that every circuit of M is a disjoint union of circuits of $M[I_r|X]$. Rather than using this to deduce a result about quotients, we deduce instead that every circuit of M is dependent in $M[I_r|X]$ and hence that every independent set in $M[I_r|X]$ is independent in M. To complete the proof that $M[I_r|X] = M$, and hence that M is binary, we now show that every circuit of $M[I_r|X]$ is a circuit of M. The result then follows from the elementary Lemma 2.1.19. Let D be a circuit of $M[I_r|X]$ and choose e in $D - B$. Then $D - e$ is contained in a basis B_1 of $M[I_r|X]$ such that $B_1 \subseteq (B \cup D) - e$. Moreover, D is the unique circuit of $M[I_r|X]$ that is contained in $B_1 \cup e$. Since B_1 is independent in $M[I_r|X]$, it is independent in M. Therefore as $|B_1| = r = r(M)$, B_1 is a basis of M. Now, from above, $C_M(e, B_1)$ is a disjoint union of circuits of

$M[I_r|X]$. But $C_M(e, B_1) \subseteq B_1 \cup e$, and D is the only circuit of $M[I_r|X]$ contained in $B_1 \cup e$. Therefore $C_M(e, B_1) = D$ and so D is a circuit of M. □

The fact that (viii) characterizes binary matroids was proved by Las Vergnas (1980). The other characterizations are older and can be found in the work of Whitney (1935), Rado (1957), Lehman (1964), Tutte (1965), and Minty (1966).

The proof of the excluded-minor characterization of binary matroids will use the last theorem and the following two lemmas. The elementary proof of the first of these is left as an exercise.

9.1.3 Lemma. *Let N be a minor of a matroid M and suppose that the set X is the intersection of a circuit and a cocircuit in N. Then X is the intersection of a circuit and a cocircuit in M.* □

9.1.4 Lemma. *Suppose that, in a matroid M, the non-empty set X is the intersection of a circuit C and a cocircuit C^*. Then M has a minor N such that X is a spanning circuit of both N and N^* and $r(N) = |X|-1 = r(N^*)$.*

Proof. By contracting the elements of $C - C^*$ and deleting the elements of $C^* - C$, we obtain a minor N_1 of M in which X is both a circuit and a cocircuit. If $E(N_1)-X$ contains an element e that is not in $\text{cl}_{N_1}(X)$, then we contract this element from N_1 to obtain the matroid N_2. Clearly X is both a circuit and a cocircuit of N_2. Now consider whether $E(N_2) - X$ contains an element that is not in $\text{cl}_{N_2}(X)$. If such an element exists, we choose one and contract it from N_2. Repeating this process of contracting suitably chosen elements one at a time, we eventually obtain a matroid N_k in which X is a circuit, a cocircuit, and a spanning set.

If X is not cospanning in N_k, then we delete a single element from $E(N_k) - \text{cl}_{N_k^*}(X)$. In the resulting matroid, X remains a spanning circuit and a cocircuit. Repeating this process of deleting suitably chosen elements one at a time, we eventually obtain the required matroid N in which X is both a spanning circuit and a cospanning cocircuit. It follows immediately that $r(N) = |X| - 1 = r(N^*)$. □

We now prove the excluded-minor characterization of binary matroids (Tutte 1958). In addition, we establish that a certain natural weakening of 9.1.2(ii) (Seymour 1976) also characterizes binary matroids.

9.1.5 Theorem. *The following statements are equivalent for a matroid M:*

(i) *M is binary.*

(ii) *If C is a circuit and C^* is a cocircuit, then $|C \cap C^*| \neq 3$.*

(iii) *M has no minor isomorphic to $U_{2,4}$.*

Proof. By the last theorem, if M is binary and C and C^* are a circuit and cocircuit respectively, then $|C \cap C^*|$ is even. Hence $|C \cap C^*| \neq 3$. Thus (i) implies (ii). To complete the proof of the theorem, we shall show that (ii) implies (iii) and that (iii) implies (i). In each case, this will be done by proving the contrapositive.

Assume that M has a $U_{2,4}$-minor N and let Z be a 3-element subset of $E(N)$. Then Z is both a circuit and a cocircuit of N. Thus, by Lemma 9.1.3, M has a circuit and a cocircuit whose intersection is Z. Hence (ii) fails. We conclude that (ii) implies (iii).

Now assume that M is non-binary. We shall prove that M has a $U_{2,4}$-minor, thereby showing that (iii) implies (i). As M is non-binary, it has a non-binary minor N of which every proper minor is binary. By Theorem 9.1.2, N has a circuit and a cocircuit whose intersection X has odd cardinality. Moreover, by Lemma 9.1.4, N has a minor N_1 such that X is a spanning circuit of both N_1 and N_1^*. As $|X|$ is odd, Theorem 9.1.2 implies that N_1 is non-binary. Therefore, by the choice of N, we must have that $N_1 = N$. Evidently $E(N) - X$ is an independent hyperplane H_0 of N. Thus, as $r(N) = |X| - 1$, it follows that $|H_0| = |X| - 2 \geq 1$. Therefore we can choose an element y in H_0. Clearly $H_0 - y$ is an independent flat of N of rank $r(N) - 2$. Let $\{H_0, H_1, H_2, \ldots, H_m\}$ be the set of hyperplanes of N containing $H_0 - y$. Then $(H_1 \cap X, H_2 \cap X, \ldots, H_m \cap X)$ is a partition of X. Furthermore, if $i \in \{1, 2, \ldots, m\}$, then $E(N) - H_i$ is a cocircuit of N. This cocircuit contains y and so does not contain X. Moreover, $E(N) - H_i$ meets X since H_0 is coindependent in N. Therefore $(E(N) - H_i) \cap X$ is a proper non-empty subset of X. As $(E(N) - H_i) \cap X$ is the intersection of a cocircuit and a circuit of N, it follows by Lemma 9.1.4 that N has a minor N_2 in which $(E(N) - H_i) \cap X$ is a spanning circuit of both N_2 and N_2^*. By the choice of N and the fact that $|(E(N) - H_i) \cap X| < |X|$, it follows that $|(E(N) - H_i) \cap X|$ is even. Thus, as $|X|$ is odd, $|H_i \cap X|$ is odd. Again, since H_i does not contain X and $|X|$ is odd, $m \geq 3$. Therefore $N/(H_0 - y)$ is a rank-2 matroid having at least four rank-1 flats, namely $\{y\}$, $H_1 - (H_0 - y)$, $H_2 - (H_0 - y), \ldots, H_m - (H_0 - y)$. Thus $N/(H_0 - y)$ has a restriction isomorphic to $U_{2,4}$. By the minimality of N again, it follows that $N \cong U_{2,4}$. □

On combining the last theorem with the Scum Theorem (3.3.7), we immediately obtain the following characterization of binary matroids in terms of their lattices of flats.

9.1.6 **Corollary.** *A rank-r matroid M is binary if and only if every rank-$(r-2)$ flat of M is contained in at most three hyperplanes.* □

Another variant of the Scum Theorem, namely Proposition 3.3.6, can be used to prove the next result (Oxley 1988), a straightforward extension of an interesting theorem of Fournier (1981) which characterizes binary matroids by a circuit double-elimination property. Fournier's result is stated here as a corollary of the next proposition, it being an immediate consequence of that proposition and Theorem 9.1.5.

9.1.7 Proposition. *Let M be a matroid and n be an integer exceeding one. Then M has no minor isomorphic to $U_{n,n+2}$ if and only if, whenever $C_1, C_2, \ldots, C_n$ are distinct circuits of M and e and f are distinct elements of $C_1 \cap C_2 \cap \cdots \cap C_n$, there is a circuit of M that is contained in $(C_1 \cup C_2 \cup \cdots \cup C_n) - \{e, f\}$.* □

9.1.8 Corollary. *(Fournier 1981) A matroid M is binary if and only if, whenever C_1 and C_2 are distinct circuits of M and e and f are elements of $C_1 \cap C_2$, there is a circuit contained in $(C_1 \cup C_2) - \{e, f\}$.* □

The list of characterizations of binary matroids given above is long but not exhaustive. It will be supplemented in the following exercises and in Section 9.3.

Exercises

1. Let X_1, X_2, and X_3 be subsets of a set E. Show that:

   (i) $(E - X_1) \triangle (E - X_2) = X_1 \triangle X_2$.

   (ii) $(E - X_1) \triangle (E - X_2) \triangle (E - X_3) = E - (X_1 \triangle X_2 \triangle X_3)$.

2. Prove that a matroid is binary if and only if, for every two distinct hyperplanes H_1 and H_2, there is a hyperplane that contains $H_1 \cap H_2$ and is contained in $H_1 \cup H_2$.

3. (Tutte 1965) Let C be a circuit of a matroid M and x and y be elements of $\mathrm{cl}_M(C)$. If both $M|(C \cup x)$ and $M|(C \cup y)$ are binary, prove that $M|(C \cup \{x, y\})$ is binary.

4. Prove that a geometric lattice $\mathcal{L}$ is the lattice of flats of a binary matroid if and only if every interval of height two in $\mathcal{L}$ contains at most three elements of height one.

5. Find all matroids for which every circuit–cocircuit intersection has odd cardinality.

6. (i) (White 1971) Prove that a matroid is binary if and only if $C_1 \triangle C_2$ is a circuit for every two distinct intersecting circuits C_1 and C_2 such that $r(C_1) + r(C_2) = r(C_1 \cup C_2) + r(C_1 \cap C_2)$.

(ii) State the corresponding characterization of binary matroids in terms of hyperplanes.

7. (Seymour 1981b) If X is a 3-element subset of a matroid M, prove that M has a $U_{2,4}$-minor whose ground set contains X if and only if M has a circuit and a cocircuit whose intersection is X.

8. Prove Proposition 9.1.7.

9. (Oxley 1984a) Call a 4-element set in a matroid M a *quad* if it is the intersection of a circuit and a cocircuit. Show that:

(i) X is a quad in M if and only if M has a minor isomorphic to one of the matroids $M(K_4)$, $\mathcal{W}^3$, Q_6, P_6, $U_{3,6}$, or R_6 (see Figures 8.10 and 8.13) in which X is a quad.

(ii) A binary matroid has a quad if and only if it has an $M(K_4)$-minor.

Now suppose M has a k-element set that is the intersection of a circuit and a cocircuit. Show that:

(iii) If $k \geq 3$, then M has a t-element set that is the intersection of a circuit and a cocircuit for some t with $\lceil k/2 \rceil \leq t \leq k-1$.

(iv)* If $k \geq 4$, then M has a quad.

(v)° If $k \geq 4$, then M has a $(k-2)$-element set that is the intersection of a circuit and a cocircuit.

10.* (Fournier 1974) Prove that a matroid is binary if and only if, whenever C_1, C_2, and C_3 are distinct circuits such that $C_1 \cap C_2 \cap C_3 \neq \emptyset$, there are distinct elements i, j, and k of $\{1,2,3\}$ such that $C_j - C_i \neq C_k - C_i$.

11.* (Greene and White in White 1971) Prove that a matroid is binary if and only if, for every two bases B_1 and B_2 and every element f of B_2, the number of elements e of B_1 such that both $(B_1 - e) \cup f$ and $(B_2 - f) \cup e$ are bases is odd. (Hint: Show that this condition on bases is equivalent to 9.1.2(ii).)

9.2 Circuit and cocircuit spaces

In this section we shall look at two vector spaces which are commonly associated with a binary matroid. In addition, we shall extend these ideas to a more general class of matroids and touch briefly on Tutte's theory of chain-groups. To begin, consider the following.

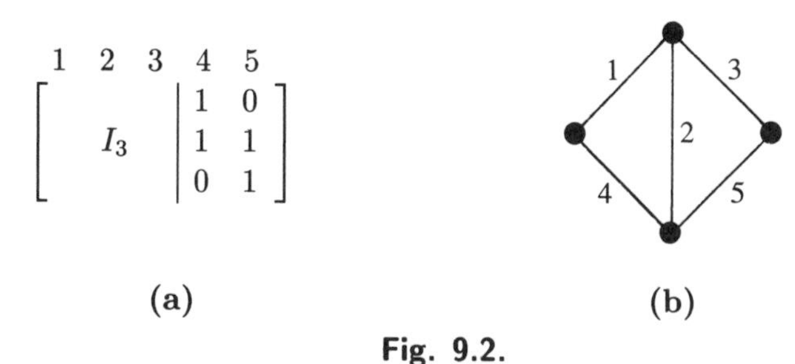

(a) (b)

Fig. 9.2.

9.2.1 Example. The matrix $[I_3|D]$ in Figure 9.2(a) is a binary representation of $M(G)$ where G is the graph shown in Figure 9.2(b). Row spaces of matrices were considered briefly in Section 2.2. Recall that the row space $\mathcal{R}(A)$ of an $m \times n$ matrix A over a field F is the subspace of $V(n, F)$ that is spanned by the rows of A. Thus, the members of the row space $\mathcal{R}[I_3|D]$ of $[I_3|D]$ are the rows of the following matrix:

$$
\begin{array}{r}
\\
\text{Row } 1 \\
\text{Row } 2 \\
\text{Row } 3 \\
\text{Row } 1 + \text{Row } 2 \\
\text{Row } 1 + \text{Row } 3 \\
\text{Row } 2 + \text{Row } 3 \\
\text{Row } 1 + \text{Row } 2 + \text{Row } 3 \\
\text{Row } 1 + \text{Row } 1
\end{array}
\quad
\begin{array}{c}
\begin{array}{ccccc} 1 & 2 & 3 & 4 & 5 \end{array} \\
\left[\begin{array}{ccccc}
1 & 0 & 0 & 1 & 0 \\
0 & 1 & 0 & 1 & 1 \\
0 & 0 & 1 & 0 & 1 \\
1 & 1 & 0 & 0 & 1 \\
1 & 0 & 1 & 1 & 1 \\
0 & 1 & 1 & 1 & 0 \\
1 & 1 & 1 & 0 & 0 \\
0 & 0 & 0 & 0 & 0
\end{array}\right]
\end{array}
$$

Viewed as incidence vectors, the rows of this matrix correspond to the sets $\{1,4\}$, $\{2,4,5\}$, $\{3,5\}$, $\{1,2,5\}$, $\{1,3,4,5\}$, $\{2,3,4\}$, $\{1,2,3\}$, and $\emptyset$. It is not difficult to check that these sets consist of all possible disjoint unions of cocircuits of $M(G)$. The reader familiar with coding theory will recognize $\mathcal{R}[I_3|D]$ as the binary linear code with generator matrix $[I_3|D]$ and parity-check matrix $[-D^T|I_2]$. This link with coding theory will be considered further in the exercises. □

This example illustrates a general phenomenon for binary matroids which will be described in the next result. Let M be a binary matroid on the set $\{1, 2, \ldots, n\}$. The *circuit space* and *cocircuit space* of M are the subspaces of $V(n, 2)$ that are generated by the incidence vectors of the circuits and cocircuits, respectively, of M.

9.2.2 Proposition. *Let A be a binary representation of a rank-r binary matroid M. Then the cocircuit space of M equals the row space of A. Moreover,* *this space has dimension equal to r and is the orthogonal subspace of the circuit space of M.*

Proof. Let $|E(M)| = n$. If A is a zero matrix, then $M \cong U_{0,n}$ and the result follows easily. Now suppose that A is not a zero matrix. By reordering the elements of M, if necessary, we may assume that the first r columns of A are linearly independent. Then, by a sequence of elementary row operations and deletions of zero rows, A can be transformed into a matrix in the form $[I_r|D]$. Moreover, the row spaces of A and $[I_r|D]$ coincide. Thus, as the rows of $[I_r|D]$ are clearly linearly independent, the row space of A has dimension equal to $r(M)$.

Let the last $n-r$ columns of $[I_r|D]$ be labelled by the set B^*. Evidently B^* is a cobasis of M. Moreover, as noted in Section 6.4, the rows of $[I_r|D]$ are the incidence vectors of the fundamental cocircuits of the elements of $E(M) - B^*$ with respect to B^*. Thus, every member of $\mathcal{R}[I_r|D]$ is in the cocircuit space of M. Furthermore, by Theorem 9.1.2 and duality, the incidence vector of every cocircuit of M is in $\mathcal{R}[I_r|D]$. We conclude that the row space of $[I_r|D]$ coincides with the cocircuit space of M. By duality, if $r < n$, then $\mathcal{R}[-D^T|I_{n-r}]$ coincides with the circuit space of M. Hence, by Proposition 2.2.23, if $r < n$, then the cocircuit space is the orthogonal subspace of the circuit space. But one easily checks that this is also true if $r = n$, so the proposition is proved. □

The following result is an easy consequence of the last proof.

9.2.3 Corollary. *Let B be a basis of an n-element rank-r binary matroid M and X be the B-fundamental-circuit incidence matrix of M. Suppose that $1 \le r \le n-1$. Then the row spaces of $[I_r|X]$ and $[X^T|I_{n-r}]$ are the cocircuit and circuit spaces, respectively, of M.* □

Another consequence of the last proposition is that one can derive a binary matroid from a family $\mathcal{A}$ of subsets of a finite set E in the following natural way. Let $\mathcal{A}'$ be the set consisting of all possible symmetric differences of members of $\mathcal{A}$ and let $\mathcal{C}$ be the set of minimal non-empty members of $\mathcal{A}'$. Then it follows immediately from Proposition 9.2.2 that $\mathcal{C}$ is the set of circuits of a binary matroid on E.

Although binary matroids are the primary focus of this chapter, we show next that many of the above ideas extend naturally to a larger class of matroids. Let F be an arbitrary field and $(v_1, v_2, \ldots, v_n)$ be a member $\underline{v}$ of $V(n, F)$. The *support* of $\underline{v}$ is $\{i : v_i \neq 0\}$. An immediate consequence of Proposition 9.2.2 is that if A is a binary representation of a matroid M, then the cocircuits of M are precisely the minimal non-empty sets that are supports of members of $\mathcal{R}(A)$. The next result captures the essence of Tutte's (1965) work on chain-groups, although the latter is set in a somewhat more general context (see Exercise 3).

9.2.4 Proposition. *Let A be an $m \times n$ matrix over a field F and $M = M[A]$. Then the set of cocircuits of M coincides with the set of minimal non-empty supports of vectors from the row space of A.*

Proof. As in the proof of Proposition 9.2.2, we may assume that A is non-zero and that it is of the form $[I_r|D]$ where $r = r(M)$. Let the columns of $[I_r|D]$ be labelled, in order, by $1, 2, \ldots, n$. If $r = n$, then $M \cong U_{n,n}$ and the result follows easily. Thus suppose that $r < n$. Then, by Theorem 2.2.8, M^* is represented by the matrix $[-D^T|I_{n-r}]$ where it too has columns labelled $1, 2, \ldots, n$.

We want to show that the circuits of M^* coincide with the minimal non-empty supports of members of $\mathcal{R}[I_r|D]$. The proof of this is straightforward, the main difficulty being notational. Let $\underline{d}\,(i)$ denote row i of D, $\underline{e}\,(j)$ denote row j of I_{n-r}, and $\underline{e}'(k)$ denote row k of I_r. Then $(\underline{e}'(i),\, \underline{d}(i))$ is the ith row of $[I_r|D]$ and

$$[-D^T|I_{n-r}] = [-\underline{d}\,(1)^T \cdots - \underline{d}\,(r)^T \,|\, \underline{e}\,(1)^T \cdots \underline{e}\,(n-r)^T].$$

Now suppose that $\{i_1, i_2, \ldots, i_s, r+j_1, r+j_2, \ldots, r+j_t\}$ is a circuit of M^* where $1 \le i_1 < i_2 < \cdots < i_s \le r$ and $1 \le j_1 < j_2 < \cdots < j_t \le n-r$. Then, for some non-zero elements $\alpha_1, \alpha_2, \ldots, \alpha_s, \beta_1, \beta_2, \ldots, \beta_t$ of F,

$$-\alpha_1 \underline{d}\,(i_1)^T - \cdots - \alpha_s \underline{d}\,(i_s)^T + \beta_1 \underline{e}\,(j_1)^T + \cdots + \beta_t \underline{e}\,(j_t)^T = \underline{0}. \quad (1)$$

Now consider the sum

$$\alpha_1(\underline{e}'(i_1), \underline{d}\,(i_1)) + \alpha_2(\underline{e}'(i_2), \underline{d}\,(i_2)) + \cdots + \alpha_s(\underline{e}'(i_s), \underline{d}\,(i_s)).$$

This is clearly a member of the row space of $[I_r|D]$. Moreover, with $\underline{0}_k$ denoting the zero vector in $V(k, F)$, we can rewrite this sum as

$$\alpha_1(\underline{e}'(i_1), \underline{0}_{n-r}) + \cdots + \alpha_s(\underline{e}'(i_s), \underline{0}_{n-r}) + \alpha_1(\underline{0}_r, \underline{d}\,(i_1)) + \cdots + \alpha_s(\underline{0}_r, \underline{d}\,(i_s)).$$

By (1), this equals

$$\alpha_1(\underline{e}'(i_1), \underline{0}_{n-r}) + \cdots + \alpha_s(\underline{e}'(i_s), \underline{0}_{n-r}) + \beta_1(\underline{0}_r, \underline{e}\,(j_1)) + \cdots + \beta_t(\underline{0}_r, \underline{e}\,(j_t)).$$

Evidently the support of this row vector is

$$\{i_1, i_2, \ldots, i_s,\ r+j_1, r+j_2, \ldots, r+j_t\}.$$

By Lemma 2.1.19, to complete the proof of the Proposition 9.2.4, it suffices to show that the support of every non-zero member of $\mathcal{R}[I_r|D]$ is dependent in M^*. The proof here essentially amounts to reversing the argument given for the first part and we omit the details. □

Exercises

1. (i) State the extension of Proposition 6.3.13 for binary matroids that arises by recognizing the non-singular diagonal matrices over $GF(2)$.

   (ii) (Bondy and Welsh 1971) Let $[I_r|A]$ be a matrix over $GF(2)$. Prove that $M[I_r|A]$ is identically self-dual if and only if A is square and $AA^T = I_r$.

2. For $r \geq 2$, consider an $r \times (2^r - 1)$ matrix A_r whose columns are all of the non-zero vectors in $V(r, 2)$. The vectors in the orthogonal subspace of $\mathcal{R}(A_r)$ form the binary *Hamming code* H_r, which arises frequently in coding theory. Note that H_r is only determined up to a permutation of the coordinates. Show that:

   (i) H_3 equals the row space of the following matrix over $GF(2)$.

$$\left[\begin{array}{c|ccc} & 0 & 1 & 1 \\ & 1 & 0 & 1 \\ A_4 & 1 & 1 & 0 \\ & 1 & 1 & 1 \end{array}\right]$$

   (ii) H_r has dimension $2^r - r - 1$.

   (iii) There is a partition of $V(2^r - 1, 2)$ into m classes where m is the number of vectors in H_r and each class contains a unique member $\underline{v}$ of H_r together with all members of $V(2^r - 1, 2)$ that differ from $\underline{v}$ in exactly one coordinate.

3. (Tutte 1965) Let R be an integral domain and E be a finite set. A *chain* on E over R is a mapping of E into R. Let N be a *chain-group* on E over R, that is, N is a set of chains on E over R that is closed under the operations of addition and multiplication by an element of R. The *support* of a chain f is $\{e \in E : f(e) \neq 0\}$. Show that:

   (i) The minimal non-empty supports of members of N form the circuits of a matroid $M\{N\}$ on E.

   (ii) The minimal sets meeting the supports of all non-zero chains form the bases of a matroid on E.

   (iii) The matroids in (i) and (ii) are duals of each other.

   (iv)* If N^* is the set of chains g on E over R such that $\sum_{e \in E} g(e)f(e) = 0$ for all f in N, then N^* is a chain-group and $M\{N^*\} = (M\{N\})^*$.

(v) If $S \subseteq E$ and $N|S$ denotes the set of restrictions to S of chains in N, then $N|S$ is a chain-group and $(M\{N|S\})^* = M|S$.

(vi) If $S \subseteq E$ and $N.S$ denotes the set of restrictions to S of chains in N whose supports are contained in S, then $N.S$ is a chain-group and $(M\{N.S\})^* = M.S$.

9.3 Other special properties

In the first section of this chapter, we proved numerous equivalent characterizations of binary matroids. In this section, we shall note a number of other special properties of such matroids. These properties are quite varied in nature, the common thread here being that each is a property that holds for binary matroids but not for matroids in general. The motivation for some of these results derives from graph theory, the matroid results being analogues or generalizations of graph-theoretic results. The first proposition is just such a result.

Recall that a graph is *Eulerian* if it contains a closed walk that traverses each edge exactly once. One of the best-known theorems in graph theory is that a non-empty connected graph is Eulerian if and only if all its vertices have even degree. To see how to extend this result to matroids, we observe that a non-empty connected graph is Eulerian if and only if its edge set can be partitioned into cycles. Moreover, it is not difficult to show that all the vertices in a graph G have even degree if and only if all the bonds in G have even cardinality (Exercise 1). Using these ideas, Welsh (1969c) generalized the above characterization of Eulerian graphs to binary matroids. His result is the equivalence of the first two parts in the next proposition. The equivalence of the second and third parts was shown independently by Brylawski (1972) and Heron (1972a). Exercise 7 of Section 6.2 sought a direct proof of the equivalence of (i) and (iii). Another proof of the equivalence of all four parts and of several other related conditions can be found in Brylawski (1982a).

9.3.1 Proposition. *The following statements are equivalent for an n-element binary matroid M:*

(i) *$E(M)$ can be partitioned into circuits.*

(ii) *Every cocircuit of M has even cardinality.*

(iii) *M^* is a binary affine matroid.*

(iv) *The n-tuple of all ones is in the circuit space of M.*

Before proving this result, we note that (i) and (ii) need not be equivalent if M is non-binary. For example, if M is the non-Fano matroid (see Figure 1.15), then $E(M)$ can be partitioned into a 3-element and a 4-element circuit, but M has a 5-element cocircuit. Hence (i) does not imply (ii). To see that (ii) does not imply (i), consider, for example, $U_{2,5}$.

Proof of Proposition 9.3.1. Since it is clear that (i) is equivalent to (iv), it suffices to show that (iii) implies (ii), that (ii) implies (iv), and that (iv) implies (iii). We leave the first two of these to the reader and prove the third (see Exercise 6, Section 1.5 for the first). Thus we suppose that (iv) holds. Then M^* has non-zero rank, r say. Let $[I_r|D]$ be a binary representation for M^*. Then, since the rows of $[I_r|D]$ span the circuit space of M, the n-tuple of all ones is a sum of rows of $[I_r|D]$. From considering the first r coordinates, we see that every row of $[I_r|D]$ is involved in this sum. Thus if $(x_1, x_2, \ldots, x_r)^T$ is a column of $[I_r|D]$, then $x_1 + x_2 + \cdots + x_r = 1$. Hence every element of M^* avoids the hyperplane $\{(x_1, x_2, \ldots, x_r) : x_1 + x_2 + \cdots + x_r = 0,\ x_i \in \{0, 1\}$ for all $i\}$ of $V(r, 2)$. Therefore M^* is a binary affine matroid, that is, (iii) holds. We conclude that (iv) implies (iii). □

An easy consequence of the last result is the following well-known graph-theoretic result.

9.3.2 Corollary. *Let G be a connected non-empty plane graph. Then G is Eulerian if and only if G^* is bipartite.* □

Next we note another property of graphs that extends to binary matroids but not to arbitrary matroids. Let C be a cycle of size exceeding one in a graph G and let e be an edge of G that is in $\text{cl}_{M(G)}(C) - C$ but is not a loop. Then it is easy to see that e is a diagonal of C. Thus there is a partition (X_1, X_2) of C such that both $X_1 \cup e$ and $X_2 \cup e$ are cycles. To see that non-binary matroids need not have this property, consider the following.

9.3.3 Example. Let k be an integer exceeding one and suppose that $M \cong U_{k,k+2}$. Let $E(M) = \{1, 2, \ldots, k+2\}$ and $C = \{1, 2, \ldots, k+1\}$. Then $k+2 \in \text{cl}(C) - C$. Evidently there is no partition (X_1, X_2) of C such that $X_1 \cup \{k+2\}$ and $X_2 \cup \{k+2\}$ are circuits of M. Note, however, that if we let $X_i = \{i\}$ for all i in $\{1, 2, \ldots, k+1\}$, then $(X_1, X_2, \ldots, X_{k+1})$ is a partition of C and, for all i, $(C - X_i) \cup \{k+2\}$ is a circuit of M. □

This example provides the clue as to how the original graph property can be generalized not only to binary matroids but also to a somewhat larger class of matroids. The proof of the following result is left as an exercise.

9.3.4 Proposition. *Let C be a circuit of a matroid M such that $|C| \geq 2$ and let e be a non-loop element of $\mathrm{cl}(C) - C$. Then, for some integer n exceeding one, there is a partition of C into non-empty subsets $X_1, X_2, \ldots, X_n$ such that each of the sets $(C - X_1) \cup e, (C - X_2) \cup e, \ldots, (C - X_n) \cup e$ is a circuit of M, and M has no other circuits that contain e and are contained in $C \cup e$. Moreover, if M has no minor isomorphic to $U_{k,k+2}$, then $n \leq k$.* □

On combining the last result with Corollary 6.5.3, we obtain the following result about the behaviour of circuits in $GF(q)$-representable matroids.

9.3.5 Corollary. *Let C be a circuit of a $GF(q)$-representable matroid M. If $e \in E(M) - C$, then there are at most q distinct circuits that contain e and are contained in $C \cup e$.* □

We know by Proposition 3.1.11 that, for an element e of a matroid M, the circuits of M/e are the minimal non-empty sets of the form $C - e$ where C is a circuit of M. In particular, if $e \in C$ and $|C| \geq 2$, then $C - e$ is a circuit of M/e. The last proposition enables us to determine the effect of contracting e on those circuits of M that do not contain e. The straightforward proof of the next result is left as an exercise.

9.3.6 Corollary. *Let C be a circuit of a matroid M and e be an element of $E(M) - C$. Then either C is a circuit of M/e, or, for some integer n exceeding one, there is a partition of C into non-empty subsets $X_1, X_2, \ldots, X_n$ such that each of the sets $C - X_1, C - X_2, \ldots, C - X_n$ is a circuit of M/e, and M/e has no other circuits contained in C. Moreover, if M has no minor isomorphic to $U_{k,k+2}$, then $n \leq k$.* □

On specializing this result to binary matroids, we immediately obtain the following.

9.3.7 Corollary. *Let C be a circuit of a binary matroid M and e be an element of $E(M) - C$. Then, in M/e, either C is a circuit, or C is a disjoint union of two circuits. In both cases, M/e has no other circuits contained in C.* □

Next we show how Proposition 9.3.4 can be extended to give a characterization of the matroids having no $U_{k,k+2}$-minor.

9.3.8 Proposition. *Let k be an integer exceeding one. Then a matroid M has no minor isomorphic to $U_{k,k+2}$ if and only if, for every circuit C and every subset S of $E(M) - C$ that is a series class of $M|(C \cup S)$, there are at most k distinct circuits of M that contain S and are contained in $C \cup S$.*

Before proving this result, we note that, on combining it with Theorem 9.1.5, we get yet another characterization of binary matroids. This result is essentially just a restatement of an observation of Bixby (1976) (see Exercise 10).

9.3.9 **Corollary.** *A matroid M is binary if and only if, for every circuit C and every subset S of $E(M)-C$ that is a series class of $M|(C \cup S)$, there are at most two distinct circuits of $M|(C \cup S)$ that contain S.* □

Proof of Proposition 9.3.8. Assume that M has no minor isomorphic to $U_{k,k+2}$. Let C be a circuit of M and S be a subset of $E(M) - C$ that is a series class of $M|(C \cup S)$. Choose e in S and contract $S - e$ from $M|(C \cup S)$ to obtain the matroid N. This matroid has C as a circuit because S is a series class of $M|(C \cup S)$. Moreover, if $D \subseteq C$, then $D \cup e$ is a circuit of N containing e if and only if $D \cup S$ is a circuit of M containing S. It follows, without difficulty, from Proposition 9.3.4 that $M|(C \cup S)$ has at most k distinct circuits that contain S.

We leave the proof of the converse to the reader. □

For the next special property of binary matroids to be noted, we look again at circuits comparing the number of these sets with the number of bases. For an arbitrary matroid M, we shall denote these numbers by $c(M)$ and $b(M)$, respectively. Welsh (1976, p. 287) suggests that 'on average a matroid has more bases than circuits' and poses the problem of determining the values of r and n such that if $\mathcal{N}$ is the set of all non-isomorphic n-element matroids of rank r, then $\sum_{M \in \mathcal{N}} b(M) \geq \sum_{M \in \mathcal{N}} c(M)$. We shall not address this problem specifically. Instead, we shall compare $b(M)$ and $c(M)$ for certain fixed matroids M and, in particular, for simple binary matroids. The following result of Oxley (1983) extends a result of Quirk and Seymour (in Welsh 1976). The proof is omitted.

9.3.10 **Proposition.** *If M is a simple binary matroid, then $b(M) \geq 2c(M)$. Moreover, equality holds here if and only if M is isomorphic to the direct sum of the Fano matroid and a free matroid.* □

The last result can be used to prove another bound which sharpens the last bound unless M is the direct sum of a free matroid and a matroid of rank less than six (Oxley 1983).

9.3.11 **Proposition.** *Let M be a rank-r simple binary matroid having no coloops. Then*

$$b(M) > \tfrac{6}{19}(r+1)\,c(M).$$ □

The constant $\frac{6}{19}$ in the last inequality does not seem to be best-possible. Indeed, the referee of Oxley's paper (1983) made the following conjecture which remains open.

9.3.12 Conjecture. *Let M be a rank-r simple binary matroid having no coloops. Then*

$$b(M) \geq \tfrac{1}{2}(r+1)\,c(M). \qquad \square$$

For simple *non-binary* matroids M, the quotient $b(M)/c(M)$ can become arbitrarily close to zero (see Exercise 5(ii)). Intuitively, the contrast between binary and non-binary matroids here is a reflection of the fact that circuits behave in a more orderly and predictable way in binary matroids than they do in matroids in general.

We opened this section by proving a generalization to arbitrary binary matroids of the well-known characterization of Eulerian graphs. In that generalization, the role of a vertex was assumed by a cocircuit. Although cocircuits are not the exact counterparts of vertices, one can often obtain a matroid analogue of a graph result by letting cocircuits take the place of vertices. To exemplify this, consider the following result for graphs that was proved by Kaugars (in Harary 1969, p. 31).

9.3.13 Proposition. *If a simple graph G is a block having minimum degree at least three, then G has a vertex v such that $G-v$ is also a block.* □

Now recall that the notions of simplicity for graphs and matroids coincide and that a graph is a block if and only if its cycle matroid is connected. Then letting cocircuits take the role of vertices, a statement for matroids that is an analogue of Proposition 9.3.13 is the following: If M is a simple connected matroid for which every cocircuit has at least three elements, then M has a hyperplane H such that $M|H$ is connected. To see that this statement does not hold for all matroids, one needs only to consider $U_{3,5}$. However, the following result of Seymour (in Oxley 1978b) asserts that the statement is true for binary matroids. The proof is left to the reader.

9.3.14 Proposition. *Let M be a simple connected binary matroid for which every cocircuit has at least three elements. Then M has a hyperplane H such that $M|H$ is connected.* □

In Chapter 6, as an indication of the computational complexity of determining whether a matroid is F-representable, we remarked that Seymour (1981a) has shown that this problem is difficult even when $F = GF(2)$. To close this chapter, we shall prove this result and note a contrasting result for graphic matroids.

We know that matroids can be specified in many ways. However, as Robinson and Welsh (1980) note, no matter which of these ways is chosen, the size of the input for a matroid problem on an n-element set is $O(2^n)$. For this reason, when discussing the computational complexity of matroid properties, it is usual to assume that the computer being used has a special subroutine or *oracle* for quickly deciding whether a given set is independent, or whether it is a basis or a circuit, or even for deciding its rank. We shall assume here that all matroids M are presented by means of an independence oracle which will determine, in unit time, whether or not any given subset of $E(M)$ is independent. Formally, an *independence oracle* is a function Ω defined for all subsets X of $E(M)$ by

$$\Omega(X) = \begin{cases} \text{YES}, & \text{if } X \in \mathcal{I}(M); \\ \text{NO}, & \text{if } X \notin \mathcal{I}(M). \end{cases}$$

To determine whether a matroid on a given set E has a certain property, one must apply the independence oracle to some sequence of subsets of E. Each application of the independence oracle to a particular subset of E is called a *probe*. Evidently, since we know we have a matroid, each probe gives more information than simply whether the probed set X is independent. For example, if $\Omega(X)$ = YES, then we deduce that $\Omega(Y)$ = YES for all subsets Y of X. Similarly, if $\Omega(X)$ = NO, we know that $\Omega(Z)$ = NO for all subsets Z of E that contain X.

9.3.15 Example. Suppose we wish to determine whether a matroid on the set $\{1,2,3,4\}$ is binary. How many probes of the independence oracle are required? We know that $U_{2,4}$ is the only non-binary matroid on a 4-element set. But, there are six different labelled matroids on the set $\{1,2,3,4\}$ that are isomorphic to the matroid $M(\mathcal{W}_2)$. To distinguish between $U_{2,4}$ and each of these matroids, one must probe all 2-element subsets of $\{1,2,3,4\}$. Hence, at least six probes are needed. We leave it to the reader to determine exactly how many probes are required. □

The last example illustrates the main idea in the proof of Seymour's result (1981a) that there is no polynomially bounded algorithm to test if a matroid presented by an independence oracle is binary.

9.3.16 Proposition. *There is no function f in $\mathbb{Z}[x]$ such that, for all positive integers n, one can determine whether an n-element matroid is binary by using at most $f(n)$ probes of an independence oracle.*

Proof. This is based on the following.

9.3.17 Example. Let r be an integer exceeding two and N_r be the vector matroid of the following matrix over $GF(2)$:

$$\begin{array}{cccc|ccccc} a_1 & a_2 & \cdots & a_r & b_1 & b_2 & b_3 & \cdots & b_r \end{array}$$
$$\left[\begin{array}{c|ccccc} & 0 & 1 & 1 & \cdots & 1 \\ & 1 & 0 & 1 & \cdots & 1 \\ I_r & 1 & 1 & 0 & \cdots & 1 \\ & \vdots & \vdots & \vdots & \ddots & \vdots \\ & 1 & 1 & 1 & \cdots & 0 \end{array}\right].$$

Evidently $N_3 \cong M(K_4)$ and $N_4 \cong AG(3,2)$. Moreover, it is straightforward to check that the set of circuits of N_r consists of all sets of the form $\{a_i, b_i, a_j, b_j\}$ for $1 \leq i < j \leq r$, together with the collection $\mathcal{D}$ of sets of the form $\{d_1, d_2, \ldots, d_r\}$ where $d_i \in \{a_i, b_i\}$ for all i and $|\{d_1, d_2, \ldots, d_r\} \cap \{b_1, b_2, \ldots, b_r\}|$ is odd. Evidently each of the circuits in $\mathcal{D}$ is also a hyperplane of N_r. Moreover, $|\mathcal{D}|$ is 2^{r-1}, the number of subsets of $\{b_1, b_2, \ldots, b_r\}$ having an odd number of elements. □

If $D \in \mathcal{D}$, let $N_r(D)$ denote the matroid obtained from N_r by relaxing the circuit–hyperplane D. It is straightforward to show that $N_r(D)$ is non-binary (Exercise 7). Now, in order to distinguish N_r from each of the matroids $N_r(D)$, one must probe each of the 2^{r-1} sets of $\mathcal{D}$. Hence the number of probes required is at least $2^{(|E(N_r)|/2)-1}$ and the proposition follows. □

In contrast to the last result, Seymour (1981a) also proved that there *is* a polynomially bounded algorithm to determine whether a matroid presented by an independence oracle is graphic. For a description of this algorithm, we refer the reader to Seymour's original paper.

Although the independence oracle is the only matroid oracle that we have formally defined here, it is clear that a basis oracle and a circuit oracle can be defined analogously. A detailed discussion of these and certain other matroid oracles and of the complexity of various matroid properties with respect to such oracles can be found in the papers of Hausmann and Korte (1978a,b, 1981), Robinson and Welsh (1980), Seymour and Walton (1981), Jensen and Korte (1982), and Truemper (1982b).

Exercises

1. Show that, in a graph G, every vertex has even degree if and only if every bond has even cardinality.

2. **(i)** In 9.3.1, prove that (ii) implies (iv).

(ii) Let M be a binary matroid. Prove that $E(M)$ can be partitioned into circuits if and only if there is a basis of the cocircuit space all of whose members have even support.

3. Give a direct proof of 9.3.7 that does not use 9.3.6.

4. (Seymour 1986a) Prove that a binary, 3-connected, vertically 4-connected matroid is internally 4-connected.

5. Let $c_{r+1}(M)$ denote the number of $(r+1)$-element circuits of a rank-r matroid M. Show that:

(i) If M is simple and binary, then $b(M) \geq (r+1)c_{r+1}(M)$.

(ii) If $M = PG(r-1, q)$, then $(q-1)^{r-1}b(M) = (r+1)c_{r+1}(M)$ for all $r \geq 2$.

6. Complete the proof of Proposition 9.3.8.

7. **(i)** Show that none of the matroids $N_r(D)$ in the proof of Proposition 9.3.16 is binary.

(ii) For fixed r, how many non-isomorphic matroids are there of the form $N_r(D)$?

8. **(i)** Let M be a matroid such that, for every circuit C and every element e not in C, there are at most two distinct circuits that contain e and are contained in $C \cup e$. Show that M need not be binary.

(ii) Assume that both a matroid M and its dual have the property that, for every circuit C and every series class S disjoint from C, there are at most three distinct circuits that contain S and are contained in $C \cup S$. Show that M need not be ternary.

9. (Lehman 1964) Let e be an element of a connected binary matroid M and let $\mathcal{C}_e$ be the set of circuits containing e. Show that the circuits of M not containing e are precisely the minimal non-empty sets of the form $C_1 \triangle C_2$ where C_1 and C_2 are in $\mathcal{C}_e$.

10. **(i)** (Bixby 1976) Prove that a matroid is binary if and only if it has no series minor isomorphic to any of the matroids $U_{r,r+2}$ for $r \geq 2$.

(ii) State the corresponding excluded-parallel minor characterization of binary matroids.

(iii) Use (i) and duality to give an alternative proof of 9.3.9.

11.* (Jensen and Korte 1982) Let M be a non-uniform matroid such that both M and M^* are paving. Prove that there is no function f in $\mathbb{Z}[x]$ such that, for all positive integers n, one can determine whether an n-element matroid has an M-minor by using at most $f(n)$ probes of an independence oracle.

10
Ternary matroids

The excluded-minor characterization of binary matroids was proved in the preceding chapter. The main task of this chapter is to prove the corresponding characterization of ternary matroids. This result was announced by Ralph Reid in 1971 but he never published his proof. The first proofs that were published are due to Bixby (1979) and Seymour (1979). Alternative proofs have since been given by Truemper (1982a), Kahn (1984), and Kahn and Seymour (1988). The last paper contains the shortest known proof, and this is the one that will be presented here.

One feature common to all the published proofs of this characterization is their reliance on Brylawski and Lucas's result (1976) that ternary matroids are uniquely $GF(3)$-representable. This result will be proved in Section 10.1 where the whole question of unique representability will be discussed.

Two other important lemmas in Kahn and Seymour's proof concern connected matroids for which one has certain information about the connectedness of a number of particular minors. These lemmas will be proved in Section 10.2 in the context of a more general discussion of matroids with such properties. With these preliminaries, the proof of the main theorem will be given in Section 10.3.

The multitude of alternative characterizations of binary matroids presented in the last chapter may lead one to hope for some similar types of results for ternary matroids. However, the best that seems to be available here are results such as Proposition 9.1.7 and Corollary 9.3.5. From these results, one can deduce certain properties of ternary matroids. Unfortunately, these properties do not characterize such matroids.

10.1 Unique representability

It is significant that the two fields for which the complete set of excluded minors has been determined have the unique representability property. The precise role that this property plays in the proof of the main theorem of this chapter will be evident when the proof is presented. Briefly, though, a key part of the argument proceeds as follows. Let M be an excluded minor for $GF(3)$-representability and a and b be distinct elements of M. Then both $M\backslash a$ and $M\backslash b$ are ternary, and the unique $GF(3)$-representability of ternary matroids guarantees that a representation for $M\backslash a, b$ can be extended to representations for *both* $M\backslash a$ and $M\backslash b$. This enables one to construct a ternary matroid M' on $E(M)$ that is very like M; the rest of the proof revolves around exploiting this similarity. Apart from unique representability, the other main idea in the proof involves connectedness. Indeed, it will be connectedness considerations that will govern the particular choice that is made for the pair $\{a, b\}$. These ideas will be developed in Section 10.2.

This section is based on an important paper of Brylawski and Lucas (1976), one of the main results of which is that ternary matroids are uniquely $GF(3)$-representable. This result will follow without difficulty from the next theorem. The technique used in the proof of this theorem is similar to that used in the proof of Theorem 6.6.3. As there, much use is made of the bipartite graph associated with a fundamental-circuit incidence matrix.

10.1.1 Theorem. *Suppose $r \geq 1$ and let $[I_r|D_1]$ and $[I_r|D_2]$ be $r \times n$ matrices over a field F with the columns of each being labelled, in order, by $e_1, e_2, \ldots, e_n$. Assume that the identity map on $\{e_1, e_2, \ldots, e_n\}$ is an isomorphism between $M[I_r|D_1]$ and $M[I_r|D_2]$. If every entry of both D_1 and D_2 is in $\{0, 1, -1\}$, then $[I_r|D_1]$ and $[I_r|D_2]$ are equivalent representations of M.*

Proof. By Proposition 6.4.1, $D_1^{\#} = D_2^{\#}$. We shall denote this matrix by D and the associated bipartite graph $G(D)$ by G. Let (B, B') be a partition of the edge set of G where B is a basis of $M(G)$. By Theorem 6.4.7, we can, by a sequence of row and column scalings, transform $[I_r|D_1]$ and $[I_r|D_2]$ into $[I_r|D_3]$ and $[I_r|D_4]$, respectively, where those entries of D_3 and D_4 that correspond to edges of B are equal to one. As the multiplicative factors in all these row and column scalings are reciprocals of non-zero matrix entries, both D_3 and D_4 retain the property that every entry is in $\{0, 1, -1\}$. We need to show that corresponding entries in D_3 and D_4 are equal. Since this is certainly true for entries that are zero or that correspond to edges in B, it remains to be checked for entries that correspond to edges in B'. If the characteristic of F is two, all such entries are one so the result certainly holds. Thus we may assume that the characteristic of F is not two.

To complete the proof, we shall use a labelling of the edges of G that is defined as follows. Firstly, we arbitrarily assign the labels $1, 2, \ldots, m$ to the edges in B. Now suppose that the labels $\{1, 2, \ldots, m+i\}$ have been

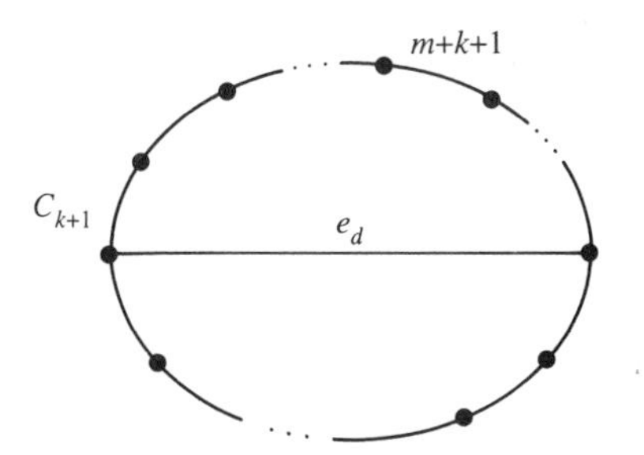

Fig. 10.1. The subgraph of G induced by $C_{k+1} \cup e_d$.

assigned but that some edges of G remain unlabelled. Choose a circuit C_{i+1} of $M(G)$ of minimum size such that $|C_{i+1}-\{1,2,\ldots,m+i\}| = 1$. As B spans $M(G)$ such a circuit certainly exists. Assign the label $m+i+1$ to the unique element of $C_{i+1} - \{1,2,\ldots,m+i\}$.

We shall argue by induction on j to show that the entries of D_3 and D_4 that correspond to the edge j of G must coincide. This is certainly true if $j \leq m$. Assume it true for $j \leq m+k$ and let $j = m+k+1$ where $k \geq 0$. Consider the cycle C_{k+1} of G. The vertices of this cycle correspond to rows and columns of D. Let D_3' and D_4' be the submatrices of D_3 and D_4, respectively, whose rows and columns are indexed by the vertices of C_{k+1}.

Suppose that D_3' has a non-zero entry d other than those corresponding to the edges of C_{k+1}. This entry corresponds in G to a diagonal e_d of C_{k+1}. Hence the subgraph of G induced by $C_{k+1} \cup e_d$ is a simple graph of the form shown in Figure 10.1. Now every edge in C_{k+1} has a label from $\{1,2,\ldots,m+k\}$. Moreover, by the definition of the labelling process, so does e_d. Thus $m+k+1$ is in a cycle with a set of edges all of which are lower labelled and this cycle has fewer edges than C_{k+1}. This contradicts the choice of C_{k+1} and enables us to conclude that the only non-zero entries of D_3' correspond to edges of C_{k+1}. Thus each row and each column of D_3' has exactly two non-zero entries.

Let d_3 and d_4 be the entries of D_3' and D_4', respectively, that correspond to the element $m+k+1$. By the induction assumption, all entries in corresponding positions in D_3' and D_4' coincide except possibly d_3 and d_4. Now all entries in D_3' and D_4' are in $\{0,1,-1\}$. Hence it follows from definition 2.2.11 of the determinant that, for $i = 3,4$,

$$\det D_i' = \varepsilon + \delta d_i$$

where both ε and δ are in $\{1,-1\}$.

By Proposition 6.4.5, $\det D_3' = 0$ if and only if $\det D_4' = 0$. Hence $d_3 = -\varepsilon/\delta$ if and only if $d_4 = -\varepsilon/\delta$. But, since each of d_3 and d_4 takes a value in $\{1, -1\}$, we must have that $d_3 = d_4$. The theorem now follows by induction. □

10.1.2 Corollary. *Ternary matroids are uniquely $GF(3)$-representable.*

Proof. Let A_1 and A_2 be $r \times n$ matrices representing a rank-r ternary matroid M over $GF(3)$ where $r \geq 1$. Then A_1 and A_2 are equivalent to matrices $[I_r|D_1]$ and $[I_r|D_2]$, respectively, over $GF(3)$ where the map that, for all i, takes column i of $[I_r|D_1]$ to column i of $[I_r|D_2]$ is an isomorphism between $M[I_r|D_1]$ and $M[I_r|D_2]$. We conclude, by the last theorem, that $[I_r|D_1]$ and $[I_r|D_2]$ are equivalent representations of M. Hence all ternary matroids of rank at least one are uniquely $GF(3)$-representable. The corollary follows immediately since all matroids of rank less than one are uniquely representable over all fields. □

Brylawski and Lucas (1976) established several other properties of uniquely representable matroids including those contained in the next four results. The reader is encouraged to fill in the proofs of these results, particularly that of the first since it is similar to the proof of Theorem 10.1.1.

10.1.3 Proposition. *If a binary matroid is representable over a field F, then it is uniquely F-representable.* □

10.1.4 Corollary. *All regular matroids are uniquely representable over every field.* □

10.1.5 Proposition. *If M is uniquely F-representable, then so too is M^*.* □

10.1.6 Proposition. *If M_1 and M_2 are uniquely F-representable, then so too are $M_1 \oplus M_2$, $P(M_1, M_2)$, $S(M_1, M_2)$, and $M_1 \oplus_2 M_2$.* □

Given an n-vertex simple graph G, it is quite common to consider the *complement* G^c of G, this being the simple graph having the same vertex set as G such that two distinct vertices are adjacent in G^c if and only if they are non-adjacent in G. One can attempt to mimic this construction for simple matroids representable over $GF(q)$ by recalling the analogy between projective geometries and complete graphs.

10.1.7 Proposition. *Let M be a simple rank-r matroid that is uniquely representable over $GF(q)$ and suppose that $k \geq r$. Let E_1 and E_2 be subsets of the set E of elements of $PG(k-1, q)$ such that*

$$PG(k-1, q)|E_1 \cong M \cong PG(k-1, q)|E_2.$$

Then

$$PG(k-1,q)\backslash E_1 \cong PG(k-1,q)\backslash E_2.$$

Proof. As M is uniquely $GF(q)$-representable, so too is $M \oplus U_{k-r,k-r}$ (Proposition 10.1.6). Let T_1 and T_2 be bases for $PG(k-1,q)/E_1$ and $PG(k-1,q)/E_2$, respectively. Then, for $i = 1,2$, $PG(k-1,q)|(E_i \cup T_i) \cong M \oplus U_{k-r,k-r}$. Moreover, the isomorphism between $PG(k-1,q)|E_1$ and $PG(k-1,q)|E_2$ can be extended to an isomorphism σ between the matroids $PG(k-1,q)|(E_1 \cup T_1)$ and $PG(k-1,q)|(E_2 \cup T_2)$ under which T_1 is mapped to T_2. Since $r(M \oplus U_{k-r,k-r}) = k = r(PG(k-1,q))$, the unique $GF(q)$-representability of $M \oplus U_{k-r,k-r}$ implies that σ is the restriction to $E_1 \cup T_1$ of an automorphism σ' of $PG(k-1,q)$. Clearly $\sigma'(E - E_1) = E - E_2$ and the result follows. □

We now define the $(GF(q),k)$-*complement* of a simple uniquely $GF(q)$-representable matroid M to be the matroid $PG(k-1,q)\backslash T$ where $M \cong PG(k-1,q)|T$ and $k \geq r(M)$. By the last proposition, this complement is well-defined up to a relabelling of the elements of its ground set. Hence, by Corollary 10.1.2, $(GF(3),k)$-complements exist for all simple ternary matroids; by Proposition 10.1.3, $(GF(2^m),k)$-complements exist for all simple binary matroids M and all $m \geq 1$; and, by Corollary 10.1.4, $(GF(q),k)$-complements exist for all simple regular matroids M and all prime powers q.

10.1.8 Example. In Table 10.1, we have listed the $(GF(2),3)$-complements of all simple binary matroids of rank three. □

Table 10.1.

Matroid	$(GF(2),3)$-complement
F_7	$U_{0,0}$
$M(K_4)$	$U_{1,1}$
$P(U_{2,3},U_{2,3})$	$U_{2,2}$
$U_{3,4}$	$U_{2,3}$
$U_{2,3} \oplus U_{1,1}$	$U_{3,3}$
$U_{3,3}$	$U_{2,3} \oplus U_{1,1}$

10.1.9 Example. Table 10.2 on the next page lists the $(GF(3),3)$-complements of certain rank-3 simple ternary matroids. A geometric representation for the matroid P_7 is shown in Figure 10.2. □

Table 10.2.

Matroid	$(GF(3),3)$-complement
$PG(2,3)$	$U_{0,0}$
$AG(2,3)$	$U_{2,4}$
$P(U_{2,4},U_{2,3})$	P_7
F_7^-	$M(K_4)$

The next example shows that a $(GF(q),k)$-complement of a uniquely $GF(q)$-representable matroid can have inequivalent $GF(q)$-representations (Brylawski and Lucas 1976). This example was looked at in a slightly different context in 6.3.12.

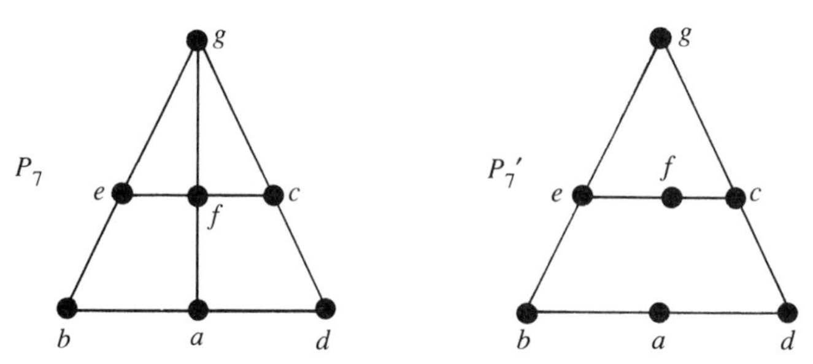

Fig. 10.2. P_7 and P_7'.

10.1.10 Example. Let P_7 and P_7' be the matroids for which geometric representations are shown in Figure 10.2. Then it is not difficult to check that the matrices A_1 and A_2 in Figure 10.3 are $GF(5)$-representations for P_7 and P_7', respectively. Evidently the identity map on $\{a,b,c,d,e,f\}$ is an isomorphism between $P_7\backslash g$ and $P_7'\backslash g$. But the matrices $A_1\backslash g$ and $A_2\backslash g$ that are obtained from A_1 and A_2 by deleting the column labelled by g are easily seen to be inequivalent representations for $P_7\backslash g$. Thus a $GF(5)$-representable matroid need not be uniquely $GF(5)$-representable.

Now let T be the six elements of $PG(2,5)$ corresponding to the columns of $A_1\backslash g$. Then $PG(2,5)|T \cong P_7\backslash g$. Consider $PG(2,5)\backslash T$. It is

$$A_1 = \begin{array}{c} \begin{matrix} a & b & c & d & e & f & g \end{matrix} \\ \left[\begin{array}{c|cccc} & 1 & 1 & 1 & 1 \\ I_3 & 1 & 2 & 2 & 1 \\ & 0 & 1 & 2 & 1 \end{array}\right] \end{array} \qquad A_2 = \begin{array}{c} \begin{matrix} a & b & c & d & e & f & g \end{matrix} \\ \left[\begin{array}{c|cccc} & 1 & 1 & 1 & 1 \\ I_3 & 1 & 2 & 2 & 1 \\ & 0 & 1 & 3 & 1 \end{array}\right] \end{array}$$

Fig. 10.3. A_1 and A_2 are $GF(5)$-representations of P_7 and P_7'.

not difficult to show that this matroid is uniquely $GF(5)$-representable (Exercise 6). However, its $(GF(5),3)$-complement, $P_7\backslash g$, is not.

This example also illustrates another disconcerting feature of matroids that are not uniquely representable. We know that P_7 is $GF(5)$-representable. If we did not have a $GF(5)$-representation for this matroid, one way that we may attempt to construct such a representation is as follows. Begin with an identity matrix representing a basis and then, one by one, adjoin columns to the matrix to correspond to the remaining elements of the matroid. These new columns should be added in such a way that, at any stage, the current matrix A is a representation for the restriction of P_7 to the set of column labels of A. We stop when we have a column for each element of P_7.

We can see from above that this naïve procedure may fail to produce a $GF(5)$-representation for P_7. In particular, the matrix $A_2\backslash g$ could be constructed as a representation for $P_7\backslash g$. But then there is no column that can be added to this matrix to give a representation for P_7.

The problems that arise here do not occur with uniquely representable matroids. For such matroids, the above procedure will always yield a representation. □

It is straightforward to extend the last example to show that, for all prime powers q exceeding three, $P_7\backslash g$ is $GF(q)$-representable but is not uniquely $GF(q)$-representable. What hope is there then for a unique representability result for such fields? In the case $q = 4$, the matroid $P_7\backslash g$ provides a clue to the answer to this question. We note that $P_7\backslash g \cong U_{2,4} \oplus_2 U_{2,4}$, so $P_7\backslash g$ is a $GF(4)$-representable matroid that is the 2-sum of two non-binary matroids. Kahn (1988) showed that such matroids are the only connected $GF(4)$-representable matroids that are not uniquely representable over $GF(4)$.

10.1.11 Proposition. *Let M be a $GF(4)$-representable matroid. Then M is uniquely $GF(4)$-representable if and only if M cannot be written as a direct sum or a 2-sum of two non-binary matroids. In particular, if M is 3-connected, it is uniquely $GF(4)$-representable.* □

For a proof of this result, we refer the reader to Kahn's paper (1988) where one can also find the following attractive conjecture which is open for all $q \geq 5$.

10.1.12 Conjecture. *For each prime power q, there is an integer $n(q)$ such that no 3-connected $GF(q)$-representable matroid has more than $n(q)$ inequivalent $GF(q)$-representations.* □

In view of Proposition 10.1.11, one may be tempted to think that by requiring a $GF(q)$-representable matroid to be sufficiently highly connected, one can guarantee unique $GF(q)$-representability. But Kahn (1988) describes examples that refute this for all $q \geq 5$.

Exercises

1. Prove Proposition 10.1.3.

2. Show that neither a deletion nor a contraction of a uniquely F-representable matroid need be uniquely F-representable.

3. Prove that if M is uniquely F-representable, so too is M^*.

4. Give two inequivalent $GF(4)$-representations of $U_{2,4} \oplus_2 U_{2,4}$.

5. Let M be a 3-connected rank-3 $GF(q)$-representable matroid having a $(q+1)$-point line as a restriction. Show that M has an $M(K_4)$-minor.

6. Show that $PG(2,5)\backslash T$ in 10.1.10 is uniquely $GF(5)$-representable.

7. Show that there are exactly three non-isomorphic 9-element rank-3 simple ternary matroids and give geometric representations for each.

8. Prove Proposition 10.1.6.

9. Show that a simple $GF(4)$-representable matroid M of rank r is uniquely $GF(4)$-representable if $|E(M)| \geq \frac{1}{3}(4^{r-1} + 14)$.

10. Let M be a ternary matroid of rank at least three and suppose that, for some field F, M is F-representable. Show that M is uniquely F-representable if F is $GF(2)$ or $GF(3)$, but that M need not be uniquely F-representable otherwise.

11. Show that the 8-element rank-4 matroid J for which a geometric representation is shown in Figure 10.4 is ternary.

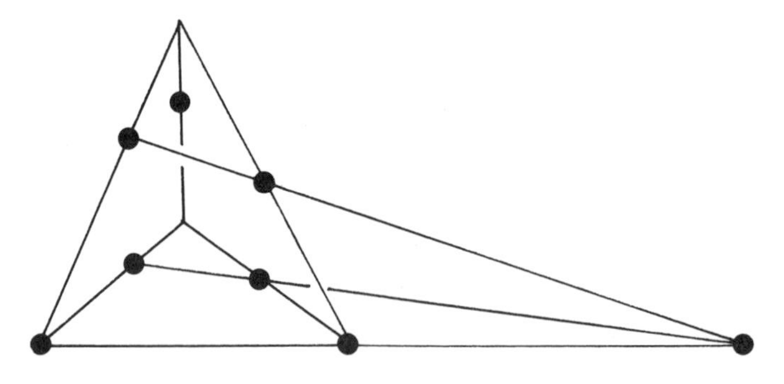

Fig. 10.4. The 8-element rank-4 matroid J.

10.2 Some connectivity results

In matroid induction arguments, it is often useful to have information about the properties of connected matroids for which certain deletions and contractions are known to be disconnected. Two results that give such information will be used in the proof of the excluded-minor theorem for ternary matroids. This section contains a brief discussion of results of this type. Various extensions of these results will be considered in the exercises and another similar result is given in Lemma 13.3.6.

Let M be a connected matroid having at least three elements. If e is an element that is in a 2-element cocircuit, then $M\backslash e$ has a coloop and is therefore disconnected. On the other hand, suppose that M is a *minimally connected matroid*, that is, M is connected but $M\backslash e$ is disconnected for all elements e. Then M has a cobasis each element of which is in a 2-cocircuit of M (Seymour 1979). The next proposition (Oxley 1981a) is an extension of this result. First, however, we give a lemma showing that in order to guarantee that a connected matroid has a 2-cocircuit, one does not need *every* single-element deletion to be disconnected.

10.2.1 Lemma. *Let M be a connected matroid having at least two elements and let $\{e_1, e_2, \dots, e_m\}$ be a circuit of M such that $M\backslash e_i$ is disconnected for all i in $\{1, 2, \dots, m-1\}$. Then $\{e_1, e_2, \dots, e_{m-1}\}$ contains a 2-cocircuit of M.*

Proof. As $M\backslash e_1$ is disconnected, Proposition 4.3.1 implies that M/e_1 is connected. Now suppose that $M/e_1\backslash e_i$ is connected for some i in $\{2, 3, \dots, m-1\}$. Then, since M is connected and $M\backslash e_i$ is disconnected, it follows that e_1 is a coloop of $M\backslash e_i$ and hence that $\{e_1, e_i\}$ is a cocircuit of M. Thus the lemma holds if $M/e_1\backslash e_i$ is connected for some i in $\{2, 3, \dots, m-1\}$ and so, in particular, it holds for $m = 3$.

Assume that the lemma is false and let k be the smallest value of m for which it fails. Since the hypotheses of the lemma are never satisfied when $m \leq 2$, we must have that $k > 2$. Moreover, from above, $k \neq 3$ and $M/e_1\backslash e_i$ is disconnected for all i in $\{2, 3, \dots, k-1\}$. As $\{e_2, e_3, \dots, e_k\}$ is a circuit of M/e_1, it follows by the choice of k that $\{e_2, e_3, \dots, e_{k-1}\}$ contains a 2-cocircuit of M/e_1. Thus the last set contains a 2-cocircuit of M; a contradiction. □

10.2.2 Proposition. *Let C be a circuit of a connected matroid M and suppose that $M\backslash e$ is disconnected for all e in C. Then either M is the circuit C or C contains at least two distinct non-trivial series classes of M.*

Proof. We may assume that $|C| \geq 3$ otherwise the hypotheses of the proposition are not satisfied. Lemma 10.2.1 implies that C contains a

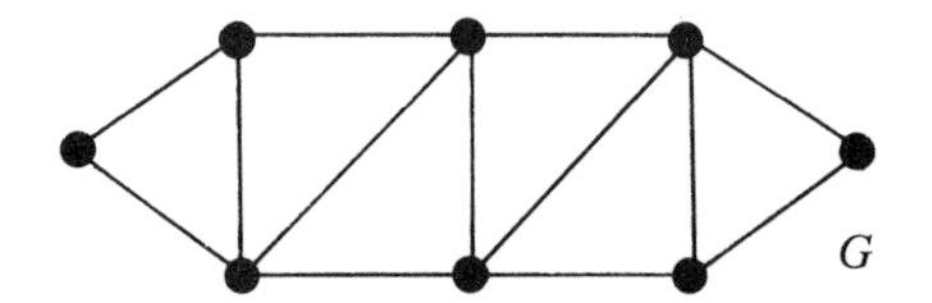

Fig. 10.5. The bound in Proposition 10.2.2 is best-possible.

2-cocircuit $\{e, f\}$ of M and that, if $|C| = 3$, then every pair of elements of C is in a 2-cocircuit.

Now suppose that every element of C is in series with e and let g be an element of $E(M) - C$. As M is connected, it has a circuit C' containing $\{e, g\}$. By Proposition 2.1.11, C' cannot have a single element in common with a cocircuit, so $C' \supseteq C$. But $g \in C' - C$, so C' properly contains C; a contradiction. We conclude that either $E(M) = C$, or $C - e$ has an element that is not in series with e. In particular, the proposition holds for $|C| = 3$. Assume it holds for $|C| < k$ and let $|C| = k$. We may also assume that $E(M) \neq C$. Thus $C - e$ contains an element h that is not in series with e.

Consider M/h. By Proposition 4.3.1, this matroid is connected. Moreover, M/h has $C - h$ as a circuit. Now, let a be an element of $C - h$ and consider $M/h\backslash a$. If this matroid is connected, then, as $M\backslash a$ is not, it follows from Proposition 8.1.10 that $\{a, h\}$ is a 2-cocircuit of M. Since $\{a, h\} \subseteq C$ and h is not in series with e, the required result holds. Thus, we may assume that $M/h\backslash a$ is disconnected for all a in $C - h$. The proposition now follows easily on applying the induction assumption. □

To see that the preceding result is best-possible, consider the cycle matroid of the plane graph G in Figure 10.5. Let C be the cycle of G bounding the infinite face. Then $M(G)\backslash e$ is disconnected for all e in C, but C contains only two distinct non-trivial series classes. It is not difficult to extend this to give examples in which C is arbitrarily large and yet still contains only two 2-cocircuits.

Using the last proposition, we can now prove the following.

10.2.3 Corollary. *Let M be a minimally connected matroid that is not a circuit. Then M has at least $r^*(M) + 1$ non-trivial series classes and therefore has at least $r^*(M) + 1$ pairwise disjoint 2-cocircuits.*

Proof. Let X be the set of elements of M that are contained in some 2-cocircuit. Then, by Proposition 10.2.2, X meets every circuit of M. Hence, by Corollary 2.1.17, X contains a cobasis B^* of M. Clearly B^* is non-empty. Let $B^* = \{e_1, e_2, \dots, e_m\}$ and C_1 be $C(e_1, E(M) - B^*)$. By Proposition 10.2.2, C_1 contains an element e_0 that is in a 2-cocircuit but is in a different series class from e_1. To complete the proof, we show that $e_0, e_1, \ldots, e_m$ are contained in distinct parallel classes of M^*. As $\{e_1, e_2, \ldots, e_m\}$ is independent in M^*, no two elements of this set are in the same parallel class. Assume that, for some $i \geq 1$, e_0 and e_i are parallel in M^*. Then $\{e_0, e_i\} \subseteq C_1$. But $C_1 \subseteq e_1 \cup (E(M) - B^*)$. Hence $i = 1$ and we have a contradiction to the fact that e_0 and e_1 are in different series classes of M. □

We now know that 2-cocircuits abound in minimally connected matroids. It follows, by duality, that a connected matroid for which every single-element contraction is disconnected has a lot of 2-circuits. Next we consider a matroid M that is both connected and simple. How many elements e can M have such that M/e is disconnected? We leave it to the reader to show (Exercise 2) that one answer to this, which can be deduced from Lemma 10.2.1, is that the number of such elements is at most $r^*(M) - 1$. A second answer is contained in the following result of Kahn and Seymour (1988), which will be used in the proof of the excluded-minor theorem for ternary matroids.

10.2.4 Lemma. *Let M be a connected simple matroid whose rank r is at least two and let $X = \{x \in E(M) : M/x$ is disconnected$\}$. Then*

(a) $|X| \leq r - 2$; *and*

(b) *if $|X| = r - 2$, then there are lines $L_0, L_1, \ldots, L_{r-2}$ of M and an ordering $x_1, x_2, \ldots, x_{r-2}$ of X such that*

 (i) $|L_i| \geq 3$ *for all i in $\{0, 1, \ldots, r-2\}$;*

 (ii) $E(M) = \bigcup_{i=0}^{r-2} L_i$; *and*

 (iii) $L_i \cap \mathrm{cl}(L_0 \cup L_1 \cup \cdots \cup L_{i-1}) = \{x_i\}$ *for all i in $\{1, 2, \ldots, r-2\}$.*

Before proving this lemma, we note that an example of a matroid satisfying (b) is shown in Figure 10.6.

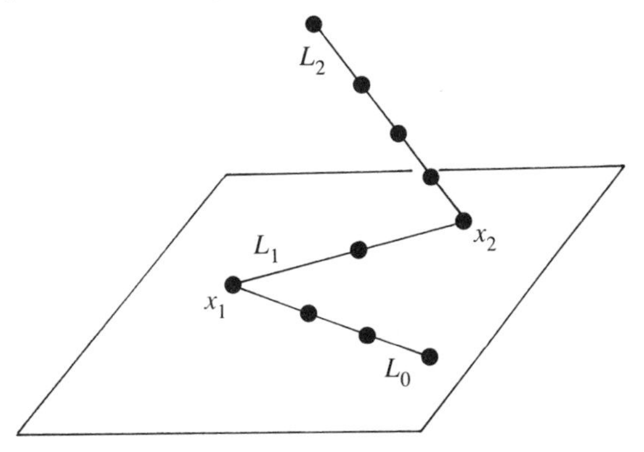

Fig. 10.6. This matroid satisfies (b) of 10.2.4.

Proof of Lemma 10.2.4. We argue by induction on r, the case $r = 2$ being trivial. Now assume the result true for $r < k$ and suppose that $r = k \geq 3$. If $|X| < k - 2$, then (a) holds and (b) does not apply. Thus we may assume that $|X| \geq k-2$. Hence X is non-empty, so, by Theorem 7.1.16, $M = P(M_1, M_2)$ for some restrictions M_1 and M_2 of M. Among such pairs, let (M_1, M_2) be chosen so that $|E(M_1)|$ is minimal. Then $X \subseteq E(M_2)$. Moreover, by Proposition 7.1.17, M_2 is connected. Let p be the basepoint of the parallel connection $P(M_1, M_2)$ and let X_2 be the set $\{e \in E(M_2) : M_2/e \text{ is disconnected}\}$. By Proposition 7.1.17 again, $X - p \subseteq X_2$. On applying the induction assumption to the simple matroid M_2, we deduce that $|X_2| \leq r(M_2) - 2$. Hence

$$r(M_2) - 2 \geq |X_2| \geq |X - p| \geq k - 3 \geq r(M_2) - 2. \tag{1}$$

Thus

$$X_2 = X - p \tag{2}$$

and

$$r(M_2) = k - 1. \tag{3}$$

As $r(P(M_1, M_2)) = k$, Proposition 7.1.15 and (3) imply that $r(M_1) = 2$, that is, M_1 is a line. Since M is connected, this line must have at least three points. Moreover, by (1), $|X_2| = r(M_2) - 2$, so, by (2), $|X| = r(M) - 2$ and hence (a) is verified. An additional consequence of applying the induction assumption to M_2 is that there are lines $L_0, L_1, \ldots, L_{k-3}$ of M_2 and an ordering $x_1, x_2, \ldots, x_{k-3}$ of X_2 satisfying the conditions of (b) with $k = r - 1$. Letting $x_{k-2} = p$ and $L_{k-2} = E(M_1)$, it is now straightforward to complete the proof of (b) and we leave the details to the reader. □

Next we turn attention to 3-connected matroids. If M is such a matroid having at least four elements, then, by Proposition 8.1.13, every single-element deletion of M is connected. Hence a smallest set whose deletion from M produces a disconnected matroid must have size at least two. By analogy with Corollary 10.2.3, we now examine the properties of those 3-connected matroids for which the deletion of every pair of distinct elements produces a disconnected matroid. From the results so far in this section, it is natural to guess that triads will abound in these matroids. The next result, due to Akkari and Oxley (1991), confirms this.

10.2.5 Proposition. *The following statements are equivalent for a matroid M having at least four elements:*

(i) *M is 3-connected and $M\backslash e,f$ is disconnected for all pairs $\{e,f\}$ of distinct elements.*

(ii) *M and all its single-element deletions are connected, but $M\backslash e,f$ is disconnected for all pairs $\{e,f\}$ of distinct elements.*

(iii) *Every pair of distinct elements of M is in a triad.*

Proof. It is straightforward to check that (iii) implies (ii). We shall complete the proof by showing that (ii) implies (i) and that (i) implies (iii).

Suppose that (ii) holds and that M is not 3-connected. Then, by Proposition 8.3.1, for some minors M_1 and M_2 of M each having at least three elements, $M = P((M_1;p),(M_2;p))\backslash p$. Now take e in $E(M_1)-p$ and f in $E(M_2)-p$. As $M\backslash e$ is connected and $M\backslash e = P((M_1\backslash e;p),(M_2;p))\backslash p$, it follows by Proposition 7.1.20 that $M_1\backslash e$ is connected. Similarly, $M_2\backslash f$ is connected and so, by Proposition 7.1.20 again, $P((M_1\backslash e;p),(M_2\backslash f;p))\backslash p$ is connected. As the last matroid is equal to $M\backslash e,f$, we have a contradiction. We conclude that (ii) implies (i).

Now assume that (i) holds. To prove that (iii) holds, we shall use the following.

10.2.6 Lemma. *If C is a circuit of M and g is an element of M not in C, then M has a triad that contains g and is contained in $C \cup g$.*

Proof. As M is 3-connected, $M\backslash g$ is connected. Now C is a circuit of $M\backslash g$ such that, for all e in C, the matroid $(M\backslash g)\backslash e$ is disconnected. Therefore, by Proposition 10.2.2, C contains a 2-cocircuit $\{a,b\}$ of $M\backslash g$. By Proposition 8.1.6, as M is 3-connected having at least four elements, it has no 2-cocircuits. Hence $\{a,b,g\}$ is a triad of M. □

Returning to the proof that (i) implies (iii), suppose that M has a subset $\{e,f\}$ that is not contained in any triads. Then $M\backslash e,f$ has no coloops, so every component of it contains a circuit having at least two elements. As $M\backslash e,f$ is disconnected, we can find two distinct components X_1 and X_2 of it. Let g be an element of X_1 and C be a circuit of $M|X_2$. By Lemma 10.2.6, M has a triad $\{g,a,b\}$ for some subset $\{a,b\}$ of C. It follows, by the dual of Proposition 3.1.11, that $M\backslash e,f$ has a cocircuit C^* that contains g and is contained in $\{g,a,b\}$. As $M\backslash e,f$ has no coloops, $|C^*| \geq 2$ so C^* meets the distinct components X_1 and X_2 of $M\backslash e,f$; a contradiction. □

Murty (1969,1970) has called a matroid a *Sylvester matroid* if every pair of distinct elements is in a triangle. The last proposition characterizes the duals of Sylvester matroids with at least four elements in terms of a connectivity condition.

Next we shall use the last proposition to prove a connectivity result of Seymour (1979) that is used in the proof of the excluded-minor theorem for ternary matroids.

10.2.7 **Corollary.** *Let M be a connected matroid having at least two elements. Suppose that, for every element e, both $M\backslash e$ and M/e are connected but, for every pair of distinct elements, e and f, both $M\backslash e, f$ and $M/e, f$ are disconnected. Then $M \cong U_{2,4}$.*

Proof. If $|E(M)| < 4$, then, as M is connected having at least two elements, it is isomorphic to one of the matroids $U_{1,2}$, $U_{1,3}$, or $U_{2,3}$. But the hypotheses of the corollary fail for each of these matroids. Hence we may assume that $|E(M)| \geq 4$. Then, by Proposition 10.2.5, M is 3-connected.

Assume that M has a 3-element subset that is both a triangle and a triad. Then, by Proposition 8.1.7, $|E(M)| \leq 4$ and so $|E(M)| = 4$. Since M has no circuits or cocircuits of size less than three, it follows in this case that $M \cong U_{2,4}$.

We may now assume that no 3-element subset of M is both a triangle and a triad. By Proposition 10.2.5 and its dual, if a and b are distinct elements of M, then they are contained in both a triangle $\{a, b, c\}$ and a triad $\{a, b, d\}$. Clearly $c \neq d$ so, by the dual of Proposition 10.2.5 again, M has a triangle T containing $\{c, d\}$. As the circuit T meets the cocircuit $\{a, b, d\}$, Proposition 2.1.11 implies that T must meet $\{a, b\}$. Hence we may assume that $T = \{a, c, d\}$. Thus a, c, and d are collinear. But, in addition, a, b, and c are collinear. Hence, as M is simple, we deduce that a, b, and d are collinear. Therefore $\{a, b, d\}$ is both a triangle and a triad of M; a contradiction. □

Further properties of minimally connected matroids will be noted in the exercises. Some properties of the analogous class of minimally 3-connected matroids were looked at in the exercises for Section 8.4.

To close this section, we prove a characterization of connectedness in terms of fundamental circuits together with a consequence of this result that will be used in the proof of the chapter's main result.

10.2.8 **Lemma.** *(Cunningham 1973, Krogdahl 1977) Let B be a basis of a matroid M and suppose that M is not isomorphic to $U_{1,1}$. Then M is connected if and only if $B \subseteq \bigcup\{C(e, B) : e \in E(M) - B\}$ and there is no partition of $E(M) - B$ into non-empty subsets X and Y such that $\bigcup\{C(e, B) : e \in X\}$ and $\bigcup\{C(e, B) : e \in Y\}$ are disjoint.*

Before proving this lemma, we note that it has the following attractive restatement when M has rank and corank at least one.

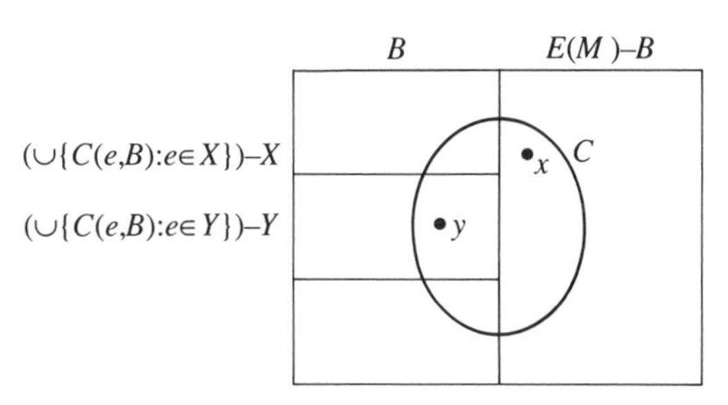

Fig. 10.7.

10.2.9 Corollary. *Let M be a matroid whose rank and corank are both positive and let B be a basis of M. Then M is connected if and only if the bipartite graph associated with the B-fundamental-circuit incidence matrix of M is connected.* □

Proof of Lemma 10.2.8. Assume that the given condition on fundamental circuits holds. Then it is straightforward to show that if a and b are distinct elements of M, there is a sequence $C_1, C_2, \ldots, C_n$ of B-fundamental circuits such that $a \in C_1$, $b \in C_n$, and $C_i \cap C_{i+1}$ is non-empty for all i in $\{1, 2, \ldots, n-1\}$. It follows immediately from Proposition 4.1.2 that M is connected.

Now assume that M is connected. If M has a loop, then $M \cong U_{0,1}$ and one easily checks that the specified condition on fundamental circuits holds. Thus, suppose that M has no loops and let X and Y be non-empty sets partitioning $E(M) - B$ such that $\bigcup\{C(e, B) : e \in X\}$ and $\bigcup\{C(e, B) : e \in Y\}$ are disjoint. As M has no loops, both $(\bigcup\{C(e, B) : e \in X\}) - X$ and $(\bigcup\{C(e, B) : e \in Y\}) - Y$ are non-empty. Let C be a circuit that meets both these sets (see Figure 10.7) and suppose that $C - B$ is minimal among such circuits. Now B cannot contain C so we can choose an element x from $C - B$. Without loss of generality, suppose that $x \in X$. Choose y in $((\bigcup\{C(e, B) : e \in Y\}) - Y) \cap C$. By the strong circuit elimination axiom, M has a circuit C' such that $y \in C' \subseteq (C \cup C(x, B)) - x$. Evidently $C' - B \subseteq (C - x) - B \subsetneq C - B$. Thus, by the choice of C, we must have that C' does not meet $(\bigcup\{C(e, B\} : e \in X\}) - X$. It follows that $C' \subseteq C - x$; a contradiction. We conclude that $\bigcup\{C(e, B) : e \in X\}$ must meet $\bigcup\{C(e, B) : e \in Y\}$. A similar argument shows that $B \subseteq \bigcup\{C(e, B) : e \in E(M) - B\}$ and we leave the details to the reader. □

The last lemma will be used in the proof of the following result, a special case of a result of Kantor (1975).

10.2.10 Lemma. *Let $[A|\underline{u}]$ and $[A|\underline{w}]$ be $r \times (n+1)$ matrices over $GF(3)$ where $\underline{u}$ and $\underline{w}$ are column vectors and $r \geq 3$. Suppose the map that, for all i,* *takes the* ith *column of* $[A|\underline{u}]$ *to the* ith *column of* $[A|\underline{w}]$ *is an isomorphism between* $M[A|\underline{u}]$ *and* $M[A|\underline{w}]$. *If* $M[A]$ *is connected of rank* r, *then* $\underline{w} = \pm\underline{u}$.

Proof. Let $\underline{v}_i$ denote the ith column of A. By Corollary 10.1.2, Proposition 6.3.9, and the fact that the field $GF(3)$ has no non-trivial automorphisms, we deduce that there is a non-singular linear transformation τ of $V(r,3)$ and a sequence $c_1, c_2, \ldots, c_n, d$ of non-zero elements of $GF(3)$ such that

$$\tau(\underline{w}) = d\,\underline{u} \tag{4}$$

and, for all i in $\{1, 2, \ldots, n\}$,

$$\tau(\underline{v}_i) = c_i\,\underline{v}_i. \tag{5}$$

As $M[A]$ has rank r, we may assume, without loss of generality, that $\{\underline{v}_1, \underline{v}_2, \ldots, \underline{v}_r\}$ is a basis of $V(r,3)$. Then, for all j in $\{r+1, r+2, \ldots, n\}$, there are elements $b_{j1}, b_{j2}, \ldots, b_{jr}$ of $GF(3)$ such that $\underline{v}_j = \sum_{k=1}^{r} b_{jk}\underline{v}_k$. Thus $c_j\underline{v}_j = \sum_{k=1}^{r} b_{jk}c_j\underline{v}_k$. But $c_j\underline{v}_j = \tau(\underline{v}_j)$ so

$$c_j\underline{v}_j = \tau\left(\sum_{k=1}^{r} b_{jk}\underline{v}_k\right) = \sum_{k=1}^{r} b_{jk}c_k\underline{v}_k.$$

Hence $\sum_{k=1}^{r} b_{jk}c_k\underline{v}_k = \sum_{k=1}^{r} b_{jk}c_j\underline{v}_k$. Therefore, for all k such that $b_{jk} \neq 0$, we have $c_k = c_j$. But $b_{jk} \neq 0$ if and only if $\underline{v}_k$ is in the fundamental circuit of $\underline{v}_j$ with respect to $\{\underline{v}_1, \underline{v}_2, \ldots, \underline{v}_r\}$. As $M[A]$ is connected, it now follows easily by Lemma 10.2.8 that $c_1 = c_2 = \cdots = c_n$. Let c be the common value of $c_1, c_2, \ldots, c_n$.

For some elements $w_1, w_2, \ldots, w_r$ of $GF(3)$, $\underline{w} = \sum_{k=1}^{r} w_k\underline{v}_k$. Thus, by (5), $\tau(\underline{w}) = \sum_{k=1}^{r} w_k c\underline{v}_k = c\underline{w}$. But, by (4), $\tau(\underline{w}) = d\underline{u}$. Therefore, $\underline{w} = (d/c)\underline{u}$. But d/c is a non-zero member of $GF(3)$. Hence $\underline{w} = \pm\underline{u}$. □

Exercises

1. (Seymour 1979) Prove that a minimally connected matroid has a cobasis of elements each of which is in a 2-cocircuit.

2. Let M be a simple connected matroid. Show that M has at most $r^*(M) - 1$ elements e for which M/e is disconnected.

3. (Oxley 1981a) Find all connected matroids M that have a triangle T such that $M\backslash e$ is disconnected for all e in T.

4. (Akkari 1988) Let X be a non-empty subset of a connected matroid M. Suppose that $M\backslash Y$ is disconnected for all non-empty proper subsets Y of X. Show that:

(i) If $M\backslash X$ is disconnected, then M/X is connected and X is independent.

(ii) If $M\backslash X$ is connected, then X is a series class of M unless M is a circuit of cardinality $|X|+1$.

5. (Oxley 1981a)

(i) Let M be a connected matroid other than a single circuit. Suppose that $A \subseteq E(M)$ and, for all e in A, $M\backslash e$ is not connected. Show that either A is independent or A contains at least $|A| - r(A) + 1$ non-trivial series classes of M.

(ii) Let G be a 2-connected graph without loops and C be a cycle of G such that, for all edges e of C, the graph $G\backslash e$ is not 2-connected. Show that either G is C, or C meets two vertices u and v of G of degree two such that the two paths joining u to v in C contain vertices of degree exceeding two.

6. (Murty 1970) Let M be a rank-r Sylvester matroid where $r \geq 2$. Show that:

(i) M is not graphic.

(ii) (Akkari 1988) If $|E(M)| \geq 4$, then M is not cographic.

(iii) $|E(M)| \geq 2^r - 1$.

(iv)* If equality is attained in (iii), then $M \cong PG(r-1, 2)$.

7. **(i)** (Oxley 1981a) Prove that a matroid M is minimally connected if and only if $|E(M)| \geq 3$, and either M is connected and has every element in a 2-cocircuit, or $M = S((M_1/q_1; p_1), (M_2/q_2; p_2))$ where both M_1 and M_2 are minimally connected matroids having at least five elements, and $\{p_1, q_1\}$ and $\{p_2, q_2\}$ are cocircuits of M_1 and M_2, respectively.

(ii) (Murty 1974) Use (i) to show that if $r \geq 3$, then a rank-r minimally connected matroid M has at most $2r - 2$ elements with this bound being attained if and only if $M \cong M(K_{2,r-1})$.

(iii) (Oxley 1984c) Let M be a minimally connected matroid having at least four elements. Show that M has at least $\frac{1}{3}(r(M) + 2)$ pairwise disjoint 2-cocircuits.

8.* (Tutte 1965) Let M be a connected matroid without coloops. Prove that M is connected if and only if, for every pair of distinct circuits C and C', there is a sequence $C_1, C_2, \ldots, C_m$ of circuits such that $C = C_1$, $C' = C_m$, and, for all i in $\{1, 2, \ldots, m-1\}$, $C_i \cap C_{i+1} \neq \emptyset$ and $r(C_i \cup C_{i+1}) = |C_i \cup C_{i+1}| - 2$.

9.* (Lemos 1989) Let M be a 3-connected matroid with $E(M) \geq 4$ and C be a circuit of M such that, for all e in C, the matroid $M\backslash e$ is not 3-connected. Show that C meets at least two triads of M.

10.* (Akkari 1988, 1991) Let M be a 3-connected matroid such that, for all $r \geq 4$, $M \not\cong M(\mathcal{W}_r)$. Suppose that C is a circuit of M such that, for all pairs $\{e, f\}$ of distinct elements of C, the matroid $M\backslash e, f$ is disconnected. Prove that every such pair $\{e, f\}$ is contained in a triad of M.

10.3 The excluded-minor characterization

Kahn's approach (1984) to Rota's conjecture (6.5.11) that the set of excluded-minors for $GF(q)$-representability is finite has been to attempt to bound the size of excluded minors without explicitly determining them. He showed that this can be done when q is three. Indeed, the main result of Kahn and Seymour (1988) provides a shorter proof of this fact. One problem with extending this approach to larger values of q is that the proof still makes essential use of unique representability. In light of this, Proposition 10.1.11 provides some hope for the case $q = 4$ of Rota's conjecture. Nevertheless, the latter still remains unproved.

We now prove the main result of this chapter, Theorem 6.5.7, first restating the result for convenience of reference.

10.3.1 Theorem. *A matroid is ternary if and only if it has no minor isomorphic to any of the matroids $U_{2,5}$, $U_{3,5}$, F_7, or F_7^*.*

Proof. We showed in Corollary 6.5.3 and Proposition 6.5.5 that the set of excluded minors for $GF(3)$-representability contains $\{U_{2,5}, U_{3,5}, F_7, F_7^*\}$. It remains to show that this set is complete. Assume that N is a minor-minimal matroid that is not ternary and let $E(N) = E$. Suppose also that N has no minor isomorphic to any of $U_{2,5}$, $U_{3,5}$, F_7, or F_7^*. Then both the rank and corank of N exceed two. The strategy of the proof is first to show that N must have rank or corank at most three and then to obtain the contradiction that N has one of $U_{2,5}$, $U_{3,5}$, F_7, or F_7^* as a minor.

The proof will be broken into a sequence of lemmas.

10.3.2 Lemma. *The matroid N is 3-connected. Moreover, for some M_1 in $\{N, N^*\}$, there are a pair of distinct elements a and b of M_1 and a ternary matroid M_2 such that*

10.3.3 *M_1 and M_2 are distinct connected matroids on a common ground set E;*

10.3.4 *$M_1\backslash a = M_2\backslash a$ and $M_1\backslash b = M_2\backslash b$;*

10.3.5 *$M_1\backslash a, b = M_2\backslash a, b$ and this matroid is connected; and*

10.3.6 *$\{a, b\}$ is not a cocircuit of M_1 or of M_2.*

Proof. The fact that N is 3-connected follows by Proposition 4.2.15 and Corollary 7.1.23. By Corollary 10.2.7, N has a pair of distinct elements a and b such that $N\backslash a, b$ or $N/a, b$ is connected. In the first case, we take $M_1 = N$; in the second case, we take $M_1 = N^*$. Then $M_1\backslash a, b$ is connected. Let $r(M_1) = r$ and suppose that $\psi : E - \{a, b\} \rightarrow V(r, 3)$ is a $GF(3)$-coordinatization of $M_1\backslash a, b$. We now extend the domain of definition of ψ to E so that $\psi|_{E-a}$ and $\psi|_{E-b}$ are $GF(3)$-coordinatizations of $M_1\backslash a$ and $M_1\backslash b$, respectively. This is possible because ternary matroids are uniquely $GF(3)$-representable. Let M_2 be the ternary matroid on E for which ψ is a coordinatization. It is now straightforward to check that 10.3.3–10.3.6 hold for the pair $\{M_1, M_2\}$. □

10.3.7 **Lemma.** *If M_1 and M_2 are arbitrary matroids having distinct elements a and b such that 10.3.3–10.3.6 hold, then at most one of M_1 and M_2 is ternary. Moreover, both $M_1\backslash a$ and $M_1\backslash b$ are connected.*

Proof. As M_1 and M_2 are distinct, their common rank r is at least three. Now assume that both M_1 and M_2 are ternary and let $\psi : E - \{a, b\} \rightarrow V(r, 3)$ be a $GF(3)$-coordinatization of $M_1\backslash a, b$. Then, by Corollary 10.1.2, ψ can be extended to coordinatizations $\phi : E \rightarrow V(r, 3)$ and $\phi' : E \rightarrow V(r, 3)$ for M_1 and M_2, respectively. By Lemma 10.2.10, there is, up to scaling, only one choice for $\phi(a)$ that makes $\phi|_{E-b}$ a coordinatization for $M_1\backslash b$. Since $\phi'|_{E-b}$ coordinatizes $M_2\backslash b$ and hence $M_1\backslash b$, we deduce that $\phi'(a) = \pm\phi(a)$. Similarly, $\phi'(b) = \pm\phi(b)$. Thus $M_1 = M_2$; a contradiction. Finally, since M_1 and $M_1\backslash a, b$ are connected but $\{a, b\}$ is not a cocircuit of M_1, both $M_1\backslash a$ and $M_1\backslash b$ must also be connected. □

The last two lemmas, in combination with the following result, will enable us to deduce a lot of important information about the minor-minimal non-ternary matroid N including the fact that it has rank or corank equal to three.

10.3.8 **Lemma.** *Let $\{M_1, M_2\}$ be a minor-minimal pair of matroids satisfying 10.3.3–10.3.6. Then their common rank r is equal to three. Moreover, there is a unique minimal set Z that is dependent in one of these matroids and independent in the other. Indeed, Z is a basis of one matroid and a circuit–hyperplane of the other.*

Proof. As $M_1 \neq M_2$, there is a subset of E that is dependent in one of the matroids but independent in the other. Let Z be a minimal such set. Then, by 10.3.4, $\{a, b\} \subseteq Z$. The next lemma will be used in identifying

$Z - \{a, b\}$ as the set of elements z for which $M_1\backslash a, b/z$ is disconnected. This will then enable us to apply Lemma 10.2.4 to $M_1\backslash a, b$.

10.3.9 Lemma. *If $x \in E - \{a, b\}$ and $M_1\backslash a, b/x$ is connected, then both M_1/x and M_2/x are connected.*

Proof. As $M_1\backslash a, b/x = M_2\backslash a, b/x$, it suffices to show that M_1/x is connected. Assume the contrary. Then $\{a, b\}$ is a separator of M_1/x that contains no coloops. Hence $(M_1/x)|\{a, b\} \cong U_{0,2}$ or $U_{1,2}$. In the first case, a and b are both parallel to x in M_1 and so, by 10.3.4, $M_1 = M_2$; a contradiction. In the second case, we obtain the contradiction that $\{a, b\}$ is a cocircuit of M_1. □

10.3.10 Lemma. *For both elements i of $\{1, 2\}$,*

$$Z = \{a, b\} \cup \{z \in E(M_i\backslash a, b) : M_i\backslash a, b/z \textit{ is disconnected}\}.$$

Proof. As $M_1\backslash a, b = M_2\backslash a, b$, it suffices to establish this lemma for $i = 1$. Suppose that $z \in Z - \{a, b\}$ but that $M_1\backslash a, b/z$ is connected. Then, by Lemma 10.3.9, M_1/z and M_2/z are connected. As $M_1/z \neq M_2/z$, it follows that 10.3.3–10.3.6 hold when we replace M_1 by M_1/z and M_2 by M_2/z. This contradicts the choice of the pair $\{M_1, M_2\}$. Hence

$$Z \subseteq \{a, b\} \cup \{z \in E(M_1\backslash a, b\} : M_1\backslash a, b/z \textit{ is disconnected}\}.$$

Now suppose that $x \in E - Z$ but that $M_1\backslash a, b/x$ is disconnected. Then, by Proposition 4.3.6, $M_1\backslash a, b, x$ is connected. Moreover, since x is in $E - Z$, $M_1\backslash x \neq M_2\backslash x$. We know that at least one of 10.3.3–10.3.6 fails for the pair $\{M_1\backslash x, M_2\backslash x\}$, although 10.3.4 and 10.3.5 clearly hold. Assume that 10.3.6 fails, that is, assume that $\{a, b\}$ is a cocircuit of $M_1\backslash x$ or $M_2\backslash x$. As $\{a, b\}$ is not a cocircuit of M_1 or M_2 , it follows that $\{a, b, x\}$ is a cocircuit of M_1 or M_2. Thus $M_1\backslash a, b$ has x as a coloop. As $M_1\backslash a, b$ is connected, we deduce that $M_1 \cong U_{1,3}$. Using this and Lemma 10.3.7, it is easy to obtain the contradiction that $M_1 = M_2$. We conclude that 10.3.6 holds for the pair $\{M_1\backslash x, M_2\backslash x\}$. Hence 10.3.3 fails for this pair, that is, $M_1\backslash x$ or $M_2\backslash x$ is disconnected. We may assume the former. Then, using the fact that both M_1 and $M_1\backslash a, b, x$ are connected, it is straightforward to again deduce that $M_1\backslash a, b$ has a coloop. From this one obtains a contradiction as above. □

We shall now complete the proof of Lemma 10.3.8. The uniqueness of the set Z follows immediately from the preceding lemma. This set is dependent in M_1 or M_2. Without loss of generality, we shall assume the former. Then Z is a circuit of M_1. Moreover, by the last two lemmas, if $x \in E - Z$, then M_1/x and M_2/x are equal, otherwise 10.3.3–10.3.6

hold for this pair. Now Z is dependent in M_1/x, so Z is dependent in M_2/x. As Z is independent in M_2, it follows that $x \in \text{cl}_2(Z)$. But x was arbitrarily chosen in $E - Z$. Thus Z spans M_2 and so is a basis of M_2. Hence $|Z| = r$. Moreover, if $y \in E - Z$, then Z spans M_2/y, so Z spans M_1/y. Hence $Z \cup y$ spans M_1. But Z is a circuit of M_1 and $|Z| = r$, so Z does not span M_1. Therefore Z is a maximal non-spanning set, that is, a hyperplane of M_1. Hence Z is a circuit–hyperplane of M_1 and a basis of M_2.

By Lemma 10.3.10, $Z - \{a, b\}$ is the set of elements of $M_1 \backslash a, b$ whose contraction disconnects the matroid. Moreover, $|Z - \{a, b\}| = r - 2$. Assume that $r \geq 4$. Then every 2-element circuit of $M_1 \backslash a, b$ is contained in $Z - \{a, b\}$. But the last set is independent in $M_1 \backslash a, b$, hence this matroid is simple. Applying Lemma 10.2.4 to it, we deduce that $Z - \{a, b\}$ contains elements x_1 and x_2 such that the line in $M_1 \backslash a, b$ spanned by these two points contains a third point. This contradicts the fact that Z is a circuit–hyperplane of M_1 containing at least four points. Thus $r \leq 3$. But, as noted at the start of the proof of Lemma 10.3.7, $r \geq 3$. Hence $r = 3$ and Lemma 10.3.8 is proved. □

To complete the proof of Theorem 10.3.1, we now focus attention again on the minor-minimal non-ternary matroid N. By Lemma 10.3.2, for some M_1 in $\{N, N^*\}$, there are distinct elements a and b of M_1 and a ternary matroid M_2 such that 10.3.3–10.3.6 hold. Moreover, by Lemma 10.3.7, $\{M_1, M_2\}$ is a minor-minimal pair of matroids satisfying 10.3.3–10.3.6. Thus, by Lemma 10.3.8, $r(M_1) = 3$. Let $\{M, M'\} = \{M_1, M_2\}$ where we use 10.3.8 to distinguish M from M': we assume that the set Z is a circuit–hyperplane of M and a basis of M'. Lemma 10.3.8 also asserts that Z is the only set that is dependent in one of M and M' but independent in the other. Hence M' is obtained from M by relaxing Z.

Let $Z = \{a, b, z\}$. Then Z is a line of M. Moreover, in M', each of $\{a, b\}$, $\{a, z\}$, and $\{b, z\}$ is a line. Now one of M and M' is ternary and the other is isomorphic to N or N^*. Hence neither M nor M' has a minor isomorphic to any of $U_{2,5}$, $U_{3,5}$, F_7, or F_7^*. Thus if c and d are elements of $E(M') - \{a, b, z\}$, then the line in M' containing $\{c, d\}$ also contains a, b, or z, otherwise $M'|\{a, b, z, c, d\} \cong U_{3,5}$. Hence, in M, the line containing $\{c, d\}$ also contains a, b, or z. Thus $M|\{a, b, z, c, d\} \cong P(U_{2,3}, U_{2,3})$ (see Figure 10.8(a)). Now one of M and M' is in $\{N, N^*\}$ and is therefore 3-connected. Thus M has a point e not on $\{a, b, z\}$ or $\text{cl}_M(\{c, d\})$. As above, both $\text{cl}_M(\{c, e\})$ and $\text{cl}_M(\{d, e\})$ meet $\{a, b, z\}$. Therefore $M|\{a, b, z, c, d, e\} \cong M(K_4)$ (see Figure 10.8(b)). If $E(M) = \{a, b, z, c, d, e\}$, then $M \cong M(K_4)$ and $M' \cong \mathcal{W}^3$. But both $M(K_4)$ and $\mathcal{W}^3$ are ternary. Therefore M has a seventh element, f say. From considering the lines through f and each of c, d, and e, we deduce that $M|\{a, b, z, c, d, e, f\} \cong F_7$ (see Figure 10.8(c)). This contradicts the fact

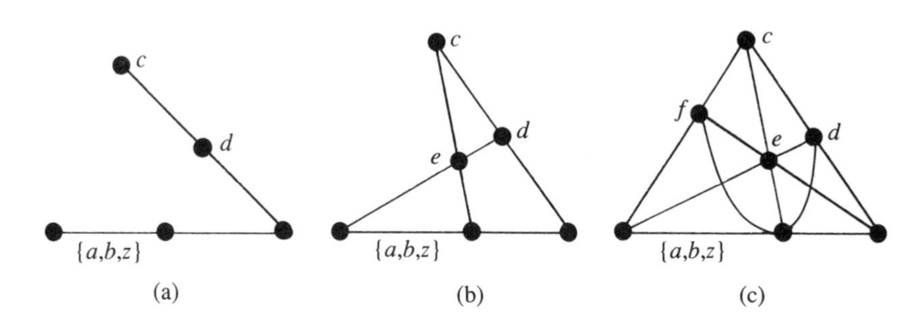

Fig. 10.8. M must have an F_7-minor.

that neither M nor M' has a minor isomorphic to any of $U_{2,5}$, $U_{3,5}$, F_7, or F_7^* and thereby completes the proof of Theorem 10.3.1. □

To close this chapter, we recall that, by Theorem 6.6.3, a matroid is regular if and only if it is both binary and ternary. In this chapter and the last, we have given proofs of the excluded-minor characterizations of ternary and binary matroids. If these proofs are combined with the proof of Theorem 6.6.3, we obtain a proof of Tutte's (1958) excluded-minor characterization of regular matroids, which we stated earlier as Theorem 6.6.4: A matroid is regular if and only if it has no minor isomorphic to $U_{2,4}$, F_7, or F_7^*. A second, more direct proof of this result will be given in Chapter 13.

Exercises

1. Let M_1 and M_2 be arbitrary matroids having distinct elements a and b such that 10.3.3–10.3.6 hold. Show that, in contrast to Lemma 10.3.7, both M_1 and M_2 can be $GF(q)$-representable for all $q \geq 4$.

2. (Oxley) Let M' be obtained from the matroid M by relaxing a circuit–hyperplane. Show that:

(i) Both M and M' are binary if and only if $M \cong U_{k-1,k} \oplus U_{1,m}$ for some positive integers k and m.

(ii)* Both M and M' are ternary if and only if either

(a) $M \cong U_{k-1,k} \oplus U_{1,m}$ for some positive integers k and m; or

(b) M can be obtained from $M(\mathcal{W}_r)$ for some $r \geq 2$ by adding elements in parallel to the spokes or adding elements in series with the rim elements.

3. (Dowling 1973a,b) Let $\{\underline{v}_1, \underline{v}_2, \ldots, \underline{v}_r\}$ be a basis of $V(r,q)$ and let $GF(q)^*$ be $GF(q) - \{0\}$, the multiplicative group of $GF(q)$.

The *Dowling geometry* $Q_r(GF(q)^*)$ is the rank-r simple matroid obtained by restricting the matroid $V(r,q)$ to the set $\{\underline{v}_1, \underline{v}_2, \ldots, \underline{v}_r\} \cup \{\underline{v}_i + c\underline{v}_j : 1 \leq i < j \leq r \text{ and } c \in GF(q)^*\}$. Show that:

(i) $Q_2(GF(q)^*) \cong U_{2,q+1}$ and $Q_r(GF(2)^*) \cong M(K_{r+1})$.

(ii) $Q_r(GF(q)^*)$ has $r\left[\frac{1}{2}(q-1)(r-1)+1\right]$ elements.

(iii) The $(GF(3),3)$-complement of $Q_3(GF(3)^*)$ is $U_{3,4}$.

(iv) For a rank-k flat X of $Q_r(GF(q)^*)$, the simple matroid associated with $Q_r(GF(q)^*)/X$ is isomorphic to $Q_{r-k}(GF(q)^*)$.

(v) If Y is a rank-k modular flat of $Q_r(GF(q)^*)$, then

$$Q_r(GF(q)^*)|Y \cong Q_k(GF(q)^*).$$

(vi) $Q_r(GF(q)^*)$ has $\frac{q^r-1}{q-1}$ hyperplanes.

(vii)* If $r \geq 3$, then $Q_r(GF(q)^*)$ is representable over $GF(q')$ if and only if $q-1$ divides $q'-1$.

4.* (Seymour 1979) Let e and f be distinct elements of a matroid M such that the following hold:

(a) for any subset X of $E(M) - \{e,f\}$, the set $X \cup e$ is a circuit if and only if $X \cup f$ is;

(b) e and f are not both loops and are not both coloops, and $\{e,f\}$ is not a circuit or a cocircuit; and

(c) $M\backslash e,f$ is connected.

Show that M has a $U_{2,5}$-minor.

5. (Bixby 1979)

(i)* Prove that a matroid is ternary if and only if it has no series minor isomorphic to any of the matroids F_7, F_7^*, $U_{2,5}$, or $U_{r,r+2}$ for $r \geq 3$.

(ii) State the dual result of (i).

11
The Splitter Theorem

In this chapter we shall state and prove Seymour's Splitter Theorem. This result, which is a generalization of Tutte's Wheels and Whirls Theorem, is a very powerful general tool for deriving matroid structure results. We shall illustrate this by noting a variety of such results that can be obtained using this theorem. In addition, we shall discuss some extensions and generalizations of the theorem. A survey of the role played by the Splitter Theorem in an analysis of matroid structure has been given by Seymour (1992). However, that paper contains no detailed proofs.

11.1 The theorem and its proof

Let $\mathcal{N}$ be a class of matroids that is closed under minors and under isomorphism. A member N of $\mathcal{N}$ is called a *splitter* for $\mathcal{N}$ if, whenever M is a member of $\mathcal{N}$ having an N-minor, either $M \cong N$, or M is 1- or 2-separable. Thus N is a splitter for $\mathcal{N}$ if and only if $\mathcal{N}$ has no 3-connected member having a proper N-minor.

Although, in general, a splitter need not be 3-connected, most interest here will be focused on 3-connected splitters. This is because a matroid that is not 3-connected can be decomposed into smaller matroids by using the operations of direct sum and 2-sum. Thus a 3-connected splitter for a class $\mathcal{N}$ is, in a very natural sense, a basic building block for $\mathcal{N}$.

An important precursor of the Splitter Theorem is the following result of K. Wagner (1960). We shall prove this result after we have stated the Splitter Theorem to indicate how the latter can be applied. H_8 denotes the 8-vertex graph shown in Figure 11.1.

11.1.1 Proposition. *Let G be a simple 3-connected graph having no K_5-minor. Then either G has no H_8-minor or $G \cong H_8$.*

Observe that, although this result is stated for graphs, by Corollary 8.2.8, it could equally well have been stated in terms of graphic matroids.

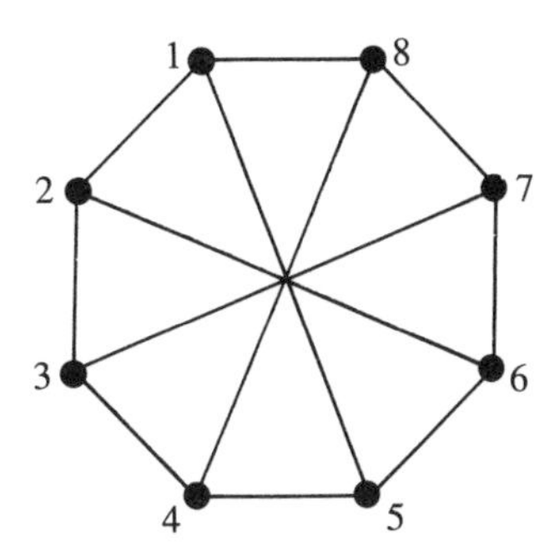

Fig. 11.1. The graph H_8.

In that context, the result asserts that $M(H_8)$ is a splitter for the class of graphic matroids with no $M(K_5)$-minor.

The following theorem is the main result of this chapter (Seymour 1980b). A matroid is *cosimple* if its dual is simple.

11.1.2 The Splitter Theorem. *Let N be a connected, simple, cosimple minor of a 3-connected matroid M. Suppose that if N is a wheel, then M has no larger wheel as a minor, while if N is a whirl, M has no larger whirl as a minor. Then either $M = N$, or M has a connected, simple, cosimple minor M_1 such that some single-element deletion or some single-element contraction of M_1 is isomorphic to N. Moreover, if N is 3-connected, so too is M_1.*

The Splitter Theorem was also proved by J.J.-M. Tan (1981) in his Ph.D. thesis. The theorem is of sufficient importance that it warrants restatement. Indeed, we shall give three reformulations of the theorem in the case that N is 3-connected. The second and third of these will be given in Section 11.2. The first is as follows.

11.1.3 Corollary. *Let $\mathcal{N}$ be a class of matroids that is closed under minors and under isomorphism. Let N be a 3-connected member of $\mathcal{N}$ having at least four elements such that if N is a wheel, it is the largest wheel in $\mathcal{N}$, while if N is a whirl, it is the largest whirl in $\mathcal{N}$. Suppose there is no 3-connected member of $\mathcal{N}$ that has N as a minor and has one more element than N. Then N is a splitter for $\mathcal{N}$.* □

On specializing the last result to graphs, we obtain the following result that was also derived by Negami (1982) independently of Seymour.

11.1.4 Corollary. *Let $\mathcal{G}$ be a class of graphs that is closed under minors and under isomorphism. Let H be a simple 3-connected graph such that if it is a wheel, it is the largest wheel in $\mathcal{G}$. Suppose that there is no simple*

3-connected graph in $\mathcal{G}$ having a single-edge deletion or a single-edge contraction isomorphic to H. Then, for every simple 3-connected graph G in $\mathcal{G}$, either G has no minor isomorphic to H, or $G \cong H$. □

Next we use this corollary to prove Wagner's result.

Proof of Proposition 11.1.1. Let $\mathcal{G}$ be the class of graphs with no minor isomorphic to K_5 and let $H = H_8$. Now suppose that G is a graph having a non-loop edge e such that $G/e = H$. As every vertex of H has degree three, G has a vertex of degree at most two and so is not 3-connected. Hence $\mathcal{G}$ does not contain a simple 3-connected graph having a single-edge contraction isomorphic to H. Suppose next that G is a simple 3-connected graph in $\mathcal{G}$ such that some single-edge deletion is isomorphic to H. Then, by symmetry, we may suppose that G is obtained from H by adding one of the edges 13 and 14 (see Figure 11.2). In each case, it is straightforward to show that $G \notin \mathcal{G}$: in the first case, we obtain a graph isomorphic to K_5 by contracting the edges 26, 78, and 45; in the second case, we obtain such a graph by contracting 23, 56, and 78. Proposition 11.1.1 now follows immediately from Corollary 11.1.4. □

Most of the remainder of this section will be devoted to proving the Splitter Theorem. The proof given here is due to C. R. Coullard and L. L. Gardner (private communication). It uses the following lemma from Seymour's proof (1980b). This lemma is also implicit in Tutte's proof of the Wheels and Whirls Theorem (1966b).

11.1.5 Lemma. *For some $n \geq 2$, let $\{x_1, x_2, \ldots, x_n\}$ and $\{y_1, y_2, \ldots, y_n\}$ be disjoint subsets X and Y of the ground set of a connected matroid M. Suppose that, for all i in $\{1, 2, \ldots, n\}$, $\{x_i, y_i, x_{i+1}\}$ is a triangle and $\{y_i, x_{i+1}, y_{i+1}\}$ is a triad, where all subscripts are to be read modulo n. Then $E(M) = \{x_1, y_1, x_2, y_2, \ldots, x_n, y_n\}$ and M is isomorphic to $M(\mathcal{W}_n)$ or $\mathcal{W}^n$.*

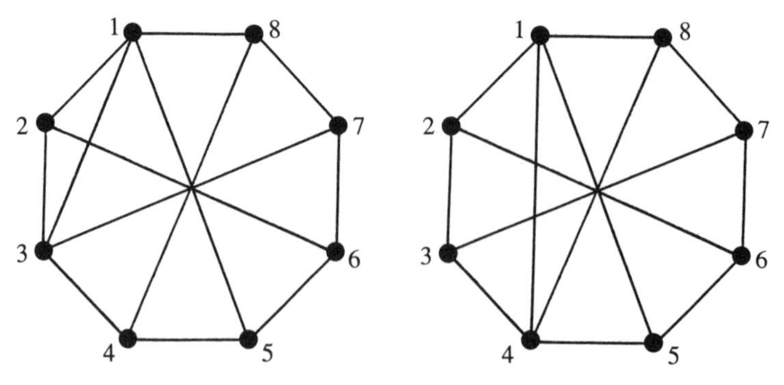

Fig. 11.2. The two simple graphs with H_8 as a single-edge deletion.

Proof. Evidently $Y \subseteq \mathrm{cl}(X)$ and $X \subseteq \mathrm{cl}^*(Y)$, so

$$r(X \cup Y) = r(X) \leq n \tag{1}$$

and

$$r^*(X \cup Y) = r^*(Y) \leq n. \tag{2}$$

On adding (1) and (2), we get that $r(X \cup Y) + r^*(X \cup Y) \leq 2n$, that is,

$$r(X \cup Y) + r^*(X \cup Y) - |X \cup Y| \leq 0. \tag{3}$$

But, for all sets Z, $r(Z) + r^*(Z) - |Z| \geq 0$. Hence equality holds in (3). Therefore, as M is connected, Proposition 4.2.6 implies that $X \cup Y = E(M)$. Moreover, as equality holds in (3), it also holds in (1), that is, $r(M) = n$.

We shall show that M is a wheel or a whirl having X as its set of spokes and Y as its rim. To this end, let $\mathcal{D}$ be the set of subsets D of $X \cup Y$ for which $|D \cap X| = 2$ such that if $D \cap X = \{x_j, x_k\}$, then $D \cap Y$ is $\{y_j, y_{j+1}, \ldots, y_{k-1}\}$ or $\{y_k, y_{k+1}, \ldots, y_{j-1}\}$ where, again, subscripts are to be read modulo n.

As a first step towards finding $\mathcal{C}(M)$, we now show that every circuit C of M contains a member of $\mathcal{D} \cup \{Y\}$. By Proposition 2.1.11, for all i in $\{1, 2, \ldots, n\}$,

$$|C \cap \{y_i, x_{i+1}, y_{i+1}\}| \neq 1. \tag{4}$$

If $C \cap X = \emptyset$, then C contains y_j for some j. By repeatedly using (4), it follows easily that $C \supseteq Y$. We now suppose that $C \cap X \neq \emptyset$ and that $C \not\supseteq Y$. Let x_k be an element of C. Then, by (4), y_{k-1} or y_k is in C. By symmetry, we may assume the latter. As $C \not\supseteq Y$, there is a non-negative integer m such that $\{y_k, y_{k+1}, \ldots, y_{k+m}\} \subseteq C$ but $y_{k+m+1} \notin C$. Then, by (4) again, $x_{k+m+1} \in C$. Hence C contains $\{x_k, y_k, y_{k+1}, \ldots, y_{k+m}, x_{k+m+1}\}$, a member of $\mathcal{D}$, and so every member of $\mathcal{C}(M)$ does indeed contain a member of $\mathcal{D} \cup \{Y\}$.

Next we show that every member of $\mathcal{D}$ is a circuit of M. To do this, it suffices, by symmetry, to show that $\{x_1, y_1, y_2, \ldots, y_t, x_{t+1}\}$ is a circuit for all t in $\{1, 2, \ldots, n-1\}$. We shall do this by induction on t. By assumption, $\{x_1, y_1, x_2\}$ is a circuit of M. Assume that $\{x_1, y_1, y_2, \ldots, y_t, x_{t+1}\}$ is a circuit for some $t < n-1$. Then, as $\{x_{t+1}, y_{t+1}, x_{t+2}\}$ is also a circuit, elimination implies that $\{x_1, y_1, y_2, \ldots, y_{t+1}, x_{t+2}\}$ contains a circuit C'. From above, C' must contain a member of $\mathcal{D} \cup \{Y\}$. Therefore $C' = \{x_1, y_1, y_2, \ldots, y_{t+1}, x_{t+2}\}$ and we conclude, by induction, that every member of $\mathcal{D}$ is a circuit of M.

If Y is a circuit of M, it follows that $\mathcal{C}(M) = \mathcal{D} \cup \{Y\}$. Then, clearly, M is isomorphic to $M(\mathcal{W}_n)$ with X being the set of spokes. If Y is not a circuit of M, then Y is an independent set having $r(M)$ elements, so Y is a basis of M. Moreover, for all x_i in X and all proper subsets Y' of Y, the set $Y' \cup x_i$ does not contain a member of $\mathcal{D}$. Hence $Y \cup x_i \in \mathcal{C}(M)$ and so $\mathcal{C}(M) = \mathcal{D} \cup \{Y \cup x_i : 1 \leq i \leq n\}$. In this case, $M \cong \mathcal{W}^n$. □

We are now ready to prove the Splitter Theorem.

Proof of Theorem 11.1.2. We first observe that it suffices to prove the theorem when N is connected, simple, and cosimple, for it follows from Proposition 8.1.10 that if N is 3-connected, then so is M_1. Now suppose that the theorem is false and that, among all pairs of matroids for which the theorem fails, (N, M) is chosen so that $|E(M) - E(N)|$ is as small as possible. Certainly $|E(M) - E(N)| > 0$. Define the sets X and Y by

$$X = \{x \in E(M) : M\backslash x \text{ has an } N\text{–minor}\}$$

and

$$Y = \{y \in E(M) : M/y \text{ has an } N\text{–minor}\}.$$

Clearly either

$$\text{(i)} \quad X \cap Y = \emptyset; \qquad \text{or} \qquad \text{(ii)} \quad X \cap Y \neq \emptyset.$$

Assume that (i) holds. In this case, we shall establish the contradiction that the hypothesis concerning wheels and whirls is violated. Evidently X or Y is non-empty. We shall assume the former noting that if X is empty, then Y is not and we may apply the argument that follows to the pair (N^*, M^*), which is also a minimal counterexample to the theorem.

11.1.6 Lemma. *Let x be an element of X. Then M has connected minors N_1 and N_2 and elements y, m, and n such that*

(a) $y \in Y - E(N_1)$, $m \in Y \cap E(N_1)$, *and* $n \in X \cap E(N_1)$;

(b) $N \cong N_1 = N_2/y\backslash x$ *and* $r(N_1) = r(N_2) - 1$; *and*

(c) $\{x, y, m\}$ *and* $\{x, y, n\}$ *are a triad and a triangle, respectively, of* N_2.

Proof. As $x \in X$, the matroid $M\backslash x$ has an N-minor N_1, say. Let N_3 be the minor of M for which $N_3\backslash x = N_1$. As the theorem fails for the pair (N, M), we must have that N_3 is disconnected, non-simple, or non-cosimple. Now, since $X \cap Y$ is empty, x is not a loop or a coloop of N_3. Therefore, by Proposition 8.1.10, N_3 is connected. Since N is cosimple, so is N_3. Hence N_3 is non-simple and so x is in a 2-circuit with some element n of N_1. Clearly $n \in X$.

As M is 3-connected, it does not have $\{x, n\}$ as a circuit. Thus there is an element y of $E(M) - E(N_3)$ and a minor N_2 of M such that $N_3 = N_2/y$ and $\{x, y, n\}$ is a triangle of N_2. Evidently $y \in Y$. By the dual of Proposition 8.1.10, as N_3 is connected, so is N_2. Consider $N_2\backslash x$. As $(N_2\backslash x)/y = N_1$, we must have that $N_2\backslash x$ is disconnected, non-simple, or non-cosimple. By (i) and the dual of Proposition 8.1.10 again, it follows that y is in a 2-cocircuit, $\{y, m\}$ say, of $N_2 \backslash x$. Clearly $m \in Y$, so $m \neq n$. As $\{x, y, n\}$ is a circuit of N_2, Proposition 2.1.11 implies that $\{y, m\}$ is not a cocircuit of N_2. Hence $\{x, y, m\}$ is a triad of N_2 and so $r(N_1) = r(N_2) - 1$. □

We now look more closely at the matroid N_2.

11.1.7 Lemma. *N_2 is simple and cosimple.*

Proof. As N_2 is connected, we need only show that N_2 has no 2-circuits and no 2-cocircuits. Assume that $\{a, b\}$ is a 2-circuit of N_2. Then $\{a, b\} \subseteq X$, so $\{a, b\} \cap Y = \emptyset$. But $N_2/y \backslash x$ equals N_1 and so is simple. Therefore $\{a, b\}$ does not contain a circuit of $N_2/y \backslash x$. Thus $x \in \{a, b\}$. Hence $|\{a, b\} \cap \{x, y, m\}|$ is non-zero. By Proposition 2.1.11 and the fact that $y \notin \{a, b\}$, we deduce that $\{a, b\} = \{x, m\}$. Hence $m \in X$; a contradiction. Thus N_2 has no 2-circuits. A dual argument establishes that N_2 has no 2-cocircuits. □

The next lemma completes the proof that case (i) cannot occur.

11.1.8 Lemma. *Both N and N_2 are wheels or both are whirls.*

Proof. Let $X_2 = \{e \in E(N_2) : N_2 \backslash e$ has an N-minor$\}$ and $Y_2 = \{e \in E(N_2) : N_2/e$ has an N-minor$\}$. Then $X_2 \subseteq X$ and $Y_2 \subseteq Y$, so $X_2 \cap Y_2 = \emptyset$. As $N_2/y \backslash x \cong N$, it follows that $x \in X_2$ and $y \in Y_2$. Moreover, $n \in X_2$ and $m \in Y_2$ since $N_2/y \backslash x \cong N_2/y \backslash n$ and $N_2 \backslash x/y \cong N_2 \backslash x/m$. Now let a be an arbitrary element of X_2. Then $N_2 \backslash a$ has an N-minor. Thus, since $r(N_2) - r(N) = 1$ and $|E(N_2) - E(N)| = 2$, the matroid $N_2 \backslash a$ has an element b that is not a loop or a coloop such that $N_2 \backslash a/b \cong N$. Hence $N_2 \backslash a$ is connected. By the choice of (N, M), it follows that $N_2 \backslash a$ is not cosimple. Thus $N_2 \backslash a$ has a 2-cocircuit, $\{c, d\}$ say. Clearly $\{c, d\} \subseteq Y_2$. Moreover, as N_2 is cosimple, $\{a, c, d\}$ is a triad of N_2. Thus every element of X_2 is in a triad with two elements of Y_2, and a dual argument shows that every element of Y_2 is in a triangle with two elements of X_2.

Now let k be the largest integer for which there are subsets $\{x_1, x_2, \ldots, x_k\}$ of X_2 and $\{y_1, y_2, \ldots, y_k\}$ of Y_2 such that $\{x_i, y_i, x_{i+1}\}$ is a triangle of N_2 and $\{y_i, x_{i+1}, y_{i+1}\}$ is a triad of N_2 for all i in $\{1, 2, \ldots, k-1\}$. By taking $x_1 = n$, $x_2 = x$, $y_1 = y$, and $y_2 = m$, we see that $k \geq 2$. Although N_2 need not be graphic, we can keep track of the triangles and triads in N_2 by using the graph in Figure 11.3. Note that in that diagram a circled vertex corresponds to a *known* triad.

Consider the element y_k. Since it is in Y_2, it is in a triangle T with two members of X_2. As T meets the triad $\{y_{k-1}, x_k, y_k\}$, it must contain x_k. Therefore T is $\{x_k, y_k, x_{k+1}\}$ for some x_{k+1} in X_2. Clearly $x_{k+1} \neq x_k$. If

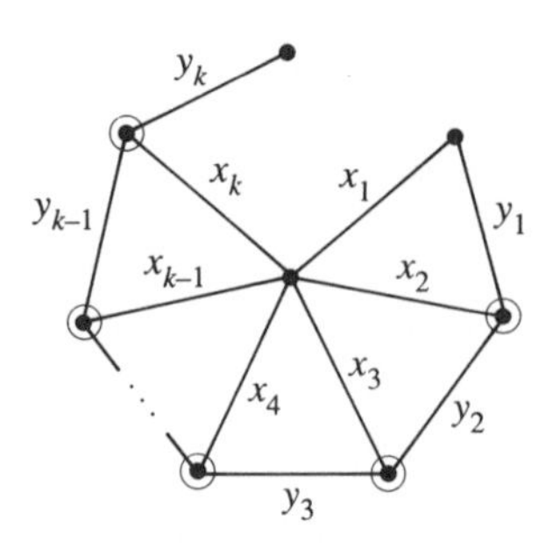

Fig. 11.3. The circled vertices correspond to known triads of N_2.

$x_{k+1} = x_i$ for some i in $\{2, 3, \ldots, k-1\}$, then the triangle $\{x_k, y_k, x_{k+1}\}$ has exactly one element in common with the triad $\{y_{i-1}, x_i, y_i\}$; a contradiction to Proposition 2.1.11. Thus $x_{k+1} \in X_2 - \{x_2, x_3, \ldots, x_k\}$.

We show next that $x_{k+1} = x_1$. Assume the contrary. As $x_{k+1} \in X_2$, there is a triad T^* which contains x_{k+1} and two members of Y_2. Since T^* meets the triangle $\{x_k, y_k, x_{k+1}\}$, $y_k \in T^*$. Hence T^* is $\{y_k, x_{k+1}, y_{k+1}\}$ for some y_{k+1} in Y_2. Clearly $y_{k+1} \neq y_k$. Moreover, if $y_{k+1} = y_i$ for some i in $\{1, 2, \ldots, k-1\}$, then the triad $\{y_k, x_{k+1}, y_{k+1}\}$ has exactly one element in common with the triangle $\{x_i, y_i, x_{i+1}\}$; a contradiction. Thus $y_{k+1} \notin \{y_1, y_2, \ldots, y_k\}$. But now the pair $(\{x_1, x_2, \ldots, x_{k+1}\}, \{y_1, y_2, \ldots, y_{k+1}\})$ contradicts the maximality of the pair $(\{x_1, x_2, \ldots, x_k\}, \{y_1, y_2, \ldots y_k\})$. Thus we do indeed have that $x_{k+1} = x_1$ and Figure 11.3 can be updated as in Figure 11.4.

This diagram suggests that $\{y_k, x_1, y_1\}$ is a triad of N_2 and we can show this by arguing as follows: x_1 is in a triad with two elements of Y_2, and the triangles $\{x_k, y_k, x_1\}$ and $\{x_1, y_1, x_2\}$ imply that these two elements must be y_k and y_1. It now follows immediately from Lemma 11.1.5 and the fact that $N \cong N_2/y\backslash x$ that both N_2 and N are wheels or both are whirls. □

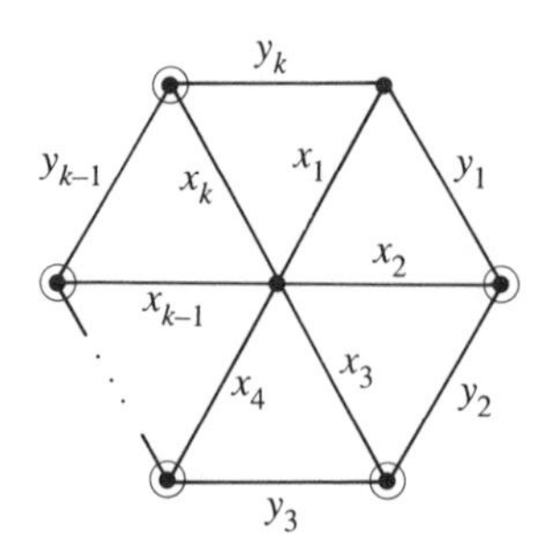

Fig. 11.4. The sequence of triangles and triads closes up.

As we now know that case (i) cannot occur, we may assume that case (ii) occurs; that is, we may suppose that, for some element e of M, both $M\backslash e$ and M/e have an N-minor. By Proposition 8.4.6, $M\backslash e$ or M/e has no non-minimal 2-separations. We shall assume the latter, noting that if the former holds, we may apply the argument that follows to $(M\backslash e)^*$, which equals M^*/e. By Proposition 8.4.6 again, $\widetilde{M/e}$ is 3-connected. Moreover, by the choice of the pair (N, M), we must have that $\widetilde{M/e} \cong N$ and $M/e \not\cong N$. Thus N is 3-connected.

As $M/e \neq \widetilde{M/e}$, there is a triangle of M containing e. Let $\{T_1, T_2, \ldots, T_k\}$ be the set of such triangles where, for all i, $T_i = \{x_i, y_i, e\}$, say. If $(T_i \cap T_j) - e \neq \emptyset$ for some distinct i and j in $\{1, 2, \ldots, k\}$, then M/e has a parallel class of size at least three and $|E(M/e)| \geq |E(N)| + 2 \geq 6$. Hence M/e has a non-minimal 2-separation; a contradiction. Thus the $2k$ elements, $x_1, y_1, x_2, y_2, \ldots, x_k, y_k$, are distinct. Moreover, each of these elements is in X. Hence, by the choice of (N, M), for all i in $\{1, 2, \ldots, k\}$, both $M\backslash x_i$ and $M\backslash y_i$ are 2-separable. Therefore, by Lemma 8.4.9, each of x_i and y_i is in a triad with e. Consider such a triad containing $\{x_1, e\}$. By Proposition 2.1.11, it must meet the disjoint sets $T_2 - e, T_3 - e, \ldots, T_k - e$. Therefore $k \leq 2$. Furthermore, by Proposition 8.1.7, $\{x_1, e, y_1\}$ is not a triad of M.

Suppose that $k = 1$. Then $M/e\backslash x_1 \cong N$ so $|E(M) - E(N)| = 2$. Moreover, by Lemma 8.4.9, $\{x_1, e, a\}$ and $\{y_1, e, b\}$ are triads of M for some elements a and b not in $\{x_1, y_1\}$. Now $a \neq b$; otherwise, by cocircuit elimination, $\{x_1, e, y_1\}$ is a triad of M; a contradiction. Thus, in $M\backslash e$, all of x_1, y_1, a, and b are in 2-cocircuits. Hence if $f \in E(M\backslash e)$, then $M\backslash e/f$ is not 3-connected. This contradicts the fact that $M\backslash e$ has an N-minor since $|E(M) - E(N)| = 2$ and N is 3-connected. We conclude that $k \neq 1$.

We may now suppose that $k = 2$. Then $|E(M) - E(N)| = 3$. Moreover, by relabelling x_2 and y_2 if necessary, we may assume that $\{x_1, e, x_2\}$ and $\{y_1, e, c\}$ are triads of M for some c in $\{x_2, y_2\}$. If $c = x_2$, then, by cocircuit elimination again, we deduce the contradiction that $\{x_1, e, y_1\}$ is a triad of M. Thus $c \neq x_2$, so $\{y_1, e, y_2\}$ is a triad of M. Now let $Z = \{x_1, y_1, x_2, y_2, e\}$. Then $\{x_1, x_2, e\}$ and $\{x_1, y_1, e\}$ span Z in M and M^*, respectively. Thus

$$r(Z) + r^*(Z) - |Z| \leq 1.$$

Hence, by Lemma 4.2.5, $(Z, E(M) - Z)$ is a 2-separation of M unless $|E(M) - Z| \leq 1$. Since M is 3-connected, it follows that $|E(M)| \leq 6$. As $|E(M) - E(N)| \geq 3$ and $|E(N)| \geq 4$, this is a contradiction and so the proof of the Splitter Theorem is complete. □

In the third section, we shall discuss several variations on the Splitter Theorem. To close this section, we observe certain ways in which this theorem cannot be extended. First, it does not remain valid if, instead of requiring that M be 3-connected, we insist only that it be connected, simple, and cosimple.

11.1.9 **Example.** Let $N = U_{2,4}$ and $M = U_{2,4} \oplus_2 U_{2,4}$. Then both N and M are connected, simple, and cosimple. Yet M has no connected, simple, cosimple minor having a single-element deletion or a single-element contraction that is isomorphic to N. □

A consequence of the requirement that N be simple and cosimple is that $|E(N)| \geq 4$. But not every 3-connected matroid has four or more elements and one may ask, in the case that N is 3-connected, whether the theorem remains valid for $|E(N)| \leq 3$. It does not, as one can see by letting $N = U_{2,3}$ and $M = M(K_5)$.

Finally, we draw attention to the fact that the matroid M_1 in the Splitter Theorem is not required to have a single-element deletion or contraction *equal* to N but only to have such a minor *isomorphic* to N. Indeed, it is not difficult to show (Exercise 7) that the theorem is no longer valid if M_1 is subject to the stronger requirement.

In one of the variants of the Splitter Theorem to be discussed in Section 11.3, we shall insist on obtaining a 3-connected minor of M that has, as a proper minor, the matroid N itself rather than just an isomorphic copy of N.

Exercises

1. Show that the graph H_8 in Figure 11.1 is isomorphic to the graph in Figure 11.5. The latter is sometimes called a *Möbius ladder.*

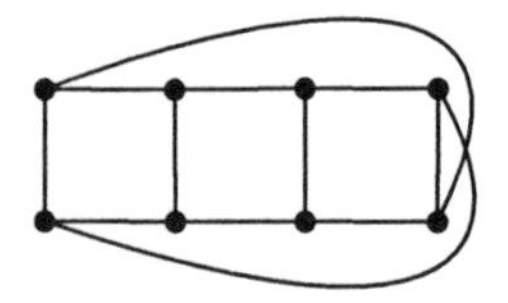

Fig. 11.5. Another drawing of H_8.

2. Let $\mathcal{N}$ be a class of matroids that is closed under isomorphism, minors, and duality. Show that if M is a splitter for $\mathcal{N}$, then so is M^*.

3. Let $\mathcal{N}_1$ and $\mathcal{N}_2$ be classes of matroids, each of which is closed under isomorphism and minors, and let $\mathcal{N}_1$ be a subclass of $\mathcal{N}_2$. Suppose that $M \in \mathcal{N}_1$. Show that if M is a splitter for $\mathcal{N}_2$, then it is a splitter for $\mathcal{N}_1$, but that the converse of this fails.

4. (Welsh 1982) Show that each of the following classes of matroids has no splitter: $GF(q)$-representable matroids; graphic matroids; and cographic matroids.

5. Give an example of a class $\mathcal{N}$ of matroids and a disconnected member N of $\mathcal{N}$ such that $\mathcal{N}$ is closed under isomorphism and under minors, and N is a splitter for $\mathcal{N}$.

6. State and prove the analogue of Corollary 11.1.3 when N is connected, simple, and cosimple.

7. Let $M = M(G)$ where G is the graph in Figure 11.6. Let $N = M(G)/\{1,2,3\}$. Show that M has no 3-connected minor M_1 having an element e such that $M_1\backslash e$ or M_1/e equals N, but that M does have such a minor M_1 and an element e for which $M_1\backslash e \cong N$.

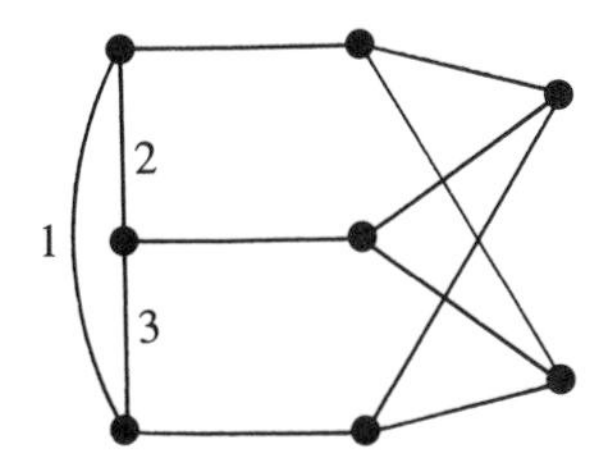

Fig. 11.6.

8. (Oxley 1989a) Show that, for each of the graphs G in Figure 11.7, $M(G)$ is a splitter for the class of graphic matroids having no minor isomorphic to $M(\mathcal{W}_5)$.

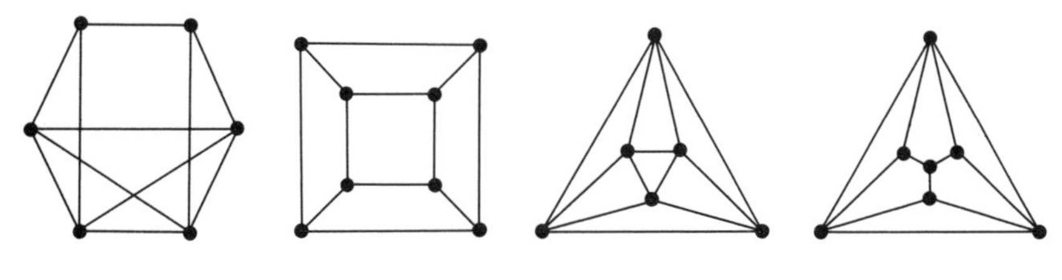

Fig. 11.7.

9. (Oxley 1990a) Show that the free extension of $AG(3,2)$ is a splitter for the class of matroids with no minor isomorphic to $\mathcal{W}^3$ or P_6.

10. (Oxley 1987c) Show that the matroid in Figure 10.4 is a splitter for the class of matroids with no minor isomorphic to $M(K_4)$, $U_{2,5}$, or $U_{3,5}$.

11. (Lemos, private communication) Let M be a connected matroid and X be a subset of $E(M)$ having at least four elements. Suppose that, for each e in X, there is a triangle T and a triad T^* such that $T \cup T^* \subseteq X$ and $e \in T \cap T^*$. Prove that M is isomorphic to a wheel or a whirl.

11.2 Applications of the Splitter Theorem

This section contains a number of examples of structure theorems for various classes of matroids and graphs. All of these results can be derived from the Splitter Theorem. We begin, however, by stating a second reformulation of the theorem in the case that N is 3-connected. The reader will find it instructive to compare this result with Corollary 4.3.7, its analogue for connected matroids.

11.2.1 Corollary. *Let M and N be 3-connected matroids such that N is a minor of M, $|E(N)| \geq 4$, and if N is a wheel, then M has no larger wheel as a minor, while if N is a whirl, then M has no larger whirl as a minor. Then there is a sequence $M_0, M_1, \ldots, M_n$ of 3-connected matroids such that $M_0 \cong N$, $M_n = M$, and, for all i in $\{0, 1, \ldots, n-1\}$, M_i is a single-element deletion or a single-element contraction of M_{i+1}.* □

We now use the last result to prove Tutte's Wheels and Whirls Theorem (1966b), which was stated earlier as Theorem 8.4.5 and is restated below for convenient reference. But first a reminder of a basic fact concerning wheels and whirls which will be used repeatedly throughout this chapter: the smallest whirl $\mathcal{W}^2$ is isomorphic to $U_{2,4}$, while the smallest 3-connected wheel $M(\mathcal{W}_3)$ is isomorphic to $M(K_4)$.

11.2.2 Corollary. *The following statements are equivalent for a 3-connected matroid M having at least one element:*

(i) *For every element e of M, neither $M\backslash e$ nor M/e is 3-connected.*

(ii) *M has rank at least three and is isomorphic to a wheel or a whirl.*

Proof. It was noted prior to Theorem 8.4.5 that (ii) implies (i). Now suppose that (i) holds and assume that M has a 3-connected wheel or a whirl as a minor. Take such a minor N of largest rank. If $N = M$, then, since $M \not\cong \mathcal{W}^2$, (ii) holds. Thus we may assume that $N \neq M$. Then, by Corollary 11.2.1, there is an element e of M such that $M\backslash e$ or M/e is 3-connected and has an N-minor. But this contradicts (i). Thus we may suppose that M has no 3-connected wheel minor and no whirl minor.

If $1 \leq |E(M)| \leq 5$, then we deduce from Table 8.1 (p. 292) that M has an element e such that $M\backslash e$ or M/e is 3-connected. Thus we may assume that $|E(M)| \geq 6$. Hence, as M is 3-connected, $r(M) \geq 2$. But if $r(M) = 2$, then $M \cong U_{2,n}$ for some $n \geq 6$ and again (i) fails. Thus $r(M) \geq 3$ and, similarly, $r^*(M) \geq 3$.

Now choose x in $E(M)$. Then, by Proposition 8.4.6, $\widetilde{M/x}$ or $\widetilde{M\backslash x}$ is 3-connected. By switching to the dual if necessary, we may assume the former. If $|E(\widetilde{M/x})| \geq 4$, then applying Corollary 11.2.1 taking N equal to $\widetilde{M/x}$, we obtain a contradiction to (i). Hence $|E(\widetilde{M/x})| \leq 3$, and, since $r(\widetilde{M/x}) \geq 2$, it follows that $r(\widetilde{M/x}) = 2$. Thus $r(M) = 3$. Now M has no $U_{2,4}$-minor and is therefore binary. Hence M is a simple binary matroid of rank three having at least six elements. Thus M has $PG(2,2)\backslash p$ as a minor for some point p. But $PG(2,2)\backslash p \cong M(\mathcal{W}_3)$ and so we have a contradiction that completes the proof. □

For the next two applications of the Splitter Theorem, we shall need to recall Tutte's result (6.6.4) that a binary matroid is regular if and only if it has no minor isomorphic to F_7 or F_7^*. The first application (Seymour 1980b) asserts that F_7 is a splitter for the class of binary matroids having no F_7^*-minor.

11.2.3 Proposition. *If M is a 3-connected binary matroid, then M has no F_7^*-minor if and only if either M is regular or $M \cong F_7$.*

Proof. If M is regular or is isomorphic to F_7, then, by Tutte's result just noted, M has no F_7^*-minor. For the converse, let $\mathcal{N}$ be the class of binary matroids having no F_7^*-minor and let $N = F_7$. Then, by Tutte's result again, it suffices to show that if N_1 is a 3-connected member of $\mathcal{N}$, then either N_1 has no N-minor, or $N_1 \cong N$. By Corollary 11.1.3, this will follow if we can show that $\mathcal{N}$ contains no 3-connected members that are single-element extensions or coextensions of N. Since $N \cong PG(2,2)$, it has no 3-connected binary single-element extensions.

Now consider the binary single-element coextensions of F_7. Two such coextensions are the vector matroids of the following matrices over $GF(2)$:

$$\left[\begin{array}{c|cccc} & 0 & 1 & 1 & 1 \\ I_4 & 1 & 0 & 1 & 1 \\ & 1 & 1 & 0 & 1 \\ & 1 & 1 & 1 & 0 \end{array}\right] \quad \text{and} \quad \left[\begin{array}{c|cccc} & 0 & 1 & 1 & 1 \\ I_4 & 1 & 0 & 1 & 1 \\ & 1 & 1 & 0 & 1 \\ & 1 & 1 & 1 & 1 \end{array}\right].$$

Clearly both these matroids are isomorphic to their duals. The first is isomorphic to $AG(3,2)$; the second will be denoted by S_8. Seymour (1985a) proved the following.

11.2.4 Lemma. *$AG(3,2)$ and S_8 are the only 3-connected binary single-element coextensions of F_7. Dually, $AG(3,2)$ and S_8 are the only 3-connected binary single-element extensions of F_7^*.*

Proof. Evidently it suffices to prove the second statement. Now F_7^* is represented by the matrix A over $GF(2)$ where

$$A = \begin{array}{c} \begin{array}{ccccccc} 1 & 2 & 3 & 4 & 5 & 6 & 7 \end{array} \\ \left[\begin{array}{c|ccc} & 0 & 1 & 1 \\ I_4 & 1 & 0 & 1 \\ & 1 & 1 & 0 \\ & 1 & 1 & 1 \end{array} \right] \end{array}.$$

Suppose that M is a 3-connected binary single-element extension of F_7^*. Then, since binary matroids are uniquely $GF(2)$-representable, M can be represented by the matrix

$$\begin{array}{c} \begin{array}{cccccccc} 1 & 2 & 3 & 4 & 5 & 6 & 7 & 8 \end{array} \\ \left[\begin{array}{c|cccc} & 0 & 1 & 1 & x_1 \\ I_4 & 1 & 0 & 1 & x_2 \\ & 1 & 1 & 0 & x_3 \\ & 1 & 1 & 1 & x_4 \end{array} \right] \end{array}$$

where each of x_1, x_2, x_3, and x_4 is in $\{0, 1\}$. We shall denote this matrix by A+8. As M is 3-connected and therefore simple, at least two of x_1, x_2, x_3, and x_4 are non-zero. By symmetry, we may assume that $(x_1, x_2, x_3, x_4)^T$ is one of $(1, 1, 1, 1)^T$, $(1, 1, 1, 0)^T$, $(1, 1, 0, 0)^T$, and $(1, 0, 0, 1)^T$. Label these columns by e_1, e_2, e_3, and e_4, respectively.

Evidently $M[A + e_1] = S_8$ and $M[A + e_2] \cong AG(3, 2)$. Moreover, as we shall show next, $M[A + e_3] \cong S_8 \cong M[A + e_4]$. The matrices $A + e_3$ and $A + e_4$ are

$$\left[\begin{array}{c|cccc} & 0 & 1 & 1 & 1 \\ I_4 & \boxed{1} & 0 & 1 & 1 \\ & 1 & 1 & 0 & 0 \\ & 1 & 1 & 1 & 0 \end{array} \right] \text{ and } \left[\begin{array}{c|cccc} & 0 & 1 & 1 & 1 \\ I_4 & 1 & 0 & 1 & 0 \\ & 1 & 1 & 0 & 0 \\ & \boxed{1} & 1 & 1 & 1 \end{array} \right].$$

On pivoting on the boxed entries in each case, we obtain the matrices

$$\left[\begin{array}{c|cccc} & 0 & 1 & 1 & 1 \\ I_4 & 1 & 0 & 1 & 1 \\ & 1 & 1 & 1 & 1 \\ & 1 & 1 & 0 & 1 \end{array} \right] \text{ and } \left[\begin{array}{c|cccc} & 0 & 1 & 1 & 1 \\ I_4 & 1 & 1 & 0 & 1 \\ & 1 & 0 & 1 & 1 \\ & 1 & 1 & 1 & 1 \end{array} \right],$$

where we recall from Section 6.6 that a pivot includes the natural column interchange. Clearly the vector matroids of the last two matrices are isomorphic to S_8 and so the lemma is proved. □

To complete the proof of Proposition 11.2.3, it suffices to note that, by the last lemma, neither $AG(3, 2)$ nor S_8 is in $\mathcal{N}$. □

Now we recall from Corollary 8.3.4 that a matroid that is not 3-connected can be built up from some of its 3-connected minors by a sequence of direct sums and 2-sums. Hence, as immediate consequences of the last proposition and duality, we have the following structural results (Seymour 1981c).

11.2.5 Corollary. *Every binary matroid that has no F_7^*-minor can be obtained from regular matroids and copies of F_7 by a sequence of direct sums and 2-sums.* □

11.2.6 Corollary. *Every binary matroid that has no F_7-minor can be obtained from regular matroids and copies of F_7^* by a sequence of direct sums and 2-sums.* □

Both these corollaries are special cases of the following result (Seymour 1981c), a third reformulation of the Splitter Theorem in the case when N is 3-connected.

11.2.7 Corollary. *Let $\mathcal{N}$ be a class of matroids that is closed under isomorphism and under minors and let N be a 3-connected member of $\mathcal{N}$ having at least four elements such that no 3-connected member of $\mathcal{N}$ with an N-minor has exactly $|E(N)|+1$ elements. Suppose that if N is a wheel, then $\mathcal{N}$ contains no larger wheels, while if N is a whirl, $\mathcal{N}$ contains no larger whirls. Then every matroid in $\mathcal{N}$ can be obtained by a sequence of direct sums and 2-sums from copies of N and members of $\mathcal{N}$ having no N-minor.* □

Our next application of the Splitter Theorem will involve showing that a certain 10-element matroid R_{10} is a splitter for the class of regular matroids. This result is an important step in Seymour's decomposition theorem for regular matroids, which will be stated in Chapter 13. The matroid R_{10} is the vector matroid of the matrix A_{10} over $GF(2)$, where

$$A_{10} = \begin{array}{c} \begin{matrix} 1 & 2 & 3 & 4 & 5 & 6 & 7 & 8 & 9 & 10 \end{matrix} \\ \left[\begin{array}{ccccc|ccccc} & & & & & 1 & 1 & 0 & 0 & 1 \\ & & & & & 1 & 1 & 1 & 0 & 0 \\ & & I_5 & & & 0 & 1 & 1 & 1 & 0 \\ & & & & & 0 & 0 & 1 & 1 & 1 \\ & & & & & 1 & 0 & 0 & 1 & 1 \end{array}\right] \end{array}.$$

This matroid has numerous attractive features (Bixby 1977, Seymour 1980b). It is clearly self-dual, although it is not identically self-dual. Moreover:

11.2.8 Lemma. *Let e be an element of R_{10}. Then $R_{10}\backslash e \cong M(K_{3,3})$ and $R_{10}/e \cong M^*(K_{3,3})$.*

Proof. Clearly the automorphism group of R_{10} is transitive on each of the sets $\{1,2,3,4,5\}$ and $\{6,7,8,9,10\}$. Moreover, pivoting on the first entry in column 6 of A_{10} and then swapping row 3 with row 4, and column 3 with column 4 gives the matrix

$$\begin{array}{ccccc|ccccc} 6 & 2 & 4 & 3 & 5 & 1 & 7 & 8 & 9 & 10 \end{array}$$
$$\left[\begin{array}{ccccc|ccccc} & & & & & 1 & 1 & 0 & 0 & 1 \\ & & & & & 1 & 0 & 1 & 0 & 1 \\ & & I_5 & & & 0 & 0 & 1 & 1 & 1 \\ & & & & & 0 & 1 & 1 & 1 & 0 \\ & & & & & 1 & 1 & 0 & 1 & 0 \end{array}\right].$$

Interchanging columns 7 and 10 now gives the matrix A_{10} with its columns relabelled. Hence R_{10} has an automorphism that interchanges 1 and 6. We conclude that the automorphism group of R_{10} is transitive. In view of this, to complete the proof of the lemma, it suffices to show that $R_{10}\backslash 10 \cong M(K_{3,3})$. To establish this, consider the matrix $A_{10}\backslash 10$. This matrix has the form $[I_5|X]$ where X is the fundamental-circuit incidence matrix of $M(K_{3,3})$ with respect to the basis $\{1,2,3,4,5\}$, the labelling of $K_{3,3}$ being as in Figure 11.8. As $M(K_{3,3})$ is binary, we conclude that $M(K_{3,3}) \cong M[I_5|X] = M[A_{10}\backslash 10] = R_{10}\backslash 10$. □

Part of the last proof showed that R_{10} has a transitive automorphism group. Another way to see this is to note that R_{10} can be represented over $GF(2)$ by the ten 5-tuples that have exactly three entries equal to one. We leave it to the reader to check this and, in addition, to show that the automorphism group of R_{10} is actually doubly transitive (Exercise 7).

On combining the last lemma with Theorem 6.6.4, we get the following result. The details of the proof are left to the reader.

11.2.9 Corollary. *R_{10} is regular.* □

As the next result shows, R_{10} is a very special member of the class of regular matroids for it is a splitter for this class (Seymour 1980b).

11.2.10 Proposition. *If M is a 3-connected regular matroid, then either M has no R_{10}-minor or $M \cong R_{10}$.*

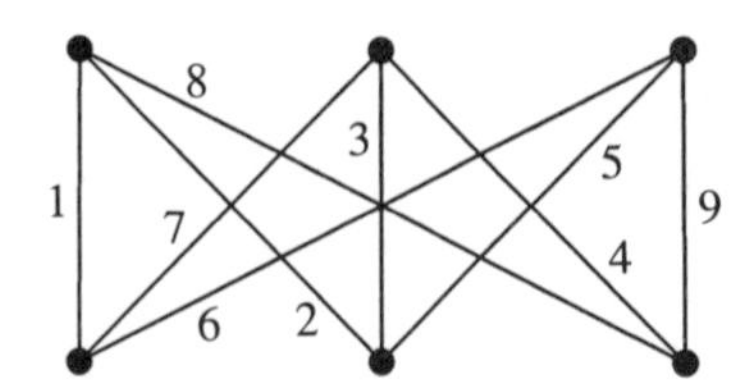

Fig. 11.8. $R_{10}\backslash 10$ is isomorphic to $M(K_{3,3})$.

Proof. By Corollary 11.1.3 and the fact that R_{10} is self-dual, it suffices to show that R_{10} has no 3-connected regular single-element extensions. Assume that R_{10} has such an extension M. Then M is represented over $GF(2)$ by a matrix of the form $A_{10} + e$. Moreover, by the symmetry of A_{10}, we may assume that e labels one of the following column vectors: $(1,0,1,x_4,x_5)^T$ for some x_4 and x_5 in $\{0,1\}$; $(1,1,0,0,0)^T$; or $(1,1,1,1,1)^T$. We leave it to the reader to check that, in the first case, $M/\{4,5\}\backslash\{9,10\} \cong F_7$; in the second case, $M/\{4,8\}\backslash\{2,7\} \cong F_7$; and in the third case, $M/\{1,3\}\backslash\{7,10\} \cong F_7$. We conclude that M is not regular; a contradiction. □

The terms 'extension' and 'coextension' have so far been used just for matroids. In our next application of the Splitter Theorem, these terms will also be used for graphs. In particular, if G is a graph and $T \subseteq E(G)$, then G will be called an *extension* of $G\backslash T$ and a *coextension* of G/T. We now look more closely at single-edge coextensions of graphs. Suppose that the graph H has been obtained from the graph G by contracting the edge e. Since contracting a loop in a graph has the same effect as deleting the loop, we shall assume that e has distinct endpoints, u and v, in G. Then, in H, the vertices u and v are identified as a new vertex w, and we shall say that G has been obtained from H by *splitting* the vertex w into the vertices u and v (Tutte 1984). Clearly the graph obtained by splitting a vertex need not be unique. For example, in Figure 11.9, both G_1 and G_2 can be obtained from K_5 by splitting w. We remark here, in case the use of similar terms causes confusion, that the notions of a 'splitter' for a class of matroids and 'splitting' a vertex in a graph are quite distinct.

The routine proof of the next result (Tutte 1961a) is left to the reader.

11.2.11 Lemma. *Let H be a simple 3-connected graph and suppose that the graph G is a single-edge coextension of H. Then G is simple and 3-connected if and only if G can be obtained from H by splitting a vertex of degree at least four into two vertices of degree at least three.* □

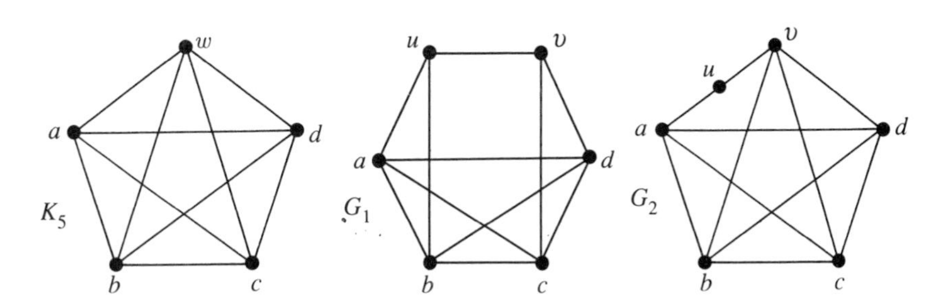

Fig. 11.9. Both G_1 and G_2 are obtained from K_5 by splitting w.

We show next how the Splitter Theorem can be used to prove a result of Hall (1943) which strengthens Kuratowski's Theorem in the case of 3-connected graphs. Recall that if G is a graph, then $\widetilde{G}$ denotes the associated simple graph.

11.2.12 Proposition. *If G is a 3-connected graph, then G has no $K_{3,3}$-minor if and only if either G is planar or $\widetilde{G} \cong K_5$.*

Proof. If $\widetilde{G} \cong K_5$, then G certainly has no $K_{3,3}$-minor. Moreover, by Proposition 5.2.6, if G is planar, it has no $K_{3,3}$-minor. To prove the converse, let $\mathcal{G}$ be the class of graphs having no $K_{3,3}$-minor and let $H = K_5$. Evidently H has no simple 3-connected single-element extensions. Moreover, by symmetry and Lemma 11.2.11, a simple 3-connected single-element coextension of K_5 is isomorphic to the graph G_1 in Figure 11.9. But $G_1 \backslash \{ab, cd\} \cong K_{3,3}$. We conclude by Corollary 11.1.4 that, for a 3-connected member G of $\mathcal{G}$, $\widetilde{G}$ has no K_5-minor or $\widetilde{G} \cong K_5$. Thus, by Proposition 5.2.6 again, either $\widetilde{G}$ is planar or $\widetilde{G} \cong K_5$. Finally, since G is planar if and only if $\widetilde{G}$ is planar, we have, as required, that either G is planar or $\widetilde{G} \cong K_5$. □

Leaving graphs for the moment, we return to matroids in general focusing again on Corollary 11.2.1 and its applications. In order to be able to use this result, we need not only a 3-connected matroid M but also some 3-connected minor N of M. The next result shows that one potential choice for N will always be a wheel or a whirl.

11.2.13 Proposition. *Every 3-connected matroid M having at least four elements has a minor isomorphic to $U_{2,4}$ or $M(K_4)$.*

Proof. If M is non-binary, it certainly has a $U_{2,4}$-minor. Thus suppose that M is binary. Then M has no whirl as a minor. If M has a wheel of rank at least three as a minor, it has an $M(K_4)$-minor. Hence we may assume that M has no such wheel as a minor. Now $|E(M)| \geq 4$. But, from Table 8.1 (p. 292), since M is binary, $|E(M)| \notin \{4,5\}$. Therefore $|E(M)| \geq 6$. As M has no 3-connected wheel or whirl as a minor, by repeated application of the Wheels and Whirls Theorem (11.2.2), we can construct a sequence $N_0, N_1, \ldots, N_k$ of 3-connected matroids with $N_0 = M$, $|E(N_k)| = 6$, and N_i a single-element deletion or a single-element contraction of N_{i-1} for all i in $\{1, 2, \ldots, k\}$. Since N_k is binary and 3-connected, it is simple and cosimple and cannot have rank or corank less than three. Thus $r(N_k) = 3 = r^*(N_k)$. Every rank-3 simple binary matroid on six elements can be obtained from $PG(2,2)$ by deleting a single element. But all such matroids are isomorphic to $M(K_4)$, so we obtain the contradiction that $N_k \cong M(K_4)$. □

We now note two corollaries of the last result. The first is immediate. The second shows that the non-empty connected matroids with no $U_{2,4}$- or $M(K_4)$-minor are precisely the cycle matroids of series–parallel networks.

11.2.14 Corollary. *Every 3-connected binary matroid having at least four elements has a minor isomorphic to $M(K_4)$.* □

11.2.15 Corollary. *Let M be a connected matroid having at least one element. Then M has no minor isomorphic to $U_{2,4}$ or $M(K_4)$ if and only if M is isomorphic to $M(G)$ for some series–parallel network G.*

Proof. If $M \cong M(G)$ where G is a series–parallel network, then M is binary and so has no $U_{2,4}$-minor. Moreover, by Corollary 5.4.12, M has no $M(K_4)$-minor. For the converse, suppose that M has no minor isomorphic to $U_{2,4}$ or $M(K_4)$. We shall argue by induction on $|E(M)|$ to show that M is isomorphic to the cycle matroid of a series–parallel network. Suppose first that $|E(M)| \in \{1,2,3\}$. Then, by Table 4.1 (p. 126), M is isomorphic to one of $U_{0,1}$, $U_{1,1}$, $U_{1,2}$, $U_{1,3}$, and $U_{2,3}$. As noted in Table 1.1 (p. 14), each of these matroids is graphic. Since none has an $M(K_4)$-minor, Corollary 5.4.12 implies that each is isomorphic to the cycle matroid of a series–parallel network.

Now suppose that $|E(M)| \geq 4$. Then by Proposition 11.2.13, M is not 3-connected. Thus by Theorem 8.3.1, $M = M_1 \oplus_2 M_2$ for some proper minors M_1 and M_2 of M. By the induction assumption, both M_1 and M_2 are graphic. Hence, by Corollary 7.1.23, so too is M. It follows by Corollary 5.4.12 again that M is isomorphic to the cycle matroid of a series–parallel network. The result now follows by induction. □

It is commonplace to use the term 'series–parallel network' both for the graph of such a network and for the cycle matroid of this graph, and we shall follow this practice. We remark here that, for some authors (for example, Seymour 1992), a series–parallel network is any matroid, connected or not, that has no $U_{2,4}$- or $M(K_4)$-minor.

Proposition 11.2.13 told us that every 3-connected matroid M with four or more elements has a minor isomorphic to $U_{2,4}$ or $M(K_4)$. We now use the Splitter Theorem to obtain some more explicit information about the minors of such a matroid M. Suppose that M has a $U_{2,4}$-minor but has no $\mathcal{W}^3$-minor. Then if $|E(M)| \geq 5$, it follows by the Splitter Theorem that M has a $U_{2,5}$- or a $U_{3,5}$-minor. The next result (Oxley 1989b) is obtained by extending this idea. The matroids Q_6 and P_6 have occurred many times earlier (see, for example, Figure 8.13).

11.2.16 Proposition. *The following statements are equivalent for a 3-connected matroid having rank and corank at least three:*

(i) *M has a $U_{2,5}$-minor.*

(ii) *M has a $U_{3,5}$-minor.*

(iii) *M has a minor isomorphic to P_6, Q_6, or $U_{3,6}$.*

Before proving this result, we note the following interesting corollary that comes from combining it with Theorem 10.3.1.

11.2.17 Corollary. *A 3-connected matroid having rank at least three is ternary if and only if it has no minor isomorphic to $U_{3,5}$, F_7, or F_7^*.* □

The proof of Proposition 11.2.16 will use the following.

11.2.18 Lemma. *Let N_1 be a 3-connected matroid having an element e such that $N_1/e \cong U_{2,k}$ for some $k \geq 5$. Then N_1 has a minor isomorphic to one of P_6, Q_6, and $U_{3,6}$.*

Proof. We argue by induction on k. If $k = 5$, then N_1 is a 3-connected coextension of $U_{2,5}$ and it is not difficult to check that N_1 is isomorphic to one of P_6, Q_6, and $U_{3,6}$. Thus the lemma is true for $k = 5$. Now assume it true for $k = n$ where $n \geq 5$ and let $k = n + 1$. Suppose that f is an element of $E(N_1/e)$. Then $N_1\backslash f/e \cong U_{2,n}$ and the induction assumption implies that, provided $N_1\backslash f$ is 3-connected, $N_1\backslash f$, and hence N_1, has a minor isomorphic to one of P_6, Q_6, and $U_{3,6}$. Thus we may assume that, for all elements f of $E(N_1) - \{e\}$, the matroid $N_1\backslash f$ is not 3-connected. But both N_1 and $N_1\backslash f/e$ are 3-connected and so it follows, by Proposition 8.1.10, that N_1 has a triad containing $\{e, f\}$. Hence every series class of $N_1\backslash e$ is non-trivial. As $r(N_1\backslash e) = 3$, it follows without difficulty that $|E(N_1\backslash e)| \leq 4$. This contradiction completes the proof of the lemma. □

Proof of Proposition 11.2.16. This is not difficult using 11.2.18 and the Splitter Theorem, and we leave the details to the reader. □

As a consequence of Theorem 9.1.5, Corollary 11.2.1, and Proposition 11.2.16, we have the following result (Oxley 1987a), the proof which is left to the reader.

11.2.19 Corollary. *A 3-connected non-binary matroid whose rank and corank exceed two has a minor isomorphic to one of $\mathcal{W}^3$, P_6, Q_6, and $U_{3,6}$.* □

On combining this with Corollary 11.2.14, we get the following.

11.2.20 Corollary. *A 3-connected matroid whose rank and corank exceed two has a minor isomorphic to one of $M(\mathcal{W}_3)$, $\mathcal{W}^3$, Q_6, P_6, and $U_{3,6}$.* □

This result (Walton 1981; Oxley 1984a) explains why the five specified 6-element matroids arise so frequently: any reasonably sized 3-connected matroid must have one of these five as a minor.

In view of the fundamental role played by wheels and whirls within the class of 3-connected matroids, it is natural to consider what can be said about the structure of a minor-closed class of matroids which avoids some small wheel or some small whirl. Previously we have sought, for a given minor-closed class, a complete list of the excluded minors for the class. We are now approaching this situation from the other end: first we specify the list of excluded minors and then we seek properties of the minor-closed class determined by this list. When these excluded minors are wheels or whirls, we already have characterizations of some of the classes of matroids that arise. By Theorem 9.1.5, if the smallest whirl $\mathcal{W}^2$ is the unique excluded minor, then the resulting class of matroids coincides with the class of binary matroids. If the excluded minors are the smallest whirl $\mathcal{W}^2$ and the smallest 3-connected wheel $M(\mathcal{W}_3)$, then, by Corollary 11.2.15, the only non-empty matroids we get are direct sums of series–parallel networks. Next we shall describe the structure of the class of matroids that have no minor isomorphic to $\mathcal{W}^2$ or $M(\mathcal{W}_4)$. Since every member of this class that is not 3-connected can be constructed from 3-connected members of the class by direct sums and 2-sums, it suffices to specify the 3-connected members of the class. In general, for a set $\{M_1, M_2, \ldots\}$ of matroids, $EX(M_1, M_2, \ldots)$ will denote the class of matroids having no minor isomorphic to any of $M_1, M_2, \ldots$.

The strategy that will be used to find the 3-connected members of $EX(\mathcal{W}^2, M(\mathcal{W}_4))$ is as follows. First we note that all 3-connected matroids with fewer than four elements are trivially in the class. Next we let M be a 3-connected member of the class having four or more elements. By Proposition 11.2.13, M has an $M(K_4)$-minor. Moreover, since M has no $M(\mathcal{W}_4)$-minor, it has no minor isomorphic to $M(\mathcal{W}_r)$ for any $r \geq 4$. Thus by Corollary 11.2.1, there is a sequence $M_0, M_1, \ldots, M_n$ of 3-connected matroids such that $M_0 \cong M(K_4)$, $M_n = M$, and, for all i in $\{1, 2, \ldots, n\}$, M_i is a single-element extension or a single-element coextension of M_{i-1}. Clearly each of $M_0, M_1, \ldots, M_n$ is binary. The unique 3-connected binary extension of $M(K_4)$ is F_7; by duality, the unique 3-connected binary coextension is F_7^*. Thus M_1 is F_7 or F_7^*. It now follows, by Lemma 11.2.4, that M_2 is $AG(3,2)$ or S_8. The fact that each of $\mathcal{W}^2$ and $M(\mathcal{W}_4)$ is self-dual means that $EX(\mathcal{W}^2, M(\mathcal{W}_4))$ is closed under duality. Moreover, as both $AG(3,2)$ and S_8 are self-dual, either M_3 or M_3^* is a binary 3-connected extension of S_8 or $AG(3,2)$. To determine the possible such extensions, we take matrices representing S_8 and $AG(3,2)$ over $GF(2)$ and consider what columns can be added to these matrices so as to avoid creating an $M(\mathcal{W}_4)$-minor. We are relying here on the unique representability of binary matroids. Continuing to analyse the sequence

$M_0, M_1, \ldots, M_n$ in this way, a pattern emerges and, from this, one can formulate and then prove the structure theorem stated below. The details of this proof can be found in Oxley (1987b).

Let r be an integer exceeding two and Z_r be the vector matroid of the following matrix over $GF(2)$:

$$\left[\begin{array}{c|ccccc|c} & & & & & a & b \\ & 0 & 1 & 1 & \cdots & 1 & 1 \\ & 1 & 0 & 1 & \cdots & 1 & 1 \\ I_r & 1 & 1 & 0 & \cdots & 1 & 1 \\ & \vdots & \vdots & \vdots & \ddots & \vdots & \vdots \\ & 1 & 1 & 1 & \cdots & 0 & 1 \end{array}\right].$$

Evidently $Z_3 \cong F_7$; $Z_4 \backslash b \cong AG(3,2)$; and $Z_4 \backslash a \cong S_8$.

11.2.21 Theorem. *Let M be a binary matroid. Then M is 3-connected and has no $M(\mathcal{W}_4)$-minor if and only if*

(i) $M \cong Z_r$, Z_r^*, $Z_r \backslash a$, *or* $Z_r \backslash b$ *for some* $r \geq 3$; *or*

(ii) $M \cong U_{0,0}$, $U_{0,1}$, $U_{1,1}$, $U_{1,2}$, $U_{1,3}$, *or* $U_{2,3}$. □

Theoretically, the technique used above of building up, an element at a time, from a wheel or a whirl could also be applied to find the structure of $EX(\mathcal{W}^2, M(\mathcal{W}_r))$ for any $r \geq 5$. But, even when $r = 5$, the number of possibilities to be considered is large and has so far defied analysis. Note, however, that the smaller class of regular matroids with no $M(\mathcal{W}_5)$-minor has been characterized (Oxley 1989a).

We now look at $EX(M(\mathcal{W}_3))$, the class of matroids obtained by excluding the smallest 3-connected wheel as a minor. As we shall see in Chapter 13, this class includes the class of gammoids and hence includes the class of transversal matroids. On the other hand, it contains no 3-connected binary matroids with more than three elements. Corollary 11.2.15 characterizes the connected binary members of $EX(M(\mathcal{W}_3))$. The next theorem (Oxley 1987c) characterizes the ternary members of $EX(M(\mathcal{W}_3))$. As in the last theorem, only the 3-connected such matroids need be specified. We note that, since the class of ternary matroids is $EX(U_{2,5}, U_{3,5}, F_7, F_7^*)$, and both F_7 and F_7^* have an $M(\mathcal{W}_3)$-minor, $EX(U_{2,5}, U_{3,5}, F_7, F_7^*) \cap EX(M(\mathcal{W}_3)) = EX(U_{2,5}, U_{3,5}, M(\mathcal{W}_3))$.

Two special matroids appear in the next result, namely the splitters for $EX(U_{2,5}, U_{3,5}, M(\mathcal{W}_3))$. One of these matroids is J, the 8-element rank-4 matroid for which a geometric representation is shown in Figure 10.4. The second such matroid is the vector matroid of the matrix A_{12} over $GF(3)$ where A_{12} is

$$\left[\begin{array}{c|cccccc} & 0 & 1 & 1 & 1 & 1 & 1 \\ & 1 & 0 & 1 & -1 & -1 & 1 \\ I_6 & 1 & 1 & 0 & 1 & -1 & -1 \\ & 1 & -1 & 1 & 0 & 1 & -1 \\ & 1 & -1 & -1 & 1 & 0 & 1 \\ & 1 & 1 & -1 & -1 & 1 & 0 \end{array}\right].$$

The last matroid is actually very well known in a slightly different context which we shall now describe. A *Steiner system* $S(t, k, v)$ is a pair $(S, \mathcal{D})$ where S is a v-element set and $\mathcal{D}$ is a collection of k-element subsets of S called *blocks* such that every t-element subset of S is contained in exactly one block. The blocks of such a Steiner system form a t-partition of S and hence, by Proposition 2.1.21, there is a paving matroid of rank $t+1$ on S that has $\mathcal{D}$ as its set of hyperplanes.

Now let E be the set of elements of $M[A_{12}]$ and $\mathcal{H}$ be its set of hyperplanes. Then the pair $(E, \mathcal{H})$ is an $S(5, 6, 12)$ (Coxeter 1958). Although, in general, there may be more than one Steiner system with a given set of parameters, t, k, and v, the Steiner system $S(5, 6, 12)$ is unique (Witt 1940). Moreover, this Steiner system has numerous attractive properties (see, for example, Cameron 1980). For example, its automorphism group is the Mathieu group M_{12}, which is 5-transitive, and the complement of every block is a block. From these observations, we deduce that both the set of circuits and the set of cocircuits of $M[A_{12}]$ equal $\mathcal{H}$, and so $M[A_{12}]$ is identically self-dual. Moreover, $M[A_{12}]$ has a 5-transitive automorphism group, that is, if $(e_1, e_2, \ldots, e_5)$ and $(f_1, f_2, \ldots, f_5)$ are ordered 5-tuples of distinct elements of $M[A_{12}]$, then there is an automorphism of $M[A_{12}]$ that, for all i, maps e_i to f_i.

We are now ready to characterize $EX(U_{2,5}, U_{3,5}, M(K_4))$. In this result and from now on, $S(5, 6, 12)$ will denote the paving matroid associated with this Steiner system.

11.2.22 Theorem. *A matroid M is 3-connected, ternary and has no $M(K_4)$-minor if and only if M is isomorphic to J, to $\mathcal{W}^r$ for some $r \geq 2$, or to a 3-connected minor of $S(5, 6, 12)$.* □

Extensions of this result to determine, say, the quaternary matroids with no $M(K_4)$-minor again run into the difficulty of large numbers of cases. However, the same technique has been used to characterize the quaternary matroids with no $M(K_4)$- or $\mathcal{W}^3$-minor (Oxley 1987c). We leave to the exercises consideration of some further results of this type.

Exercises

1. Let M be a binary matroid having no $M(K_4)$-minor. Show that M can be obtained by a sequence of direct sums and 2-sums starting with matroids on at most three elements.

2. Show that, for all prime powers q, there is at least one Steiner system $S(2, q+1, q^2+q+1)$ and at least one $S(2, q, q^2)$.

3. Show, without using Theorem 11.2.21, that $M(\mathcal{W}_4)$, $AG(3,2)$, and S_8 are the only 8-element binary 3-connected matroids.

4. Show that the hyperplanes of $AG(3,2)$ are the blocks of an $S(3,4,8)$ on $E(AG(3,2))$.

5. Show that a non-ternary 3-connected matroid for which the rank and corank exceed two has a minor isomorphic to one of F_7, F_7^*, P_6, Q_6, and $U_{3,6}$.

6. Characterize the ternary matroids with no $\mathcal{W}^3$-minor.

7. (Seymour 1980b) Show that:

   (i) R_{10} can be represented over $GF(2)$ by the ten 5-tuples that have exactly three entries equal to one.

   (ii) R_{10} has a doubly transitive automorphism group (p. 501).

8. Let $(S, \mathcal{D})$ be a Steiner system $S(t,k,v)$. If $x \in S$, let $\mathcal{D}_x = \{H - x : x \in H \in \mathcal{D}\}$. Show that:

   (i) $(S - x, \mathcal{D}_x)$ is an $S(t-1, k-1, v-1)$. This Steiner system is said to be *derived* from $(S, \mathcal{D})$.

   (ii) $\mathcal{D}_x$ is the set of hyperplanes of the matroid M/x where M is the matroid on S that has $\mathcal{D}$ as its set of hyperplanes.

9. (Acketa 1988) Show that the only 3-connected binary paving matroids are $U_{0,0}$, $U_{0,1}$, $U_{1,1}$, $U_{1,2}$, $U_{1,3}$, $U_{2,3}$, $M(K_4)$, F_7, F_7^*, and $AG(3,2)$.

10. (Oxley 1984a) Prove that a 3-connected matroid has a quad (Exercise 9, Section 9.1) if and only if both its rank and corank exceed two.

11. (Oxley 1988) Consider the following circuit elimination property: whenever C_1, C_2, and C_3 are circuits, none of which is contained in the union of the others, and e_1, e_2, and e_3 are distinct elements of $C_1 \cap C_2 \cap C_3$, there is a circuit contained in $(C_1 \cup C_2 \cup C_3) - \{e_1, e_2, e_3\}$. Show that:

(i) A matroid has this property if and only if it has no minor isomorphic to any of $U_{3,6}$, $\mathcal{W}^3$, P_6, or Q_6.

(ii) A 3-connected matroid whose rank and corank exceed two has the property if and only if it is binary.

12. (Seymour 1981c) Show that:

(i) Every binary matroid with no $AG(3,2)$- or S_8-minor can be obtained from regular matroids and copies of F_7 and F_7^* by a sequence of direct sums and 2-sums.

(ii) If G_1 is as shown in Figure 11.9, then every binary matroid with no F_7- or $M(G_1)$-minor can be obtained from copies of $M(K_5)$ and from binary matroids with no F_7- or $M(K_5)$-minor by a sequence of direct sums and 2-sums.

13. A graph is *outerplanar* if it is isomorphic to a plane graph in which every vertex is on the boundary of the infinite face. Show that:

(i) (Chartrand and Harary 1967) The following statements are equivalent for a graph G:

(a) G is outerplanar.

(b) G has no subgraph homeomorphic from K_4 or $K_{2,3}$.

(c) G has no minor isomorphic to K_4 or $K_{2,3}$.

(ii) (Seymour 1992) G has no $K_{2,3}$-minor if and only if every block of G is either outerplanar or has fewer than five vertices.

14. (Robertson and Seymour 1984) Let G be a 3-connected simple graph with no $(K_5\backslash e)$-minor and at least four vertices. Show that G is a wheel, or G is isomorphic to $K_{3,3}$ or $(K_5\backslash e)^*$.

15. (Oxley 1989b)

(i)* Let M be a 3-connected matroid having a $U_{2,5}$-minor. Show that either M is uniform or M has a minor isomorphic to P_6 or Q_6.

(ii) Deduce the following extension of 11.2.9: A non-binary 3-connected matroid whose rank and corank exceed two is either uniform or has a minor isomorphic to $\mathcal{W}^3$, P_6, or Q_6.

(iii) Show the following are equivalent for a 3-connected matroid M:

(a) $M \in EX(M(K_4), P_6, Q_6) - EX(M(K_4), U_{2,5}, U_{3,5})$.

(b) M is a non-binary member of $EX(\mathcal{W}^3, P_6, Q_6)$ having rank and corank exceeding two.

(c) M is uniform having rank and corank exceeding two.

16. (i)* (Seymour 1980b) Show that $M(K_5)$ is a splitter for the class of regular matroids with no minor isomorphic to $M(K_{3,3})$.

(ii) (Walton and Welsh 1980) Deduce that $M(K_5)$ is a splitter for each of the classes $EX(U_{2,4}, F_7, M(K_{3,3}))$ and $EX(U_{2,4}, F_7^*, M(K_{3,3}))$.

17.* (Oxley 1991) Show that the 3-connected ternary paving matroids are precisely the 3-connected minors of $PG(2,3)$, $S(5,6,12)$, R_8, and T_8 where the last two matroids were introduced in Exercise 8 of Section 2.2.

11.3 Variations on the Splitter Theorem

The Splitter Theorem has been extended in several directions, some of which we shall discuss in this section. Most of the results here will not be proved. We begin by noting that the restriction on excluding wheels and whirls can be weakened so that instead of applying to all such matroids, it applies only to the smallest 3-connected wheels and whirls (Coullard 1985). The result here is stated in the case when N is 3-connected since the more general case, when N is connected, simple, and cosimple, is covered by the Splitter Theorem.

11.3.1 Proposition. *Let N be a 3-connected proper minor of a 3-connected matroid M such that $|E(N)| \geq 4$ and M is not a wheel or a whirl. Suppose that if $N \cong \mathcal{W}^2$, then M has no $\mathcal{W}^3$-minor, while if $N \cong M(\mathcal{W}_3)$, then M has no $M(\mathcal{W}_4)$-minor. Then M has a 3-connected minor M_1 and an element e such that $M_1 \backslash e$ or M_1 / e is isomorphic to N.*

Proof. This can be found in Coullard and Oxley (1992). □

We noted at the end of Section 11.1 that the Splitter Theorem cannot be strengthened to require that the minor M_1 of M be a single-element extension or coextension of N itself. Indeed, if M and N are as given in Exercise 7 of Section 11.1, then, for every 3-connected minor M_2 of M having N as a minor, $|E(M_2) - E(N)| \geq 3$. This example prompts the question as to, in general, how large the gap can be between N and a next largest 3-connected minor of M having N as a minor. Truemper (1984a) answered this question by showing that this gap has size at most three.

This result was strengthened slightly by Bixby and Coullard (1986) who proved the following result.

11.3.2 Theorem. *Let N be a 3-connected proper minor of a 3-connected matroid M. Then M has a 3-connected minor M_1 and an element e such that N is a cosimple matroid associated with $M_1\backslash e$ or a simple matroid associated with M_1/e, and $|E(M_1) - E(N)| \leq 3$.* □

In fact, Bixby and Coullard proved a somewhat stronger result than this since they analysed more closely what happens when $|E(M_1)-E(N)|$ takes the values 2 and 3. For an explicit statement of their result, the reader is referred to Coullard and Oxley (1992). Alternatively, this result can be deduced from Theorem 11.3.6, which will be stated later in the section.

In the case when N has no circuits or cocircuits of size less than four, Bixby and Coullard (1986) improved the bound in the last theorem by one.

11.3.3 Proposition. *Let N be a 3-connected minor of a 3-connected matroid M and suppose that N has no circuits or cocircuits with fewer than four elements. Then there is a sequence $M_0, M_1, \ldots, M_n$ of 3-connected matroids such that $M_0 = N$, $M_n = M$, and, for all i in $\{1, 2, \ldots, n\}$, $|E(M_i)-E(M_{i-1})| \leq 2$ and there is an element e_i of M_i such that M_{i-1} is a cosimple matroid associated with $M_i\backslash e_i$ or a simple matroid associated with M_i/e_i.* □

Next we consider a variant of Theorem 11.3.2 in which we seek a 3-connected minor M_1 of M that not only has an N-minor but also contains some nominated element e of $E(M) - E(N)$. We leave it to the reader to show (Exercise 9) using Proposition 4.3.6 that such an M_1 can be found with $|E(M_1)| - |E(N)| \leq 1$. Note that this allows the N-minor of M_1 to be an isomorphic copy of N. If we require this N-minor to be N itself, then we have the following result, also due to Bixby and Coullard (1987).

11.3.4 Proposition. *Let N be a 3-connected minor of a 3-connected matroid M. Suppose that $|E(N)| \geq 4$ and $e \in E(M) - E(N)$. Then M has a 3-connected minor M_1 such that $e \in E(M_1)$, N is a minor of M_1, and $|E(M_1) - E(N)| \leq 4$.* □

Returning to the situation of seeking a 3-connected single-element extension or coextension of some isomorphic copy N_1 of N, we now add the requirement that N_1 uses some nominated element e of N in the same way that e is used in N. To make the last idea precise, we introduce the following concept. Let M_1 and M_2 be matroids and e be an element that is in both $E(M_1)$ and $E(M_2)$. Then M_1 and M_2 are *e-isomorphic* if there is an isomorphism between M_1 and M_2 under which e is fixed. The following result is due to Tseng and Truemper (1986).

11.3.5 Theorem. *Let N be a 3-connected proper minor of a 3-connected matroid M. Suppose that $|E(N)| \geq 4$ and e is an element of N. Then M has a 3-connected minor M_1 which has a minor N_1 that is e-isomorphic to N such that either*

(i) $|E(M_1) - E(N_1)| = 1$; *or*

(ii) $N_1 \cong M(\mathcal{W}_n)$ *for some* $n \geq 3$ *and* $M_1 \cong M(\mathcal{W}_{n+1})$; *or*

(iii) $N_1 \cong \mathcal{W}^n$ *for some* $n \geq 2$ *and* $M_1 \cong \mathcal{W}^{n+1}$. □

An extension of the last result in which the restriction on $|E(N)|$ is dropped will be noted in Exercise 10. Proposition 11.3.4 is actually a simplified version of another result of Bixby and Coullard (1987), which we state next. The detailed information provided by this theorem makes it useful in proving a number of results for 3-connected matroids. We shall say that a matroid M *uses* a set T if $T \subseteq E(M)$.

11.3.6 Theorem. *Let N be a 3-connected minor of a 3-connected matroid M. Suppose that $|E(N)| \geq 4$, $e \in E(M) - E(N)$, and M has no 3-connected proper minor that both uses e and has N as a minor. Then, for some (N_1, M_1) in $\{(N, M), (N^*, M^*)\}$, one of the following holds where $|E(M) - E(N)| = n$:*

(i) $n = 1$ *and* $N_1 = M_1 \backslash e$.

(ii) $n = 2$, $N_1 = M_1 \backslash e/f$, *and N_1 has an element x such that* $\{e, f, x\}$ *is a triangle of* M_1.

(iii) $n = 3$, $N_1 = M_1 \backslash e, g/f$, *and N_1 has an element x such that* $\{e, f, x\}$ *is a triangle of M_1 and* $\{f, g, x\}$ *is a triad of M_1. Moreover,* $M_1 \backslash e$ *is 3-connected.*

(iv) $n = 3$, $N_1 = M_1 \backslash e, g/f = M_1 \backslash e, f/g = M_1 \backslash f, g/e = M_1 \backslash e, f, g$, *and* $\{e, f, g\}$ *is a triad of M_1. Moreover, N_1 has distinct elements x and y such that* $\{e, g, x\}$ *and* $\{e, f, y\}$ *are triangles of* M_1.

(v) $n = 4$, $N_1 = M_1 \backslash e, g/f, h$ *and N_1 has an element x such that* $\{e, f, x\}$ *and* $\{g, h, x\}$ *are triangles of M_1, and* $\{f, g, x\}$ *is a triad of M_1. Moreover,* $M_1 \backslash e$ *and* $M_1 \backslash e/f$ *are 3-connected.* □

Although the last result applies to all matroids and not just graphic ones, Bixby and Coullard use graphs to depict what happens in (iii)–(v) (see Figure 11.10). Note that a vertex is circled if it corresponds to a

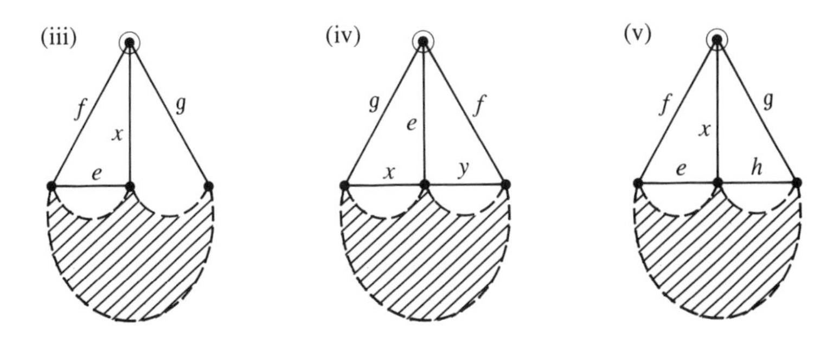

Fig. 11.10. A graphic depiction of 11.3.6(iii)–(v).

known triad in the matroid; all cycles shown are indeed circuits of the matroid; and the shaded part of the diagram corresponds to the rest of the matroid.

The applications of the last theorem that we shall describe next are concerned with relating certain minors in a matroid to particular elements of the matroid. Interest in results of this type began with some results for non-binary matroids, the first of which was proved by Bixby (1974).

11.3.7 Proposition. *Let M be a 2-connected matroid having a $U_{2,4}$-minor and suppose that $e \in E(M)$. Then M has a $U_{2,4}$-minor using e.*

Proof. Let N be a $U_{2,4}$-minor of M and suppose that $e \notin E(N)$. Then by Corollary 4.3.7, M has a 2-connected minor M_1 such that $M_1\backslash e$ or M_1/e is isomorphic to N. As N is self-dual, we may assume the former. But then either M_1 is a 5-point line, or M_1 is a parallel extension by e of a 4-point line. In both cases, it follows easily that M has a $U_{2,4}$-minor using e. □

It is not difficult to see that without the requirement that M is 2-connected, the last result may fail. However, Seymour (1981b) showed that if one imposes a stronger condition on the connectivity of M, then one can strengthen the conclusion as follows.

11.3.8 Proposition. *Let M be a 3-connected matroid having a $U_{2,4}$-minor and suppose that e and f are distinct elements of M. Then M has a $U_{2,4}$-minor using $\{e, f\}$.* □

We remark here that the last result, which can be verified straightforwardly using Theorem 11.3.6, is a basic tool in the proof of Kahn's result for unique representability over $GF(4)$ (Proposition 10.1.11).

The last two results prompt several questions including whether the analogue of these results holds for 4-connected matroids, and whether $\{U_{2,4}\}$ is the only set of matroids with the properties noted there. To formalize these questions, we introduce the following definition (Seymour 1985a). Let t be a positive integer. A class $\mathcal{N}$ of matroids is *t-rounded* if every member of $\mathcal{N}$ is $(t+1)$-connected and the following condition holds: If M is a $(t+1)$-connected matroid having an $\mathcal{N}$-minor and X is a subset of $E(M)$ with at most t elements, then M has an $\mathcal{N}$-minor using X.

Restated using this terminology, Propositions 11.3.7 and 11.3.8 assert that $\{U_{2,4}\}$ is both 1- and 2-rounded. In general, the task of checking whether a given class of matroids is t-rounded is potentially infinite. However, Seymour (1977a, 1985a) showed that in the two most frequently studied cases, when t is 1 or 2, this task is finite.

11.3.9 Theorem. *Let t be 1 or 2 and $\mathcal{N}$ be a collection of $(t+1)$-connected matroids. Then $\mathcal{N}$ is t-rounded if and only if the following condition holds: If M is a $(t+1)$-connected matroid having an $\mathcal{N}$-minor N such that $|E(M) - E(N)| = 1$, and X is a subset of $E(M)$ with at most t elements, then M has an $\mathcal{N}$-minor using X.*

Proof. For $t = 1$, the proof is similar to that of Proposition 11.3.7. We leave the details here to the reader, as we do for the case when $t = 2$, a proof of which can be derived from Theorem 11.3.6. □

Using this theorem it is not difficult to show that each of the following collections of matroids is 2-rounded: $\{U_{2,4}, M(K_4)\}$, $\{U_{2,4}, F_7, F_7^*, S_8\}$ (Seymour 1985a); $\{U_{2,4}, M(\mathcal{W}_4)\}$ (Oxley and Reid 1990); $\{U_{3,6}, P_6, Q_6, \mathcal{W}^3\}$, and $\{\mathcal{W}^3, P_6, Q_6\}$ (Oxley 1984a, 1989b). Since a 2-rounded set is also 1-rounded (Exercise 2), all these collections are 1-rounded as are, for example, $\{U_{2,5}, U_{3,5}, F_7, F_7^*\}$, $\{U_{2,4}, F_7, F_7^*\}$, and $\{U_{2,4}, F_7, F_7^*, M^*(K_5), M^*(K_{3,3}), M^*(K'_{3,3})\}$ (Seymour 1977a). In the last of these sets, $K'_{3,3}$ denotes the graph that is obtained from $K_{3,3}$ by adding an edge joining two non-adjacent vertices.

We began this discussion of roundedness with some questions arising from Propositions 11.3.7 and 11.3.8. We now answer these questions. Firstly, suppose that N is a matroid having at least four elements. Then $\{N\}$ is 2-rounded if and only if $N \cong U_{2,4}$, while $\{N\}$ is 1-rounded if and only if $N \cong U_{2,4}$, Q_6, or $M(\mathcal{W}_2)$ (Oxley 1984b). Secondly, we observe that Kahn (1985) and Coullard (1986) independently showed that $\{U_{2,4}\}$ is not 3-rounded thereby disproving a conjecture of Seymour (1981b). In light of their result and Proposition 11.3.8, it is natural to try to characterize when a set of three elements in a 4-connected non-binary matroid is not in a $U_{2,4}$-minor. This problem is still open although a partial answer to it is provided by the following result for non-binary 3-connected matroids

(Oxley 1987a). A proof of this result can be obtained by using Theorem 11.3.6 and Proposition 11.3.8.

11.3.10 Proposition. *Let $\{x, y, z\}$ be contained in a 3-connected matroid M that has a $U_{2,4}$-minor. If M has no $U_{2,4}$-minor using $\{x, y, z\}$, then M has a $\mathcal{W}_3$-minor that uses $\{x, y, z\}$ as its rim or its spokes.* □

To conclude this section, we follow Seymour (1978, 1992) in sketching how the type of structural results considered above can be used to prove a matroid extension of the edge form of Menger's Theorem. The statement of Menger's Theorem given in Chapter 8 is sometimes called the vertex form of the theorem. The companion edge form (Menger 1927) asserts that if u and v are distinct vertices in a graph G, then the minimum number of edges whose removal from G leaves u and v in different components is equal to the maximum number of edge-disjoint paths joining u and v. Now suppose that we add a new edge e to G joining u and v and let the resulting graph be $G + e$. Then, for $M = M(G + e)$, the edge form of Menger's Theorem asserts that M has the following property.

11.3.11 *The minimum size of a cocircuit of M containing e is one more than the maximum size of a set $\mathcal{P}$ of circuits in M such that each member of $\mathcal{P}$ contains e, and any two distinct members of $\mathcal{P}$ meet in $\{e\}$.*

This formulation of Menger's Theorem suggests the problem of determining all pairs (M, e) such that e is a non-loop element of a matroid M for which 11.3.11 holds. One difficulty with this problem is that the class of matroids satisfying 11.3.11 is not minor-closed. For example, although 11.3.11 holds for $(\mathcal{W}^3, e)$ when e is a spoke of $\mathcal{W}^3$, it does not hold for $(\mathcal{W}^2, e)$ when e is an arbitrary element of $\mathcal{W}^2$.

This difficulty suggests that one should seek another reformulation of Menger's Theorem, one which will yield a matroid condition that is preserved under the taking of minors. We shall now describe such a reformulation due to Seymour (1977b). This was obtained be generalizing the statement of the max-flow min-cut theorem, the latter being essentially a digraph version of Menger's Theorem.

Let e be an element of an arbitrary matroid M and suppose that a non-negative integer capacity $c(f)$ is assigned to each element f of $E(M) - e$. Assume that

11.3.12 *M has a multiset $\{C_1, C_2, \dots, C_k\}$ of circuits each of which contains e such that every element f of $E(M) - e$ is in at most $c(f)$ members of this multiset.*

Then it is not difficult to show (see Exercise 6(i)) that the following does not hold.

11.3.13 *M has a cocircuit C^* containing e such that $\sum_{f\in C^*-e} c(f) < k$.*

The matroid M is said to have the *integer max-flow min-cut* ($\mathbb{Z}^+$-*MFMC*) *property* with respect to the element e if, for all positive integers k and all integer-valued functions c on $E(M) - e$, exactly one of 11.3.12 and 11.3.13 holds.

It is straightforward to show (see Exercises 5(i) and 6(ii)) that the edge form of Menger's Theorem is equivalent to the assertion that if e is an element of a graphic matroid M, then M has the $\mathbb{Z}^+$-MFMC property with respect to e. We note that e can even be a loop here since, in that case, 11.3.13 fails while 11.3.12 holds. If M is an arbitrary matroid having the $\mathbb{Z}^+$-MFMC property with respect to an element e and N is a minor of M using e, then we leave it to the reader to check (Exercise 6(iii)) that N has the $\mathbb{Z}^+$-MFMC property with respect to e. Thus we have obtained a reformulation of Menger's Theorem of the type we were seeking. We are now faced with the problem of determining precisely when a matroid has the $\mathbb{Z}^+$-MFMC property with respect to a specified element. This problem, which is very difficult, was solved by Seymour (1977b).

11.3.14 Theorem. *A matroid M has the $\mathbb{Z}^+$-MFMC property with respect to an element e if and only if M has no $U_{2,4}$- or F_7^*-minor using e.* □

As an alternative to his original 'long and cumbersome' proof of this theorem, Seymour (1992) has outlined a proof which relies mainly on the type of structural results that we have been considering in this section. In particular, this proof uses Proposition 11.3.7; Gallai (1959) and Minty's (1966) result that every regular matroid has the $\mathbb{Z}^+$-MFMC property with respect to each of its elements; and the following structural theorem of Tseng and Truemper (1986).

11.3.15 Theorem. *Let e be an element of a 3-connected, internally 4-connected binary matroid M and suppose that $e \in E(M)$. Then exactly one of the following holds:*

(i) *There is an F_7^*-minor of M using e.*

(ii) *M is regular.*

(iii) $M \cong F_7$. □

The original proof of this theorem used Theorem 11.3.5. Bixby and Rajan (1989) have given a shorter proof that relies on some further structural results of Truemper (1986). Seymour (1978, 1981c) discusses numerous results and problems related to 11.3.14 and proves several other interesting structural results. Both of the last two papers are rich in open conjectures.

Finally, we note that on combining Theorem 11.3.14 with Corollary 11.2.5, we immediately obtain the following.

11.3.16 Corollary. *A matroid M has the $\mathbb{Z}^+$-MFMC property with respect to every element if and only if M can be obtained from regular matroids and copies of F_7 by a sequence of direct sums and 2-sums.* □

Exercises

1. Find all non-binary 3-connected matroids M such that, for some element e, either $M\backslash e$ or M/e is isomorphic to $M(\mathcal{W}_3)$.

2. Prove that if a class of matroids is 2-rounded, then it is also 1-rounded.

3. Show that both of the classes $\{Q_6\}$ and $\{M(\mathcal{W}_2)\}$ are 1-rounded but not 2-rounded.

4. (Oxley and Reid 1990) Show that $\{U_{2,4}, M(\mathcal{W}_n)\}$ is 2-rounded if and only if $n \in \{3, 4\}$.

5. **(i)** Show that if a matroid M has the $\mathbb{Z}^+$-MFMC property with respect to an element e, then e is a loop or 11.3.11 holds.

   (ii) Deduce that 11.3.11 holds if M is a regular matroid and e is a non-loop element of M.

   (iii) Show that the converse of (i) fails.

6. **(i)** By considering the set of pairs (f, C_i) such that f is an element of M that is in C_i, prove that if 11.3.12 holds, then 11.3.13 fails.

   (ii) Let e be a non-loop edge of a graph G. Suppose that c is a non-negative integer-valued function defined on $E(G)-e$ and k is a positive integer such that 11.3.13 does not hold for $M = M(G)$. Use the following technique to show that M satisfies 11.3.12: for each edge f in $E(G) - e$ with $c(f) > 0$, replace f by $c(f)$ edges in parallel; then delete all those edges f with $c(f) = 0$; finally use the fact that 11.3.11 holds in the cycle matroid of the resulting graph.

   (iii) Prove that if M is an arbitrary matroid having the $\mathbb{Z}^+$-MFMC property with respect to an element e and N is a minor of M using e, then N also has the $\mathbb{Z}^+$-MFMC property with respect to e.

7. Prove that a 3-connected matroid M has every element in both a triangle and a triad if and only if M is isomorphic to a wheel of rank at least three or a whirl of rank at least two.

8. (Tan 1981; Truemper 1984a) Let N be a 3-connected minor of a 3-connected matroid M such that $|E(N)| \geq 4$. Show that there is a sequence $N_0, N_1, \ldots, N_n$ of 3-connected matroids, each a minor of its successor, such that $N_0 \cong N$, $N_n = M$, and, for all i in $\{0, 1, \ldots, n-1\}$, $|E(N_{i+1}) - E(N_i)| \leq 2$ with equality only if N_{i+1} and N_i are both wheels or are both whirls.

9. Let N be a 3-connected minor of a 3-connected matroid M and suppose that $e \in E(M) - E(N)$. Show that M has a 3-connected minor M_1 such that M_1 has an N-minor, $e \in E(M_1)$, and $|E(M_1)| - |E(N)| \leq 1$.

10. (Coullard in Oxley and Row 1989) Prove the following extension of Theorem 11.3.5: Let N be a 3-connected proper minor of a 3-connected matroid M and suppose that $e \in E(N)$. Then, provided M is neither a wheel nor a whirl of rank at least three, there is an element f of M such that some member M_0 of $\{M\backslash f, M/f\}$ is 3-connected and has a minor N_0 that is e-isomorphic to N.

11. **(i)** (Oxley 1987a) Let M be a 3-connected non-binary matroid such that, for some element e, both $M\backslash e$ and M/e are binary. Prove that $M \cong U_{2,4}$.

(ii) Let M be a 3-connected binary matroid having an $M(K_4)$-minor. Prove that if M has an element e such that neither $M\backslash e$ nor M/e has an $M(K_4)$-minor, then $M \cong M(K_4)$.

12.* (Reid 1988) Prove that there is no matroid M such that $\{M\}$ is 3-rounded.

13. (Coullard and Oxley 1992) Let n be an integer exceeding two. Consider the following assertion: Every 3-connected matroid M with a $\mathcal{W}^2$- and an $M(\mathcal{W}_n)$-minor has a 3-connected non-binary minor N and an element e such that $N\backslash e$ or N/e is isomorphic to $M(\mathcal{W}_n)$.

(i)* Prove the assertion is false for all $n \geq 5$.

(ii)* Prove the assertion is true for $n = 3$.

(iii)° Is the assertion true or false for $n = 4$?

(iv) Find an 8-element 3-connected non-ternary matroid M having an $M(\mathcal{W}_3)$-minor such that M has no proper minor that is 3-connected, non-ternary, and has an $M(\mathcal{W}_3)$-minor.

14.* (Reid 1991) Let $\{e, f, g\}$ be a circuit of a 3-connected binary matroid M and N be a 3-connected minor of M that uses e. Prove that M has a 3-connected minor M_1 using $\{e, f, g\}$ such that M_1 has an N-minor and $|E(M_1)| \leq |E(N)| + 2$.

12

Submodular functions and matroid union

A number of matroid constructions were discussed in Chapter 7. In this chapter, we shall consider several more such constructions. The common thread here is the association of these constructions to submodular functions where, for a set E, a function f from 2^E into $\mathbb{R}$ is *submodular* if $f(X) + f(Y) \geq f(X \cup Y) + f(X \cap Y)$ for all subsets X and Y of E. Such a function is *increasing* if $f(X) \leq f(Y)$ whenever $X \subseteq Y \subseteq E$. In Section 12.1, we shall prove a result of Edmonds and Rota (1966) which associates a matroid on E to every increasing submodular function on 2^E. Section 12.2 considers several applications of submodular functions, one of which is in the proof of Hall's Marriage Theorem, a result that specifies precisely when a family of sets has a transversal.

The most important of the matroid operations we shall consider here is that of matroid union. This operation will be discussed in Section 12.3 along with several of its attractive applications. Finally, in Section 12.4, we shall consider the problem of sticking two matroids together across a common restriction. Much of the discussion there will be focused on a generalization of the operation of parallel connection.

The treatment given here of the topics in Sections 12.1–12.3 is similar to that found in Chapters 7 and 8 of Welsh (1976) and the reader is referred to that book for a more detailed exposition of some of these topics. We remark here that several authors including Crapo and Rota (1970) refer to submodular functions as *semimodular functions.*

12.1 Submodular functions

For any matroid M on a set E, the rank function is an increasing submodular function on 2^E. Moreover, $\mathcal{C}(M) = \{C \subseteq E\colon C$ is minimal and non-empty such that $r(C) < |C|\}$. The next result (Edmonds and Rota 1966) generalizes this observation.

12.1.1 **Proposition.** *Let f be an increasing submodular function from 2^E into $\mathbb{Z}$. Let $\mathcal{C}(f) = \{C \subseteq E$: C is minimal and non-empty such that $f(C) < |C|\}$. Then $\mathcal{C}(f)$ is the collection of circuits of a matroid on E.*

Proof. Evidently $\mathcal{C}(f)$ satisfies (C1) and (C2). Now suppose that C_1 and C_2 are distinct members of $\mathcal{C}(f)$ such that $e \in C_1 \cap C_2$. If $i \in \{1,2\}$, then $|C_i - e| \leq f(C_i - e)$. Since f is increasing and $C_i \in \mathcal{C}_f$, it follows that $|C_i - e| \leq f(C_i) < |C_i|$. Thus

$$f(C_i) = |C_i| - 1. \tag{1}$$

To complete the proof that $\mathcal{C}(f)$ satisfies (C3), we need only show that

$$f((C_1 \cup C_2) - e) < |(C_1 \cup C_2) - e|, \tag{2}$$

for it follows from this that $(C_1 \cup C_2) - e$ contains a member of $\mathcal{C}(f)$. Since f is increasing and submodular,

$$f((C_1 \cup C_2) - e) \leq f(C_1 \cup C_2) \leq f(C_1) + f(C_2) - f(C_1 \cap C_2).$$

Moreover, as $C_1 \cap C_2 \subsetneqq C_1$, it follows that $f(C_1 \cap C_2) \geq |C_1 \cap C_2|$. Thus, by (1),

$$\begin{aligned} f((C_1 \cup C_2) - e) &\leq |C_1| - 1 + |C_2| - 1 - f(C_1 \cap C_2) \\ &\leq |C_1| - 1 + |C_2| - 1 - |C_1 \cap C_2| \\ &= |C_1 \cup C_2| - 2 = |(C_1 \cup C_2) - e| - 1. \end{aligned}$$

Hence (2) holds and the proposition is proved. □

Let $M(f)$ denote the matroid on E that has $\mathcal{C}(f)$ as its set of circuits. We shall say that $M(f)$ is *induced* by f. The next result is a straightforward consequence of the preceding proposition.

12.1.2 **Corollary.** $\mathcal{I}(M(f)) = \{I \subseteq E : |I'| \leq f(I')$ *for all non-empty subsets* I' *of* $I\}$. □

12.1.3 **Example.** Let G be a graph and E be its edge set. For all subsets X of E, denote $V(G[X])$ by $V(X)$ and let

$$f_{-1}(X) = |V(X)| - 1.$$

Certainly f_{-1} is integer-valued and one easily shows that it is increasing. Moreover, if $X, Y \subseteq E$, then

$$\begin{aligned} f_{-1}(X) + f_{-1}(Y) &= |V(X)| - 1 + |V(Y)| - 1 \\ &= |V(X) \cup V(Y)| - 1 + |V(X) \cap V(Y)| - 1. \end{aligned}$$

Fig. 12.1. The subdivisions of these graphs are the circuits of $B(G)$.

But $V(X) \cup V(Y) = V(X \cup Y)$ and $V(X) \cap V(Y) \supseteq V(X \cap Y)$. Hence

$$f_{-1}(X) + f_{-1}(Y) \geq f_{-1}(X \cup Y) + f_{-1}(X \cap Y).$$

Thus f_{-1} is submodular. Therefore f_{-1} induces a matroid on E, and it is not difficult to check that this matroid is precisely $M(G)$.

Next consider the function f_0 defined, for all subsets X of E, by

$$f_0(X) = |V(X)| = f_{-1}(X) + 1.$$

It is an immediate consequence of the next result, the proof of which is left as an exercise, that f_0 is integer-valued, increasing, and submodular.

12.1.4 Lemma. *Let f be an integer-valued, increasing, submodular function on 2^E and, for all $X \subseteq E$, let $a(X) = f(X)+1$. Then a is also integer-valued, increasing, and submodular.* □

In order to characterize the circuits of the matroid $B(G)$ induced on E by f_0, we shall use the following elementary graph-theoretic result. Recall that $\omega(G)$ denotes the number of connected components of the graph G.

12.1.5 Lemma. *A graph G has a unique cycle if and only if $|E(G)| = |V(G)| - \omega(G) + 1$.* □

Every circuit of $B(G)$ is easily derived from one of the graphs in Figure 12.1. Specifically:

12.1.6 Proposition. *A subset C of $E(G)$ is a circuit of $B(G)$ if and only if $G[C]$ is a subdivision of one of the three graphs in Figure 12.1.*

Proof. It is routine to check that if $G[C]$ is a subdivision of one of the specified graphs, then $|C| > |V(C)| = f_0(C)$, so C is dependent in $B(G)$.

Now let C be a circuit in $B(G)$. Then C is a minimal non-empty set such that $|V(C)| < |C|$. Hence $|C| \neq 1$, so, for all x in C, the set $C - x$ is non-empty, and therefore

$$|V(C - x)| \geq |C - x| = |C| - 1.$$

It follows that, for all such x,

$$|V(C)| = |V(C-x)| = |C| - 1. \tag{3}$$

Now assume that C has a proper non-empty subset X such that $G[X]$ is a component of $G[C]$. Then, as every proper subset of C is independent,

$$|V(C)| = |V(X)|+|V(C-X)| = f_0(X)+f_0(C-X) \geq |X|+|C-X| = |C|.$$

This contradiction implies that $G[C]$ is connected.

By the choice of C, every vertex of $G[C]$ has degree at least two. But $2|C| = \sum_{v \in V(G)} d(v)$. Therefore, by (3),

$$|V(C)| = |C| - 1 = \tfrac{1}{2}[\sum_{v \in V(C)} d(v) - 2].$$

Hence, as $d(v) \geq 2$ for all v, it follows that every vertex of $G[C]$ has degree two except that there is either (i) exactly one vertex of degree four, or (ii) exactly two vertices of degree three. In case (i), it is easy to show that $G[C]$ is a subdivision of the first graph in Figure 12.1. In the second case, $G[C]$ certainly has a path joining the two vertices of degree three, and from this it is straightforward to show that $G[C]$ is a subdivision of the second or third graph in Figure 12.1.

We conclude that every circuit in $B(G)$ is a subdivision of one of the graphs in Figure 12.1. Moreover, as the edge set of every graph that is a subdivision of one of these three graphs is dependent in $B(G)$, every such edge set is a circuit in $B(G)$. □

We call $B(G)$ the *bicircular matroid* of G. Some of the attractive properties of such matroids will be considered in the exercises. As examples, note that $B(K_4) \cong U_{4,6}$, while if G is the graph consisting of two vertices joined by n parallel edges for some $n \geq 2$, then $B(G) \cong U_{2,n}$. □

The increasing submodular functions considered so far have been allowed to take arbitrary integer values on the empty set. If we now insist that $f(\emptyset) = 0$ for such a function f, then we can specify the rank function of $M(f)$ in terms of f (Edmonds and Rota 1966). Sometimes such a function f is known as an *integer polymatroid*, or as an *integer polymatroid function*. Detailed discussion of polymatroids and their properties can be found in Welsh (1976, Chapter 18) and Lovász and Plummer (1986, Chapter 11).

12.1.7 Proposition. *Let f be an integer polymatroid on 2^E. If $X \subseteq E$, then its rank $r_f(X)$ in $M(f)$ is given by*

$$r_f(X) = \min\{f(Y) + |X - Y| : Y \subseteq X\}.$$

Proof. Let $a(X) = \min\{f(Y) + |X - Y| : Y \subseteq X\}$ for all subsets X of E. We need to show that, for all such X,

$$a(X) = r_f(X). \tag{4}$$

We leave it to the reader to show this when $X \in \mathcal{I}(M(f))$ and to show that a is increasing.

Now suppose that $X \notin \mathcal{I}(M(f))$ and let B be a basis of X. Then, as (4) holds for all members of $\mathcal{I}(M(f))$ and a is increasing,

$$r_f(X) = |B| = r_f(B) = a(B) \leq a(X).$$

To complete the proof of (4), it remains to show that

$$a(X) \leq |B| = r_f(X). \tag{5}$$

Let $X - B = \{e_1, e_2, \ldots, e_m\}$ and suppose $i \in \{1, 2, \ldots, m\}$. Then $B \cup e_i \notin \mathcal{I}(M(f))$. As B is in $\mathcal{I}(M(f))$, there is a subset I_i of B such that $f(I_i \dot\cup e_i) < |I_i \dot\cup e_i|$. Evidently $I_i \in \mathcal{I}(M(f))$, so $f(I_i) \geq |I_i|$ and therefore

$$f(I_i \cup e_i) = f(I_i) = |I_i|. \tag{6}$$

We now argue, by induction on j, that, for all j in $\{1, 2, \ldots, m\}$,

$$f(\bigcup_{i=1}^{j} I_i \cup \{e_1, e_2, \ldots, e_j\}) \leq |\bigcup_{i=1}^{j} I_i|. \tag{7}$$

By (6), this is true for $j = 1$. Assume it true for $j \leq k$ and let $j = k + 1$. Then, as f is submodular,

$$\begin{aligned} f(\bigcup_{i=1}^{k+1} I_i \cup \{e_1, e_2, \ldots, e_{k+1}\}) &\leq f(\bigcup_{i=1}^{k} I_i \cup \{e_1, e_2, \ldots, e_k\}) \\ &\quad + f(I_{k+1} \cup e_{k+1}) - f((\bigcup_{i=1}^{k} I_i) \cap I_{k+1}) \\ &\leq |\bigcup_{i=1}^{k} I_i| + |I_{k+1}| - |(\bigcup_{i=1}^{k} I_i) \cap I_{k+1}| \quad (8) \end{aligned}$$

where the last step follows by the induction hypothesis and the fact that $(\bigcup_{i=1}^{k} I_i) \cap I_{k+1} \in \mathcal{I}(M(f))$. From (8), we deduce that (7) holds for $j = k + 1$. It follows by induction that (7) holds for $j = m$, that is,

$$f(\bigcup_{i=1}^{m} I_i \cup (X - B)) \leq |\bigcup_{i=1}^{m} I_i|.$$

Therefore

$$f(\bigcup_{i=1}^{m} I_i \cup (X - B)) + |B - (\bigcup_{i=1}^{m} I_i)| \leq |B|.$$

Thus $a(X) \leq |B|$, so (5) holds and the proposition is proved. □

A consequence of the last result is that the function a, being the rank function of $M(f)$, also induces $M(f)$, that is, $M(a) = M(f)$. The problem of characterizing those matroids that are induced by a unique integer polymatroid was solved independently by Duke (1981) and Dawson (1983).

Next we shall describe a useful geometric interpretation of integer polymatroids and of the matroids they induce (Helgason 1974, McDiarmid 1975c, Lovász 1977). The key to this description is the following.

12.1.8 Example. Let M be a matroid on a set S and let E be a set, each member of which labels some flat of M, where the same flat can receive several different labels. For all $X \subseteq E$, define $f(X)$ to be the rank in M of the union of those sets labelled by members of X. Then it is routine to check that f is an integer polymatroid on 2^E. Indeed, the same conclusion holds more generally when the members of E label arbitrary subsets of S. □

The feature of this example which makes it particularly noteworthy is that every integer polymatroid is representable in this way as a collection of flats of some matroid. Formally:

12.1.9 Theorem. *Let f be a function defined on the set of subsets of a set E. Then f is an integer polymatroid on 2^E if and only if, for some matroid M, there is a function φ from E into the set of flats of M such that $f(X) = r_M(\bigcup_{x \in X} \varphi(x))$ for all $X \subseteq E$.*

The key idea in the proof of this theorem is that of freely adding a point on an element of an integer polymatroid, an operation which we shall see is effectively the same as that of freely adding a point on a flat of a matroid. Suppose f is an integer polymatroid on 2^E. If $x \in E$ and $x' \notin E$, then we can extend the domain of definition of f to include all subsets of $E \cup x'$ by letting

$$f(X \cup x') = \begin{cases} f(X) & \text{if } f(X \cup x) = f(X); \\ f(X) + 1 & \text{if } f(X \cup x) > f(X). \end{cases}$$

It is not difficult to check that this extended function is an integer polymatroid on the set of subsets of $E \cup x'$; we say that this new integer polymatroid has been obtained from f by *freely adding* x' to x. If we repeat this construction by freely adding a new element y' to some member

y of E, then one can show that the order in which these two extensions are performed is irrelevant. In particular, by freely adding one element to each member of E, we obtain a well-defined integer polymatroid f' on the set of subsets $E \dot{\cup} E'$ where E and E' are in one-to-one correspondence, and the restriction of f' to the set of subsets of E is the original integer polymatroid f on 2^E. We leave it to the reader to prove the following result of Edmonds (1970).

12.1.10 Proposition. *Let f be an integer polymatroid on 2^E. Extend this to an integer polymatroid on the set of subsets of $E \dot{\cup} E'$ by adding one element freely on each member of E. Then the restriction of this extended function to the set of subsets of E' is the rank function of a matroid and this matroid is isomorphic to $M(f)$.* □

This proposition is used in the proof of Theorem 12.1.9, which we now sketch.

Proof of Theorem 12.1.9. It suffices to show that, for an integer polymatroid f on 2^E, there is a matroid M having the specified properties. For each element y of E, replace y by a set A_y consisting of $f(y)$ new elements. Let A be $\bigcup_{y \in E} A_y$ and k be the function defined on all subsets X of A by $k(X) = f(\{y : A_y \cap X \neq \emptyset\})$. Then it is not difficult to check that k is an integer polymatroid on 2^A. Moreover, Proposition 12.1.10 implies that if we freely add a set A' of elements, one to each member of A, we obtain a matroid M on A'. For all y in E, let F_y be the closure in M of those elements that were added on members of A_y. Then one easily shows that $f(X) = r_M(\bigcup_{x \in X} F_x)$ for all $X \subseteq E$. The theorem follows by taking $\varphi(y) = F_y$ for all y in E. □

We conclude this section with a generalization of Proposition 12.1.7 that was proved by McDiarmid (1973) (see also Welsh 1976, pp. 116–117). Its proof is outlined in Exercise 9. Let $\mathcal{E}$ be a collection of subsets of the set E such that $\mathcal{E}$ is a lattice under set inclusion. In such a case, we call $\mathcal{E}$ a *lattice of subsets* of E. A function t from $\mathcal{E}$ into $\mathbb{R}$ is *submodular on* $\mathcal{E}$ if $t(X) + t(Y) \geq t(X \vee Y) + t(X \wedge Y)$ for all X and Y in $\mathcal{E}$.

12.1.11 Proposition. *Let $\mathcal{E}$ be a lattice of subsets of a set E such that $\mathcal{E}$ is closed under intersection and contains $\emptyset$ and E. Suppose that f is a non-negative, integer-valued, submodular function on $\mathcal{E}$ for which $f(\emptyset) = 0$. Let $\mathcal{I}(\mathcal{E}, f) = \{X \subseteq E : f(X) \geq |X \cap T| \text{ for all } T \text{ in } \mathcal{E}\}$. Then $\mathcal{I}(\mathcal{E}, f)$ is the collection of independent sets of a matroid on E. The rank of a subset X of E in this matroid is* $\min\{f(T) + |X - T| : T \in \mathcal{E}\}$. □

Exercises

1. Let $E = \{1, 2, 3\}$ and f be the function defined for all X in 2^E by

$$f(X) = \begin{cases} 1.5, & \text{if } X \text{ is } \{1,2\} \text{ or } \{1,3\}; \\ 2\ , & \text{if } X = E; \\ |X|, & \text{otherwise.} \end{cases}$$

   (i) Show that f is increasing and submodular.

   (ii) Find $\mathcal{C}(f)$.

   (iii) Deduce that 12.1.1 may fail for functions whose range is $\mathbb{Q}$.

2. Let G be an undirected graph having vertex set V. Assign to every edge e of E a positive integral weight. For all $X \subseteq V$, let $f_1(X)$ be obtained by summing the weights of all those edges with at least one end in X, and let $f_2(X)$ be the corresponding sum over all edges with exactly one end in X. Show that both f_1 and f_2 are submodular.

3. Prove Lemma 12.1.4.

4. Consider the non-Fano matroid labelled as in Figure 1.15. Consider the integer polymatroid defined on the set of subsets of $\{\{1,2,3\},$ $\{1,4,7\}, \{1,5,6\}\}$ as in Example 12.1.8. Show that this polymatroid is isomorphic to the corresponding integer polymatroid defined on the set of subsets of $\{\{1,5,6\}, \{2,4\}, \{4,6\}\}$.

5. (Dawson 1983) Let f be an integer polymatroid on 2^E and let r be the rank function of $M(f)$. A subset X of E is *f-balanced* if $f(X) = r(X)$. Show that:

   (i) every circuit of $M(f)$ is f-balanced;

   (ii) a union of f-balanced sets is f-balanced;

   (iii) a set is f-balanced if and only if its closure is f-balanced;

   (iv) if $X \cup Y \in \mathcal{I}(M(f))$ and X and Y are f-balanced, then $X \cap Y$ is f-balanced.

6. (Dawson 1983) Let f be an integer polymatroid on the set of subsets of a set E, and let cl denote the closure operator of $M(f)$. Show that $f(\text{cl}(X)) = f(X)$ for all $X \subseteq E$.

7. (Matthews 1977) Let G be a connected graph with more than one edge and assume that G is not a cycle. Prove that $B(G)$ is a connected matroid if and only if G has no vertices of degree one.

8. (Oxley, Prendergast, and Row 1982) Let f be an arbitrary integer-valued function defined on a set E. If $X \subseteq E$, define $a(X)$ to be $\max\{f(x) : x \in X\}$ if X is non-empty, and $\min\{f(x) : x \in E\}$ otherwise. Show that:

(i) a is submodular and increasing, and hence $M(a)$ is a matroid.

(ii) The class of matroids $M(a)$ arising in this way coincides with the class of transversal matroids having a nested presentation (Section 1.6, Exercise 12).

9. Let E, $\mathcal{E}$, f, and $\mathcal{I}(\mathcal{E}, f)$ be as defined in 12.1.11 and let a be the function defined on 2^E by $a(X) = \min\{f(T) + |X - T| : T \in \mathcal{E}\}$.

(i) Show that, for all subsets X, Y, Z, and W of E,

$$|X - Z| + |Y - W| \geq |(X \cup Y) - (Z \cup W)| + |(X \cap Y) - (Z \cap W)|.$$

(ii) Use (i) to show that a is submodular.

(iii) Now show that a is increasing and finish the proof of 12.1.11.

12.2 The theorems of Hall and Rado

Let $\mathcal{A}$ be a family $(A_j : j \in J)$ of subsets of a set S. When does $M[\mathcal{A}]$ have rank $|J|$? Equivalently, under what circumstances does $M[\mathcal{A}]$ have a transversal? This problem is often referred to as the *marriage problem* for, historically, it was originally formulated as follows. Let J be the set of unmarried men in a primitive village and, for each j in J, let A_j be the set of unmarried women in the village each of whom would be an acceptable wife for man j. The problem of finding necessary and sufficient conditions under which acceptable marriages can be arranged for every man in J is precisely the problem of finding necessary and sufficient conditions for $(A_j : j \in J)$ to have a transversal. We begin this section by noting P. Hall's (1935) solution to the marriage problem. Then we present a generalization of Hall's Theorem due to Rado (1942). Both this and Hall's original result will be obtained here as corollaries of a still more general result due to Welsh (1971a). Several applications of these results will also be noted in this section including a formula for the rank function of an arbitrary transversal matroid.

If $K \subseteq J$, then we shall denote $\bigcup_{j \in K} A_j$ by $A(K)$. Evidently if $(A_j : j \in J)$ has a transversal, then $|A(K)| \geq |K|$ for all $K \subseteq J$. Hall's Theorem asserts that the converse of this is also true.

12.2.1 Hall's Marriage Theorem. *Let $(A_j : j \in J)$ be a family of subsets of a set S. Then $(A_j : j \in J)$ has a transversal if and only if, for all $K \subseteq J$,*

$$|A(K)| \geq |K|.$$

12.2.2 Rado's Theorem. *Let $(A_j : j \in J)$ be a family of subsets of a set S and let M be a matroid on S having rank function r. Then $(A_j : j \in J)$ has a transversal that is independent in M if and only if, for all $K \subseteq J$,*

$$r(A(K)) \geq |K|.$$

In the last result, just as in Hall's Theorem, an obvious set of necessary conditions is also sufficient. This phenomenon is quite common in transversal theory. If we let the matroid M in Rado's Theorem be the free matroid on S, then we immediately obtain the Marriage Theorem. The next result was obtained by Welsh (1971a) by extracting the essential features from a proof of Hall's Theorem due to Rado (1967). A *system of representatives* for a family $(A_j : 1 \leq j \leq n)$ of subsets of a set S is a sequence $(e_1, e_2, \ldots, e_n)$ such that $e_j \in A_j$ for all j in $\{1, 2, \ldots, n\}$. Such a sequence differs from a transversal in that its members need not be distinct. The subset of E corresponding to this sequence will be denoted by $\{e_1, e_2, \ldots, e_n\}$.

12.2.3 Theorem. *Let $\mathcal{A}$ be a family $(A_j : j \in J)$ of subsets of a set S and let f be a non-negative, integer-valued, increasing, submodular function on 2^S. Then $\mathcal{A}$ has a system of representatives $(e_j : j \in J)$ such that*

$$f(\{e_j : j \in K\}) \geq |K| \text{ for all } K \subseteq J \quad (1)$$

if and only if

$$f(A(K)) \geq |K| \text{ for all } K \subseteq J. \quad (2)$$

Proof. Let $J = \{1, 2, \ldots, n\}$. If $(e_j : j \in J)$ is a system of representatives for $\mathcal{A}$ satisfying (1), then $f(A(K)) \geq f(\{e_j : j \in K\}) \geq |K|$ for all subsets K of J, so (2) holds.

Now assume that (2) holds. If, for each j in J, the set A_j contains a unique element a_j, then $(a_j : j \in J)$ is the required system of representatives for $\mathcal{A}$. Thus we may assume, without loss of generality, that $|A_1| \geq 2$.

12.2.4 Lemma. *For some element x of A_1, the family $(A_1 - x, A_2, \ldots, A_n)$ satisfies* (2).

Proof. Assume that no such element x exists. Then if x_1 and x_2 are distinct elements of A_1, there are non-empty subsets K_1 and K_2 of $\{2, 3, \ldots, n\}$ such that

$$f((A_1 - x_1) \cup A(K_1)) < |K_1| + 1,$$

and

$$f((A_1 - x_2) \cup A(K_2)) < |K_2| + 1.$$

As f is submodular,

$$\begin{array}{rcl} |K_1| + |K_2| & \geq & f((A_1 - x_1) \cup A(K_1)) + f((A_1 - x_2) \cup A(K_2)) \\ & \geq & f((A_1 - x_1) \cup A(K_1) \cup (A_1 - x_2) \cup A(K_2)) \\ & + & f\left([(A_1 - x_1) \cup A(K_1)] \cap [(A_1 - x_2) \cup A(K_2)]\right) \\ & \geq & f(A_1 \cup A(K_1 \cup K_2)) + f(A(K_1) \cap A(K_2)). \end{array}$$

Thus, as $A(K_1) \cap A(K_2) \supseteq A(K_1 \cap K_2)$,

$$|K_1| + |K_2| \geq f(A_1 \cup A(K_1 \cup K_2)) + f(A(K_1 \cap K_2)). \tag{3}$$

But $\mathcal{A}$ satisfies (2) and $1 \notin K_1 \cup K_2$, so

$$f(A_1 \cup A(K_1 \cup K_2)) \geq 1 + |K_1 \cup K_2|$$

and

$$f(A(K_1 \cap K_2)) \geq |K_1 \cap K_2|.$$

Hence, by (3),

$$|K_1| + |K_2| \geq 1 + |K_1 \cup K_2| + |K_1 \cap K_2| = 1 + |K_1| + |K_2|.$$

This contradiction completes the proof of the lemma. □

By this lemma, we may successively delete elements from A_1, maintaining a family of sets satisfying (2), until we arrive at a singleton set. We repeat this process for $A_2, A_3, \ldots, A_n$ reducing each to a singleton set while maintaining a family of sets that satisfies (2). But we have already noted that the required system of representatives exists when all the sets in the family are singletons. Hence the theorem is proved. □

Clearly Rado's Theorem follows from the last result by taking f to be the rank function r of M. Now let d be a non-negative integer not exceeding $|J|$. Then we obtain the following extension of Rado's Theorem, due to Perfect (in Mirsky 1971), by letting $f(X) = r(X) - d$ for all $X \subseteq S$.

12.2.5 Corollary. *Let $\mathcal{A}$ be a family $(A_j : j \in J)$ of subsets of a set S and let M be a matroid on S having rank function r. Then $\mathcal{A}$ has a partial transversal that has size $|J| - d$ and is independent in M if and only if, for all subsets K of J,*

$$r(A(K)) \geq |K| - d.$$ □

Next we use the last result to find the rank function of an arbitrary transversal matroid.

12.2.6 Proposition. *Let $\mathcal{A}$ be a family $(A_j : j \in J)$ of subsets of a set S. If $X \subseteq S$, then its rank in $M[\mathcal{A}]$ is given by*

$$r(X) = \min\{|A(K) \cap X| - |K| + |J| : K \subseteq J\}.$$

Proof. Consider the family $\mathcal{A}_X = (A_j \cap X : j \in J)$. We shall apply the last result to this family of sets taking M to be the free matroid on S. Evidently $r(X) = \max\{|Y| : Y \text{ is a partial transversal of } \mathcal{A}_X\}$. Hence, from Corollary 12.2.5,

$$r(X) = \max\{|J| - d : d \geq 0 \text{ and } |A(K) \cap X| \geq |K| - d \text{ for all } K \subseteq J\}.$$

If $K = \emptyset$, the condition that $|A(K) \cap X| \geq |K| - d$ is equivalent to the condition that $d \geq 0$. Hence

$$\begin{aligned} r(X) &= \max\{|J| - d : |A(K) \cap X| \geq |K| - d \text{ for all } K \subseteq J\} \\ &= \max\{|J| - d : |A(K) \cap X| - |K| + |J| \geq |J| - d \text{ for all } K \subseteq J\} \\ &= \min\{|A(K) \cap X| - |K| + |J| : K \subseteq J\}. \end{aligned}$$ □

As an immediate consequence of this, we have the following.

12.2.7 Corollary. *Let $\mathcal{A}$ be a family $(A_j : j \in J)$ of subsets of a set S. A subset X of S is independent in $M[\mathcal{A}]$ if and only if, for all $K \subseteq J$,*

$$|A(K) \cap X| - |K| + |J| \geq |X|.$$ □

We can obtain a different characterization of the independent sets in $M[\mathcal{A}]$ by reference to the bipartite graph $\Delta[\mathcal{A}]$ associated with $\mathcal{A}$ where we recall that $\Delta[\mathcal{A}]$ was introduced in Section 1.6. A subset X of S is independent in $M[\mathcal{A}]$ if and only if there is a matching W in $\Delta[\mathcal{A}]$ such that every vertex in X is incident with an edge of W. This can be expressed in terms of a new family of sets as follows. For all $Y \subseteq X$, let $N(Y)$ be those members of J that are adjacent in $\Delta[\mathcal{A}]$ to some member of Y. Let $\mathcal{N}_X = (N(x) : x \in X)$. Then X is independent in $M[\mathcal{A}]$ if and only if $\mathcal{N}_X$ has a transversal.

12.2.8 Example. Let $\mathcal{A}$ be the family of subsets of $\{x_1, x_2, \ldots, x_6\}$ whose members are $A_1 = \{x_1, x_2, x_5, x_6\}$, $A_2 = \{x_1, x_3\}$, $A_3 = \{x_2, x_6\}$, $A_4 = \{x_5\}$, and $A_5 = \{x_3, x_4, x_6\}$. Then $\Delta[\mathcal{A}]$ is as shown in Figure 12.2. As $\{x_2 1, x_3 2, x_4 5, x_6 3\}$ is a matching in $\Delta[\mathcal{A}]$, $\{x_2, x_3, x_4, x_6\}$ is independent in $M[\mathcal{A}]$. Equivalently, $\mathcal{N}_X$ has a transversal $(1, 2, 5, 3)$ where

$$\mathcal{N}_X = (N(x_2), N(x_3), N(x_4), N(x_6)) = (\{1, 3\}, \{2, 5\}, \{5\}, \{1, 3, 5\}).$$ □

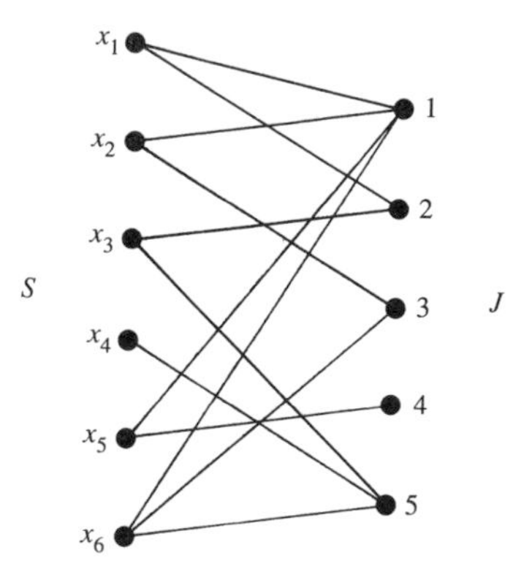

Fig. 12.2. The bipartite graph $\Delta[\mathcal{A}]$ associated with $\mathcal{A}$.

Using Hall's Theorem, we can now specify when X is independent in terms of the family $\mathcal{N}_X$. The straightforward proof is left to the reader.

12.2.9 Proposition. *Let $\mathcal{A}$ be a family $(A_j : j \in J)$ of subsets of a set S. For a subset X of S, let $\mathcal{N}_X = (N(x) : x \in X)$ where $N(x) = \{j : x \in A_j\}$. Then X is independent in $M[\mathcal{A}]$ if and only if*

$$|N(Y)| \geq |Y| \text{ for all } Y \subseteq X. \qquad \square$$

As an immediate consequence of this result one can show that bicircular matroids arise in a very natural way as transversal matroids (Matthews 1977).

12.2.10 Corollary. *Let G be a graph and, for each v in $V(G)$, let A_v be the set of edges incident with v. Then $B(G)$ is equal to the transversal matroid of the family $(A_v : v \in V(G))$.* □

Our final application of Rado's Theorem will use the theory of submodular functions to define a class of matroids that contains the class of transversal matroids. Let Δ be a bipartite graph with vertex classes S and J and suppose that M is a matroid on J. Let $\Delta(\mathcal{I}) = \{X \subseteq S : X$ can be matched in Δ onto a member of $\mathcal{I}(M)\}$.

12.2.11 Example. Let Δ be the bipartite graph shown in Figure 12.3(a) and suppose that the matroid M on J is the cycle matroid of the graph G in Figure 12.3(b). Then it is not difficult to check that $\Delta(\mathcal{I}) = \{X \subseteq S : |X| \leq 2\}$. □

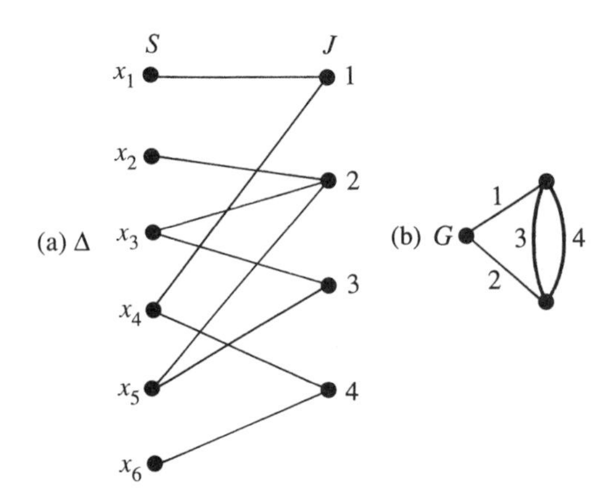

Fig. 12.3. The matroid induced from $M(G)$ by Δ is $U_{2,6}$.

In the last example, $\Delta(\mathcal{I})$ is the set of independent sets of a matroid on S. Moreover, if $\mathcal{I}(M) = 2^J$, that is, if M is the free matroid on J, then $\Delta(\mathcal{I})$ is again the set of independent sets of a matroid on S, this matroid being the transversal matroid on S associated with Δ. These observations are special cases of the following basic result (Perfect 1969).

12.2.12 Theorem. *Let Δ be a bipartite graph with vertex classes S and J and let M be a matroid on J. Suppose that $\Delta(\mathcal{I})$ is the set of subsets X of S that are matched in Δ onto a member of $\mathcal{I}(M)$. Then $\Delta(\mathcal{I})$ is the set of independent sets of a matroid $\Delta(M)$ on S.*

The matroid $\Delta(M)$ is said to be *induced from M by Δ*. We shall prove Theorem 12.2.12 by showing that $\Delta(M)$ is the matroid induced on S by a submodular function. This proof follows Welsh (1976). As before, for all $X \subseteq S$, we let $N(X)$ be the set of vertices in J that are adjacent in Δ to some member of X. Define $f : 2^S \to \mathbb{Z}^+ \cup \{0\}$ by

$$f(X) = r_M(N(X)).$$

12.2.13 Lemma. *The function f is non-negative, integer-valued, increasing, and submodular. A subset X of S is independent in $M(f)$ if and only if $X \in \Delta(\mathcal{I})$. Hence $\Delta(M)$ is the matroid $M(f)$.*

Proof. Evidently f is non-negative and integer-valued. Moreover, it is straightforward to check that f is increasing and submodular. Now X is independent in $M(f)$ if and only if, for all subsets Y of X,

$$r_M(N(Y)) \geq |Y|.$$

But, by Rado's Theorem, this occurs if and only if the family $(N(e) : e \in X)$ has a transversal that is independent in M. Since the latter occurs precisely when Δ has a matching of X onto a member of $\mathcal{I}(M)$, the lemma follows and hence so does Theorem 12.2.12. □

On combining the last lemma with Proposition 12.1.7, we obtain the rank function of $\Delta(M)$.

12.2.14 Corollary. *If $X \subseteq S$, its rank in $\Delta(M)$ is equal to $\min\{r_M(N(Y)) + |X - Y| : Y \subseteq X\}$.* □

A straightforward consequence of Theorem 12.2.12 is that any function σ on the ground set of a matroid can be used to induce a matroid on the range of σ (Nash-Williams 1966).

12.2.15 Corollary. *Let M be a matroid on a set J and suppose that σ is a function from J into a set S. Then there is a matroid $\sigma(M)$ on S whose set of independent sets is $\{\sigma(I) : I \in \mathcal{I}(M)\}$.* □

In general, an induced matroid $\Delta(M)$ retains very few of the properties of the original matroid M. It need not have the same rank as M, although clearly $r(\Delta(M)) \leq r(M)$. In Example 12.2.11, M is graphic and cographic and hence is regular, binary, and ternary. However, $\Delta(M) \cong U_{2,6}$ and hence has none of these five properties. M is isomorphic to its dual, but $\Delta(M)$ and its dual do not even have the same rank. Although, in general, F-representability of M does not imply F-representability of $\Delta(M)$, Piff and Welsh (1970) were able to prove the following.

12.2.16 Proposition. *Let Δ be a bipartite graph with vertex classes S and J and let M be a matroid on J. Then there is an integer $n(M)$ such that if M is representable over a field F that has at least $n(M)$ elements, then the induced matroid $\Delta(M)$ is F-representable.*

Before proving this result, we note two immediate consequences of it, both of which were stated without proof in earlier chapters.

12.2.17 Corollary. *Let M be a transversal matroid and F be a field. Then there is an integer $n(M)$ such that M is representable over every extension field of F having at least $n(M)$ elements.* □

12.2.18 Corollary. *A transversal matroid is representable over fields of every characteristic.* □

On combining Corollary 12.2.17 with Proposition 6.7.10 and Exercise 5 of Section 6.7, we obtain the following.

12.2.19 Corollary. *A transversal matroid is algebraic over all fields.* □

The next lemma contains the core of the proof of Proposition 12.2.16.

12.2.20 Lemma. *Let M be a matroid on a fixed set J and let $\sigma : J \to S$ be a surjection. Then there is an integer $n(M)$ such that if M is representable over a field F that has at least $n(M)$ elements, the matroid $\sigma(M)$ is also F-representable.*

Proof. The surjection σ will be called *simple* if $|J| = |S| + 1$. As every surjection is the composition of at most $|J| - 1$ simple surjections, it will suffice to prove the proposition in the case that σ is simple. Thus assume this and let $\psi : J \to V(r, F)$ be an F-coordinatization of M.

Let $J = \{j_1, j_2, \ldots, j_m\}$, $S = \{s_2, s_3, \ldots, s_m\}$, and suppose that $\sigma(j_i) = s_i$ for all i in $\{2, 3, \ldots, m\}$, while $\sigma(j_1) = s_2$. Now define $\psi_1 : S - s_2 \to V(r, F)$ by $\psi_1(s_i) = \psi(j_i)$ for all i in $\{3, 4, \ldots, m\}$. We shall show that there are elements c_1 and c_2 of F such that if $\psi_1(s_2) = c_1\psi(j_1) + c_2\psi(j_2)$, then the extended map $\psi_1 : S \to V(r, F)$ is an F-coordinatization of $\sigma(M)$. This is certainly true if $\{j_1\}$, $\{j_2\}$, or $\{j_1, j_2\}$ is a circuit of M. Hence we may assume that $\{j_1, j_2\}$ is independent in M. Thus $\{s_2\}$ is independent in $\sigma(M)$.

Now let $\mathcal{B}'$ be the set of bases of $\sigma(M)/s_2$ and suppose $B \in \mathcal{B}'$. Then $B \cup s_2$ is independent in $\sigma(M)$, so $B \cup s_2 = \sigma(I)$ for some I in $\mathcal{I}(M)$. Therefore $\langle \psi_1(B) \rangle$ is a proper subspace of $\langle \psi(I) \rangle$. Hence $\langle \psi(j_1), \psi(j_2) \rangle \cap \langle \psi_1(B) \rangle$ is a 0- or 1-dimensional subspace of the 2-dimensional space $\langle \psi(j_1), \psi(j_2) \rangle$. Now if $F = GF(q)$, then $V(2, F)$ has $q + 1$ subspaces of dimension one, while if F is infinite, $V(2, F)$ has infinitely many such subspaces. Thus if $|F| \geq |\mathcal{B}'|$, then $\langle \psi(j_1), \psi(j_2) \rangle$ contains a 1-dimensional subspace that is not in $\bigcup\{\langle \psi_1(B) \rangle : B \in \mathcal{B}'\}$. Let $c_1\psi(j_1) + c_2\psi(j_2)$ be a non-zero member of such a subspace and define $\psi_1(s_2) = c_1\psi(j_1) + c_2\psi(j_2)$. Now there are only finitely many non-isomorphic matroids M on the fixed set J and only finitely many simple surjections σ with domain J. Therefore the maximum value of $|\mathcal{B}'|$ over all such M and all such σ exists and is finite. We take this maximum to be the required integer $n(M)$.

To complete the proof, we need to show that ψ_1 is an F-coordinatization for M, that is, if $X \subseteq S$, then X is independent in $\sigma(M)$ if and only if $\psi_1(X)$ is linearly independent in $V(r, F)$. This is easily seen to be true if $s_2 \notin X$, so assume that $s_2 \in X$. Suppose that X is independent in $\sigma(M)$. Then X is contained in a basis $B \,\dot\cup\, s_2$ of $\sigma(M)$. From above, as $B \subseteq S - s_2$, the set $\psi_1(B)$ is linearly independent. Moreover, the choice of $\psi_1(s_2)$ guarantees that $\psi_1(B \cup s_2)$ is linearly independent. Thus $\psi_1(X)$ is linearly independent. Now suppose that X is a dependent set in $\sigma(M)$ containing s_2. Then it is not difficult to show that $\psi_1(X)$ is linearly dependent in $V(r, F)$ and this finishes the proof of the lemma. □

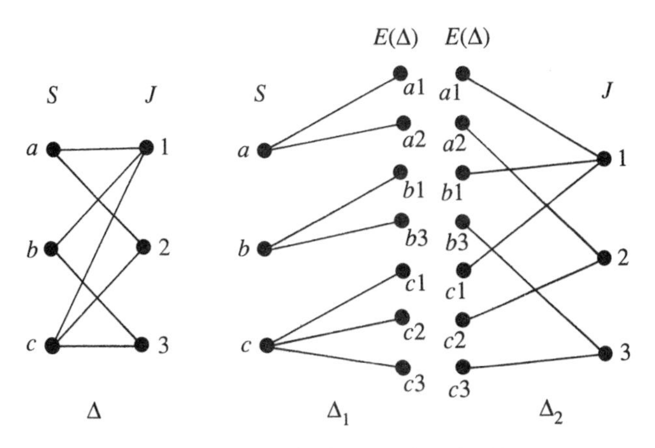

Fig. 12.4. $\Delta(M) = \Delta_1(\Delta_2(M))$.

Proof of Proposition 12.2.16. We shall assume that Δ has no isolated vertices since the general case follows easily from this case. Now let Δ_1 be the simple bipartite graph with vertex classes S and $E(\Delta)$ in which a member v of S is joined to a member e of $E(\Delta)$ precisely where e is incident with v in Δ. Let Δ_2 be defined similarly with vertex classes $E(\Delta)$ and J. An example of this construction is shown in Figure 12.4.

Now, clearly, $\widetilde{\Delta_2(M)} \cong \widetilde{M}$. Thus if M is F-representable, so too is $\Delta_2(M)$. Moreover, $\Delta(M) = \Delta_1(\Delta_2(M))$ and clearly Δ_1 corresponds to a surjection. The proposition now follows immediately from Lemma 12.2.20. □

We shall show next that both the construction of an induced matroid from a bipartite graph and the construction of a strict gammoid given in Section 2.4 are special cases of an operation which induces a matroid from an arbitrary directed graph. Lemma 2.4.3, which provided the key step in the proof that the dual of a transversal matroid is a strict gammoid, is also of fundamental importance in the proof that this more general construction does indeed produce a matroid.

Let G be a directed graph having vertex set V, and let M be a matroid on V. Let $L(G, M)$ be the collection of subsets of V that can be linked in G to an independent set of M. Thus, for example, if $B_0 \subseteq V$ and M is the direct sum of a free matroid on B_0 and a rank-0 matroid on $V - B_0$, then $L(G, M)$ is precisely the set of independent sets of a matroid, namely the strict gammoid $L(G, B_0)$. This observation is a special case of the following result, a proof of which is outlined in Exercise 10.

12.2.21 Proposition. *If G is a directed graph having vertex set V, and $L(G, M)$ is the set of subsets of V that are linked to an independent set in M, then $L(G, M)$ is the set of independent sets of a matroid on V.* □

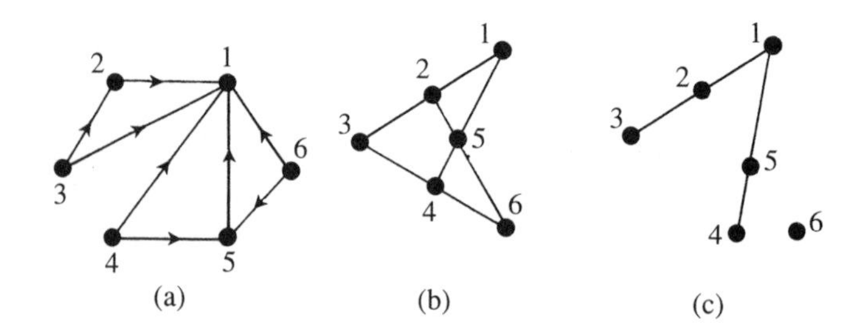

Fig. 12.5. (a) G. (b) M. (c) The induced matroid $L(G, M)$.

By analogy with strict gammoids, we use $L(G, M)$ to denote both the matroid obtained here and its collection of independent sets.

12.2.22 Example. Let G be the directed graph shown in Figure 12.5(a) and M be the matroid on $\{1, 2, \dots, 6\}$ for which a geometric representation is shown in Figure 12.5(b). Then it is not difficult to check that Figure 12.5(c) is a geometric representation for $L(G, M)$. For example, $\{3, 4, 6\}$ is independent in $L(G, M)$ though not in M because $\{3, 4, 6\}$ is linked to the independent set $\{3, 4, 1\}$ in G. □

We call $L(G, M)$ the matroid *induced from M by G*. Its rank function was determined by McDiarmid (1975a) using the theory of submodular functions. Although G has thus far been required to be a directed graph, we can easily extend the definition of $L(G, M)$ to include the possibility that G is undirected. In this case, one replaces each non-loop edge of G by two oppositely directed edges, deletes each loop from G, and then finds the matroid induced from M by the resulting directed graph.

Next we note how two of the matroid constructions discussed in Chapter 7 can be achieved by this operation of inducing by a graph (Mason 1977). We begin with the principal extension $M +_F e$ of M in which the element e is freely added to the flat F. In the following result, a straightforward consequence of Proposition 12.2.21, the graph G is as illustrated in Figure 12.6(a).

12.2.23 Corollary. *Let F be a flat in a matroid M and let G be the graph with vertex set $E(M) \dot\cup e$ and edge set $\{ef : f \in F\}$. Then $M +_F e$ is the matroid $L(G, M \oplus U_{0,1})$, where $M \oplus U_{0,1}$ is formed from M by adjoining e as a loop.* □

The free coextension of M by the element e can also be obtained as an induced matroid. Let M' be an isomorphic copy of M for which the ground set E' is disjoint from the ground set E of M. For each x in E, let x' denote the corresponding element of E'. Now form the bipartite

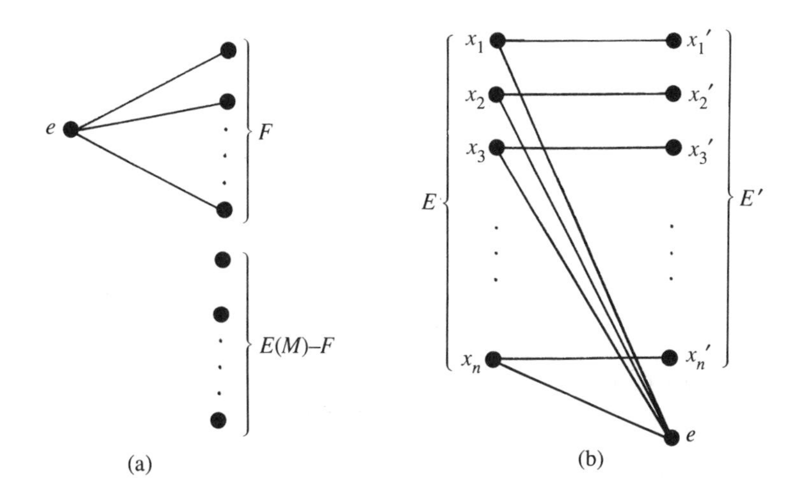

Fig. 12.6. (a) G in 12.2.23. (b) H in 12.2.24.

graph H by taking the vertex set to be $E \cup E' \cup \{e\}$ and the edge set to be $\{xx' : x \in E\} \cup \{xe : x \in E\}$ (see Figure 12.6(b)). Let M_0 be the matroid obtained from M' by adjoining e as a coloop and adjoining the elements of E as loops. We leave it to the reader to check the following.

12.2.24 Proposition. *The restriction of $L(H, M_0)$ to $E \cup e$ is the free coextension of M.* □

In the next section, we give another important example of a matroid operation that can be achieved by using this technique of inducing by a graph. We close this section by using the above ideas to describe the geometric significance of the operation of inducing a matroid from a bipartite graph Δ. Suppose that Δ has vertex classes S and J, and let M be a matroid on J. Then it is not difficult to show that, by adding edges to Δ, we can find a bipartite graph Γ with $\Gamma(M) = \Delta(M)$ such that, for every x in S, the set $N_\Gamma(x)$ of neighbours of x in Γ forms a flat in M (see Exercise 9). Using this fact, one can prove the following extension of Corollary 12.2.23.

12.2.25 Proposition. *Let Δ be a bipartite graph with vertex classes S and J and let M be a matroid on J. Let M^+ be the matroid that is obtained from M by adjoining all of the elements of S as loops. Then the matroids obtained by restricting $L(\Delta, M^+)$ to J and S are M and $\Delta(M)$, respectively.* □

Geometrically, the matroid $L(\Delta, M^+)$ is constructed from M by a sequence of single-element extensions with each element x of S being freely added to the flat $\mathrm{cl}_M(N_\Gamma(x))$ of M. Since transversal matroids are precisely the matroids that are induced by bipartite graphs from free matroids, this construction can be used to give a geometric description of transversal matroids (Ingleton 1971b, Brylawski 1975d). Before making the details of this description more explicit, we comment briefly on terminology. A matroid of the form $L(\Delta, M^+)$ where M is free is called a *principal transversal matroid* or a *fundamental transversal matroid.* Such matroids have several attractive features some of which will be explored in the exercises.

If $|J| = m$, a free matroid on J can be represented geometrically by m affinely independent points in $\mathbb{R}^{m-1}$. So, for instance, when $m = 3$ and $m = 4$, the elements of J are the vertices of a triangle and a tetrahedron respectively. In general, such a collection of affinely independent points is called a *simplex.* A *vertex* of the simplex is any of the points of J; a *face* of the simplex is any flat of the affine matroid on J.

In view of the above discussion, a consequence of Proposition 12.2.25 is that a rank-r matroid M is transversal if and only if there is a simplex in $\mathbb{R}^{r-1}$ such that every element e of M is freely added to a face F_e of the simplex so that M has no dependencies except those that are forced by the faces F_e. Clearly the last statement needs to be made more precise. This is done in the next proposition where, for ease of statement, attention is confined to loopless non-empty matroids. Little generality is lost by this restriction since the addition or deletion of loops does not alter whether or not a matroid is transversal.

12.2.26 Proposition. *For $r \geq 1$, a loopless rank-r matroid M is transversal if and only if there are disjoint finite subsets J and S of $\mathbb{R}^{r-1}$ such that*

(i) *$|J| = r$ and the elements of J form a simplex;*

(ii) *M is isomorphic to the affine matroid N on S; and*

(iii) *every element e of S is freely added to a face F_e of the simplex J so that, for every flat X of N containing e, the closure of X in the affine matroid on $S \cup J$ contains F_e.* □

The simplex J in the last result is called the *spanning simplex* of the matroid. The last two propositions are illustrated in the following.

12.2.27 Example. Let Δ be the bipartite graph shown in Figure 12.7(a) where $S = \{a, b, \ldots, h\}$ and $J = \{1, 2, 3, 4\}$. Let M_J be the free matroid on J and M_J^+ be the matroid obtained from M_J by adjoining the elements of S as loops. A geometric representation for the principal transversal matroid $L(\Delta, M_J^+)$ is shown in Figure 12.7(b) where we note that the element c of this matroid is free on the flat spanned by $\{1, 2, 3\}$. The restriction

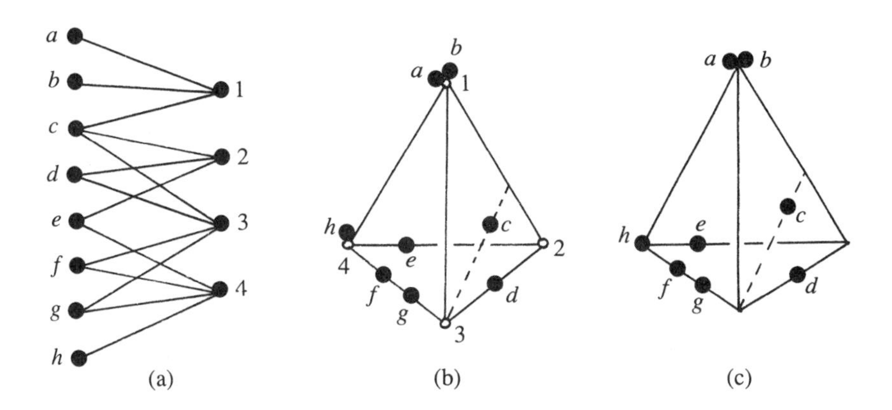

Fig. 12.7. (a) Δ. (b) $L(\Delta, M_J^+)$. (c) $L(\Delta, M_J^+)|S = \Delta(M_J)$.

of $L(\Delta, M_J^+)$ to J is a simplex, whereas the restriction of $L(\Delta, M_J^+)$ to S equals the matroid $\Delta(M_J)$. A geometric representation for the latter restriction is shown in Figure 12.7(c). Of course, $\Delta(M_J)$ is precisely the transversal matroid on S that is associated with the bipartite graph Δ.□

Finally, we note that, for many matroids M, Proposition 12.2.26 provides a quick way to show that M is not transversal. For instance, $M(K_4)$ is not transversal. If it were, then, since each of its six elements lies on the intersection of two 3-point lines, each such element must be parallel to a vertex of the associated simplex. But this simplex has only three vertices and $M(K_4)$ has no non-trivial parallel classes.

Exercises

1. (Ore 1955) Show that a family $(A_j : j \in J)$ of subsets of a set S has a partial transversal of size $|J| - d$ if and only if $|A(K)| \geq |K| - d$ for all subsets K of J.

2. Let $(A_j : j \in J)$ be a family of subsets of a set S and let $\mathcal{I}$ be the set of subsets K of J for which $(A_j : j \in K)$ has a transversal. Prove that $\mathcal{I}$ is the set of independent sets of a matroid on J.

3. Prove Proposition 12.2.9 and Corollary 12.2.10.

4. Find $L(G, M)$ if M is the matroid in Figure 12.5(b) and G is the undirected graph underlying the directed graph in Figure 12.5(a).

5. Let Δ be a bipartite graph with vertex classes S and J and M be a matroid on J. Show that:

(i) the characteristic set of M is a subset of the characteristic set of $\Delta(M)$, and these two sets need not be equal;

(ii) if M is transversal, then $\Delta(M)$ need not be transversal.

6. (Matthews 1977) Prove that bicircular matroids are precisely the loopless transversal matroids that have a presentation in which the intersection of any three members is empty.

7. Let $\mathcal{A}$ be a family $(A_j : j \in J)$ of subsets of a set S and m be a positive integer. Use Theorem 12.2.3 to show that:

(i) $\mathcal{A}$ has m disjoint transversals if and only if $|A(K)| \geq m|K|$ for all $K \subseteq J$;

(ii) $\mathcal{A}$ has a system of representatives in which no element occurs more than m times if and only if $m|A(K)| \geq |K|$ for all $K \subseteq J$.

8. Let M be a rank-r transversal matroid.

(i) Show that M has at most r non-trivial parallel classes.

(ii) Extend (i) by determining the maximum number of rank-k flats in M each of which has a spanning circuit.

(iii) Show that, for all $k \geq 3$, exactly one of the two non-isomorphic parallel extensions of $\mathcal{W}^k$ is transversal.

9. Let Δ be a bipartite graph with vertex classes S and J. Let M be a matroid on J and x be in S.

(i) Suppose that $y \in \text{cl}_M(N_\Delta(x)) - N_\Delta(x)$ and let Δ_1 be obtained from Δ by adding the edge xy. Show that $\Delta_1(M) = \Delta(M)$.

(ii) Deduce that there is a bipartite graph Δ' that can be obtained by adding edges to Δ such that $\Delta'(M) = \Delta(M)$ and, for every x in S, the set of neighbours of x in Δ' forms a flat in M.

10. Let M be a matroid on the vertex set V of a directed graph G and let $\widehat{G}$ be the bipartite graph corresponding to G (see Section 2.4). Let $\widehat{M}$ be an isomorphic copy of M on the set $\widehat{V}$. Prove Proposition 12.2.21 by showing, using Lemma 2.4.3, that the maximal members of $L(G, M)$ coincide with the cobases of the matroid induced by $\widehat{G}$ from the dual of $\widehat{M}$.

11. For a graph G, let Δ be the bipartite graph with vertex classes $E(G)$ and $V(G)$ and edge set $\{ev : e \in E(G), v \in V(G), e \text{ is incident with } v \text{ in } G\}$. Let M be the free matroid on $V(G)$.

(i) Show that $\Delta(M)$ is the bicircular matroid $B(G)$.

(ii) Use this to show that (i)–(iii) of 12.2.26 give a geometric characterization of bicircular matroids if one adds to (iii) the requirement that, for all e in S, the face F_e of the simplex to which e is freely added has rank one or two.

12. (Brualdi 1969) Use Hall's Theorem to prove that $\mathcal{B}$ is the collection of bases of a matroid on E if and only if $\mathcal{B}$ satisfies (B1) and the following *bijective basis exchange axiom*:

(B2)″ *If B_1 and B_2 are in $\mathcal{B}$, then there is a bijection $\alpha : B_1 \to B_2$ such that $(B_1 - b) \cup \alpha(b) \in \mathcal{B}$ for all b in B_1.*

13. (Zaslavsky 1987b, 1989) A graph that is a subdivision of the third graph in Figure 12.1 is a *Θ-graph.* A *biased graph* (G, Ψ) consists of a graph G and a set Ψ of cycles of G such that if C_1 and C_2 are in Ψ and $G[C_1 \cup C_2]$ is a Θ-graph, then all cycles in $G[C_1 \cup C_2]$ are in Ψ. The members of Ψ are called *balanced cycles.* Let (G, Ψ) be a biased graph and $\mathcal{I} = \{I \subseteq E(G) : G[I]$ contains no balanced cycles and has no component with more than one cycle$\}$. Show that:

(i) $\mathcal{I}$ is the collection of independent sets of a matroid $B(G, \Psi)$ on $E(G)$.

(ii) If Ψ is the set of all cycles of G, then $B(G, \Psi) = M(G)$.

(iii) If Ψ is empty, then $B(G, \Psi) = B(G)$.

(iv) If Ψ' is the set of all even-length cycles of G, then (G, Ψ') is a biased graph. In this case, $B(G, \Psi')$ is Doob's (1973) *even-cycle matroid.*

14. (Las Vergnas 1970) Prove that the dual of a principal transversal matroid is a principal transversal matroid.

15. Prove that a transversal matroid M is a principal transversal matroid if and only if M has a cobasis B^* such that the family $(C^*(e, B^*) : e \in E(M) - B^*)$ is a presentation for M.

16. (Hall 1935) Let $(A_j : j \in J)$ and $(D_j : j \in J)$ be families of subsets of a set S. Prove that there is a sequence $(x_j : j \in J)$ that is a system of representatives for both of these families if and only if $|\{j \in J : D_j \cap A(K) \neq \emptyset\}| \geq |K|$ for all $K \subseteq J$.

17. (Mendelsohn and Dulmage 1958) Let $(A_j : j \in J)$ be a family of subsets of a set S. Suppose $J_1 \subseteq J$ and $S_1 \subseteq S$. Prove that the following are equivalent.

(a) $(A_j : j \in J_1)$ has a transversal and S_1 is a partial transversal of $(A_j : j \in J)$.

(b) For some J_2 with $J_1 \subseteq J_2 \subseteq J$, there is a set S_2 such that $S_1 \subseteq S_2 \subseteq S$ and S_2 is a transversal of $(A_j : j \in J_2)$.

12.3 Matroid union and its applications

In this section we consider a way of joining matroids which, unlike the previous such operations we have considered, is defined irrespective of the number of elements common to the matroids. This important operation was introduced by Nash-Williams (1966). We begin by defining it for two matroids on a common ground set, although eventually we shall extend the definition to include any finite set of matroids having arbitrary ground sets.

12.3.1 Theorem. *Let M_1 and M_2 be matroids on a set E having rank functions r_1 and r_2, respectively. Let*

$$\mathcal{I} = \{I_1 \cup I_2 : I_1 \in \mathcal{I}(M_1),\ I_2 \in \mathcal{I}(M_2)\}.$$

Then $\mathcal{I}$ is the set of independent sets of a matroid $M_1 \vee M_2$ on E. Moreover, if $X \subseteq E$, its rank in $M_1 \vee M_2$ is

$$\min\{r_1(Y) + r_2(Y) + |X - Y| : Y \subseteq X\}.$$

Proof. Construct the bipartite graph Δ as follows. Let ϕ_1 and ϕ_2 be bijections from E onto disjoint sets E_1 and E_2. Then, for $i = 1, 2$, the function ϕ_i induces an isomorphic copy of M_i on E_i. Hence on $E_1 \dot\cup E_2$, we have a matroid isomorphic to $M_1 \oplus M_2$. Let Δ have E as one vertex class and $E_1 \cup E_2$ as the other. Join each element e of E to $\phi_1(e)$ and $\phi_2(e)$ and to no other vertices in $E_1 \cup E_2$ (see Figure 12.8).

Now consider the matroid N on E induced from $M_1 \oplus M_2$ by Δ. Since $M_1 \oplus M_2$ has rank function $r_1 + r_2$, Corollary 12.2.14 implies that the rank function of N is as specified in the theorem. Moreover, one easily checks that $\mathcal{I}(N) = \{I_1 \cup I_2 : I_1 \in \mathcal{I}(M_1),\ I_2 \in \mathcal{I}(M_2)\}$. □

The matroid $M_1 \vee M_2$ is called the *union* of M_1 and M_2, or sometimes the *join* or *sum* of M_1 and M_2. Thus, for example, if M_1 and M_2 are as shown in Figure 12.9, then $M_1 \vee M_2 \cong U_{6,7}$. Notice that both $M_1 \vee M_1$ and $M_2 \vee M_2$ are also isomorphic to $U_{6,7}$.

It is straightforward to extend 12.3.1 to show that if $(M_1, M_2, \ldots, M_n)$ is an arbitrary finite family of matroids on a set E, then there is a matroid $M_1 \vee M_2 \vee \cdots \vee M_n$ on E such that

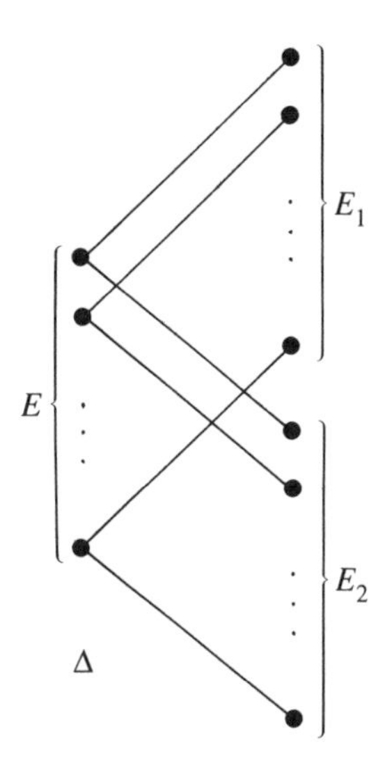

Fig. 12.8. $M_1 \vee M_2$ is induced from $M_1 \oplus M_2$ by Δ.

12.3.2 $\mathcal{I}(M_1 \vee M_2 \vee \cdots \vee M_n) = \{I_1 \cup I_2 \cup \cdots \cup I_n : I_i \in \mathcal{I}(M_i) \text{ for } 1 \leq i \leq n\}$.

Moreover, if M_i has rank function r_i, then the rank of X in $\vee_{i=1}^n M_i$ is

12.3.3 $\min\{\sum_{i=1}^n r_i(Y) + |X - Y| : Y \subseteq X\}$.

12.3.4 Example. If $m = m_1 + m_2 + \cdots + m_k$, then

$$U_{m_1,n} \vee U_{m_2,n} \vee \cdots \vee U_{m_k,n} = \begin{cases} U_{m,n} & \text{if } m < n, \\ U_{n,n} & \text{if } m \geq n. \end{cases}$$

Hence a union of uniform matroids is uniform. However, a union of graphic matroids need not even be binary. For instance, if G_1 and G_2 are as shown in Figure 12.10, then $M(G_1) \vee M(G_2) \cong U_{2,4}$. □

So far the operation of matroid union has only been defined for matroids that share a common ground set. We shall show next that the

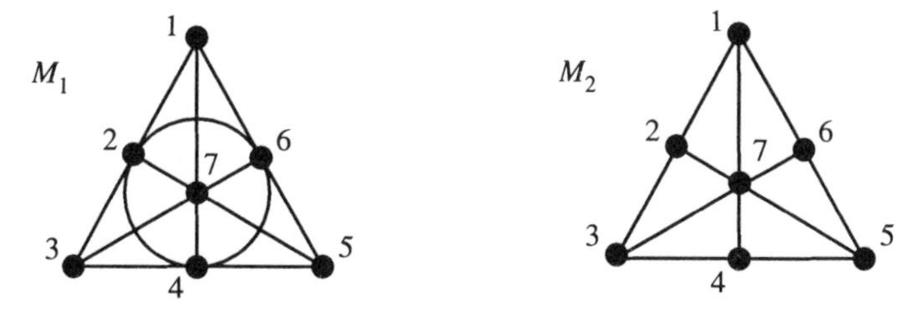

Fig. 12.9. Each of $M_1 \vee M_2$, $M_1 \vee M_1$, and $M_2 \vee M_2$ is $U_{6,7}$.

Fig. 12.10. $M(G_1) \vee M(G_2)$ is non-binary.

definition can be extended in a natural way to cover the case when this condition does not hold. Let $M_1, M_2, \ldots, M_n$ be matroids having ground sets $E_1, E_2, \ldots, E_n$, respectively, and let $E = E_1 \cup E_2 \cup \cdots \cup E_n$. For all i in $\{1, 2, \ldots, n\}$, define M_i^+ to be the matroid on E such that $M_i^+|E_i = M_i$ and $M_i^+|(E - E_i)$ has rank zero. Thus M_i^+ is formed by adjoining a set of loops to M_i, these loops being labelled by the elements of $E - E_i$. As $M_1^+, M_2^+, \ldots, M_n^+$ have the same ground set, $M_1^+ \vee M_2^+ \vee \cdots \vee M_n^+$ is well-defined, and we take $M_1 \vee M_2 \vee \cdots \vee M_n$ to be equal to this matroid. Thus $E(M_1 \vee M_2 \vee \cdots \vee M_n) = E$ and $\mathcal{I}(M_1 \vee M_2 \vee \cdots \vee M_n)$ is as specified in 12.3.2. Moreover, the rank of a subset X of E in $\vee_{i=1}^n M_i$ is

12.3.5 $$\min\{|X - Y| + \textstyle\sum_{i=1}^n r_i(Y \cap E_i) : Y \subseteq X\}$$

where r_i is the rank function of M_i.

The operation of matroid union has now been defined for any finite collection of matroids. Two important special cases of this operation that we have already met are direct sum and series connection.

12.3.6 Proposition. *Let M_1 and M_2 be matroids.*

(i) *If $E(M_1) \cap E(M_2) = \emptyset$, then $M_1 \vee M_2 = M_1 \oplus M_2$.*

(ii) *If $E(M_1) \cap E(M_2) = \{p\}$, then $M_1 \vee M_2 = S(M_1, M_2)$.*

Proof. This is left as an exercise. □

Transversal matroids have the following attractive characterization in terms of the matroid union operation. The straightforward proof of this result is omitted.

12.3.7 Proposition. *A matroid is transversal if and only if it is a union of rank-1 matroids.* □

As an immediate consequence of this proposition we have the following.

12.3.8 Corollary. *A union of transversal matroids is transversal. In particular, a series connection of transversal matroids is transversal.* □

Evidently every matroid M can be written as a union of itself with a rank-0 matroid. We call M *reducible* if it can be written as the union of two matroids neither of which is equal to M; otherwise M is *irreducible*. Welsh (1971b) raised the following problem that has so far defied solution.

12.3.9 Problem. *Characterize all irreducible matroids.* □

To get some feeling for the sorts of difficulty involved with this problem, the reader is urged to check that $M(K_4)$ is irreducible, although P_6, Q_6, $U_{3,6}$, and $\mathcal{W}^3$ are all reducible. Currently the best partial result towards 12.3.9 is the following result of Cunningham (1979), which generalizes earlier results of his (1977) and Lovász and Recski (1973). For a short proof of this result, the reader is referred to Duke (1987).

12.3.10 Proposition. *A binary matroid M is irreducible if and only if M and all of its single-element deletions are connected.* □

There are a number of other unsolved problems associated with decomposing matroids as unions of other matroids. These problems and the progress towards their solutions have been surveyed by Recski (1985).

The remainder of this section discusses various applications of the operation of matroid union. We begin by proving a packing result and a covering result due to Edmonds (1965a,b). The original proofs of these results preceded the discovery of the operation of matroid union and were difficult. The easy proofs given below (Harary and Welsh 1969) exemplify the power of the union operation. In each of the next five results, k will denote an arbitrary positive integer.

12.3.11 Theorem. *A matroid M has k disjoint bases if and only if, for every subset X of $E(M)$,*

$$kr(X) + |E(M) - X| \geq kr(M).$$

12.3.12 Theorem. *A matroid M has k independent sets whose union is $E(M)$ if and only if, for every subset X of $E(M)$,*

$$kr(X) \geq |X|.$$

Proof of Theorem 12.3.11. M has k disjoint bases if and only if $\vee_{i=1}^{k} M$ has rank at least $kr(M)$. By 12.3.3, the latter occurs if and only if $kr(X) + |E(M) - X| \geq kr(M)$ for all $X \subseteq E(M)$. □

Proof of Theorem 12.3.12. M has k independent sets whose union is $E(M)$ if and only if M has k bases whose union is $E(M)$. But the latter occurs if and only if $\vee_{i=1}^{k} M$ has rank at least $|E(M)|$. The result now follows using 12.3.3 again. □

In conjunction with his proof of Theorem 12.3.12, Edmonds (1965a) described an algorithm that, for fixed k and a given matroid M, will either produce a partition of $E(M)$ into k independent sets or establish that no such partition exists. This algorithm, known as the *matroid partitioning algorithm*, proceeds in polynomial time provided that one can decide in polynomial time whether or not any given set is independent.

An immediate consequence of Theorem 12.3.12 is a characterization of when a finite subset of a vector space can be partitioned into k linearly independent sets. This result was originally proved by Horn (1955) using an intricate algebraic argument which does not appear to generalize to arbitrary matroids.

Part of the motivation for Theorems 12.3.11 and 12.3.12 derived from the fact that the corresponding results for graphs had been proved just a few years earlier. These two graph results are derived here as consequences of Theorems 12.3.11 and 12.3.12. Both results were originally proved by Tutte (1961b), with the first also being independently derived by Nash-Williams (1961).

If π is a partition of the vertex set $V(G)$ of a graph G, then $E_\pi(G)$ will denote the set of edges of G that join vertices in different classes of π. In addition, $|\pi|$ will denote the number of classes in π.

12.3.13 Proposition. *A connected graph G has k edge-disjoint spanning trees if and only if, for every partition π of $V(G)$,*

$$|E_\pi(G)| \geq k(|\pi| - 1).$$

Proof. Suppose that the specified condition holds for all partitions π of $V(G)$. By Theorem 12.3.11, to show that $M(G)$ has k disjoint bases, we need to show that, for all subsets X of $E(G)$,

$$kr(X) + |E(G) - X| \geq kr(M(G)). \tag{1}$$

For such a subset X, let G_X be the subgraph of G having vertex set $V(G)$ and edge set X. Let π be the partition of $V(G)$ in which the classes are the vertex sets of connected components of G_X. Then, as $E(G) - X \supseteq E_\pi(G)$,

$$|E(G) - X| \geq k(|\pi| - 1). \tag{2}$$

On combining (2) with (1), we see that (1) holds provided that

$$|\pi| - 1 \geq r(M(G)) - r(X). \tag{3}$$

But $r(M(G)) = |V(G)| - 1$ and $r(X) = |V(G_X)| - \omega(G_X) = |V(G)| - |\pi|$, so (3) holds with equality and therefore G has k edge-disjoint spanning trees. The straightforward proof of the converse is left as an exercise. □

12.3.14 Proposition. *Let G be a graph. Then $E(G)$ can be partitioned into k disjoint forests if and only if, for all subsets X of $V(G)$,*

$$|E(G[X])| \leq k(|X| - 1).$$

Proof. Apply Theorem 12.3.12 to $M(G)$. □

Next we shall show how to use the operation of matroid union to determine necessary and sufficient conditions for two matroids to have a common k-element independent set (Edmonds 1970). Although the solution to this problem is relatively straightforward, the corresponding problem for three matroids remains unsolved and, indeed, is notoriously difficult (see Welsh 1976 and Garey and Johnson 1979).

12.3.15 Theorem. *Let M_1 and M_2 be matroids with rank functions r_1 and r_2 and a common ground set E. Then there is a k-element subset of E that is independent in both M_1 and M_2 if and only if, for all subsets X of E,*

$$r_1(X) + r_2(E - X) \geq k.$$

Proof. By Proposition 2.1.9, $r_2(E - X) = |E - X| + r_2^*(X) - r_2^*(M_2)$. Thus, $r_1(X) + r_2(E - X) \geq k$ for all $X \subseteq E$ if and only if, for all such X,

$$r_1(X) + r_2^*(X) + |E - X| \geq k + r_2^*(M_2). \tag{4}$$

By Theorem 12.3.1, inequality (4) holds for all $X \subseteq E$ if and only if

$$r(M_1 \vee M_2^*) \geq k + r_2^*(M_2). \tag{5}$$

To complete the proof, we shall show that (5) holds if and only if M_1 and M_2 have a common k-element independent set. If the latter holds and I is such a set, then I is independent in M_1 and $E - I$ is spanning in M_2^*. Thus $E - I$ contains a basis B_2^* of M_2^* and so

$$r(M_1 \vee M_2^*) \geq |I \,\dot{\cup}\, B_2^*| = k + r_2(M_2^*),$$

that is, (5) holds. Conversely, if (5) holds and D_2^* is a basis of M_2^*, then D_2^* is contained in a basis B of $M_1 \vee M_2^*$. Thus $B = I_1 \,\dot{\cup}\, D_2^*$ where $I_1 \in \mathcal{I}(M_1)$. Now

$$|I_1| + r_2(M_2^*) = |I_1 \,\dot{\cup}\, D_2^*| = |B| = r(M_1 \vee M_2^*) \geq k + r_2(M_2^*).$$

Hence $|I_1| \geq k$. Moreover, $I_1 \subseteq E - D_2^*$, so $I_1 \in \mathcal{I}(M_2)$. Thus I_1 contains a k-element subset that is independent in both M_1 and M_2, and the theorem follows. □

As an immediate consequence of the last theorem, we have the following result that specifies the size of a largest common independent set in two matroids.

12.3.16 Corollary. *Let M_1 and M_2 be matroids with rank functions r_1 and r_2 and a common ground set E. Then*

$$\max\{|I| : I \in \mathcal{I}(M_1) \cap \mathcal{I}(M_2)\} = \min\{r_1(T) + r_2(E-T) : T \subseteq E\}. \quad \square$$

The problem of finding a maximum-sized set that is independent in both M_1 and M_2 can be solved in polynomial time assuming that one has polynomial-time subroutines for testing independence of sets in M_1 and M_2. For descriptions of this *matroid intersection algorithm* and of algorithms that solve the corresponding weighted version of this problem, the reader is referred to Lawler (1976), Edmonds (1979), and Faigle (1987). The fundamental idea in the matroid intersection algorithm is similar to that used in the algorithm for finding a maximum-sized matching in a bipartite graph: construction of an augmenting path.

Theorem 12.3.15 will now be used to prove a result of Perfect (1968), which generalizes Ford and Fulkerson's theorem (1958) characterizing when two families of sets have a common transversal.

12.3.17 Proposition. *Let $(A_j : j \in J)$ and $(D_j : j \in J')$ be two families, $\mathcal{A}$ and $\mathcal{D}$, of subsets of a set E. Then there is a k-element subset of E that is a partial transversal of both $\mathcal{A}$ and $\mathcal{D}$ if and only if, for all $K \subseteq J$ and all $K' \subseteq J'$,*

$$|A(K) \cap D(K')| \geq |K| + |K'| - |J| - |J'| + k.$$

Proof. Evidently $\mathcal{A}$ and $\mathcal{D}$ have a common k-element partial transversal if and only if $M[\mathcal{A}]$ and $M[\mathcal{D}]$ have a common k-element independent set. By Theorem 12.3.15, this occurs if and only if, for all $X \subseteq E$,

$$r_1(X) + r_2(E-X) \geq k, \tag{6}$$

where r_1 and r_2 are the rank functions of $M[\mathcal{A}]$ and $M[\mathcal{D}]$. By Corollary 12.2.7, inequality (6) holds for all $X \subseteq E$ if and only if, for all such X,

$$\begin{aligned}&\min\{|A(K) \cap X| - |K| + |J| : K \subseteq J\}\\ &+ \min\{|D(K') \cap (E-X)| - |K'| + |J'| : K' \subseteq J'\} \geq k.\end{aligned}$$

In turn, this holds for all $X \subseteq E$ if and only if

$$\begin{aligned}&|A(K) \cap X| + |D(K') \cap (E-X)| \geq |K| + |K'| - |J| - |J'| + k\\ &\text{for all } X \subseteq E, \text{ all } K \subseteq J, \text{ and all } K' \subseteq J'.\end{aligned} \tag{7}$$

Now the left-hand side of (7) equals

$$|(A(K)\cap X)\cup(D(K')\cap(E-X))|+|A(K)\cap X\cap D(K')\cap(E-X)|,$$

which equals $|(A(K)\cap X)\cup(D(K')\cap(E-X))|$. Thus $M[\mathcal{A}]$ and $M[\mathcal{D}]$ have a common k-element independent set if and only if

$$|(A(K)\cap X)\cup(D(K')\cap(E-X))|\geq|K|+|K'|-|J|-|J'|+k$$
$$\text{for all } X\subseteq E,\text{ all } K\subseteq J,\text{ and all } K'\subseteq J'. \tag{8}$$

Letting $X=D(K')$, it follows that if $M[\mathcal{A}]$ and $M[\mathcal{D}]$ have a common k-element independent set, then

$$|A(K)\cap D(K')|\geq|K|+|K'|-|J|-|J'|+k$$
$$\text{for all } K\subseteq J \text{ and all } K'\subseteq J'. \tag{9}$$

Finally, if (9) holds, then so does (8) since

$$\begin{aligned}A(K)\cap D(K') &= (A(K)\cap D(K')\cap X)\cup(A(K)\cap D(K')\cap(E-X))\\ &\subseteq (A(K)\cap X)\cup(D(K')\cap(E-X)).\end{aligned}$$ □

12.3.18 Corollary. *(Ford and Fulkerson 1958) Two families $(A_j : j\in J)$ and $(D_j : j\in J)$ of subsets of a set E have a common transversal if and only if, for all subsets K and K' of J,*

$$|A(K)\cap D(K')|\geq|K|+|K'|-|J|.$$ □

A number of other results can be derived using this operation of matroid union. Some of these will be considered in the exercises. For a detailed general account of results in transversal theory such as 12.3.17, 12.3.18, and the theorems of Hall and Rado, we refer the reader to Mirsky's book (1971).

Exercises

1. Show that if M_1 and M_2 are matroids on a set E and $T\subseteq E$, then $(M_1\vee M_2)|T=(M_1|T)\vee(M_2|T)$.

2. For matroids M_1 and M_2 on a set E, let $M_1\wedge M_2$ be $(M_1^*\vee M_2^*)^*$. Show that $\mathcal{S}(M_1\wedge M_2)=\{X_1\cap X_2 : X_1\in\mathcal{S}(M_1),\ X_2\in\mathcal{S}(M_2)\}$.

3. Let f_1 and f_2 be non-negative, integer-valued, increasing, submodular functions on 2^E. Show that:

   (i) (Pym and Perfect 1970) $M(f_1+f_2)=M(f_1)\vee M(f_2)$;

(ii) (McDiarmid in Welsh 1976) if f_1 and f_2 are not increasing, then (i) may fail when $|E| = 2$.

4. Let T be a subset of the ground set of a matroid M. Show that:

(i) if $M|T$ is irreducible and $r(T) = r(M)$, then M is irreducible;

(ii) the converse of (i) does not hold;

(iii) the assumption that $r(T) = r(M)$ is needed in (i);

(iv) F_7 is irreducible.

5. In a matroid M having ground set E, prove that

$$\max\{|X| - 2r(X) : X \subseteq E\} = \min\{|B_1^* \cap B_2^*| : B_1^*, B_2^* \in \mathcal{B}^*(M)\}.$$

6. Prove 12.3.18 directly from 12.3.15.

7. (Welsh 1976) Prove that two families $(A_j : j \in J)$ and $(D_j : j \in J)$ of subsets of a set E have a common transversal containing a subset X of E if and only if, for all subsets K and K' of J,

$$\begin{aligned} |(A(K) \cap D(K')) - X| &+ |A(K) \cap X| + |D(K') \cap X| \\ &\geq |K| + |K'| + |X| - |J|. \end{aligned}$$

[Hint: Let M be the matroid on E whose bases are the bases of $M[(D_j : j \in J)]$ containing X. Then the required common transversal exists if and only if $(A_j : j \in J)$ has a transversal that is a basis of M. Now use Rado's Theorem.]

8. (Pym and Perfect 1970) Show that if M_1 and M_2 are matroids on a set E and $X \subseteq E$, then the rank of X in $M_1 \vee M_2$ is $\max\{r_1(Y) + r_2(X - Y) : Y \subseteq X\}$.

9. (Piff and Welsh 1970) Let $M_1, M_2, \ldots, M_m$ be F-representable matroids on a set E. Prove that there is an integer n such that $\vee_{i=1}^m M_i$ is representable over every extension of F with at least n elements.

10. (Brualdi and Scrimger 1968, Brualdi 1969) A matroid M is *base-orderable* if, given any two bases B_1 and B_2, there is a bijection $\alpha : B_1 \to B_2$ such that, for every element e of B_1, both $(B_1 - e) \cup \alpha(e)$ and $(B_2 - \alpha(e)) \cup e$ are bases. M is *strongly base-orderable* if, given any two bases B_1 and B_2, there is a bijection $\beta : B_1 \to B_2$ such that, for every subset X of B_1, both $(B_1 - X) \cup \beta(X)$ and $(B_2 - \beta(X)) \cup X$ are bases. Prove that the classes of base-orderable and strongly base-orderable matroids are closed under each of the following operations:

(i) restriction;

(ii) the taking of duals;

(iii) the taking of minors;

(iv) union;

(v) truncation.

11. Show that:

(i) All gammoids and hence all transversal matroids are strongly base-orderable.

(ii) $M(K_4)$ is not base-orderable.

(iii) P_8 is base-orderable but not strongly base-orderable.

(iv) The matroid in Figure 10.4 is not base-orderable.

(v) P_7 is strongly base-orderable but is not a gammoid.

12. (McDiarmid 1975b) Let X and Y be independent sets in a matroid M and let (X_1, X_2) be a partition of X. By considering the matroid $(M/X_1) \vee (M/X_2)$, prove that there is a partition (Y_1, Y_2) of Y such that both $X_1 \cap Y_1$ and $X_2 \cap Y_2$ are empty, and both $X_1 \cup Y_1$ and $X_2 \cup Y_2$ are independent.

13.* (Brualdi 1970) Let Δ be a bipartite graph with vertex classes S_1 and S_2, and k be a non-negative integer. For $i = 1, 2$, let M_i be a matroid on S_i having rank function r_i. Use 12.3.15 to show that the following are equivalent:

(a) there is a k-element subset of M_1 that can be matched into an independent set of M_2;

(b) $r_1(S_1 - X) + r_2(N_\Delta(X)) \geq k$ for all $X \subseteq S_1$.

12.4 Amalgams and the generalized parallel connection

Let M_1 and M_2 be matroids having a common restriction N. In such circumstances one often wants to stick these matroids together across N. We have already seen with incompatible extensions (Example 7.2.3) that this may not be possible. In this section, we investigate some circumstances under which it is possible, concentrating in particular on the operation of generalized parallel connection. The properties of this operation are discussed in the last part of the section beginning with Proposition 12.4.13. The first part of the section is based on some unpublished results of

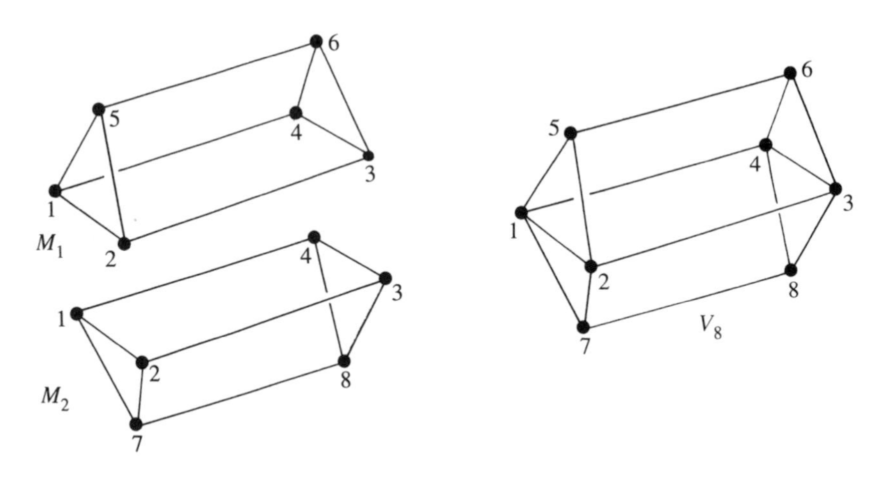

Fig. 12.11. V_8 is an amalgam of M_1 and M_2.

A. W. Ingleton and on an unpublished exposition of these results due to J. H. Mason.

Suppose that the matroids M_1 and M_2 have ground sets E_1 and E_2, rank functions r_1 and r_2, and closure operators cl_1 and cl_2. Let $E_1 \cup E_2 = E$. We are assuming that $M_1|T = M_2|T = N$ where $E_1 \cap E_2 = T$. The rank function of this common restriction of M_1 and M_2 will be denoted by r. If M is a matroid on E such that $M|E_1 = M_1$ and $M|E_2 = M_2$, then M is called an *amalgam* of M_1 and M_2. We call M_0 the *free amalgam* of M_1 and M_2 if M_0 is an amalgam of M_1 and M_2 that is freer than all other amalgams, that is, for any other amalgam M, every independent set in M is independent in M_0.

We have seen already that there may be no amalgam of M_1 and M_2. The next example shows that, even when there is an amalgam, there may be no free amalgam.

12.4.1 Example. Consider the three rank-4 matroids for which geometric representations are shown in Figure 12.11. It is clear that the Vámos matroid, V_8, is an amalgam of M_1 and M_2. Another amalgam of these two matroids is the vector matroid of the matrix A over $\mathbb{R}$ where

$$A = \begin{array}{c} \begin{array}{cccccccc} 1 & 2 & 4 & 5 & 7 & 3 & 6 & 8 \end{array} \\ \left[\begin{array}{c|ccc} & 1 & 1 & 1 \\ & 1 & 0 & 0 \\ I_5 & 1 & 1 & 1 \\ & 0 & 1 & 0 \\ & 0 & 0 & 1 \end{array}\right] \end{array}.$$

Assume that the free amalgam M_0 of M_1 and M_2 exists. Then, as $\{1,2,4,5,7\}$ is independent in $M[A]$, it is independent in M_0, so $r(M_0) \geq 5$. Moreover, as $\{5,6,7,8\}$ is independent in V_8, it is independent in M_0. Therefore $\{5,6,7,8\}$ is contained in a 5-element independent set I in M_0. By symmetry, we may assume that I is $\{5,6,7,8,1\}$. But $\{5,6,1,4\}$ is a circuit of M_1 and hence of M_0, and $\{7,8,1,4\}$ is a circuit of M_2 and hence of M_0. Thus, eliminating the element 4, we get that $\{5,6,7,8,1\}$ is dependent in M_0. This contradiction shows that there is no free amalgam of M_1 and M_2. □

We observe next that if M is an arbitrary amalgam of M_1 and M_2, then, by submodularity of the rank function, for all $X \subseteq E$,

$$r_M(X) \leq \eta(X)$$

where

$$\eta(X) = r_1(X \cap E_1) + r_2(X \cap E_2) - r(X \cap T). \tag{1}$$

Now let

$$\zeta(X) = \min\{\eta(Y) : Y \supseteq X\}.$$

Then, for all $X \subseteq E$,

$$\zeta(X) \geq r_M(X). \tag{2}$$

Throughout the rest of this section, where ambiguity does not arise, we shall often drop the subscripts on the various rank functions involved.

12.4.2 Proposition. *If ζ is a submodular function on 2^E, then it is the rank function of a matroid on E. Moreover, this matroid is the free amalgam of M_1 and M_2.*

Proof. It is not difficult to show that ζ satisfies (R1)–(R3). Thus ζ is the rank function of a matroid M on E. If $X \subseteq E_1$ and $Y \supseteq X$, then, by (1),

$$\eta(Y) = r_1(Y \cap E_1) + r_2(Y \cap E_2) - r(Y \cap T) \geq r_1(Y \cap E_1) \geq r_1(X).$$

Hence $\zeta(X) \geq r_1(X)$. As $\eta(X) = r_1(X)$, it follows that $\zeta(X) = r_1(X)$ and so $M|E_1 = M_1$. Similarly, $M|E_2 = M_2$. Thus M is indeed an amalgam of M_1 and M_2. Moreover, by (2) and Proposition 7.3.11, M is the free amalgam of M_1 and M_2. □

When ζ is submodular, the matroid on E that has ζ as its rank function is called the *proper amalgam* of M_1 and M_2. The last result established that every proper amalgam is a free amalgam. The next example

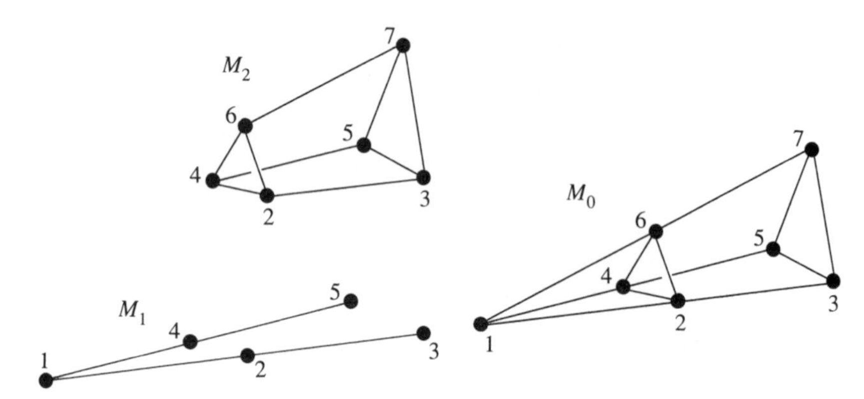

Fig. 12.12. M_0 is the free amalgam of M_1 and M_2 but is not proper.

shows that some free amalgams are not proper. To verify that this example has the stated properties, we shall use the following result, the proof of which is left to the reader.

12.4.3 Proposition. *A given matroid M is the proper amalgam of $M|E_1$ and $M|E_2$ if and only if, for every flat F of M,*

$$r(F) = r(F \cap E_1) + r(F \cap E_2) - r(F \cap T). \qquad \square$$

12.4.4 Example. In Figure 12.12, the matroid M_0 is clearly an amalgam of M_1 and M_2. In fact, it is the free amalgam. We recall here from Example 1.5.5 that 1, 6, and 7 must be collinear in order for M_0 to be a matroid. To see that M_0 is not the proper amalgam of M_1 and M_2, it suffices to note that the rank equation in Proposition 12.4.3 fails when F is the flat $\{1, 6, 7\}$ of M_0. $\square$

The remainder of our discussion of general amalgams will be concerned with finding conditions to guarantee the existence of the proper amalgam of M_1 and M_2. We remark here that while Proposition 12.4.3 gives necessary and sufficient conditions for a *given* matroid to be this proper amalgam, it says nothing about when the proper amalgam exists.

Let $\mathcal{L}(M_1, M_2)$ denote the set of subsets X of E such that $X \cap E_1$ and $X \cap E_2$ are flats of M_1 and M_2, respectively. Clearly every flat of any amalgam of M_1 and M_2 is in $\mathcal{L}(M_1, M_2)$. Moreover, $\mathcal{L}(M_1, M_2)$ is a lattice of subsets of E in which $X \wedge Y = X \cap Y$. The next result shows that ζ is uniquely determined by the values taken by η on members of $\mathcal{L}(M_1, M_2)$.

12.4.5 Proposition. *For all $X \subseteq E$,*

$$\zeta(X) = \min\{\eta(Y) : Y \in \mathcal{L}(M_1, M_2) \text{ and } Y \supseteq X\}.$$

This result follows without difficulty from the following.

12.4.6 Lemma. *If $Y \subseteq E$ and Z is the least member of $\mathcal{L}(M_1, M_2)$ containing Y, then $\eta(Z) \le \eta(Y)$.*

Proof. For all $X \subseteq E$, let $\phi_1(X) = \mathrm{cl}_1(X \cap E_1) \cup (X \cap E_2)$ and $\phi_2(X) = (X \cap E_1) \cup \mathrm{cl}_2(X \cap E_2)$. Then

$$\phi_1(Y) \cap E_2 = (\phi_1(Y) \cap T) \cup (Y \cap E_2)$$

and

$$Y \cap T \subseteq (\phi_1(Y) \cap T) \cap (Y \cap E_2).$$

Using these facts and the submodularity of r_2, we get that

$$r_2(\phi_1(Y) \cap E_2) + r_2(Y \cap T) \le r_2(\phi_1(Y) \cap T) + r_2(Y \cap E_2),$$

so

$$r_2(\phi_1(Y) \cap E_2) - r_2(\phi_1(Y) \cap T) \le r_2(Y \cap E_2) - r_2(Y \cap T).$$

But also $r_1(\phi_1(Y) \cap E_1) = r_1(Y \cap E_1)$. Hence

$$\eta(\phi_1(Y)) \le r_1(Y \cap E_1) + r_2(Y \cap E_2) - r_2(Y \cap T) = \eta(Y).$$

Now applying the above argument with the subscripts 1 and 2 interchanged and Y replaced by $\phi_1(Y)$, we get that

$$\eta(\phi_2(\phi_1(Y))) \le \eta(\phi_1(Y)).$$

Thus, by applying the operations ϕ_1 and ϕ_2 alternately, we get, after a finite number of steps, a set Z_0 such that $\eta(Z_0) \le \eta(Y)$ and $\phi_1(Z_0) = Z_0 = \phi_2(Z_0)$. It is not difficult to check that $Z_0 \in \mathcal{L}(M_1, M_2)$ and that, in fact, $Z_0 = Z$. The lemma follows. □

The next proposition is Ingleton's main result on when the proper amalgam of two matroids is guaranteed to exist.

12.4.7 Proposition. *Suppose that η is submodular on $\mathcal{L}(M_1, M_2)$. Then the proper amalgam of M_1 and M_2 exists.*

Proof. Suppose $X_1, X_2 \subseteq E$. By Proposition 12.4.5, for $i = 1, 2$, there is a member Y_i of $\mathcal{L}(M_1, M_2)$ such that $X_i \subseteq Y_i$ and $\zeta(X_i) = \eta(Y_i)$. Evidently $X_1 \cap X_2 \subseteq Y_1 \wedge Y_2$ and $X_1 \cup X_2 \subseteq Y_1 \vee Y_2$. Therefore

$$\begin{aligned} \zeta(X_1 \cap X_2) + \zeta(X_1 \cup X_2) &\le \eta(Y_1 \wedge Y_2) + \eta(Y_1 \vee Y_2) \\ &\le \eta(Y_1) + \eta(Y_2) = \zeta(X_1) + \zeta(X_2). \end{aligned}$$

Thus ζ is submodular on 2^E and so, by Proposition 12.4.2 and the remarks following it, the proper amalgam of M_1 and M_2 exists. □

Next we seek conditions that will ensure that η is submodular on $\mathcal{L}(M_1, M_2)$. If M is an arbitrary matroid and X and Y are subsets of $E(M)$, we define the *modular defect* $\delta(X,Y)$ of (X,Y) to be $r(X)+r(Y)-r(X\cup Y)-r(X\cap Y)$. A subset T of $E(M)$ is said to be *fully embedded* in M if $\delta(F_1\cap T, F_2\cap T)\le\delta(F_1,F_2)$ for all flats F_1 and F_2 of M. Thus, in Example 12.4.4, if $T=\{2,3,4,5\}$, then T is not fully embedded in M_0. To see this, let $F_1=\{1,4,5,6,7\}$ and $F_2=\{1,2,3,6,7\}$. Then $\delta(F_1,F_2)=0$. But $\delta(F_1\cap T,F_2\cap T)=1$.

It is not difficult to check that loops and parallel elements have no effect on whether or not a set is fully embedded.

12.4.8 Lemma. *Let M be a matroid and suppose that $T\subseteq T'\subseteq E(M)$. Assume that every element of $T'-T$ is either a loop or is parallel to an element of T. Then T' is fully embedded if and only if T is fully embedded.* □

The main examples of proper amalgams will be derived from the following result.

12.4.9 Proposition. *If T is fully embedded in M_1, then η is submodular on $\mathcal{L}(M_1,M_2)$ and hence the proper amalgam of M_1 and M_2 exists.*

Proof. Suppose X and Y are in $\mathcal{L}(M_1,M_2)$. Then, as T is fully embedded in M_1, it follows that $\delta(X\cap T, Y\cap T)\le\delta(X\cap E_1, Y\cap E_1)$. Now, by Lemma 12.4.6 and the fact that $X\wedge Y=X\cap Y$, we have that

$$\eta(X)+\eta(Y)-\eta(X\wedge Y)-\eta(X\vee Y)\ge\eta(X)+\eta(Y)-\eta(X\cap Y)-\eta(X\cup Y).$$

On expanding the right-hand side of the last line using (1) and then rearranging terms, we obtain

$$\begin{aligned}[\delta(X\cap E_1,Y\cap E_1) \; - \; &\delta(X\cap T,Y\cap T)]+[r_2(X\cap E_2)+r_2(Y\cap E_2)\\ &-r_2((X\cup Y)\cap E_2)-r_2((X\cap Y)\cap E_2)].\end{aligned}$$

Each quantity enclosed in square brackets is non-negative, hence η is indeed submodular on $\mathcal{L}(M_1,M_2)$. It follows by Proposition 12.4.7 that the proper amalgam of M_1 and M_2 exists. □

The next result is our main application of Proposition 12.4.9.

12.4.10 Theorem. *If either of the following conditions holds, then T is fully embedded in M_1 and so the proper amalgam of M_1 and M_2 exists:*

(i) *T is a modular flat of M_1.*

(ii) *$M_1|T$ is a modular matroid.*

Proof. Part (ii) is immediate from the definition. To prove (i), we shall use the following lemma which holds for an arbitrary matroid having rank function r. The proof is left to the reader.

12.4.11 Lemma. *If F and T are flats with T modular and if $F \cap T \subseteq W \subseteq T$, then*

$$r(F \cup W) + r(F \cap W) = r(F) + r(W).$$ □

Now suppose that T is a modular flat of M_1. To complete the proof of 12.4.10, we shall show that T is fully embedded in M_1. Let X and Y be flats of M_1. Then, it is not difficult to show using Lemma 8.2.10 that

$$\begin{aligned} r(X \cup Y) &- r([X \cap (Y \cup T)] \cup [Y \cap (X \cup T)]) \\ \leq r(X) &- r(X \cap (Y \cup T)) + r(Y) - r(Y \cap (X \cup T)). \end{aligned}$$

Moreover, by Lemma 12.4.11,

$$r(X \cap (Y \cup T)) = r(X \cap Y) + r(X \cap T) - r(X \cap Y \cap T)$$

and

$$r(Y \cap (X \cup T)) = r(X \cap Y) + r(Y \cap T) - r(X \cap Y \cap T).$$

Also $[X \cap (Y \cup T)] \cup [Y \cap (X \cup T)] = (X \cap Y) \cup [(X \cup Y) \cap T]$ and, by submodularity,

$$r((X \cap Y) \cup [(X \cup Y) \cap T]) \leq r(X \cap Y) + r((X \cup Y) \cap T) - r(X \cap Y \cap T).$$

On combining these observations, it is not difficult to show that $\delta(X, Y) \geq \delta(X \cap T, Y \cap T)$. Hence T is fully embedded in M_1. □

On combining the last theorem with Lemma 12.4.8, we get the following.

12.4.12 Corollary. *Suppose that $\mathrm{cl}_1(T)$ is a modular flat of M_1 and every non-loop element of $\mathrm{cl}_1(T) - T$ is parallel to some element of T. Then T is fully embedded in M_1, so the proper amalgam of M_1 and M_2 exists.* □

Now recall that, if M is a matroid, then $\widetilde{M}$ denotes the simple matroid associated with M. We shall denote the matroid $\widetilde{M_1|T}$ by $\widetilde{T}$. In this notation, the hypothesis of the last corollary is that $\widetilde{T}$ is a modular flat of $\widetilde{M_1}$. When this condition holds, we call the proper amalgam of M_1 and M_2 the *generalized parallel connection* of M_1 and M_2. This matroid will be denoted by $P_N(M_1, M_2)$ where we recall that $N = M_1|T = M_2|T$. Brylawski (1975b) has studied the generalized parallel connection in detail when both M_1 and M_2 are simple. The extension of his work to the general case is straightforward. Note that, by Proposition 12.4.3, for every flat F of $P_N(M_1, M_2)$,

$$\zeta(F) = r(F) = r(F \cap E_1) + r(F \cap E_2) - r(F \cap T). \tag{3}$$

12.4.13 Proposition. *The flats of $P_N(M_1, M_2)$ are precisely the members of $\mathcal{L}(M_1, M_2)$, that is, F is a flat of $P_N(M_1, M_2)$ if and only if $F \cap E_1$ is a flat of M_1 and $F \cap E_2$ is a flat of M_2.*

Proof. Since $P_N(M_1, M_2)$ is an amalgam of M_1 and M_2, every flat of it is certainly in $\mathcal{L}(M_1, M_2)$. Now suppose X is in $\mathcal{L}(M_1, M_2)$ and let Z be its closure $\mathrm{cl}_P(X)$ in $P_N(M_1, M_2)$. Then Z is also in $\mathcal{L}(M_1, M_2)$ and, since ζ is the rank function of $P_N(M_1, M_2)$, we have $\zeta(X) = \zeta(Z)$. For $i = 1, 2$, let $X_i = X \cap E_i$ and $Z_i = Z \cap E_i$. By (3), (1), and the fact that $Z \cap T = Z_1 \cap T$, we have

$$\zeta(Z) = r(Z_1) + r(Z_2) - r(Z_1 \cap T) = \eta(Z).$$

But $\zeta(X) \leq \eta(X)$. Hence $\eta(Z) \leq \eta(X)$, that is,

$$r(Z_1) + r(Z_2) - r(Z_1 \cap T) \leq r(X_1) + r(X_2) - r(X_1 \cap T). \tag{4}$$

As $\widetilde{T}$ is a modular flat of $\widetilde{M_1}$, it follows easily that

$$r(X_1) + r(T) = r(X_1 \cap T) + r(X_1 \cup T)$$

and

$$r(Z_1 \cap T) + r(Z_1 \cup T) = r(Z_1) + r(T).$$

Adding the last two equations and eliminating $r(T)$, we get

$$r(X_1) - r(X_1 \cap T) = r(Z_1) - r(Z_1 \cap T) + [r(X_1 \cup T) - r(Z_1 \cup T)].$$

Therefore, as $X_1 \cup T \subseteq Z_1 \cup T$,

$$r(X_1) - r(X_1 \cap T) \leq r(Z_1) - r(Z_1 \cap T). \tag{5}$$

But we also know, since $X_2 \subseteq Z_2$, that

$$r(X_2) \leq r(Z_2). \tag{6}$$

If we now combine (5) and (6) with (4), we deduce that equality holds in all three of those statements. In particular, as X_2 and Z_2 are flats of M_2 of the same rank and $X_2 \subseteq Z_2$, we have $X_2 = Z_2$. Thus

$$X_1 \cap T = X \cap T = X_2 \cap T = Z_2 \cap T = Z \cap T = Z_1 \cap T.$$

Hence, since (5) holds with equality, $r(X_1) = r(Z_1)$, so $X_1 = Z_1$. We conclude that $X = Z$, that is, X is a flat of $P_N(M_1, M_2)$. □

As an alternative to proceeding through the derivation of $P_N(M_1, M_2)$ as a particular proper amalgam, one can simply define $P_N(M_1, M_2)$ to be the matroid on $E_1 \cup E_2$ whose flats are those subsets X of $E_1 \cup E_2$ such that $X \cap E_1$ is a flat of M_1 and $X \cap E_2$ is a flat of M_2. The fact that this construction does indeed produce a matroid is proved directly by Brylawski (1975b, pp. 19–20) in the case that both M_1 and M_2 are simple. The extension of this argument to the general case is straightforward.

The generalized parallel connection is a very natural construction and has numerous attractive properties. The 11-point matroid in Figure 1.14(b) was obtained by sticking together two copies of F_7 along a 3-point line. We are now able to identify this matroid as $P_N(F_7, F_7)$ where $N \cong U_{2,3}$. When $|T| = 1$, $P_N(M_1, M_2)$ is just the parallel connection $P(M_1, M_2)$. Indeed, the parallel connection of M_1 and M_2 with respect to the basepoint p can always be thought of as a generalized parallel connection provided $M_1|\{p\} = M_2|\{p\}$. The last condition fails precisely when p is a loop in exactly one of M_1 and M_2.

The proofs of the following properties of the generalized parallel connection are not difficult and are left to the reader.

12.4.14 Proposition. *The generalized parallel connection has the following properties:*

(i) $P_N(M_1, M_2)|E_1 = M_1$ *and* $P_N(M_1, M_2)|E_2 = M_2$.

(ii) *If* $\widetilde{T}$ *is a modular flat in* $\widetilde{M_2}$ *as well as in* $\widetilde{M_1}$, *then* $P_N(M_2, M_1) = P_N(M_1, M_2)$.

(iii) $\widetilde{E}_2$ *is a modular flat of the simple matroid associated with* $P_N(M_1, M_2)$.

(iv) *If* $e \in E_1 - T$, *then* $P_N(M_1, M_2)\backslash e = P_N(M_1\backslash e, M_2)$.

(v) *If* $e \in E_1 - \mathrm{cl}_1(T)$, *then* $P_N(M_1, M_2)/e = P_N(M_1/e, M_2)$.

(vi) *If* $e \in E_2 - T$, *then* $P_N(M_1, M_2)\backslash e = P_N(M_1, M_2\backslash e)$.

(vii) *If* $e \in E_2 - \mathrm{cl}_2(T)$, *then* $P_N(M_1, M_2)/e = P_N(M_1, M_2/e)$.

(viii) *If* $e \in T$, *then* $P_N(M_1, M_2)/e = P_{N/e}(M_1/e, M_2/e)$.

(ix) $P_N(M_1, M_2)/T = (M_1/T) \oplus (M_2/T)$. □

In the preceding proposition, we did not identify $P_N(M_1, M_2)\backslash e$ when $e \in T$; nor did we identify $P_N(M_1, M_2)/e$ when $e \in \mathrm{cl}_1(T) - T$ or

$e \in \text{cl}_2(T) - T$. In the first case, $\widetilde{T\backslash e}$ need not be modular in $\widetilde{M_1\backslash e}$; in the third case, N remains a restriction of M_1 but not necessarily of M_2/e. In the second case, e is a loop or is parallel to some element of T, so we can determine $P_N(M_1, M_2)/e$ from (iv) or (viii), respectively.

We saw in Theorem 7.1.16 that if p is an element of a connected matroid M and if M/p is disconnected, then M can be written as a parallel connection. The next result generalizes this.

12.4.15 Proposition. *Let M be a matroid on a set E and suppose that, for some subset T of E, the matroid $M/T = M_1 \oplus M_2$. If $\widetilde{T}$ is a modular flat of the simple matroid associated with $M\backslash E(M_2)$, then*

$$M = P_{M|T}(M\backslash E(M_2), M\backslash E(M_1)).$$

Proof. A proof for simple matroids can be found in Brylawski (1975b). As an alternative, one can argue using the modular short circuit axiom (6.9.9). Although the argument is non-trivial, the details are omitted. □

It was noted in Chapter 7 that the operation of parallel connection is closely related to the operation of 2-sum for graphs. The generalized parallel connection is similarly related to the more general graph operation of *clique-sum* (Robertson and Seymour 1984). In the latter, one sticks two graphs together along a common complete subgraph and then deletes all identified edges. An example of this operation is shown in Figure 12.13. This operation is formally described as follows. If G_1 and G_2 are graphs each having a K_n-subgraph for some positive integer n, then to form an *n-sum* of G_1 and G_2, one first pairs the vertices of the chosen K_n-subgraph of G_1 with distinct vertices of the chosen K_n-subgraph of G_2. The paired vertices are then identified, as are the corresponding pairs of edges. Finally, all identified edges are deleted.

To see the relationship between this operation and generalized parallel connection, observe that if G_1' is the chosen K_n-subgraph of G_1, then, by Proposition 6.9.11, $E(G_1')$ is a modular flat T of $\widetilde{M(G_1)}$. Thus, if G is the resulting n-sum of G_1 and G_2, then $M(G) \cong P_N(M(G_1), M(G_2))\backslash T$. This idea of deleting the elements of T from the generalized parallel connection is what distinguishes the 2-sum of two matroids from their parallel connection where, in this case, $|T| = 1$. The same idea was used by Seymour (1980b) in defining the 3-sum of two binary matroids M_1 and M_2, a generalization of the operation of 3-sum for graphs. Suppose $E(M_1) \cap E(M_2) = T$ and let $M_1|T$ and $M_2|T$ be 3-circuits. By Corollary 6.9.6, T is a modular flat in both $\widetilde{M_1}$ and $\widetilde{M_2}$. Now assume that both $E(M_1)$ and $E(M_2)$ have more than six elements and that T does not contain a cocircuit of M_1 or M_2. Then the *3-sum* $M_1 \oplus_3 M_2$ of M_1 and M_2 is $P_N(M_1, M_2)\backslash T$. This operation features prominently in Seymour's

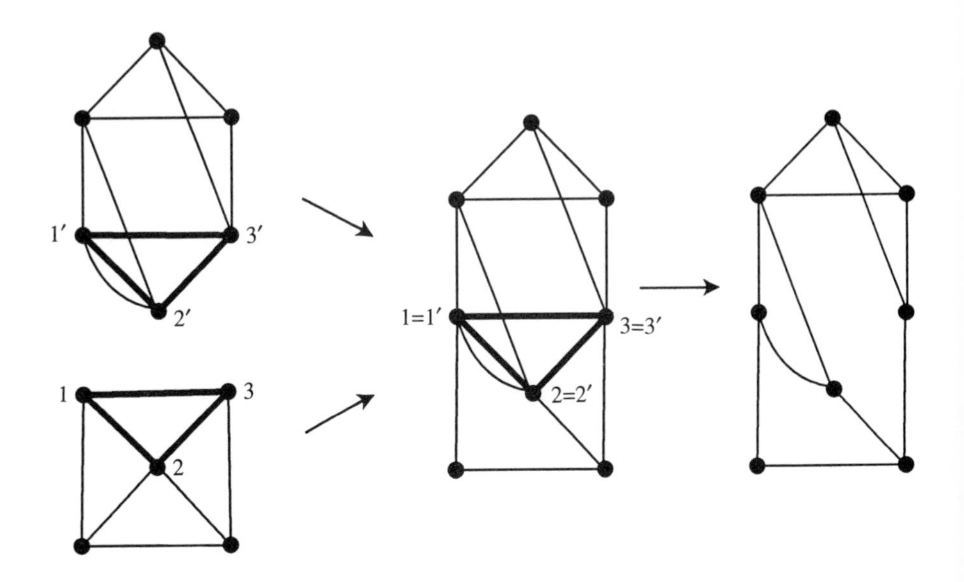

Fig. 12.13. Forming a 3-sum of two graphs.

decomposition theorem for regular matroids, which will be presented in the next chapter. We remark that the requirement that both $|E(M_1)|$ and $|E(M_2)|$ exceed six is equivalent to the requirement that $|E(M_1 \oplus_3 M_2)|$ exceeds both $|E(M_1)|$ and $|E(M_2)|$.

Seymour actually defined the 3-sum of binary matroids M_1 and M_2 as a special case of the following operation. If $E(M_1) = E_1$ and $E(M_2) = E_2$, then it is routine to check that there is a binary matroid $M_1 \triangle M_2$ on $E_1 \triangle E_2$, the circuits of which are the minimal non-empty subsets of $E_1 \triangle E_2$ that are of the form $X_1 \triangle X_2$ where X_i is a disjoint union of circuits of M_i for $i = 1, 2$. If $E_1 \cap E_2$ is a triangle of both M_1 and M_2 that does not contain a cocircuit of M_1 or M_2, and both $|E_1|$ and $|E_2|$ exceed six, then the reader can check that $M_1 \triangle M_2$ is precisely $M_1 \oplus_3 M_2$.

A fundamental feature of the matroid operations of direct sum and 2-sum is their link with the presence of 1- and 2-separations: M can be written as a direct sum of two of its proper minors if and only if M has an exact 1-separation; and M can be written as a 2-sum of two of its proper minors if and only if M has an exact 2-separation. In the case of 3-sums, the situation is similar but somewhat more complicated. Firstly, we are confined to binary matroids. Secondly, we need to impose an additional condition to guarantee that the parts of a 3-sum are isomorphic to minors of the whole. Specifically:

12.4.16 Proposition. *If M is a 3-connected binary matroid and M is a 3-sum of M_1 and M_2, then M_1 and M_2 are isomorphic to proper minors of M.*

Proof. We refer the reader to Seymour (1980b) for the proof, which is non-trivial. □

There is a third way in which the behaviour of 3-sums differs from that of 1- and 2-sums: being expressible as a 3-sum is not equivalent to having an exact 3-separation, although, as the next result shows, these properties are very similar.

12.4.17 Proposition. *Let (X_1, X_2) be an exact 3-separation of a binary matroid M with $|X_1|, |X_2| \geq 4$, and let Z be a 3-element set that is disjoint from $E(M)$. Then there are binary matroids M_1 and M_2 on $X_1 \cup Z$ and $X_2 \cup Z$, respectively, such that $M = M_1 \oplus_3 M_2$. Conversely, if M is a 3-sum of M_1 and M_2, then $(E(M_1) - E(M_2), E(M_2) - E(M_1))$ is an exact 3-separation of M, and both $|E(M_1) - E(M_2)|$ and $|E(M_2) - E(M_1)|$ exceed three.*

Proof. We shall assume that M is simple since the only additional difficulties posed by the general case are notational. The argument we shall give is due to Seymour (private communication). Let $r(M) = r$. Then M can be viewed as a restriction of $PG(r-1, 2)$. As (X_1, X_2) is an exact 3-separation of M,

$$r(X_1) + r(X_2) = r(M) + 2. \tag{7}$$

Let cl_P denote the closure operator of the matroid $PG(r-1, 2)$. Then, by (7), as $PG(r-1, 2)$ is a modular matroid, $r(\mathrm{cl}_P(X_1) \cap \mathrm{cl}_P(X_2)) = 2$. Thus $\mathrm{cl}_P(X_1) \cap \mathrm{cl}_P(X_2)$ is a 3-point line $\{a_1, a_2, a_3\}$ of $PG(r-1, 2)$. Now, for each i in $\{1, 2, 3\}$, adjoin a new element z_i to $PG(r-1, 2)$ in parallel with a_i. Let $Z = \{z_1, z_2, z_3\}$ and P' be the extension of $PG(r-1, 2)$ by Z. Then one easily checks that $M = M_1 \oplus_3 M_2$ where $M_j = P'|(X_j \cup Z)$ for $j = 1, 2$.

We omit the proof of the converse (see Seymour 1980b, (2.9)). □

The last two results are used in the proof of Seymour's decomposition theorem for regular matroids. To close this chapter, we comment briefly on how various classes of matroids behave under generalized parallel connection. We noted in Example 7.1.24 that the class of transversal matroids is not closed under parallel connection. Hence this class is certainly not closed under generalized parallel connection; nor is the class of cographic matroids (Exercise 7). Moreover, these are not the only well-known classes that are not closed under generalized parallel connection.

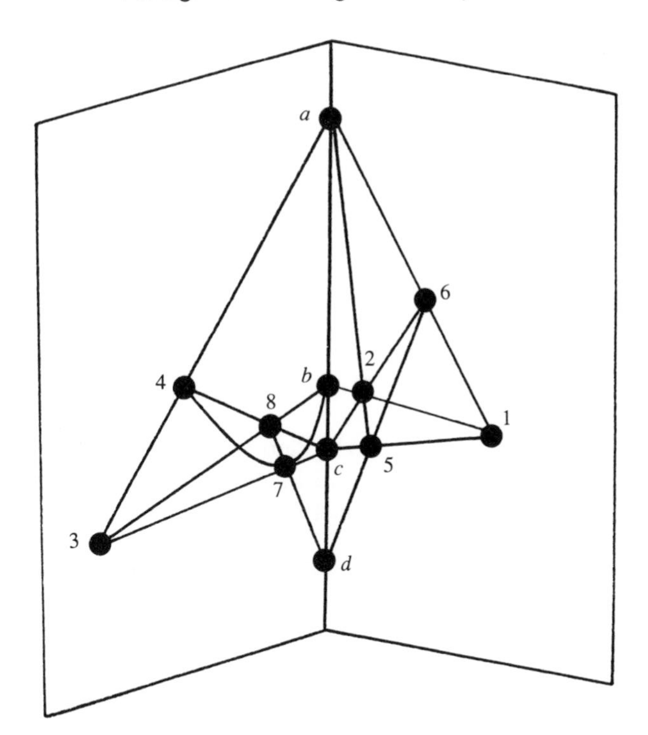

Fig. 12.14. A generalized parallel connection across a 4-point line.

12.4.18 Example. Consider T_8, the vector matroid of the matrix $[I_4|J_4 - I_4]$ over $GF(3)$. This matroid is F-representable if and only if F has characteristic three (Exercise 5 of Section 6.5). If the columns of $[I_4|J_4 - I_4]$ are labelled, in order, $1, 2, \ldots, 8$, then $T_8 = M|\{1, 2, \ldots, 8\}$ where M is the 12-element matroid for which a geometric representation is shown in Figure 12.14. Letting $M_1 = M\backslash\{1, 2, 5, 6\}$, $M_2 = M\backslash\{3, 4, 7, 8\}$, and $N = M|\{a, b, c, d\}$, it is not difficult to check that $M = P_N(M_1, M_2)$. It is also straightforward to show that M_1 is F-representable unless F has characteristic two and that $M_2 \cong M_1$. We conclude that if F has characteristic other than two or three, then both M_1 and M_2 are F-representable yet $P_N(M_1, M_2)$, since it has T_8 as a restriction, is not. □

In contrast to the above, we have the following.

12.4.19 Proposition. *The classes of graphic, regular, binary, and ternary matroids are all closed under the operation of generalized parallel connection.* □

For a proof of this result and for a discussion of various other properties of the generalized parallel connection, we refer the reader to the work of Brylawski (1975b, 1986a). Moreover, further results on amalgams in general can be found in Bachem and Kern (1988), Nešetřil, Poljak and Turzík (1981, 1985), and Poljak and Turzík (1982, 1984).

Exercises

1. (Poljak and Turzík 1984) Let M_1 and M_2 be matroids on the sets E_1 and E_2, respectively, and let M be an amalgam of these two matroids. Show that:

(i) if $X \subseteq E_1 \cup E_2$, then $M|X$ is an amalgam of $M_1|X$ and $M_2|X$;

(ii) if $Y \subseteq E_1 \cap E_2$, then M/Y is an amalgam of M_1/Y and M_2/Y.

2. Let M_1 and M_2 be the matroids for which geometric representations are shown in Figure 12.15. Show that M_1 and M_2 have an amalgam isomorphic to F_7 but that their free amalgam is isomorphic to F_7^-.

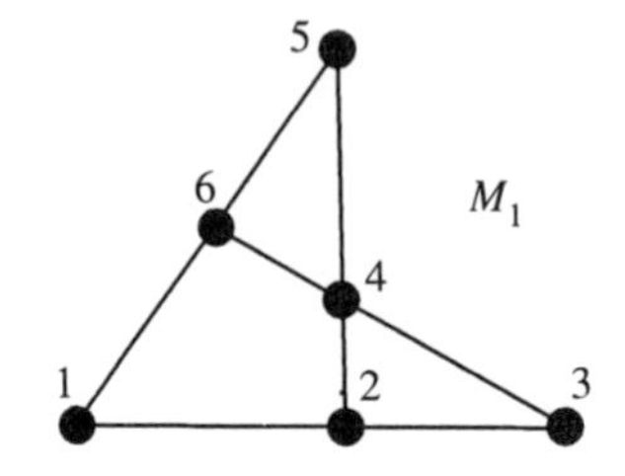

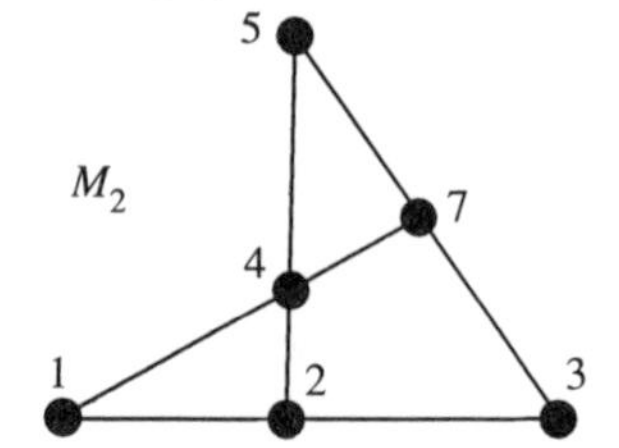

Fig. 12.15.

3. (Poljak and Turzík 1984) Let M_1 and M_2 be F-representable matroids on the sets E_1 and E_2, respectively. Show that if the matroids $M_1|(E_1 \cap E_2)$ and $M_2|(E_1 \cap E_2)$ are equal and are uniquely F-representable, then there is an F-representable matroid that is an amalgam of M_1 and M_2.

4. Show that if a set T is fully embedded in a loopless matroid M, then T is a flat or a spanning set of M.

5. Let $M = P_N(M_1, M_2)$ where $N = M_1|T = M_2|T$. Let cl_1, cl_2, and cl_M denote the closure operators of M_1, M_2, and M, respectively. If $X \subseteq E(M)$, show that:

(i) $\text{cl}_M(X) = \text{cl}_1(X \cap E_1) \cup \text{cl}_2(X \cap E_2)$; and

(ii) $r(X) = r(X \cap E_1) + r(X \cap E_2) - r(T \cap [\text{cl}_1(X \cap E_1) \cup \text{cl}_2(X \cap E_2)])$.

6. (Seymour 1980b) Let M_1 and M_2 be binary matroids. Show that:

(i) if $M = M_1 \Delta M_2$, then $M^* = M_1^* \Delta M_2^*$;

(ii) if $M = M_1 \oplus_3 M_2$, then $M^* \neq M_1^* \oplus_3 M_2^*$.

7. (Brylawski 1975b) Show that the generalized parallel connection of two copies of $M(K_5\backslash e)$ across an $M(K_4)$ has an $M(K_{3,3})$-minor. Deduce that the class of cographic matroids is not closed under the operation of generalized parallel connection.

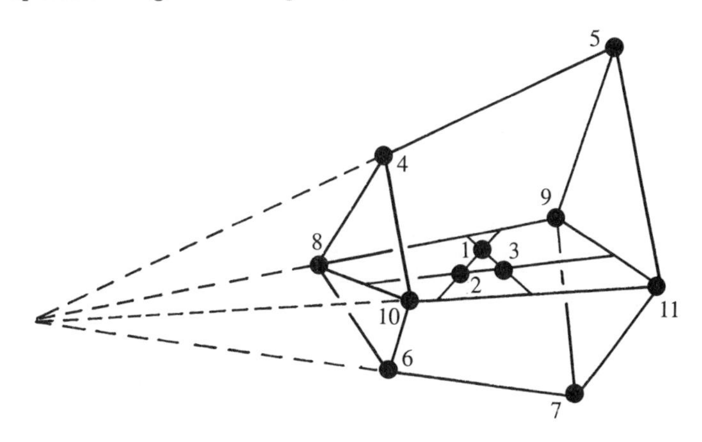

Fig. 12.16. The converse of Proposition 12.4.7 fails.

8. (Mason) Let M be the matroid for which a geometric representation is shown in Figure 12.16. Let $E_1 = \{1, 2, \ldots, 9\}$ and $E_2 = \{1, 2, \ldots, 7\} \cup \{10, 11\}$. Show that the converse of Proposition 12.4.7 fails by showing that M is the proper amalgam of $M|E_1$ and $M|E_2$, but η is not submodular on $\mathcal{L}(M|E_1, M|E_2)$.

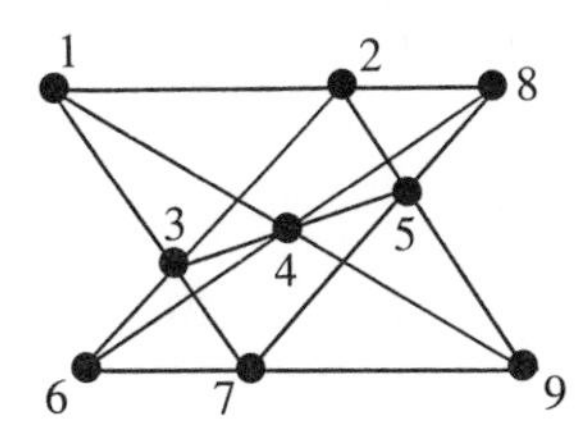

Fig. 12.17. The Pappus matroid.

9. (Ingleton) Let M be the Pappus matroid labelled as in Figure 12.17. Let $E_1 = \{1, 2, \ldots, 8\}$ and $E_2 = \{1, 2, \ldots, 7, 9\}$. Show that the converse of Proposition 12.4.9 fails by showing that M is the proper amalgam of $M|E_1$ and $M|E_2$, and η is submodular on $\mathcal{L}(M|E_1, M|E_2)$, but $\{1, 2, \ldots, 7\}$ is not fully embedded in M_1 or M_2.

10. (T.J. Reid, private communication) Let $P = P_N(M_1, M_2)$ where M_1 and M_2 are n-connected matroids each having at least $2n-2$ elements for some $n \geq 2$. Prove that P is n-connected if and only if N has at least $n-1$ elements. [Hint: Use Exercise 5(ii) of Section 8.3.]

11. Let $F_8 = P_N(F_7, F_7^-)\backslash E(N)$ where $N \cong U_{2,3}$ (see Figure 12.18 or Exercise 5 of Section 6.4). Show that:

(i) F_8 is self-dual but is not identically self-dual.

(ii) F_8 can be obtained from the 3-sum of two copies of F_7 by relaxing two intersecting circuit–hyperplanes.

(iii) F_8 is not a (minimal) excluded minor for F-representability for any field F.

(iv) F_8 is non-algebraic even though there are fields, for example $GF(2)$, over which both F_7 and F_7^- are algebraic.

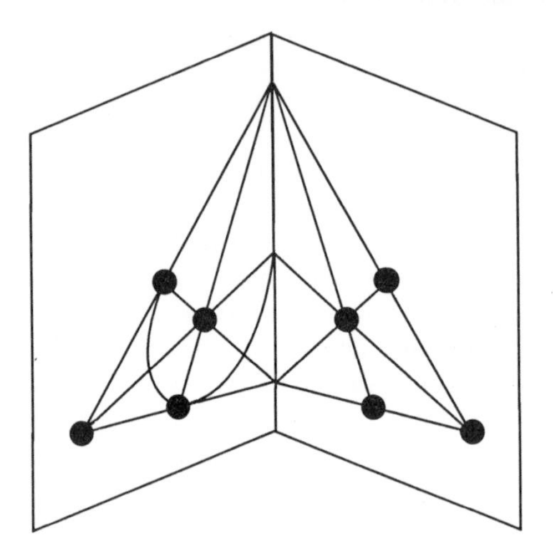

Fig. 12.18. The matroid F_8.

13
Regular matroids

A regular matroid was defined in Chapter 1 to be a matroid that can be represented over $\mathbb{R}$ by a totally unimodular matrix, the latter being a matrix for which the determinant of every square submatrix is in $\{0, 1, -1\}$. We proved in Chapter 6 that a matroid is regular if and only if it is representable over every field. The last result certainly provides adequate motivation for an investigation of regular matroids. But there are several other reasons why this class of matroids has been so frequently studied. One of these is the association of regular matroids with totally unimodular matrices. Such matrices are of fundamental importance in combinatorial optimization because, as Hoffman and Kruskal (1956) showed, there is a basic link between total unimodularity and integer linear programming, the details of which can be found in Schrijver's book (1986, Chapter 19). Another important reason for studying the class of regular matroids is that this class occupies an intermediate position between the important classes of graphic and binary matroids. There are a number of results for graphs which, although they cannot be generalized to the class of all binary matroids, can be extended to the class of regular matroids.

In Section 13.1, we present a surprisingly simple proof of Tutte's excluded-minor characterization of the class of regular matroids (Gerards 1989). Section 13.2 gives an alternative characterization of regular matroids, this a constructive one due to Seymour (1980b). In Section 13.3, the focus switches to graphic matroids; most of that section is devoted to Wagner's (1985) proof of Tutte's excluded-minor characterization of the class of graphic matroids. The chapter concludes with a brief section reviewing some additional properties of regular and graphic matroids.

13.1 Proof of the excluded-minor theorem

The main purpose of this section is to present Gerards' (1989) proof of Tutte's (1958) excluded-minor characterization of the class of regular matroids. We noted earlier that one proof of this result can be obtained by combining certain results from Chapters 6, 9, and 10. That proof, while it is shorter than Tutte's original, very difficult proof, is still quite long. Gerards' proof is the shortest and most elementary one known of Tutte's theorem. It puts one of the most celebrated results of matroid theory within relatively easy reach. Although this result was stated earlier, we restate it now for ease of reference.

13.1.1 Theorem. *A matroid is regular if and only if it has no minor isomorphic to $U_{2,4}$, F_7, or F_7^*.*

This theorem is actually a combination of the excluded-minor characterization of binary matroids (Theorem 6.5.4) and the following result.

13.1.2 Theorem. *A binary matroid is regular if and only if it has no minor isomorphic to F_7 or F_7^*.*

To see this, we note that it is an immediate consequence of the elementary Proposition 6.4.5 that every regular matroid is binary. In order to prove Theorem 13.1.2, we shall reformulate it in matrix terms. First recall from Section 6.4 that if D is a matrix, then $D^{\#}$ denotes the matrix that is obtained from D by replacing every non-zero entry by a one. Let Y be a $(0,1)$-*matrix*, that is, a matrix for which every entry is in $\{0,1\}$. A *signing* of Y is a real matrix Z having every entry in $\{0,1,-1\}$ such that $Z^{\#} = Y$. Thus every signing of Y can be obtained from Y by changing some of its entries from 1 to -1. Now let $[I_r|X]$ be a $GF(2)$-representation for a binary matroid M. Then it follows by Proposition 6.4.1 and Lemma 2.2.21 that M is regular if and only if $[I_r|X]$ has a signing that is totally unimodular. But, clearly, the latter is true if and only if X has a totally unimodular signing. It now follows without difficulty that Theorem 13.1.2 is equivalent to the next result. The matrix X_F in this theorem is

$$\begin{bmatrix} 1 & 1 & 1 & 0 \\ 1 & 1 & 0 & 1 \\ 1 & 0 & 1 & 1 \end{bmatrix}.$$

Thus $[I_3|X_F]$ is a $GF(2)$-representation of F_7.

13.1.3 Theorem. *The following are equivalent for a $(0,1)$-matrix $[I_r|X]$.*

(i) *X has no totally unimodular signing.*

(ii) *When viewed as a matrix over $GF(2)$, $[I_r|X]$ can be transformed into $[I_3|X_F]$ or $[I_4|X_F^T]$ by a sequence of the following operations: deleting rows or columns; permuting rows or columns; and pivoting.*

The proof of this theorem, which occupies the rest of this section, will require a number of preliminaries. One of the main tools used in the proof will be the simple bipartite graph $G(X)$ that is associated with a $(0,1)$-matrix X. This was introduced in Section 6.4.

13.1.4 Lemma. *Let G be a simple, connected, bipartite graph such that, whenever two distinct vertices from the same vertex class are deleted, a disconnected graph results. Then G is a path or a cycle.*

Proof. Assume that G is neither a path nor a cycle. Then it has a spanning tree T that is not a path. Evidently T has at least three vertices of degree one. Therefore T has two degree-1 vertices u and v that are in the same vertex class of G. But deleting u and v from G results in a connected graph; a contradiction. □

The next two lemmas note some elementary properties of totally unimodular matrices.

13.1.5 Lemma. *Let D be an $n \times n$ matrix $[d_{ij}]$ with entries in $\{0,1,-1\}$. If $G(D^{\#})$ is a cycle, then D is totally unimodular if and only if the number of negative entries in D is congruent to n modulo two.*

Proof. As $G(D^{\#})$ is a cycle, by permuting rows and permuting columns, D can be transformed into a matrix D_1 such that

$$D_1^{\#} = \begin{bmatrix} 1 & 0 & 0 & \cdots & 0 & 0 & 1 \\ 1 & 1 & 0 & \cdots & 0 & 0 & 0 \\ 0 & 1 & 1 & \cdots & 0 & 0 & 0 \\ 0 & 0 & 1 & \cdots & 0 & 0 & 0 \\ \vdots & \vdots & \vdots & \ddots & \vdots & \vdots & \vdots \\ 0 & 0 & 0 & \cdots & 1 & 1 & 0 \\ 0 & 0 & 0 & \cdots & 0 & 1 & 1 \end{bmatrix}.$$

By 2.2.16, these permutations can only affect the determinant of a square submatrix of D by changing its sign. Therefore we may assume that $D = D_1$. Now a straightforward induction argument shows that if D' is a proper square submatrix of D, then D' has a row or column with at most one non-zero entry. Using this observation iteratively we deduce that such a submatrix D' has determinant in $\{0,1,-1\}$. Thus D is totally unimodular if and only if det $D \in \{0,1,-1\}$. By scaling, in turn, column 1, row 2, column 2, row 3, ..., column $(n-1)$, and row n, we can transform D into the matrix D_2 where

$$D_2 = \begin{bmatrix} 1 & 0 & 0 & \cdots & 0 & 0 & d_{1n} \\ 1 & 1 & 0 & \cdots & 0 & 0 & 0 \\ 0 & 1 & 1 & \cdots & 0 & 0 & 0 \\ 0 & 0 & 1 & \cdots & 0 & 0 & 0 \\ \vdots & \vdots & \vdots & \ddots & \vdots & \vdots & \vdots \\ 0 & 0 & 0 & \cdots & 1 & 1 & 0 \\ 0 & 0 & 0 & \cdots & 0 & 1 & 1 \end{bmatrix}$$

and $d_{1n} \in \{1, -1\}$. Moreover, $|\det D_2| = |\det D|$ and, modulo two, D_2 and D have the same number of negative entries. But, by expanding along the first row of D_2, we find that $\det D_2 = 1 + (-1)^{n+1} d_{1n}$. Hence $|\det D| = |1 + (-1)^{n+1} d_{1n}|$. As $d_{1n} \in \{1, -1\}$, it is straightforward to show that $\det D \in \{0, 1, -1\}$ if and only if the number of negative entries of D is congruent to n modulo two. □

The next lemma is due to Camion (1963). The proof that we shall give is Seymour's (in Gerards 1989).

13.1.6 Lemma. *Let D_1 and D_2 be totally unimodular matrices. If $D_1^\# = D_2^\#$, then D_2 can be obtained from D_1 by multiplying some rows and columns by -1.*

Proof. Clearly $G(D_1^\#) = G(D_2^\#)$. We shall call an edge of this bipartite graph *even* if the corresponding entries in D_1 and D_2 are the same; all other edges are called *odd.* We shall show first that every cycle C of $G(D_1^\#)$ has an even number of odd edges. The proof of this is by induction on the number n of diagonals of C, the core of the proof being contained in the case $n = 0$ which we now present. Let X_1 and X_2 be the submatrices of D_1 and D_2, respectively, that correspond to C, where we assume that C has no diagonals. Then $G(X_1^\#) = G(X_2^\#) = C$. An edge of C is odd if and only if exactly one of the corresponding entries of X_1 and X_2 is equal to -1. Thus the set of odd edges of C is the symmetric difference of the set of edges of $G(X_1^\#)$ for which the corresponding entry of X_1 is -1 and the set of edges of $G(X_1^\#)$ for which the corresponding entry of X_2 is -1. But, by the preceding lemma, the cardinalities of the last two sets are equal modulo two. Thus C has an even number of odd edges; that is, a cycle in $G(D_1^\#)$ without diagonals has an even number of odd edges. We leave it to the reader to complete the straightforward induction argument that every cycle in $G(D_1^\#)$ has an even number of odd edges.

Next we define a new partition (V_1, V_2) of the vertex set of $G(D_1^\#)$. This part of the argument is similar to the argument that is used to show that a graph is bipartite if all its cycles have even length. We begin by choosing a set W of vertices of $G(D_1^\#)$ such that W contains exactly one vertex from each component of $G(D_1^\#)$. For each vertex u of $G(D_1^\#)$, let

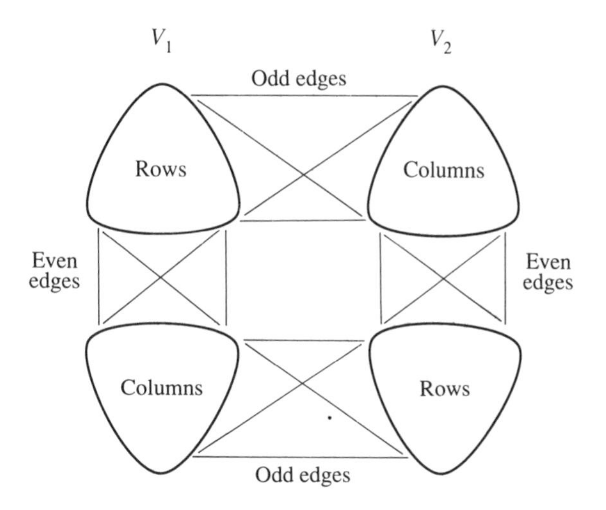

Fig. 13.1. The positions of the even and odd edges in $G(D_1^{\#})$.

S_u be the set of shortest paths from u to the unique vertex of W that is in the same component of $G(D_1^{\#})$ as u. If P_1 and P_2 are two members of S_u, then the symmetric difference of their edge sets is a disjoint union of cycles and therefore contains an even number of odd edges. Thus both P_1 and P_2 have an even number of odd edges, or both have an odd number of odd edges. Hence either every member of S_u contains an even number of odd edges, in which case we put u in V_1; or every member of S_u contains an odd number of odd edges and we put u in V_2. Certainly (V_1, V_2) is a partition of the vertex set of $G(D_1^{\#})$. Moreover, it is straightforward to show that the odd edges of $G(D_1^{\#})$ are precisely those edges that join vertices in V_1 to vertices in V_2 (see Figure 13.1).

If we now multiply by -1 the rows and columns of D_2 that correspond to vertices in V_2, then, when the resulting matrix D_2' is compared with D_1, we find that every edge is even. Hence $D_2' = D_2$. □

To complete the preliminaries needed to prove Theorem 13.1.3, we now note some properties of the pivoting operation. This operation was introduced in Section 2.2 and modified in Section 6.6 to include the natural column interchange as part of the operation. Let the matrix A_1 be as shown in Figure 13.2 where Z is a matrix, α is a non-zero scalar, and $\underline{x}$ and $\underline{y}$ are column vectors. Recall that a pivot on α seeks to transform column $r+1$ of A_1 into the first natural basis vector. Once this is achieved by elementary row operations, columns 1 and $r+1$ of the resulting matrix A_2 are interchanged. Thus the effect of the whole operation is to transform A_1 into A_3 (see Figure 13.2). Moreover, if we pivot on the entry α^{-1} of A_3, then it is easy to show that we obtain A_1.

$$\left[\begin{array}{c|c|c} I_r & \alpha & \underline{x}^T \\ \hline & \underline{y} & Z \end{array}\right] \qquad \left[\begin{array}{c|c|c|c} \alpha^{-1} & 0 & 1 & \alpha^{-1}\underline{x}^T \\ \hline -\alpha^{-1}\underline{y} & I_{r-1} & 0 & Z-\alpha^{-1}\underline{y}\,\underline{x}^T \end{array}\right]$$

$$A_1 \qquad\qquad\qquad A_2$$

$$\left[\begin{array}{c|c|c} I_r & \alpha^{-1} & \alpha^{-1}\underline{x}^T \\ \hline & -\alpha^{-1}\underline{y} & Z-\alpha^{-1}\underline{y}\,\underline{x}^T \end{array}\right]$$

$$A_3$$

Fig. 13.2. A pivot on the entry α transforms A_1 into A_3.

13.1.7 Lemma.

(i) *If $G(A_1^{\#})$ is connected, then so is $G(A_3^{\#})$.*

(ii) *If $\left[\begin{array}{c|c} \alpha & \underline{x}^T \\ \hline \underline{y} & Z \end{array}\right]$ is square, then its determinant is $\alpha^{-1}\det(Z-\alpha^{-1}\underline{y}\,\underline{x}^T)$.*

Proof. (i) Suppose that $G(A_3^{\#})$ is disconnected. Then, by permuting rows and permuting columns, A_3 can be transformed into a block diagonal matrix D_3. As A_1 can be obtained from A_3 by pivoting on the entry α^{-1} of the latter, a matrix that is a row and column permutation of A_1 can be obtained from D_3 by pivoting. Since a pivot in D_3 clearly produces another matrix in block diagonal form, we conclude that $G(A_1^{\#})$ is disconnected; a contradiction.

(ii) This follows from 2.2.15 and 2.2.17 by noting which row operations must be applied to obtain A_2 from A_1 in Figure 13.2. □

When we consider pivoting in the rest of the proof of Theorem 13.1.3, it will be convenient to suppress the identity matrices in A_1 and A_3. Hence when we pivot on the entry α of A_1' in Figure 13.3, we get the matrix A_3'.

$$\left[\begin{array}{c|c} \alpha & \underline{x}^T \\ \hline \underline{y} & Z \end{array}\right] \qquad \left[\begin{array}{c|c} \alpha^{-1} & \alpha^{-1}\underline{x}^T \\ \hline -\alpha^{-1}\underline{y} & Z-\alpha^{-1}\underline{y}\,\underline{x}^T \end{array}\right]$$

$$A_1' \qquad\qquad A_3'$$

Fig. 13.3.

Proof of Theorem 13.1.3. Suppose first that X has a totally unimodular signing and assume that $[I_3|X_F]$ or $[I_4|X_F^T]$ can be obtained from $[I_r|X]$ by a sequence of the operations specified in (ii). Then it follows by 2.2.16 and Lemma 2.2.20 that $[I_3|X_F]$ or $[I_4|X_F^T]$ has a totally unimodular signing. Hence $[I_3|X_F]$ has a totally unimodular signing. This implies that F_7 is $\mathbb{R}$-representable. One can deduce that this is not so from Proposition 6.4.8, although it is straightforward to show this directly using only some elementary ideas from Section 6.4. We conclude that if (i) fails, then so does (ii).

Now suppose that X has no totally unimodular signing, but that every proper submatrix of X does have such a signing. Then one easily checks that $G(X)$ is connected but is not a path or a cycle; otherwise X has a totally unimodular signing. It follows by Lemma 13.1.4 that, by permuting columns, X or X^T can be transformed into a matrix Y of the form $[\underline{x}|\underline{y}|Z]$ where $\underline{x}$ and $\underline{y}$ are column vectors and $G(Z)$ is connected. By assumption, both $[\underline{x}|Z]$ and $[\underline{y}|Z]$ have totally unimodular signings. Moreover, these signings can be chosen so that Z is signed the same way in both cases. To see this, suppose that $[\underline{x}_1|Z_1]$ and $[\underline{y}_2|Z_2]$ are totally unimodular signings of $[\underline{x}|Z]$ and $[\underline{y}|Z]$, respectively. Then, by Lemma 13.1.6, Z_1 can be obtained from Z_2 by multiplying some rows and columns by -1. Performing the same row and column scalings in $[\underline{y}_2|Z_2]$, we transform this matrix into $[\underline{y}_1|Z_1]$. Thus some matrix Y that is obtained from X or X^T by permuting columns has a signing $[\underline{x}_1|\underline{y}_1|Z_1]$ such that

(i) $G(Z_1^{\#})$ is connected; and

(ii) both $[\underline{x}_1|Z_1]$ and $[\underline{y}_1|Z_1]$ are totally unimodular.

The rest of the proof will show that we can transform Y into X_F by a sequence of pivots, row and column deletions, and row and column permutations.

By assumption, $[\underline{x}_1|\underline{y}_1|Z_1]$ is not totally unimodular. By Lemma 13.1.7(i) and Lemma 2.2.20, if we pivot in $[\underline{x}_1|\underline{y}_1|Z_1]$ on an entry of Z_1, then we maintain a matrix of the same form as $[\underline{x}_1|\underline{y}_1|Z_1]$ for which (i) and (ii) above both hold. We remark that we do this pivot over $\mathbb{R}$ so as to maintain a signing. By Lemma 2.2.20 again, if the resulting matrix is viewed modulo two, then we get exactly the same matrix as we would by viewing $[\underline{x}_1|\underline{y}_1|Z_1]$ over $GF(2)$ and doing the pivot over $GF(2)$.

Now let $\mathcal{Z}$ be the set of matrices that can be obtained from $[\underline{x}_1|\underline{y}_1|Z_1]$ by performing a sequence of pivots in Z_1. Let $[\underline{x}_3|\underline{y}_3|Z_3]$ be a member of $\mathcal{Z}$ having a square submatrix W whose determinant is not in $\{0, 1, -1\}$ such that every square submatrix of a member of $\mathcal{Z}$ that is smaller than W has determinant in $\{0, 1, -1\}$.

13.1.8 Lemma. *W is a submatrix of $[\underline{x}_3|\underline{y}_3]$ and can be obtained from $\left[\begin{smallmatrix}1 & 1\\ 1 & -1\end{smallmatrix}\right]$ by multiplying some rows or columns of $[\underline{x}_3|\underline{y}_3|Z_3]$ by -1.*

Proof. Clearly W meets both of the columns $\underline{x}_3$ and $\underline{y}_3$. If W also meets Z_3, then we pivot in $[\underline{x}_3|\underline{y}_3|Z_3]$ on an entry of Z_3 that is also in W. Let $[\underline{x}_4|\underline{y}_4|Z_4]$ be the resulting matrix. Using Lemma 13.1.7(ii) and the fact that every non-zero entry of Z_3 is in $\{1, -1\}$, we deduce that $[\underline{x}_4|\underline{y}_4|Z_4]$ has a square submatrix W' such that $|\det W'| = |\det W|$ but W' is smaller than W. This contradiction to the choice of W implies that W

itself is a submatrix of $[\underline{x}_3|\underline{y}_3]$. Thus W is a 2×2 matrix and the lemma follows without difficulty. □

By this lemma, we may assume that $\left[\begin{smallmatrix} 1 & 1 \\ 1 & -1 \end{smallmatrix}\right]$ occurs as a submatrix of $[\underline{x}_3|\underline{y}_3]$. By permuting rows, we may also suppose that this submatrix occupies rows 1 and 2 of $[\underline{x}_3|\underline{y}_3]$. Let $Y_3 = [\underline{x}_3|\underline{y}_3|Z_3]$. Now $G(Z_3^{\#})$ is connected and so has a shortest path P linking the vertices corresponding to rows 1 and 2. If P has length two, then Y_3 has $\left[\begin{smallmatrix} 1 & 1 & a \\ 1 & -1 & b \end{smallmatrix}\right]$ as a submatrix for some a and b in $\{1, -1\}$. But if $a = b$, then $|\det\left[\begin{smallmatrix} 1 & a \\ -1 & b \end{smallmatrix}\right]| = 2$; while, if $a = -b$, then $|\det\left[\begin{smallmatrix} 1 & a \\ 1 & b \end{smallmatrix}\right]| = 2$. In each case we obtain a contradiction to the fact that both $[\underline{y}_3|Z_3]$ and $[\underline{x}_3|Z_3]$ are totally unimodular. Thus P must have length exceeding two. Therefore, by permuting rows of Y_3 and columns of Z_3, we can obtain a $k \times (k+1)$ matrix of the following form for some $k \geq 3$ where each starred entry is in $\{1, -1\}$ and the entries in the submatrix D are unspecified:

$$\left[\begin{array}{cc|cccccc} 1 & 1 & * & 0 & 0 & \cdots & 0 & 0 & 0 \\ 1 & -1 & 0 & 0 & 0 & \cdots & 0 & 0 & * \\ \hline & & * & * & 0 & \cdots & 0 & 0 & 0 \\ & & 0 & * & * & \cdots & 0 & 0 & 0 \\ & D & 0 & 0 & * & \cdots & 0 & 0 & 0 \\ & & \vdots & \vdots & \vdots & \ddots & \vdots & \vdots & \vdots \\ & & 0 & 0 & 0 & \cdots & * & * & 0 \\ & & 0 & 0 & 0 & \cdots & 0 & * & * \end{array}\right]. \tag{1}$$

Next we note that each of the starred entries may be assumed to be equal to 1. To achieve this, we scale, in turn, column 3, row 3, column 4, ..., column k, and row k. This makes all the starred entries equal to 1 except for the two in the last column. By scaling this column and then row 2, we can make these last two starred entries equal to 1, although we may need to interchange columns 1 and 2 to preserve the matrix $\left[\begin{smallmatrix} 1 & 1 \\ 1 & -1 \end{smallmatrix}\right]$ in the top left corner.

Now, in (1), we pivot on the entry in row 3 and column 4 and then delete that row and column to obtain the $(k-1) \times k$ matrix

$$\left[\begin{array}{cc|cccccc} 1 & 1 & 1 & 0 & 0 & \cdots & 0 & 0 & 0 \\ 1 & -1 & 0 & 0 & 0 & \cdots & 0 & 0 & 1 \\ \hline & & -1 & 1 & 0 & \cdots & 0 & 0 & 0 \\ & & 0 & 1 & 1 & \cdots & 0 & 0 & 0 \\ & D' & 0 & 0 & 1 & \cdots & 0 & 0 & 0 \\ & & \vdots & \vdots & \vdots & \ddots & \vdots & \vdots & \vdots \\ & & 0 & 0 & 0 & \cdots & 1 & 1 & 0 \\ & & 0 & 0 & 0 & \cdots & 0 & 1 & 1 \end{array}\right].$$

Arguing as above, we see that, by multiplying some rows and columns of this matrix by -1 and then interchanging columns 1 and 2 if necessary, we can get a matrix that has the same form as (1) but is smaller. Repeating this process, we eventually obtain a matrix of this form having three rows. This matrix is

$$\begin{bmatrix} 1 & 1 & 1 & 0 \\ 1 & -1 & 0 & 1 \\ c & d & 1 & 1 \end{bmatrix} \tag{2}$$

where c and d are in $\{0, 1, -1\}$. Moreover, deleting either of the first two columns of this matrix gives a totally unimodular matrix. The submatrices $\left[\begin{smallmatrix} 1 & 1 \\ d & 1 \end{smallmatrix}\right]$ and $\left[\begin{smallmatrix} -1 & d \\ 1 & 1 \end{smallmatrix}\right]$ imply that $d \neq -1$ and $d \neq 1$. Thus $d = 0$. The submatrix obtained by deleting column 2 has determinant equal to $c - 2$. Therefore $c = 1$. We conclude that, when viewed over $GF(2)$, the matrix in (2) is X_F. This completes the proof that (i) implies (ii) thereby finishing the proof of Theorem 13.1.3. □

To conclude this section, we remark that Lovász and Schrijver (in Gerards 1989) have observed that there is another proof of the excluded-minor characterization of ternary matroids that follows the lines of the above proof.

Exercises

1. Let $[I_r|D]$ be a real matrix. Prove that $M[I_r|D]$ is regular if and only if every $r \times r$ submatrix of $[I_r|D]$ has determinant in $\{0, 1, -1\}$.
2. Show that pivoting on the entry α^{-1} of the matrix A_3 in Figure 13.2 gives the matrix A_1.
3. Show that a matrix with every entry in $\{0, 1, -1\}$ is totally unimodular if every column contains exactly one 1 and exactly one -1.
4. Let A_{10} be the $GF(2)$-representation for R_{10} that is given in Figure 13.4 at the start of the next section. Show that R_{10} is regular by giving a totally unimodular signing of A_{10}.
5. Show that a 3-connected binary matroid of rank at least four is regular if and only if it has no minor isomorphic to F_7^*.

13.2 Seymour's decomposition theorem

The excluded-minor theorem (13.1.1) provides one characterization of the class of regular matroids. In this section we shall discuss a different characterization of this class: a decomposition theorem for members of the class which was proved by Seymour (1980b). This theorem has a number of important consequences one of which is that it leads to a polynomial-time algorithm to test whether a given real matrix is totally unimodular.

The matroid R_{10} will play a fundamental role in this decomposition theorem where we recall from Chapter 11 that R_{10} is the matroid for which the matrix A_{10} in Figure 13.4 is a $GF(2)$-representation. It was proved in Proposition 11.2.10 that R_{10} is a splitter for the class of regular matroids, that is, if a 3-connected regular matroid M has an R_{10}-minor, then $M \cong R_{10}$. The next proposition follows immediately on combining that result with the fact that every matroid that is not 3-connected can be built up from 3-connected matroids by a sequence of direct sums and 2-sums (Corollary 8.3.4).

13.2.1 Proposition. *Every regular matroid can be obtained from copies of R_{10} and from 3-connected regular matroids without R_{10}-minors by a sequence of direct sums and 2-sums.* □

This result is the first step towards Seymour's theorem. It leads one naturally to an examination of the class $\mathcal{N}$ of 3-connected regular matroids without R_{10}-minors. Two prominent and familiar subclasses of $\mathcal{N}$ are the classes of 3-connected graphic and 3-connected cographic matroids. Moreover, the matroid R_{12} that is represented over $GF(2)$ by the matrix A_{12} in Figure 13.4 is a member of $\mathcal{N}$ which is neither graphic nor cographic. Indeed, Seymour proved the following result using a long argument in which the main difficulties are graph-theoretic.

$$A_{10} = \left[\begin{array}{c|ccccc} & 1 & 1 & 0 & 0 & 1 \\ & 1 & 1 & 1 & 0 & 0 \\ I_5 & 0 & 1 & 1 & 1 & 0 \\ & 0 & 0 & 1 & 1 & 1 \\ & 1 & 0 & 0 & 1 & 1 \end{array}\right] \qquad A_{12} = \left[\begin{array}{c|cccccc} & 1 & 1 & 1 & 0 & 0 & 0 \\ & 1 & 1 & 0 & 1 & 0 & 0 \\ I_6 & 1 & 0 & 0 & 0 & 1 & 0 \\ & 0 & 1 & 0 & 0 & 0 & 1 \\ & 0 & 0 & 1 & 0 & 1 & 1 \\ & 0 & 0 & 0 & 1 & 1 & 1 \end{array}\right]$$

Fig. 13.4. $GF(2)$-representations for R_{10} and R_{12}.

13.2.2 Theorem. *Let M be a 3-connected regular matroid. Then either M is graphic or cographic, or M has a minor isomorphic to one of R_{10} and R_{12}.* □

We shall not attempt to give the proof of this theorem here, commenting only that one of the results it uses is Tutte's (1959) excluded-minor characterization of graphic matroids. Theorem 13.2.2 motivates an examination of the regular matroids that have an R_{12}-minor, and Seymour (1980b) established the following result for such matroids.

13.2.3 Proposition. *Every regular matroid with an R_{12}-minor has an exact 3-separation (X_1, X_2) in which both $|X_1|$ and $|X_2|$ exceed five.*

Proof. First observe that R_{12} has exactly two triangles and that these triangles are disjoint (Exercise 1); let Y_1 be their union and Y_2 be $E(R_{12}) - Y_1$. Then one can show that every regular matroid with an R_{12}-minor has an exact 3-separation (Z_1, Z_2) with $Y_1 \subseteq Z_1$ and $Y_2 \subseteq Z_2$. The details of this argument are not short and are omitted. □

The last three results are the key steps in the proof of the following fundamental result of Seymour (1980b), the main theorem of this section.

13.2.4 Theorem. *Every regular matroid M can be constructed by means of direct sums, 2-sums, and 3-sums starting with matroids each of which is isomorphic to a minor of M, and each of which is either graphic, cographic, or isomorphic to R_{10}.*

Proof. First we remark that by Propositions 4.2.23 and 7.1.19, respectively, if N is the direct sum or 2-sum of two non-empty matroids N_1 and N_2, then each of N_1 and N_2 is isomorphic to a proper minor of N.

Now let M be a regular matroid that is minor-minimal with the property that it cannot be constructed by direct sums and 2-sums from graphic matroids, cographic matroids, and copies of R_{10}. Then M is 3-connected. Moreover, M is neither graphic nor cographic. Thus, by Theorem 13.2.2, M has a minor isomorphic to R_{10} or R_{12}. In the first case, by Proposition 13.2.1, $M \cong R_{10}$; a contradiction. In the second case, by Proposition 13.2.3, M has an exact 3-separation (X_1, X_2) in which both $|X_1|$ and $|X_2|$ exceed five. Thus, by Propositions 12.4.16 and 12.4.17, M is a 3-sum of two of its proper minors, and the theorem follows without difficulty. □

The converse of the last theorem, namely that taking the direct sum, 2-sum, or 3-sum of two regular matroids produces a regular matroid, comes from combining Proposition 4.2.15, Corollary 7.1.23, and Proposition 12.4.19.

Theorem 13.2.4 is similar to the decomposition theorems considered in Chapter 11, but differs in that 3-sums as well as direct sums and 2-sums are used to stick the matroids together. Motivated by theorems of this type, Truemper (1984b, 1985a,b, 1986, 1988, 1990, 1992a,b) developed a general decomposition theory for matroids. A detailed discussion of this theory and its many applications is given in Truemper's (1992c) monograph *Matroid decomposition.*

We have already noted numerous interesting properties of R_{10}. Two further such properties are contained in the next result (Bixby 1977), which can be proved without difficulty using Theorem 13.2.2.

13.2.5 Corollary. *Let M be a regular matroid that is neither graphic nor cographic. Then the following statements are equivalent:*

(i) $M \cong R_{10}$.

(ii) $|E(M)| \leq 10$.

(iii) $r(M) \leq 5$ *and M is simple.* □

An immediate consequence of Theorem 13.2.4 is the following.

13.2.6 Corollary. *If M is a 3-connected, internally 4-connected, regular matroid, then M is graphic, cographic, or isomorphic to R_{10}.* □

Theorem 13.2.4 has a multitude of other interesting and important applications, some of which appear in a series of papers by Seymour (1980c, 1981c,d, 1985b, 1986b). We shall conclude this section by looking briefly at one such application that relates to problems in combinatorial optimization. We have already commented on the important role played by totally unimodular matrices in linear programming. In view of this, one of the applications of Theorem 13.2.4 that has attracted the most attention is an algorithm that tests in polynomial time whether a given real matrix is totally unimodular. Since we have put very little emphasis here on the well-developed subject of matroid algorithms, we shall be content to follow Bixby (1981) and Welsh (1982) in sketching the main ideas of this algorithm. A thorough description and justification of the algorithm has been given by Schrijver (1986). The reader who is interested in a detailed general account of the algorithmic aspects of matroid theory is referred to the books by Lawler (1976) and Recski (1989) and the survey papers of Bixby (1981) and Faigle (1987).

The main parts of the algorithm for testing whether a real matrix is totally unimodular are polynomial subroutines that will

(i) test whether a vector matroid M of a matrix is (a) graphic, (b) cographic, and (c) isomorphic to R_{10}; and

(ii) find 1-, 2-, and 3-separations in such a matroid.

Evidently one can easily check if $M \cong R_{10}$. Several authors including Tutte (1960) and Bixby and Cunningham (1980) have given algorithms to test whether a given vector matroid M is graphic. Applying such an algorithm to M^* will test whether M is cographic. Finally, Cunningham and Edmonds (in Cunningham 1973) have given a polynomial algorithm for finding k-separations in a matroid for any fixed positive integer k.

We conclude this discussion with some brief comments on Cunningham and Edmonds' algorithm. It relies on the next result which relates the problem of finding k-separations to a matroid intersection problem. As noted in Section 12.3, problems of the latter type can be solved in polynomial time using the matroid intersection algorithm.

13.2.7 Proposition. *Let M be a matroid and k be a positive integer. If X_1 and X_2 are disjoint k-element subsets of $E(M)$, then M has a k-separation (Y_1, Y_2) with $X_1 \subseteq Y_1$ and $X_2 \subseteq Y_2$ if and only if $M/X_1\backslash X_2$ and $M/X_2\backslash X_1$ do not have a common t-element independent set where $t = r(M) + k - r(X_1) - r(X_2)$.*

Proof. Let $M_1 = M/X_1\backslash X_2$, $M_2 = M/X_2\backslash X_1$, and r_1 and r_2 be their rank functions. Then, for all $X \subseteq E(M) - (X_1 \cup X_2)$ and all i in $\{1, 2\}$, $r_i(X) = r(X \cup X_i) - r(X_i)$. Now, by Theorem 12.3.15, M_1 and M_2 do not have a common t-element independent set if and only if, for some $T \subseteq E(M) - (X_1 \cup X_2)$,

$$r_1(T) + r_2(E(M) - (X_1 \cup X_2) - T) \leq t - 1,$$

that is,

$$r(T \cup X_1) + r(E(M) - (T \cup X_1)) \leq t - 1 + r(X_1) + r(X_2).$$

But, as $t = r(M) + k - r(X_1) - r(X_2)$, the last inequality is equivalent to the inequality

$$r(T \cup X_1) + r(E(M) - (T \cup X_1)) \leq r(M) + k - 1. \tag{1}$$

Now $T \cup X_1 \supseteq X_1$ and $E(M) - (T \cup X_1) \supseteq X_2$. Hence both $T \cup X_1$ and $E(M) - (T \cup X_1)$ have at least k elements. Therefore (1) holds if and only if $(T \cup X_1, E(M) - (T \cup X_1))$ is a k-separation of M. The proposition follows without difficulty. □

In view of this proposition, we can determine whether M has a k-separation for some fixed k by testing all pairs (X_1, X_2) of disjoint k-element subsets of $E(M)$ to determine whether $M/X_1\backslash X_2$ and $M/X_2\backslash X_1$ have a common independent set of size $r(M)+k-r(X_1)-r(X_2)$. This matroid intersection problem can be solved in polynomial time for each choice of (X_1, X_2). Moreover, the number of choices for (X_1, X_2) is polynomial in $|E(M)|$. Therefore, for a fixed k, we have a polynomial algorithm for finding k-separations. On combining this algorithm with those discussed earlier, we get the desired polynomial algorithm for testing whether a real matrix is totally unimodular.

Exercises

1. (i) Show that R_{12} has exactly two triangles and that these triangles are disjoint.

   (ii) Let X be the union of these two triangles and Y be $E(R_{12})-X$. Show that (X, Y) is an exact 3-separation of R_{12} and use this to deduce that R_{12} is isomorphic to a 3-sum of $M^*(K_{3,3})$ and $M(K_5 - e)$.

2. Show that a 5-connected regular matroid is either graphic or cographic.

3. Find all maximal subsets X of $E(R_{12})$ such that if x and y are in X, then there is an automorphism of R_{12} mapping x to y.

4. (Seymour 1980b) Suppose that $M[A_{12} + e]$ is a simple regular matroid. Show that the column vector labelled by e is one of $(1,1,0,0,0,0)^T$, $(1,1,0,0,1,1)^T$, $(0,0,0,0,1,1)^T$, or $(0,0,1,1,0,0)^T$, and that, while the first three cases give isomorphic matroids, the fourth does not.

5.* Show that R_{10} is the unique splitter for the class of regular matroids.

13.3 The excluded-minor characterization of graphic matroids

The main purpose of this section is to prove the following excluded-minor characterization of the class of graphic matroids (Tutte 1959) which was stated earlier as Theorem 6.6.5(i).

13.3.1 Theorem. *A matroid is graphic if and only if it has no minor isomorphic to any of the matroids $U_{2,4}$, F_7, F_7^*, $M^*(K_5)$, and $M^*(K_{3,3})$.*

This theorem is actually a combination of three results: the excluded-minor characterization of regular matroids; the assertion that every graphic matroid is regular (Proposition 5.1.3); and the following theorem of Tutte (1959).

13.3.2 Theorem. *A regular matroid is graphic if and only if it has no minor isomorphic to $M^*(K_5)$ or $M^*(K_{3,3})$.*

Alternatively, one can deduce Theorem 13.3.1 by combining three other results: the excluded-minor characterization of binary matroids; the assertion that every graphic matroid is binary (Proposition 5.1.2); and the following result, also due to Tutte (1965).

13.3.3 Theorem. *A binary matroid is graphic if and only if it has no minor isomorphic to* F_7, F_7^*, $M^*(K_5)$, *or* $M^*(K_{3,3})$.

In this section, we shall present Wagner's (1985) proof of the last theorem. This proof is shorter and conceptually simpler than the original, difficult proof of Tutte. However, it is still quite long occupying, after a few short remarks, all of the rest of this section. The proof makes extensive use of Whitney's 2-Isomorphism Theorem (5.3.1). Another proof of Theorem 13.3.3 has been given by Seymour (1980c). It is based on some of the ideas used in the proof of his decomposition theorem for regular matroids.

We observe here that by dualizing Theorems 13.3.1, 13.3.2, and 13.3.3, one can deduce excluded-minor characterizations of the class of cographic matroids within the classes of matroids, regular matroids, and binary matroids, respectively. The first of these was stated earlier as (ii) of Theorem 6.6.5. The other two are now stated for reference.

13.3.4 Corollary. *A regular matroid is cographic if and only if it has no minor isomorphic to* $M(K_5)$ *or* $M(K_{3,3})$. □

13.3.5 Corollary. *A binary matroid is cographic if and only if it has no minor isomorphic to* F_7, F_7^*, $M(K_5)$, *or* $M(K_{3.3})$. □

The proof of Theorem 13.3.3 will require some preliminary lemmas that are of independent interest. The first of these is due to Cunningham (1981) and Seymour (1980b). Recall that the cosimple matroid $\underset{\sim}{N}$ associated with a matroid N is obtained from N by deleting all its coloops and contracting all but one element from every non-trivial series class.

13.3.6 Lemma. *Let* M *be a 3-connected matroid having at least four elements. Then* M *has an element* e *such that* $\underset{\sim}{M\backslash e}$ *is 3-connected.*

Proof. Let M be a matroid for which the result does not hold such that $|E(M)|$ is as small as possible. Then one easily checks that M is not a wheel or a whirl. Thus, by Tutte's Wheels and Whirls Theorem (8.4.5), M has an element e such that $M\backslash e$ or M/e is 3-connected. In the former case, the lemma holds. Thus we may assume that M/e is 3-connected. Hence, by the choice of M, there is an element f such that the cosimple matroid associated with $M/e\backslash f$ is 3-connected. Therefore $E(M/e\backslash f)$ has a subset X such that $M/e\backslash f/X$ is 3-connected and every element of X is in series with some element of $M/e\backslash f/X$. Now every 2-cocircuit of $M/e\backslash f$ is a 2-cocircuit of $M\backslash f$. Thus if $M\backslash f/X$ is 3-connected, then it is isomorphic to $\underset{\sim}{M\backslash f}$, contradicting the fact that $\underset{\sim}{M\backslash f}$ is not 3-connected. Hence $M\backslash f/X$ is not 3-connected. Thus, by the dual of Proposition

8.1.10, e is a loop, a coloop, or a series element of $M\backslash f/X$. In the third case, we obtain the contradiction that $M\backslash f$ is 3-connected since it is isomorphic to $M\backslash f/X/e$. In the second case, e is a coloop of $M\backslash f$, so $\{e, f\}$ contains a cocircuit of M; a contradiction to Proposition 8.1.6. Finally, in the first case, $M\backslash f$ must have a circuit C with at least three elements such that $e \in C \subseteq X \cup e$. Thus $C - e$ is a circuit of $M/e\backslash f$ that is contained in X. Hence if $x \in C - e$, then $M/e\backslash f$ has a 2-cocircuit C^* containing x and some element y of $M/e\backslash f/X$. But then $|(C-e) \cap C^*| = 1$, a contradiction to Proposition 2.1.11. □

The next lemma (Löfgren 1959) will be extensively used in the proof of Theorem 13.3.3. A set P of edges in a graph G is called a *hypopath* of G if P is the edge set of a path in some graph that is 2-isomorphic to G. Throughout this section, we shall maintain our practice of frequently not distinguishing between a subset of the edge set of a graph and the subgraph induced by that subset.

13.3.7 Lemma. *Let M be a binary matroid having an element e such that $M\backslash e = M(G)$ for some graph G. Let C be a circuit of M containing e and let $P = C - e$. Then M is graphic if and only if P is a hypopath of G.*

Proof. If M is graphic, then C is a cycle in some graph G_1 such that $M(G_1) \cong M$. Thus $M(G_1\backslash e) \cong M\backslash e$ and so $M(G_1\backslash e) \cong M(G)$. Therefore, by Whitney's 2-Isomorphism Theorem (5.3.1), G is 2-isomorphic to $G_1\backslash e$. As P is a path of the latter, P is a hypopath of G.

Conversely, suppose that P is a hypopath of G. Then P is a path of some graph G_2 that is 2-isomorphic to G. Let $G_2 + e$ be the graph that is obtained from G_2 by adding an edge labelled by e joining the end-vertices of P. Then C is a cycle of $G_2 + e$. Thus both $M(G_2 + e)$ and M are binary matroids having C as a circuit, and $M\backslash e = M(G_2 + e)\backslash e$. By constructing the B-fundamental-circuit incidence matrices for $M(G_2 + e)$ and M with respect to some basis B that contains P, we easily deduce that $M = M(G_2 + e)$. □

We now begin the proof of Theorem 13.3.3. In one direction the proof is straightforward and this will be done first. The proof of the converse will require considerably more effort.

Proof of Theorem 13.3.3. Since every minor of a graphic matroid is graphic, to show that a graphic matroid has no minor isomorphic to F_7, F_7^*, $M^*(K_5)$, or $M^*(K_{3,3})$, it suffices to show that none of these four matroids is graphic. Proposition 2.3.3 noted that the last two matroids are non-graphic. One can deduce that neither F_7 nor F_7^* is graphic from the fact that neither is regular. Alternatively, one can show this directly as follows: F_7 is not graphic because there is no simple graph on four vertices having more edges than K_4. On the other hand, if F_7^* is graphic, then there is a connected graph G for which $M(G) \cong F_7^*$, $|E(G)| = 7$, and $|V(G)| = 5$. But G has average vertex degree less than three, so $M^*(G)$, which is isomorphic to F_7, has a circuit of size less than three; a contradiction.

Now to prove the converse, suppose that M is a non-graphic binary matroid for which every proper minor is graphic. Then, by Proposition 4.2.15, Corollary 7.1.23, and Theorem 8.3.1, M is 3-connected. Moreover, $|E(M)| \geq 5$. By Lemma 13.3.6, M has an element e such that $\widetilde{M\backslash e}$ is 3-connected. Throughout the rest of the proof of Theorem 13.3.3, G will denote a graph without isolated vertices for which $M\backslash e = M(G)$. In addition, C will denote a fixed circuit of M containing e, and P will denote $C - e$. The proof will concentrate for some time on this set P and its subsets. In particular, if such a subset X is contained in the edge set of a graph H, then when we refer to the vertex degrees of X in H, we mean the vertex degrees in the induced graph $H[X]$.

13.3.8 Lemma. *P is not a hypopath of G. Moreover, P has at least four odd-degree vertices in any graph that is 2-isomorphic to G.*

Proof. Lemma 13.3.7 implies that P is not a hypopath of G. Since P clearly contains no cycles of G, it must contain more than two vertices of odd degree in any graph 2-isomorphic to G. Since every graph has an even number of odd-degree vertices, the lemma follows. □

Much of the remainder of the proof of Theorem 13.3.3 will involve playing off the last lemma against the following result.

13.3.9 Lemma. *Let H be the graph $G\backslash X/Y$ where P avoids X. If $P - Y$ and $X \cup Y$ are non-empty, then there is a graph that is 2-isomorphic to H in which $P - Y$ has exactly two vertices of odd degree. In particular, if $H = G/f$ where $f \in P$, then $P - f$ is a hypopath of H.*

Proof. As $X \cup Y$ is non-empty, it follows by the choice of M that there is a graph H' such that $M(H') = M\backslash X/Y$. Since M is binary, Corollary 9.3.7 implies that $(P - Y) \cup e$, which equals $C - Y$, is a disjoint union of circuits of $M(H')$. Thus every vertex of $(P - Y) \cup e$ has even degree in H'. Hence $P - Y$ has exactly two vertices of odd degree in $H'\backslash e$. Now $M(H'\backslash e) = M(H')\backslash e = M\backslash X/Y\backslash e = (M\backslash e)\backslash X/Y = M(G)\backslash X/Y$ $= M(H)$. Thus, by Whitney's 2-Isomorphism Theorem, the graph $H'\backslash e$ is 2-isomorphic to H. We conclude that the first assertion of the lemma holds. If $H = G/f$, then, as $C - f$ is a circuit of the graphic matroid M/f, it follows by Lemma 13.3.7 that $P - f$ is a hypopath of H. □

Since M is 3-connected having at least five elements and $M(G) = M\backslash e$, the graph G is simple and has at least four edges. Thus, by Corollary 8.2.8, $M\backslash e$ is a 3-connected matroid if and only if G is a 3-connected graph. In order to determine the possibilities for M, we shall first determine the possibilities for G. Recalling that $\underset{\sim}{M\backslash e}$ is 3-connected, we can divide the argument here into the following two cases:

(i) G is not 3-connected, but $\underset{\sim}{M(G)}$ is 3-connected; and

(ii) G is 3-connected.

To determine the possibilities for G, we shall use six lemmas in the first case and five in the second.

Consider case (i). We know that $\underset{\sim}{M(G)}$ is obtained from $M(G)$ by contracting all but one edge from each non-trivial series class. The following definition will be useful in giving an explicit description of the non-trivial series classes of $M(G)$. A *series path* in a graph H is a path X in H of length at least two that is maximal with the property that $V(X)$ meets $V(H[E(H) - X])$ in the set of end-vertices of X.

13.3.10 Lemma. *If G is not 3-connected, then every non-trivial series class of $M(G)$ is the edge set of a series path in G. Thus $\underset{\sim}{M(G)} \cong M(\underset{\sim}{G})$ where $\underset{\sim}{G}$ is obtained from G by contracting all but one edge from each series path. Moreover, every graph that is 2-isomorphic to G can be obtained from G by a sequence of twistings, where each of these twistings takes place about two vertices that are in the same series path.*

Proof. Let $\{u, v\}$ be a vertex cut in G such that no series path has u and v among its vertices. If f is an edge of G that is in a non-trivial series class of $M(G)$, then $\{u, v\}$ is a vertex cut of G/f. By repeating this process, we find that if G_1 is the graph that is obtained by contracting all but one edge from each non-trivial series class of $M(G)$, then $\{u, v\}$ is a vertex cut of G_1 and so $|V(G_1)| \geq 4$. Thus, by Corollary 8.2.8, $M(G_1)$, which is isomorphic to $\underset{\sim}{M(G)}$, is not 3-connected. This contradiction implies that if two vertices form a vertex cut in G, then these two vertices are in some series path of G. The lemma follows from this without difficulty. □

Next we bound the number of series paths in G.

13.3.11 Lemma. *If G is not 3-connected, then it has at most three series paths. Moreover, each of these series paths contains exactly two edges, exactly one of which is in P.*

Proof. First suppose that G has a series path S_1 containing more than one edge from P, and let f_1 be in $S_1 \cap P$. Then, by Lemma 13.3.9, there

Fig. 13.5.

is a graph G_1 that is 2-isomorphic to G/f_1 in which $P - f_1$ has exactly two vertices of odd degree. Therefore, by the last lemma, there is a graph G_1' that is 2-isomorphic to G such that $G_1'/f_1 = G_1$. Now, by Lemma 13.3.8, P has at least four vertices of odd degree in G_1'. But $P - f_1$ has only two vertices of odd degree in G_1. Therefore both the ends v_1 and v_2 of f_1 must be odd-degree vertices of P in G_1', and P has exactly four odd-degree vertices in G_1'. Thus, in G_1', the series path S_1 contains a path with edges $v_1v_2, v_2v_3, \ldots, v_{n-1}v_n$ for some $n \geq 4$ where P contains v_1v_2 and $v_{n-1}v_n$ but none of $v_2v_3, v_3v_4, \ldots,$ and $v_{n-2}v_{n-1}$ (see Figure 13.5, noting that the boldface edges there are those in P). Clearly $v_1, v_2,$ and v_{n-1} are three of the four odd-degree vertices of P in G_1'. If, in G_1', we twist the path from v_1 to v_{n-1} about its end-vertices, we get a graph that is 2-isomorphic to G in which P has only two odd-degree vertices. This contradiction to Lemma 13.3.8 establishes that no series path of G contains more than one edge of P.

Next we use a similar argument to that just given to show that every series path in G has at most one edge that is not in P. Assume to the contrary that P has a series path S_2 that contains at least two edges not in P, one of which is f_2. Arguing as above using Lemmas 13.3.8 and 13.3.9, we deduce that there is a graph G_2' that is 2-isomorphic to G such that G_2'/f_2 is 2-isomorphic to G/f_2, and $P - f_2$ has exactly two odd-degree vertices in G_2'/f_2. Now $P - f_2 = P$ because $f_2 \notin P$. Therefore, since P has at least four vertices of odd degree in G_2', it follows that P has exactly four such vertices and that two of these are the ends of f_2. Thus, in G_2', the series path S_2 contains a path as in Figure 13.5 where $v_1v_2 = f_2$ and, this time, the boldface edges are those not in P. As before, $v_1, v_2,$ and v_{n-1} are three of the four odd-degree vertices of P, and twisting the path from v_1 to v_{n-1} about its end-vertices produces a contradiction. We conclude that every series path contains at most one edge not in P. On combining this with the conclusion of the first paragraph, we deduce that every series path in G contains exactly one edge in P and exactly one edge not in P.

Next we suppose that G has more than three series paths. Choose one such series path S_3 and let f_3 be an edge in $S_3 \cap P$. Evidently G/f_3 has at least three series paths and, by Lemma 13.3.10, the unique degree-2 vertex of each of these series paths is a vertex of odd degree in $P - f_3$ in every graph that is 2-isomorphic to G/f_3. This contradiction to Lemma 13.3.9 establishes that G has at most three series paths and thereby completes the proof of Lemma 13.3.11. □

To determine the possibilities for G when it is not 3-connected, we shall rely largely on information about series paths. The next lemma gives more such information.

13.3.12 Lemma. *If G is not 3-connected, then either G has exactly three series paths or some edge of P is not in any series path.*

Proof. Suppose that G has at most two series paths and that these series paths contain all the edges of P. Then, since P is not a hypopath of G, the latter has exactly two series paths. Moreover, as P has four odd-degree vertices in every graph 2-isomorphic to G, these two series paths do not share a common vertex. It follows that if $f \in S - P$ where S is one of the series paths, then, in every graph 2-isomorphic to G/f, there are four odd-degree vertices in P. This contradicts Lemma 13.3.9. □

By this lemma, the following three cases, (I), (II)(a), and (II)(b), exhaust all the possibilities when G is not 3-connected:

(I) G has a 4-cycle formed by two series paths;

(II) G has no such 4-cycle; and

- **(a)** some edge of P is not in any series path; or
- **(b)** G has exactly three series paths and every edge of P is in one of these series paths.

These three cases will be treated, in order, by the next three lemmas, the second of which has the longest proof.

13.3.13 Lemma. *If G is not 3-connected and G has a 4-cycle that is formed by two series paths, then $G \cong K_{2,3}$.*

Proof. As $\underset{\sim}{G}$ has a 2-cycle and $M(\underset{\sim}{G})$ is 3-connected, by Proposition 8.1.6 and Table 8.1 (p.292), $M(\underset{\sim}{G})$ is isomorphic to $U_{1,2}$ or $U_{1,3}$. Since the 4-cycle in G is formed by two distinct series paths, $|E(G)| \neq 4$ so $|E(G)| \geq 5$. Thus, either $G \cong K_{2,3}$ or $G \cong K_4\backslash e$. In the latter case, M is a rank-3 simple binary matroid having six elements, so $M \cong M(K_4)$; a contradiction. Therefore $G \cong K_{2,3}$. □

13.3.14 Lemma. *Suppose that G is not 3-connected, G has no 4-cycle formed by two series paths, and P has an edge that is not in any series path. Then G is 2-isomorphic to the graph shown in Figure 13.6.*

Proof. Let S be a series path of G and choose f in $S \cap P$. By Lemma 13.3.9, $P - f$ is a hypopath of G/f. Therefore, by Lemma 13.3.10, the

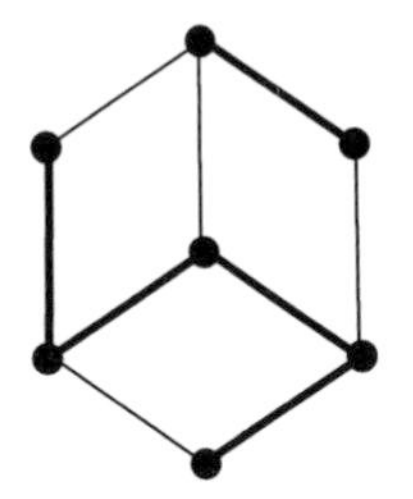

Fig. 13.6. Edges of P are marked in boldface.

edges of P that are not in a series path of G/f form a path P_1 of G/f and hence of G. Moreover, every series path of G/f contains an end-vertex of P_1. As f was arbitrarily chosen as an edge of P that is in a series path, we deduce that every series path of G contains an end-vertex of P_1. Since P is not a hypopath of G, there are at least two series paths S_1 and S_2 of G that contain the same end-vertex, say v_1, of P_1. Let w_1 and w_2 be the internal vertices of S_1 and S_2, respectively. Evidently v_1 must be an end-vertex of both S_1 and S_2. If the other end-vertices, u_1 and u_2, of S_1 and S_2 are also the same, then these two series paths form a 4-cycle. Thus we may assume that $u_1 \neq u_2$. Moreover, if v_2 is the end-vertex of P_1 other than v_1, then we may also suppose that $u_1 \neq v_2$ (see Figure 13.7). Note that possibly $u_2 = v_2$. Also u_1 or u_2 may be an internal vertex of P_1. By Lemma 13.3.11, $S_1 - P$ contains a unique edge f_1. Contracting f_1 from G will identify w_1 with one of v_1 and u_1.

Assume that S_1 and S_2 are the only series paths of G. Then, in every graph that is 2-isomorphic to G/f_1, the vertices u_1, v_2, and w_2 of P all have odd degree unless $u_2 = v_2$. Thus, by Lemma 13.3.9, we may assume that $u_2 = v_2$. But then P is a hypopath of G; a contradiction. We conclude that G has more than two series paths, so, by Lemma 13.3.11, G has exactly three such paths.

Let S_3 be the third series path of G and w_3 be its internal vertex. Then, in every graph that is 2-isomorphic to G/f_1, the vertices w_2 and

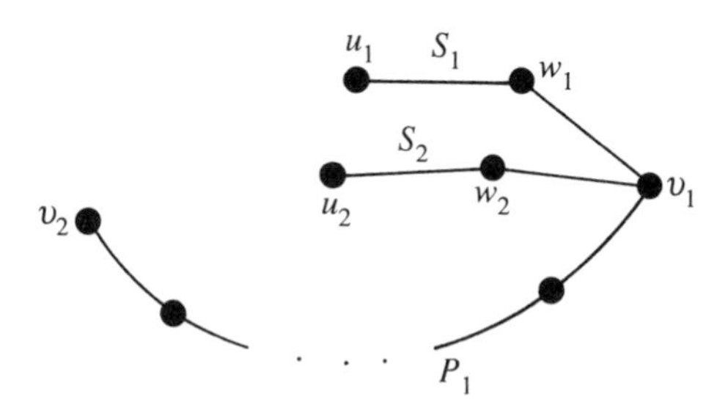

Fig. 13.7. The series paths S_1 and S_2 and the path P_1.

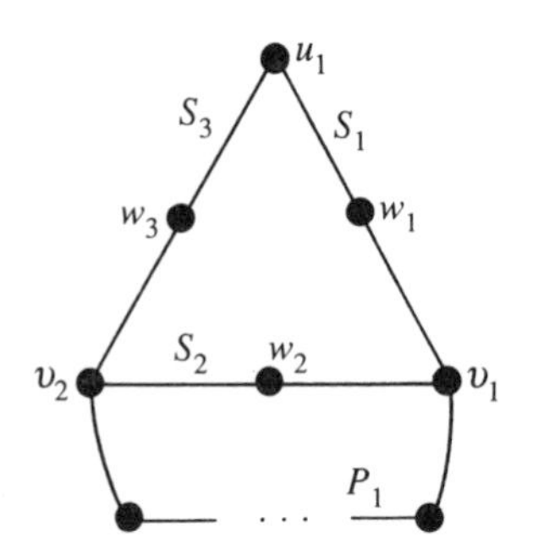

Fig. 13.8. The relationship between S_1, S_2, S_3, and P_1.

w_3 of P both have odd degree. Therefore, by Lemma 13.3.9, S_3 has u_1 as an end-vertex, otherwise u_1 will also be an odd-degree vertex of P in G/f_1. Since S_3 has v_1 or v_2 as an end-vertex, but G has no 4-cycle formed by two series classes, S_3 must have v_2 as an end-vertex. But now $u_2 = v_2$ otherwise, in every graph that is 2-isomorphic to G/f_1, one of v_2 and u_1 is an odd-degree vertex of P.

We now know that the relationship between S_1, S_2, S_3, and P_1 is as shown in Figure 13.8 where possibly u_1 is a vertex of P_1. Since $\underset{\sim}{G}$ is simple, $|P_1| > 1$. Assume that u_1 is not a vertex of P_1. Because $\underset{\sim}{G}$ is 3-connected, $\underset{\sim}{G} - \{v_1, v_2\}$ is connected. Therefore there is a path in G from u_1 to some vertex in P_1 other than v_1 or v_2. Moreover, we may assume that this path P_2 is internally disjoint from P_1. Thus G has the graph H in Figure 13.9 as a subgraph. Moreover, the edges of P consist of the edges of P_1 together with exactly one of the edges from each of S_1, S_2, and S_3.

If H is a proper subgraph of G, then, by Lemma 13.3.9, P is a hypopath of H. This is a contradiction since $\underset{\sim}{H} \cong K_4$ and, as is not difficult to see, P has more than two odd-degree vertices in any graph that is 2-isomorphic to H. We conclude that $H = G$. Thus, as G has only three series paths, P_1 has just two edges and P_2 just one edge. Therefore G is 2-isomorphic to the graph shown in Figure 13.6.

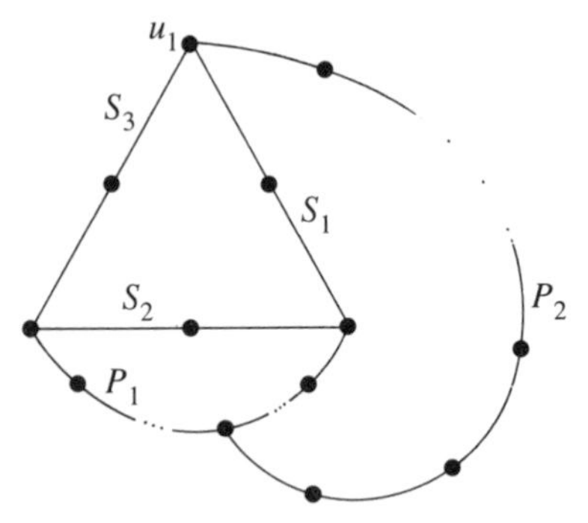

Fig. 13.9. The case when u_1 is not a vertex of P_1.

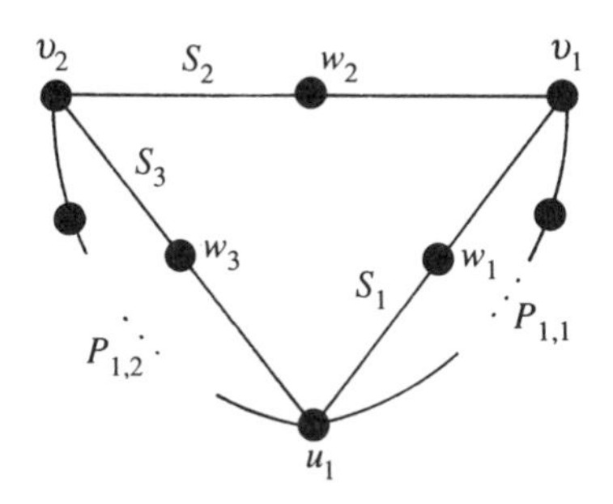

Fig. 13.10. The case when u_1 is a vertex of P_1.

It remains to consider the case when u_1 is a vertex of P_1 as in Figure 13.10. For $i = 1, 2$, let $P_{1,i}$ be the segment of P_1 with end-vertices v_i and u_1. Since $\underset{\sim}{G}$ is simple, $|P_{1,i}| > 1$. Now $\underset{\sim}{G} - \{u_1, v_1\}$ is connected, so there is a path in G from an internal vertex of $P_{1,1}$ to a vertex of $V(P_{1,2}) - u_1$. Moreover, we may assume that this path P_3 is internally disjoint from P_1. If possible, we choose P_3 so that it avoids v_2; otherwise there is a path P_4 from an internal vertex of $P_{1,2}$ to v_1 that is internally disjoint from both P_1 and P_3. Hence G has one of the graphs H_1 and H_2 in Figure 13.11 as a subgraph. Thus either H_1/P_3 or H_2/P_3 is a proper minor of G containing P. But it is not difficult to check that P has more than two odd-degree vertices in any graph that is 2-isomorphic to H_1/P_3 or H_2/P_3. This contradiction to Lemma 13.3.9 implies that u_1 is not a vertex of P_1, thereby finishing the proof of the lemma. □

13.3.15 Lemma. *Suppose that G is not 3-connected, has no 4-cycle formed by two series paths, and has exactly three series paths. If every edge of P is in a series path of G, then G is 2-isomorphic to the graph shown in Figure 13.12.*

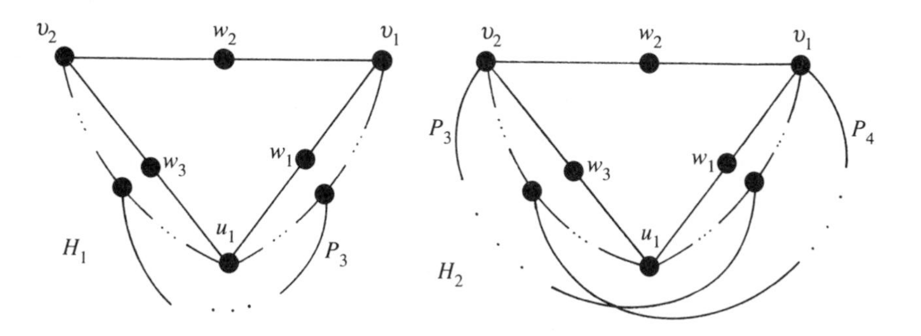

Fig. 13.11. The graphs H_1 and H_2.

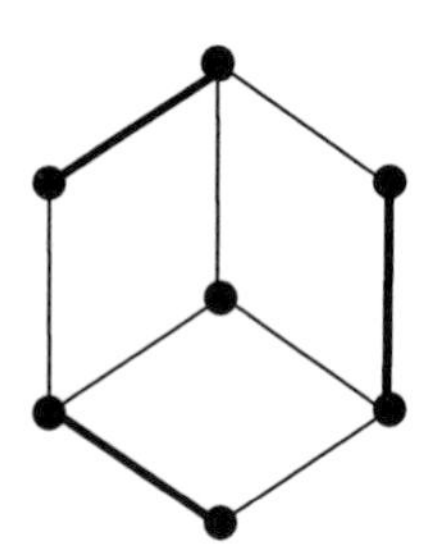

Fig. 13.12. Edges of P are marked in boldface.

Proof. Choose f in P. Then, by Lemma 13.3.9, $P - f$ is a hypopath of G/f. Using this and Lemma 13.3.10, we deduce that each pair of series paths has at least one vertex in common.

Now assume that the three series paths of G form a 6-cycle C'. If $\{u, v, w\}$ is the set of end-vertices of these series paths, and V' is the set of vertices of G not in C', then, as $M(\underset{\sim}{G})$ is 3-connected, none of u, v, and w has degree 2 in $\underset{\sim}{G}$. Therefore, as $\underset{\sim}{G}$ is simple and no proper subset of $\{u, v, w\}$ is a vertex cut of G, there is a minor H of G that is 2-isomorphic to the graph in Figure 13.12. If $H = G$, then the lemma holds. Therefore we may assume that H is a proper minor of G. But then Lemma 13.3.9 gives a contradiction since it is clear that H is not 2-isomorphic to a graph in which P has exactly two odd-degree vertices. We conclude that if the three series paths of G form a 6-cycle, then G is 2-isomorphic to the graph in Figure 13.12.

We may now assume that the three series paths of G all meet at a single vertex v but are otherwise disjoint. In that case, if S is a series path of G and f is an edge of $S - P$, it is not difficult to see that $P - f$ must have more than two odd-degree vertices in every graph that is 2-isomorphic to G/f. This contradiction to Lemma 13.3.9 finishes the proof of Lemma 13.3.15. □

This completes the determination of all the possibilities for G in case (i), that is, when G is not 3-connected. Next we consider case (ii), that is, when G is 3-connected. In the argument that follows, an edge f of G is called *deletable* if the cosimple matroid associated with $M(G\backslash f)$ is 3-connected; f is called *contractible* if the simple matroid associated with $M(G/f)$ is 3-connected. By Proposition 8.4.6, every edge of G is either deletable or contractible.

We note that if G is 3-connected and f is a contractible edge of it, then every graph 2-isomorphic to G/f actually equals G/f. The next two lemmas relate the set P to the behaviour of contractible and deletable edges. If v is a vertex of G, then C_v^* will denote the set of edges incident with v.

13.3.16 Lemma. *If G is 3-connected, then every contractible edge of G joins two odd-degree vertices of P.*

Proof. If f is a contractible edge of G, then, by Lemma 13.3.9 and the remarks preceding this lemma, $P - f$ has exactly two odd-degree vertices in G/f. But, by Lemma 13.3.8, P has at least four odd-degree vertices in G. Thus f must join two odd-degree vertices of P. □

13.3.17 Lemma. *If G is 3-connected and f is a deletable edge not in P, then G has a degree-3 vertex v that meets f and exactly one edge of P. Moreover, if v is the only such vertex of G and $C_v^* = \{f, g, h\}$, then both ends of g and both ends of h have odd degree in P.*

Proof. Since G is a 3-connected graph, every series path in $G\backslash f$ contains exactly two edges. Now $M\backslash f$ is graphic. Thus, by Lemma 13.3.7, P is a hypopath of $G\backslash f$. But P is not a hypopath of G. Therefore $G\backslash f$ must have a series path S that contains exactly one edge of P. The vertex of S that has degree two in $G\backslash f$ has degree three in G and meets f. Taking this vertex to be v, we see that the first assertion of the lemma holds.

Now suppose that v is the only degree-3 vertex of G that meets both f and exactly one edge of P. Then S is the only series path in $G\backslash f$. Since P is a hypopath of $G\backslash f$ but not of G, we must have that P is a path of the graph obtained from $G\backslash f$ by twisting S about its end-vertices. Thus, in G, both the end-vertices of S are odd-degree vertices of P. □

For each circuit C' of M that contains e, let $O_{C'}$ denote the set of odd-degree vertices in $G[C' - e]$. In particular, O_C is the set of odd-degree vertices of P.

13.3.18 Lemma. *If G is 3-connected and C' is a circuit of M containing e, then a vertex v of G is in $O_{C'}$ if and only if $C_v^* \cup e$ is a cocircuit of M. Hence $O_{C'} = O_C$.*

Proof. As G is 3-connected, for all v in $V(G)$, the set C_v^* is a cocircuit of $M(G)$. Thus, since $M(G) = M\backslash e$, either C_v^* or $C_v^* \cup e$ is a cocircuit of M. Now M is binary, so every circuit meets every cocircuit in an even number of elements. Therefore, since $v \in O_{C'}$ if and only if $|(C' - e) \cap C_v^*|$ is odd, it follows that $v \in O_{C'}$ if and only if $|C' \cap (C_v^* \cup e)|$ is even. The lemma is now immediate. □

13.3.19 Lemma. *If G is 3-connected, then every edge of G is incident with a vertex of O_C.*

Proof. Suppose f is an edge of G that is not incident with a vertex of O_C. Since M is 3-connected, $\{e, f\}$ does not contain a cocircuit of M.

Therefore M has a circuit C' that contains e but not f. Let $P' = C' - e$. We now apply Lemmas 13.3.16 and 13.3.17 using P' in place of P. First we note, by Lemma 13.3.16, that if f is contractible, then it meets a vertex of $O_{C'}$. But if f is not contractible, then it is deletable and therefore, by Lemma 13.3.17, it again meets a vertex of $O_{C'}$. Hence, in both cases, f meets a vertex of $O_{C'}$. But, by Lemma 13.3.18, $O_{C'} = O_C$ and so we have a contradiction. □

With these preliminaries, we can now determine the possibilities for G when G is 3-connected.

13.3.20 Lemma. *If G is 3-connected, then G is isomorphic to $\mathcal{W}_3$ or $\mathcal{W}_4$.*

Proof. By the dual of Lemma 13.3.6, G has a contractible edge f. If P has more than four odd-degree vertices in G, then $P - f$ has more than two odd-degree vertices in G/f; a contradiction to Lemma 13.3.9. Therefore, by Lemma 13.3.8, P has exactly four odd-degree vertices in G, that is, $|O_C| = 4$.

We show next that G has at most one vertex that is not in O_C. Assume, to the contrary, that $V(G) - O_C$ has size at least two and let u_1 and u_2 be distinct members of this set. The 3-connectedness of G implies that both u_1 and u_2 have degree at least three. By Lemma 13.3.19 and the fact that $|O_C| = 4$, it follows that O_C contains two distinct vertices v_1 and v_2 each of which is adjacent to both u_1 and u_2. As P contains no cycles, at least one of the edges u_1v_1, u_1v_2, u_2v_1, and u_2v_2 is not in P. Let $f = u_1v_1$ and assume, without loss of generality, that $f \notin P$. Because $u_1 \notin O_C$, Lemma 13.3.16 implies that f is not contractible. Therefore f is deletable. Hence, by Lemma 13.3.17, v_1 has degree three and is adjacent to at least two vertices of O_C. Since v_1 is adjacent to u_1 and u_2, neither of which is in O_C, we have a contradiction. We conclude that G has at most one vertex not in O_C. Therefore $|V(G)|$ is 4 or 5 and, as G is 3-connected and simple, it follows that $G \cong \mathcal{W}_3$ or $\mathcal{W}_4$. □

We have now determined all the possibilities for the graph G in both cases (i) and (ii). It is not difficult to use this information to determine all the possibilities for the binary minor-minimal non-graphic matroid M where we recall that $M\backslash e = M(G)$ and $P \cup e$ is a circuit of M. The next two lemmas complete the proof of Theorem 13.3.3.

13.3.21 Lemma. *If G is not 3-connected, then M is isomorphic to F_7^* or $M^*(K_5)$.*

Proof. By Lemmas 13.3.13, 13.3.14, and 13.3.15, G is isomorphic to $K_{2,3}$ or one of the graphs in Figures 13.6 and 13.12. In the first case, M is a 3-connected binary rank-4 matroid having seven elements. Thus

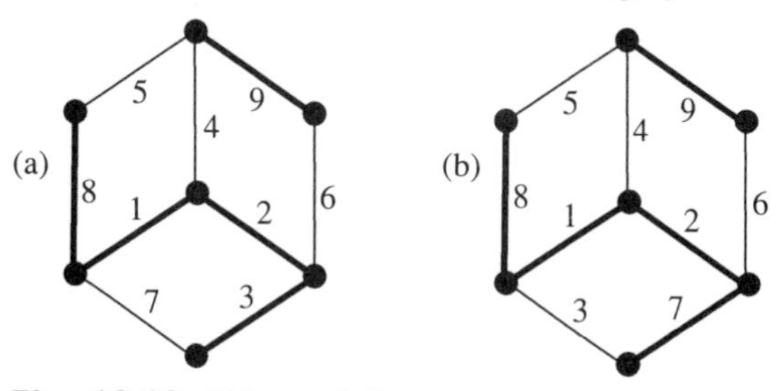

Fig. 13.13. Edges of P are marked in boldface.

M^* is a simple binary rank-3 matroid having seven elements. Hence $M^* \cong PG(2,2) \cong F_7$, so $M \cong F_7^*$.

If G is 2-isomorphic to either of the graphs in Figures 13.6 and 13.12, then with the elements labelled as in (a) and (b), respectively, of Figure 13.13, we find that M is represented over $GF(2)$ by the following matrix:

$$\begin{array}{c} \begin{matrix} 1 & 2 & 3 & 4 & 5 & 6 & 7 & 8 & 9 & e \end{matrix} \\ \left[\begin{array}{c|cccc} & 1 & 1 & 0 & 0 \\ & 1 & 0 & 1 & 0 \\ I_6 & 1 & 0 & 0 & 1 \\ & 0 & 1 & 1 & 0 \\ & 0 & 1 & 0 & 1 \\ & 0 & 0 & 1 & 1 \end{array}\right]. \end{array}$$

Thus $M \cong M^*(K_5)$. □

13.3.22 Lemma. *If G is 3-connected, then M is isomorphic to F_7 or $M^*(K_{3,3})$.*

Proof. By Lemma 13.3.20, G is isomorphic to $\mathcal{W}_3$ or $\mathcal{W}_4$. In the first case, M is simple and binary having rank three and seven elements. Thus $M \cong F_7$. In the second case, $M\backslash e$ is represented by the following matrix over $GF(2)$:

$$\begin{array}{c} \begin{matrix} 1 & 2 & 3 & 4 & 5 & 6 & 7 & 8 \end{matrix} \\ \left[\begin{array}{c|cccc} & 1 & 0 & 0 & 1 \\ I_4 & 1 & 1 & 0 & 0 \\ & 0 & 1 & 1 & 0 \\ & 0 & 0 & 1 & 1 \end{array}\right]. \end{array}$$

Therefore a binary representation for M can be obtained by adjoining a column to this matrix. By the symmetry of the matrix and the simplicity of M, we may assume that this column is one of $(1,0,1,0)^T$, $(1,1,1,0)^T$, or $(1,1,1,1)^T$. In the first case, M is isomorphic to a single-element deletion of $M(K_5)$ and hence is graphic; a contradiction. In the second case, one easily checks that $M/8\backslash 1 \cong F_7$, so M is not minor-minimal non-graphic; another contradiction. In the third case, $M \cong M^*(K_{3,3})$.

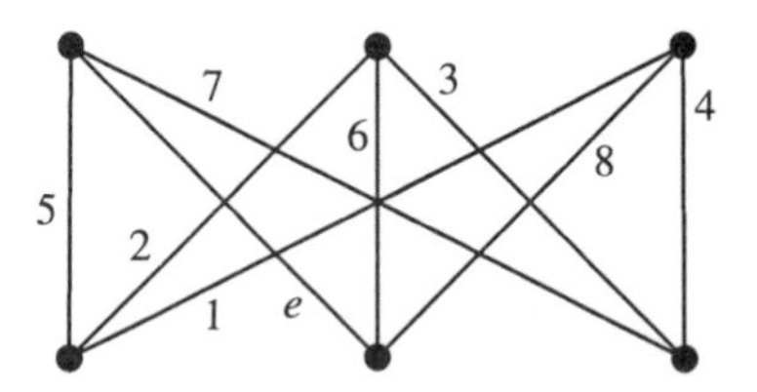

Fig. 13.14. $M^* = M(K_{3,3})$ under this labelling of $K_{3,3}$.

The last claim follows from the fact that, if $K_{3,3}$ is labelled as in Figure 13.14, then $M(K_{3,3})$ is represented over $GF(2)$ by the matrix:

$$\begin{array}{c} \begin{matrix} 5 & 6 & 7 & 8 & e & 1 & 2 & 3 & 4 \end{matrix} \\ \left[\begin{array}{c|cccc} & 1 & 1 & 0 & 0 \\ & 0 & 1 & 1 & 0 \\ I_5 & 0 & 0 & 1 & 1 \\ & 1 & 0 & 0 & 1 \\ & 1 & 1 & 1 & 1 \end{array}\right]. \end{array}$$

This completes the proof of Lemma 13.3.22 and thereby that of Theorem 13.3.3. □

Exercises

1. (Oxley 1979) Use the fact that $M(K_5)$ and $M^*(K_{3,3})$ are maximal regular (Exercise 3, Section 6.6) to prove that a regular matroid is cographic if and only if it has no parallel minor isomorphic to $M(K_5)$ and no series minor isomorphic to $M(K_{3,3})$.

2. Let M be a cyclically 3-connected non-empty matroid. Prove that M has a series class S such that $M\backslash S$ is cyclically 3-connected.

3.* (Graver 1966, Welsh 1969b) Prove that a matroid M is graphic if and only if M is binary and its cocircuit space has a basis B such that every element of $E(M)$ is in at most two members of B.

4.* (Asano, Nishizeki, and Seymour 1984) Let $\{e, f, g\}$ be a circuit of a 3-connected non-graphic matroid M. Prove that M has a minor using $\{e, f, g\}$ that is isomorphic to $U_{2,4}$, F_7, or $M^*(K_{3,3})$.

13.4 Further properties of regular and graphic matroids

In this section we survey some additional properties of regular and graphic matroids. Most of the proofs of these results will be omitted. We begin by noting some extensions of Theorems 13.1.2 and 13.3.2 due to Bixby

(1976, 1977). Recall from Chapter 5 that a matroid N is a parallel minor of a matroid M if $N = M/Y\backslash X$ for some subsets X and Y of $E(M)$ such that, in M/Y, every element of X is parallel to some element not in X. If N^* is a parallel minor of M^*, then N is a series minor of M.

13.4.1 Theorem. *The following are equivalent for a binary matroid M:*

(i) *M is regular.*

(ii) *M has no parallel minor isomorphic to F_7 or F_7^*.*

(iii) *M has no series minor isomorphic to F_7 or F_7^*.*

Proof. This is left to the reader (see Exercise 4). □

The next result extends Propositions 5.4.3 and 5.4.4. It is taken from a paper of Bixby (1977) that was the first to point out the special role that R_{10} plays within the class of regular matroids. The full realization of the importance of R_{10} within this class emerged with Seymour's decomposition theorem (13.2.4). The simple graphs $K'_{3,3}$, $K''_{3,3}$, and $K'''_{3,3}$ are obtained from $K_{3,3}$ by adding one, two, and three edges, respectively, within the same vertex class (see Figure 5.14).

13.4.2 Theorem. *The following are equivalent for a regular matroid M:*

(i) *M is graphic.*

(ii) *M has no series minor isomorphic to R_{10}, $M^*(K_5)$, $M^*(K_{3,3})$, $M^*(K'_{3,3})$, $M^*(K''_{3,3})$, or $M^*(K'''_{3,3})$.*

(iii) *M has no parallel minor isomorphic to $M^*(K_5)$ or $M^*(K_{3,3})$.* □

13.4.3 Corollary. *The following are equivalent for a regular matroid M:*

(i) *M is cographic.*

(ii) *M has no series minor isomorphic to $M(K_5)$ or $M(K_{3,3})$.*

(iii) *M has no parallel minor isomorphic to R_{10}, $M(K_5)$, $M(K_{3,3})$, $M(K'_{3,3})$, $M(K''_{3,3})$, or $M(K'''_{3,3})$.* □

Now let M be a simple rank-r matroid. We know that if M is binary, then it is isomorphic to a restriction of $PG(r-1,2)$. Hence $|E(M)| \leq 2^r-1$ with this bound being attained if and only if $M \cong PG(r-1,2)$. If M is graphic, then $M \cong M(G)$ for some $(r+1)$-vertex graph G. Therefore $|E(M)| \leq \binom{r+1}{2}$, the bound being attained if and only if $M \cong M(K_{r+1})$. These observations for binary and graphic matroids prompt one to seek a similar result when M is regular. Heller (1957) proved that, in this case,

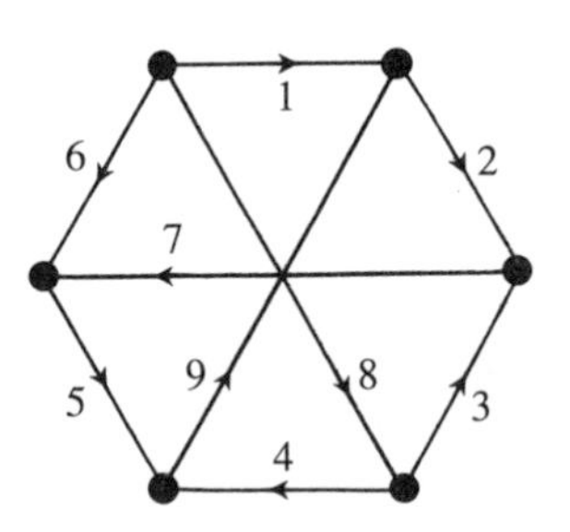

Fig. 13.15. An orientation of $K_{3,3}$.

both the bound on $|E(M)|$ and the extremal matroids are the same as when M is graphic; that is, a simple rank-r regular matroid M has at most $\binom{r+1}{2}$ elements with equality holding if and only if $M \cong M(K_{r+1})$. This result can be proved using Seymour's decomposition theorem (13.2.4) along with the bound for graphic matroids noted above and the corresponding bound for cographic matroids (Exercise 8). However, there are a number of more elementary proofs of the following stronger result of Murty (1976) (see, for example, Baclawski and White 1979 or Bixby and Cunningham 1987).

13.4.4 Proposition. *Let M be a simple rank-r binary matroid having no F_7-minor. Then $|E(M)| \leq \binom{r+1}{2}$ with this bound being attained if and only if $M \cong M(K_{r+1})$.* □

Next we consider a property of graphs that extends to regular matroids but not to all binary matroids. As we shall see, this property can actually be used to give another characterization of the class of regular matroids. Let M be a matroid on the set $\{1, 2, \ldots, n\}$ such that $\mathcal{C}(M) = \{C_1, C_2, \ldots, C_m\}$. The *circuit incidence matrix* $A(\mathcal{C})$ of M is the $m \times n$ matrix $[a_{ij}]$ in which a_{ij} is 1 or 0 depending on whether j is or is not in C_i. The *cocircuit incidence matrix* $A(\mathcal{C}^*)$ of M is the circuit incidence matrix of M^*. Evidently each of $A(\mathcal{C})$ and $A(\mathcal{C}^*)$ is only determined to within a permutation of its rows and a permutation of its columns. Following White (1987c), we call the matroid M *signable* if there are signings $A'(\mathcal{C})$ and $A'(\mathcal{C}^*)$ of $A(\mathcal{C})$ and $A(\mathcal{C}^*)$, respectively, such that, over $\mathbb{R}$, every row of $A'(\mathcal{C})$ is orthogonal to every row of $A'(\mathcal{C}^*)$.

Given a graph G, an orientation of G can be used in a natural way to give signings of the circuit and cocircuit incidence matrices of $M(G)$. For instance, let $G \cong K_{3,3}$ and suppose G is oriented as in Figure 13.15. Let C and C^* be the circuit $\{1, 2, 3, 4, 5, 6\}$ and the cocircuit $\{1, 4, 7, 8, 9\}$ of $M(G)$. The corresponding rows, $\underline{v}_C$ and $\underline{v}_{C^*}$, of $A(\mathcal{C})$ and $A(\mathcal{C}^*)$ are $(1, 1, 1, 1, 1, 1, 0, 0, 0)$ and $(1, 0, 0, 1, 0, 0, 1, 1, 1)$. Now arbitrarily assign orientations to C and C^* and change the signs of those entries of

$\underline{v}_C$ and $\underline{v}_{C^*}$ for which these directions disagree with the directions on the corresponding edges of G. For example, if C is oriented clockwise and C^* from left to right, the signed rows corresponding to C and C^* are $(1, 1, -1, 1, -1, -1, 0, 0, 0)$ and $(1, 0, 0, -1, 0, 0, -1, 1, 1)$, respectively. By repeating this process for all cycles and bonds of G, we obtain signings $A'(\mathcal{C})$ and $A'(\mathcal{C}^*)$ of $A(\mathcal{C})$ and $A(\mathcal{C}^*)$. Moreover, it is not difficult to show that, for all graphs, these signings have the property that the rows of $A'(\mathcal{C})$ are orthogonal to those of $A'(\mathcal{C}^*)$ (Exercise 6). Hence every graphic matroid is signable. As the dual of a signable matroid is clearly signable, every cographic matroid is also signable. The following result (Minty 1966) identifies the class of signable matroids. Our proof follows White (1987c).

13.4.5 Proposition. *A matroid is regular if and only if it is signable.*

Proof. Let M be a signable matroid and suppose $A'(\mathcal{C})$ and $A'(\mathcal{C}^*)$ are signings of the circuit and cocircuit incidence matrices of M such that, over $\mathbb{R}$, the rows of $A'(\mathcal{C})$ are orthogonal to those of $A'(\mathcal{C}^*)$. Clearly this orthogonality condition implies that a circuit and cocircuit of M cannot meet in an odd number of elements. Therefore, by Theorem 9.1.2, M is binary. Moreover, $A(\mathcal{C}^*)$ is a $GF(2)$-representation of M. To see this, let B be a basis of M and consider the submatrix A_1 of $A(\mathcal{C}^*)$ whose rows are the incidence vectors of the $(E(M) - B)$-fundamental cocircuits of M. Then some column permutation of A_1 has the form $[I_r|X]$ where X is the B-fundamental-circuit incidence matrix of M. Thus A_1 is a $GF(2)$-representation of M. Since every row of $A(\mathcal{C}^*)$ is in the cocircuit space of M and is therefore a linear combination of rows of A_1, we conclude that $A(\mathcal{C}^*)$ is indeed a $GF(2)$-representation of M.

Next we show that M is equal to the matroid M_1 for which $A'(\mathcal{C}^*)$ is an $\mathbb{R}$-representation. If C is the circuit $\{i_1, i_2, \dots, i_m\}$ of M, let the row of $A'(\mathcal{C})$ corresponding to C be $(\alpha_1, \alpha_2, \dots, \alpha_n)$. Then α_j is 1 or -1 if j is in $\{i_1, i_2, \dots, i_m\}$, and α_j is 0 otherwise. Since $(\alpha_1, \alpha_2, \dots, \alpha_n)$ is orthogonal to every row of $A'(\mathcal{C}^*)$, if we sum, over all j in $\{i_1, i_2, \dots, i_m\}$, the product of α_j and the jth column of $A'(\mathcal{C}^*)$, then we get the zero vector. Thus C is dependent in M_1, that is, every circuit of M is dependent in M_1. But, since $A(\mathcal{C}^*)$ is a $GF(2)$-representation for M, it follows easily that every circuit of M_1 is dependent in M. We conclude, by Lemma 2.1.19, that $\mathcal{C}(M) = \mathcal{C}(M_1)$ and therefore $M = M_1$.

As M is both binary and $\mathbb{R}$-representable, it follows, by Theorem 6.6.3, that M is regular. Hence if M is signable, then it is regular. We leave the proof of the converse of this to the reader (Exercise 7). □

Signable matroids were introduced by Minty (1966) in an attempt to develop a theory of orientation for matroids. Indeed, Minty used the word 'orientable' rather than 'signable' to describe such matroids. However, Rockafellar (1969) suggested that, in view of Minty's result that all signable matroids are regular, 'a much broader theory of orientation ought to be possible'. Such a theory was introduced and developed by Bland and Las Vergnas (1978) and Folkman and Lawrence (1978). This theory provides a natural context for the generalization of many results concerning directed graphs, convex polyhedra, and linear programs. Indeed, the theory is now rich enough to fill an entire book (Björner, Las Vergnas, Sturmfels, White, and Ziegler 1992). We shall not go into the details of this theory here. Instead, we shall be content to indicate how it extends the notion of signability discussed above.

Let M be a matroid. Clearly M is signable if and only if there are signings $A'(\mathcal{C})$ and $A'(\mathcal{C}^*)$ of the circuit and cocircuit incidence matrices of M such that, for all circuits C and all cocircuits C^*,

$$|\{e \in E(M) : \underline{v}'_C(e) = \underline{v}'_{C^*}(e) \neq 0\}| = |\{e \in E(M) : \underline{v}'_C(e) = -\underline{v}'_{C^*}(e) \neq 0\}|$$

where, for example, $\underline{v}'_C(e)$ is the entry of $A'(\mathcal{C})$ corresponding to the circuit C and the element e. The matroid M is *orientable* if there are signings $A'(\mathcal{C})$ and $A'(\mathcal{C}^*)$ of $A(\mathcal{C})$ and $A(\mathcal{C}^*)$ such that, whenever a circuit C and a cocircuit C^* have non-empty intersection, both $\{e \in E(M) : \underline{v}'_C(e) = \underline{v}'_{C^*}(e) \neq 0\}$ and $\{e \in E(M) : \underline{v}'_C(e) = -\underline{v}'_{C^*}(e) \neq 0\}$ are non-empty. Evidently every signable matroid is orientable. Indeed, Bland and Las Vergnas (1978) proved that the class of signable matroids coincides with the class of binary orientable matroids. The next corollary follows immediately on combining this result with the last proposition.

13.4.6 Corollary. *The following statements are equivalent for a matroid M:*

(i) *M is signable.*

(ii) *M is regular.*

(iii) *M is binary and orientable.* □

Transversal matroids share with regular matroids the property of being representable over fields of all characteristics. But exactly how do these two important classes of matroids overlap? To conclude this chapter, we shall answer this and a related question. Unlike the class of regular matroids, the class of transversal matroids is not closed under contraction (see Example 3.2.9). However, as Bondy (1972) showed, this class is closed under the more restrictive operation of series contraction.

13.4.7 Proposition. *A matroid is transversal if and only if all of its series minors are transversal.*

Proof. An outline of a proof due to Piff (1972) is given in Exercise 9. □

Bondy used the last result as a step towards characterizing those matroids that are both graphic and transversal (see also Las Vergnas 1970). Subsequently, de Sousa and Welsh (1972) established that such matroids are precisely the matroids that are both binary and transversal. Clearly these matroids are also the matroids that are both regular and transversal. If $k \geq 3$, we shall denote by C_k^2 the graph that is obtained from a k-edge cycle by replacing each edge by two edges in parallel. Thus, for example, C_3^2 is isomorphic to the graph G_2 in Figure 1.20.

13.4.8 Theorem. *The following statements are equivalent for a matroid M:*

(i) *M is graphic and transversal.*

(ii) *M is regular and transversal.*

(iii) *M is binary and transversal.*

(iv) *M has no series minor isomorphic to any of the matroids $U_{2,4}$, $M(K_4)$, and $M(C_k^2)$ for $k \geq 3$.* □

The smallest minor-closed class of matroids that contains the class of regular transversal matroids is the class of regular gammoids, where we recall from Chapter 3 that a gammoid is a minor of a transversal matroid. The next result, which characterizes the class of regular gammoids, follows immediately from a result of Brylawski (1971, 1975b) and Ingleton (1971b) which identified the class of binary gammoids.

13.4.9 Theorem. *The following statements are equivalent for a matroid M:*

(i) *M is a graphic gammoid.*

(ii) *M is a regular gammoid.*

(iii) *M is a binary gammoid.*

(iv) *M has no minor isomorphic to $U_{2,4}$ or $M(K_4)$.*

(v) *M is a direct sum of series–parallel networks.* □

The equivalence of (iii) and (iv) in 13.4.8 and 13.4.9 was generalized by Oxley (1986b, 1987c) in results that characterize the classes of ternary transversal matroids and ternary gammoids. One may wonder about similar characterizations of the class of all transversal matroids and the class of all gammoids. These problems look very difficult. Ingleton (1977) has noted that there are infinitely many minor-minimal non-gammoids and infinitely many series-minor-minimal non-transversal matroids. Moreover, he described as 'probably futile' the task of completely characterizing the class of gammoids by excluded minors.

Exercises

1. Update the Venn diagram in Figure 6.27 by indicating where the classes of transversal matroids and gammoids fit.

2. (Bondy 1972, Las Vergnas 1970) Let G be a graph. Use Theorem 13.4.8 to prove that $M(G)$ is transversal if and only if no subgraph of G is a subdivision of K_4 or C_k^2 for any $k \geq 3$.

3. Characterize those graphs whose cycle matroids are strict gammoids.

4. **(i)** Show that a binary matroid has an F_7-minor if and only if it has F_7 as a parallel minor.

   (ii) Show that a binary matroid with no F_7-minor has an F_7^*-minor if and only if it has F_7^* as a parallel minor.

   (iii) Use (i) and (ii) to prove Theorem 13.4.1.

   (iv) Prove that a binary matroid is regular if and only if it has no parallel minor isomorphic to F_7 and no series minor isomorphic to F_7^*.

5. Prove that the following are equivalent for a binary matroid M:

   (a) M is graphic.

   (b) M has no series minor isomorphic to R_{10}, $M^*(K_5)$, $M^*(K_{3,3})$, $M^*(K'_{3,3})$, $M^*(K''_{3,3})$, $M^*(K'''_{3,3})$, F_7, or F_7^*.

   (c) M has no parallel minor isomorphic to $M^*(K_5)$, $M^*(K_{3,3})$, F_7, or F_7^*.

6. Use the signings of the circuit and cocircuit matrices described before Proposition 13.4.5 to prove that every graphic matroid is signable.

7. Let $[I_r|A]$ be a totally unimodular matrix.

   (i) Use the pivoting operation (without the natural column interchange) to show that, for each cocircuit C^* of $M[I_r|A]$, the row space of $[I_r|A]$ contains a signing of the incidence vector of C^*.

   (ii) Deduce that if a matroid is regular, then it is signable.

8. (Jaeger 1979) Let M be a simple rank-r cographic matroid. Prove that $|E(M)| \leq 3(r-1)$ and characterize those matroids for which equality holds.

9. Let M be the strict gammoid $L(G, B)$ and suppose that x and y are parallel in M. Show that $M\backslash x$ is a strict gammoid and hence prove Proposition 13.4.7. [Hint: There are two similar cases. If $|\{x, y\} \cap B| = 0$, then G has a vertex whose removal separates $\{x, y\}$ from B. Modify G to obtain a new graph G' for which $L(G, B) = L(G', B)$ and $L(G' - x, B) = M\backslash x$.]

10. (Walton 1981) Prove that if M is a simple rank-r binary matroid having no minor isomorphic to $AG(3, 2)$ or S_8, then $|E(M)| \leq \binom{r+1}{2} + 1$ with equality being attained if and only if $M \cong F_7$.

11.* (Lindström 1984b) Prove that $U_{1,2}$ is the only connected regular matroid that is identically self-dual.

12.* (J. Edmonds in Bondy 1972) Prove that a transversal matroid is binary if and only if it has a presentation for which the associated bipartite graph is a forest.

13.* (Matthews 1977) Let G be a loopless graph. Prove that $M(G)$ is bicircular if and only if G has no subgraph that is a subdivision of K_4, C_3^2, or one of the three graphs in Figure 13.16.

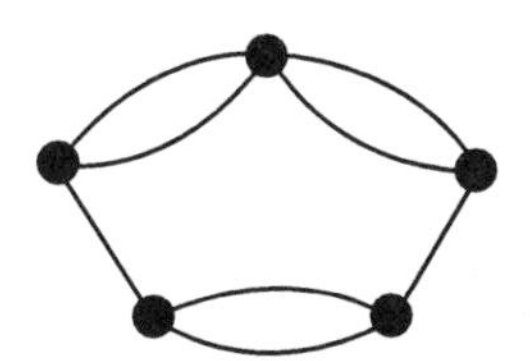
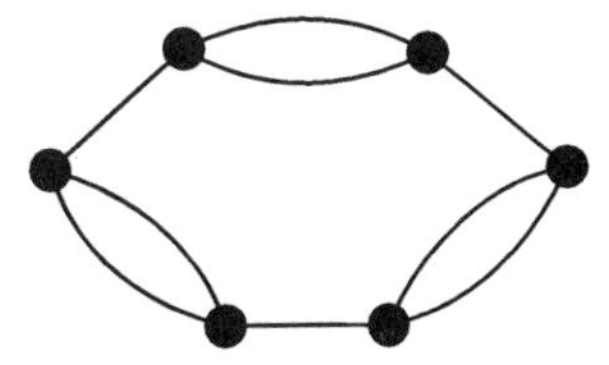

Fig. 13.16.

14.* (Tutte 1965) Let M be a connected binary matroid having rank at least three. Prove that M has an F_7-minor or M has a cocircuit C^* such that $M\backslash C^*$ is disconnected.

14
Unsolved problems

The coverage of matroid theory in this book has been far from encyclopaedic. Many important areas of the subject have been ignored completely or have received just a cursory treatment. However, most of the areas that are not fully covered here are treated in detail in one of the numerous books on matroids that have appeared since 1985. The first two volumes of White's trilogy on matroid theory (White 1986, 1987a) have been referred to frequently in earlier chapters. The third volume (White 1992) considers a variety of applications of matroids, while a related monograph (Björner, Las Vergnas, Sturmfels, White, and Ziegler 1992) is entirely devoted to orientable matroids.

Greedoids are another generalization of matroids that have received considerable attention in the last decade. One of the many equivalent ways to define a *greedoid* is as a finite set E and a collection $\mathcal{I}$ of subsets of E satisfying (I1) and (I3). A detailed treatment of greedoids and their properties can be found in Björner and Ziegler (1992) and in Korte, Lovász, and Schrader (1991). Recent developments in the structural theory of matroids are the subject of a monograph on matroid decompositions by Truemper (1992c), while Recski (1989), in addition to developing the basic theory of matroids, devotes much attention to the applications of matroids in electrical network theory and statics. To conclude this brief survey of recent or forthcoming books in matroid theory, we note two very valuable reference resources: Kung's (1986) 'Source book' an annotated reprinting of eighteen matroid papers which have had a major influence on the development of the subject; and Brown's (1980) *Reviews in graph theory*, which reprints, from the volumes of *Mathematical reviews* for 1940 through 1978, reviews of all the papers in graph theory and related areas including matroid theory.

The primary purpose of this chapter is to bring together, in one place, a number of unsolved problems. Several of these have appeared in earlier chapters and, for these, we have noted the earlier reference since the background to the problem is usually described there. For many of the problems that have not appeared earlier, some details of the background are provided. Several of the books listed above contain further details relating to a number of these problems.

While the selection of problems listed here is partially influenced by the author's preference, it has also been dictated to a large extent by the topics covered in earlier chapters. Where possible, problems have been grouped by topic although the reader will note that some problems actually belong in more than one section.

14.1 Representability: linear and algebraic

Problems in linear representability tend to be among the oldest and hardest in matroid theory. One of the most important problems in the subject is to settle the following conjecture of Rota (1971), noted earlier as 6.5.11.

14.1.1 Conjecture. *If q is a prime power, then the complete set of excluded minors for representability over $GF(q)$ is finite.*

In the first open case of this conjecture, when $q = 4$, a more concrete problem, which was stated earlier as 6.5.9, is the following.

14.1.2 Problem. *Is $\{U_{2,6}, U_{4,6}, F_7^-, (F_7^-)^*, P_6, P_8\}$ the complete set of excluded minors for $GF(4)$-representability?*

The obvious follow-up to a negative answer to the last problem would be to find the actual complete set of excluded minors.

Kahn (1988) has described the next conjecture (10.1.12) as being 'in many ways similar to Rota's conjecture'.

14.1.3 Conjecture. *For each prime power q, there is an integer $n(q)$ such that no 3-connected $GF(q)$-representable matroid has more than $n(q)$ inequivalent representations.*

The next problem (6.5.12) has received considerable attention from projective geometers.

14.1.4 Problem. *Determine all of the finite fields $GF(q)$ over which $U_{r,n}$ is representable.*

A proof of the following conjecture of Hirschfeld (in Fenton and Vámos 1982) would mean that the odd-order fields over which $U_{r,n}$ is representable are completely determined.

14.1.5 Conjecture. *Suppose $n \geq r + 2$ and q is odd. Then $U_{r,n}$ is $GF(q)$-representable if and only if $n \leq q + 1$.*

An interesting special case of 14.1.4 is the following.

14.1.6 Problem. *If $U_{r,n}$ is $GF(q)$-representable and $q' > q$, is $U_{r,n}$ representable over $GF(q')$?*

The next problem, due to Brylawski (private communication), relates to characteristic sets (see Section 6.8) and is not as well known as most of the others stated above.

14.1.7 Problem. *Let M be a ternary matroid. Is one of the following statements true for M?*

(i) *M is representable only over fields of characteristic three.*

(ii) *M is representable over fields of all characteristics.*

(iii) *M is representable over fields of all characteristics except two.*

The classes of binary and ternary matroids are minor-closed, as are their intersection and their union. The former is the class of regular matroids.

14.1.8 Problem. *Find the complete set of excluded minors for the class of matroids that is the union of the classes of binary and ternary matroids.*

Consider the minor-closed class $\mathcal{Q}$ of matroids that are representable over $\mathbb{Q}$ and all the fields $GF(p)$ for p an odd prime. Results of Kung (1990) and Kung and Oxley (1988) suggest that $\mathcal{Q}$ is very similar to the class of regular matroids. In particular, the maximum number of elements in a rank-r simple member M of $\mathcal{Q}$ is r^2 with equality being attained if and only if M is isomorphic to the Dowling geometry $Q_r(GF(3)^*)$. The corresponding result for regular matroids, noted earlier as 13.4.4, asserts that $M(K_{r+1})$, or equivalently $Q_r(GF(2)^*)$, is the unique rank-r simple regular matroid with the largest number of elements. The next two problems were raised by Kung and Oxley (1988).

14.1.9 Problem. *Find the excluded-minor characterization of $\mathcal{Q}$.*

14.1.10 Problem. *Is there an analogue of Seymour's regular matroids decomposition theorem (13.2.4) for $\mathcal{Q}$?*

A popular conjecture is that $\mathcal{Q}$ is precisely the class of matroids that are representable over all the fields $GF(p)$ for p an odd prime, that is:

14.1.11 Conjecture. *If a matroid M is representable over $GF(p)$ for all odd primes p, then M is representable over $\mathbb{Q}$.*

Turning briefly to algebraic representability, we recall from 6.7.15 that the main unsolved problem in this subject is the following.

14.1.12 Problem. *Is the dual of an algebraic matroid also algebraic?*

The intractability of this problem prompted Lindström (1988b) to propose the following question (6.7.16).

14.1.13 Problem. *If a matroid is algebraic over a field of characteristic k, is its dual also algebraic over a field of this characteristic?*

Several other related unsolved problems for algebraic matroids are discussed by Lindström (1985c, 1988b), Gordon (1988), and Lemos (1988).

14.2 Unimodal conjectures

Let $a_1, a_2, \ldots, a_n$ be a finite sequence of real numbers. This sequence is *unimodal* if $a_j \geq \min\{a_i, a_k\}$ for all i, j, and k such that $1 \leq i \leq j \leq k \leq n$. Equivalently, $(a_i : 1 \leq i \leq n)$ is unimodal if, for some t with $1 \leq t \leq n$, the members of the sequence are ordered so that $a_1 \leq a_2 \leq \cdots \leq a_t \geq a_{t+1} \geq \cdots \geq a_n$. A somewhat stronger property of a sequence is that it be *logarithmically concave* or *log concave* for short, that is, $a_k^2 \geq a_{k-1}a_{k+1}$ for all k in $\{2, 3, \ldots, n-1\}$.

In this section, we look at three important sequences that are associated with a matroid, each of which is conjectured to be log concave. Let i_k denote the number of k-element independent sets of a matroid. Welsh (1971b) proposed the following.

14.2.1 Conjecture. *For all matroids M, the sequence $(i_k : 0 \leq k \leq r(M))$ is unimodal.*

Three successive strengthenings of this conjecture were proposed by Mason (1972b):

14.2.2 Conjecture. *If M is a matroid, then, for all k in $\{1, 2, \ldots, r(M)-1\}$:*

(i) $i_k^2 \geq i_{k-1}i_{k+1}$;

(ii) $i_k^2 \geq \frac{k+1}{k} i_{k-1}i_{k+1}$;

(iii) $i_k^2 \geq \left(\frac{k+1}{k}\right)\left(\frac{i_1-k+1}{i_1-k}\right) i_{k-1}i_{k+1}$.

Progress on these conjectures has been very limited. Dowling (1980) proved (i) for $k \leq 7$ and proposed a further extension of it; Hamidoune and Salaün (1989) proved (iii) for $k = 3$ and conjectured another variant of it; Seymour (1975) proved (iii) for those k such that M has no circuits of cardinality $3, 4, \ldots,$ or $k-1$; and Mahoney (1985) proved (i) when M is the cycle matroid of an outerplanar graph.

The number of rank-k flats in a matroid M is denoted by W_k, and the numbers $W_0, W_1, \ldots, W_{r(M)}$ are called the *Whitney numbers of M of the second kind.* Rota (1971) conjectured that this sequence is unimodal.

14.2.3 Conjecture. *If M is a matroid, then*

$$W_k \geq \min\{W_{k-1}, W_{k+1}\} \text{ for all } k \text{ in } \{1, 2, \ldots, r(M)-1\}.$$

In the same paper that he proposed 14.2.2, Mason (1972b) offered the following analogous strengthenings of 14.2.3.

14.2.4 Conjecture. *If M is a rank-r matroid, then, for all k in $\{1, 2, \ldots, r-1\}$,*

(i) $W_k^2 \geq W_{k-1}W_{k+1}$;

(ii) $W_k^2 \geq \frac{k+1}{k} W_{k-1}W_{k+1}$;

(iii) $W_k^2 \geq \left(\frac{k+1}{k}\right)\left(\frac{W_1-k+1}{W_1-k}\right) W_{k-1}W_{k+1}$.

The following special case of (ii) has been called the *points–lines–planes conjecture* (Mason 1972b, Welsh 1976).

14.2.5 Conjecture. *If W_1 points in Euclidean 3-space determine W_2 lines and W_3 planes, then*

$$W_2^2 \geq \tfrac{3}{2} W_1 W_3.$$

For W_1, W_2, and W_3 defined as in 14.2.5, Purdy (1986) proved the existence of a positive constant c for which $W_2^2 \geq cW_1W_3$. There has also been some progress on certain special cases of 14.2.4(iii). Stonesifer (1975) proved that it holds when $k = 2$ for all graphic matroids, and his result was extended by Seymour (1982) who showed, again in the case $k = 2$, that 14.2.4(iii) holds for all matroids in which no line has five or more points. Incidentally, for Seymour (1982), the points–lines–planes conjecture is the case $k = 2$ of 14.2.4(iii). Aigner and Schoene (1987) have verified 14.2.4(i) for binary matroids in the case $k = 2$, and Aigner (1987) expresses optimism that the log concavity of the Whitney numbers of the second kind 'may be established in the near future'. Further support for 14.2.3 is provided by some striking results of Dowling and Wilson (1974, 1975). These and other results on Whitney numbers are surveyed by Aigner (1987). The last paper contains results not only for Whitney numbers of the second kind but also for Whitney numbers of the first kind, which we shall define in a moment. The latter results are also relevant here since the sequence of Whitney numbers of the first kind is also conjectured to be unimodal.

For a matroid M having ground set E, the *characteristic* (or *chromatic*) *polynomial* of M is defined by

$$p(M;\lambda) = \sum_{X \subseteq E} (-1)^{|X|} \lambda^{r(M)-r(X)}.$$

The *Whitney numbers of the first kind* are the coefficients of this polynomial. More precisely, they are the numbers $w_0, w_1, \ldots, w_{r(M)}$ where

$$p(M;\lambda) = \sum_{k=0}^{r(M)} w_k \lambda^{r(M)-k}.$$

Surveys of the properties of characteristic polynomials can be found in Welsh (1976) and Zaslavsky (1987a). In particular, we note that it can be shown that if M has a loop, then $p(M;\lambda)$ is identically zero, whereas if M is loopless, then $w_0 = 1$ and $w_0, w_1, \ldots, w_{r(M)}$ alternate in sign. Moreover, if G is a graph, then $p(M(G);\lambda)$ is closely related to the chromatic polynomial $\chi(G;\lambda)$ of the graph G where $\chi(G;\lambda)$ is the number of ways to colour the vertices of G so that no two adjacent vertices are coloured alike. In particular, if G has $\omega(G)$ connected components, then

$$\chi(G;\lambda) = \lambda^{\omega(G)} p(M(G);\lambda).$$

Implicit in Rota (1971) and explicit in Heron (1972b) is the following generalization of an earlier graph-theoretic conjecture of Read (1968).

14.2.6 Conjecture. *If M is a loopless matroid, then*

$$|w_k| \geq \min\{|w_{k-1}|, |w_{k+1}|\} \textit{ for all } k \textit{ in } \{1, 2, \ldots, r(M)-1\}.$$

Welsh (1976) proposed that Read's conjecture could be extended further and that the Whitney numbers of the first kind are actually log concave.

14.2.7 Conjecture. *Let M be a matroid. Then*

$$w_k^2 \geq w_{k-1}w_{k+1} \textit{ for all } k \textit{ in } \{1, 2, \ldots, r(M)-1\}.$$

Progress towards the last two conjectures has been summarized by Aigner (1987). Probably the most significant partial result asserts that if M is simple and not free, then $|w_0| < |w_1| < \cdots < |w_{\lfloor (r+1)/2 \rfloor}|$ (Heron 1972b).

14.3 The critical problem

The characteristic polynomial of a matroid is one of a number of important matroid isomorphism invariants. The universal such invariant is a two-variable polynomial called the Tutte polynomial and a detailed discussion of this polynomial, its properties, and its applications can be found in Brylawski and Oxley (1992). We shall not attempt to list all of the many interesting unsolved problems related to the Tutte polynomial but instead we concentrate on a selection of such problems that relate to the characteristic polynomial. For details of the background to these problems and of progress made towards their solutions, we refer the reader to Brylawski and Oxley's (1992) paper and to the items in that paper's extensive bibliography.

Let M be a loopless rank-r matroid representable over $GF(q)$ and let $\psi : E(M) \to V(r,q)$ be a $GF(q)$-coordinatization of M. The *critical exponent* $c(M;q)$ of M is defined to be the least natural number j for which $V(r,q)$ has j hyperplanes $H_1, H_2, \ldots, H_j$ such that $(\bigcap_{i=1}^{j} H_i) \cap \psi(E(M))$ is empty. If M has a loop and is representable over $GF(q)$, then $c(M;q)$ is taken to be ∞. Ostensibly, $c(M;q)$ depends upon the particular coordinatization ψ. However, a fundamental result of Crapo and Rota (1970) establishes that this is not the case. In particular, if M is $GF(q)$-representable, then

$$c(M;q) = \begin{cases} \infty & \text{if } M \text{ has a loop;} \\ \min\{j \in \mathbb{Z}^+ \cup \{0\} : p(M;q^j) > 0\} & \text{otherwise.} \end{cases}$$

Numerous problems from various diverse branches of combinatorics can be expressed in terms of evaluating a certain critical exponent. For instance, the Four Colour Theorem (Appel and Haken 1976) is equivalent to the assertion that $c(M(G);2) \le 2$ for all loopless planar graphs G.

One of the best-known conjectures in this area is the following.

14.3.1 Tutte's Tangential 2-Block Conjecture. *(Tutte 1966c) The only binary matroids M such that $c(M;2) = 3$ but $c(N;2) \le 2$ for all loopless proper minors N of M are F_7, $M(K_5)$, and $M^*(P_{10})$.*

Seymour (1981d) has shown that this conjecture is equivalent to the following conjecture for cographic matroids which is also due to Tutte (1966b) and is a variant of his celebrated 5-Flow Conjecture (Tutte 1954).

14.3.2 Conjecture. *Let M be a loopless cographic matroid for which $p(M;4) = 0$. Then M has a minor isomorphic to $M^*(P_{10})$.*

The first unsolved case of the following conjecture of Brylawski (1975d) occurs when $q = 4$.

14.3.3 Conjecture. *If M is a loopless $GF(q)$-representable matroid and M has no $M(K_4)$-minor, then $c(M;q) \leq 2$.*

A weaker conjecture of Brylawski (1975d) which is also open for all $q \geq 4$ is as follows.

14.3.4 Conjecture. *If M is a loopless $GF(q)$-representable gammoid, then $c(M;q) \leq 2$.*

The next conjecture, due to Walton and Welsh (1980), was motivated by Tutte's 5-Flow Conjecture.

14.3.5 Conjecture. *If M is a loopless binary matroid having no minor isomorphic to $M(K_5)$, then $c(M;2) \leq 3$.*

The best partial result towards the last conjecture is due to Kung (1987) who proved that, for a matroid M satisfying the hypotheses of the conjecture, $c(M;2) \leq 8$.

A longstanding graph-theoretic conjecture of Lovász (in Bollobás 1978, p. 290) asserts that complete graphs are the only simple graphs for which the chromatic number drops by two whenever two adjacent vertices are deleted. A matroid analogue of this, which has received far less attention than the other conjectures in this section is the following.

14.3.6 Conjecture. *Let M be a simple rank-r $GF(q)$-representable matroid such that, for all rank-$(r-2)$ flats X of M, $c(M|X;q) = c(M;q) - 2$. Then $M \cong PG(r-1,q)$.*

14.4 From graphs to matroids

The title of this section is the same as that of a section in Welsh's (1971b) paper which surveyed what were then the major unsolved problems in matroid theory. Now, as then, many important matroid problems are motivated by corresponding problems for graphs. Some of these were discussed in the last section. The following conjecture of Welsh (private communication) was motivated by Dirac's theorem (1952b) on Hamiltonian cycles in graphs.

14.4.1 Conjecture. *If M is a simple regular connected matroid and every cocircuit has at least $\frac{1}{2}(r(M)+1)$ elements, then M has a circuit of size $r(M)+1$.*

Of course, this conjecture holds for graphic matroids by Dirac's result. In view of Seymour's decomposition theorem for regular matroids (13.2.4), a natural starting point for the conjecture would be to verify it for cographic matroids. The generalization of the conjecture to all binary matroids is not true since it fails for F_7^*. However, in view of Corollary 11.2.5, it seems reasonable to expect the generalization to hold for all binary matroids with no F_7^*-minor.

The *reconstruction conjecture* of Kelly and Ulam (in Kelly 1942) is another famous problem for graphs that suggests numerous interesting matroid problems. The graph conjecture is that if G and H are graphs having at least three vertices and $\alpha : V(G) \to V(H)$ is a bijection such that $G - v \cong H - \alpha(v)$ for all v in $V(G)$, then $G \cong H$. There is a corresponding *edge reconstruction conjecture* for graphs with at least four edges. Brylawski (1974, 1975a) has considered matroid analogues of these conjectures. Let M be a matroid on a set E and suppose that N is another matroid on E such that $M \backslash e \cong N \backslash e$ for all e in E. If all such matroids N are isomorphic to M, then M is said to be *reconstructible*. The matroids $U_{n-1,n}$ and $U_{n,n}$ are not reconstructible since each has every single-element deletion isomorphic to $U_{n-1,n-1}$. Thus any matroid analogue of the edge reconstruction conjecture for graphs would need to make an exception for circuits and free matroids. Brylawski (1974) conjectured that these are the only exceptions, that is, all matroids other than circuits and free matroids are reconstructible. However, soon after, he (Brylawski 1975a) found a counterexample to this conjecture. Indeed, as noted in Section 6.1, if M^* is a projective plane of order p^2 where p is a prime exceeding two, then M is not reconstructible (Brylawski 1986a). In spite of these examples, the following question remains unanswered.

14.4.2 Problem. *If M is a binary matroid that is neither a circuit nor a free matroid, then is M reconstructible?*

How should one formulate a matroid analogue of the (vertex) reconstruction conjecture for graphs? We have seen several instances earlier in the book when the matroid analogue of a vertex has been taken to be a cocircuit. This will be done again here. Let M be a matroid and N be another matroid for which there is a bijection $\alpha : \mathcal{H}(M) \to \mathcal{H}(N)$ such that $M|H \cong N|\alpha(H)$ for all H in $\mathcal{H}(M)$. If all such matroids N are isomorphic to M, then M is said to be *hyperplane reconstructible*. Since there is such a bijection between the sets of hyperplanes of the non-isomorphic matroids Q_6 and $U_{2,4} \oplus_2 U_{2,4}$, not all matroids are hyperplane reconstructible (Brylawski 1974). However, the following conjecture (Brylawski 1974) is still open.

14.4.3 Conjecture. *All binary matroids are hyperplane reconstructible.*

Since the vertex bonds in a loopless 3-connected graph G are exactly the cocircuits C^* of $M(G)$ for which $M(G) \backslash C^*$ is connected, Brylawski also proposed that a binary matroid of fixed rank and corank could be reconstructed from its connected hyperplanes. However, Welsh (private communication) has noted that this fails since if G is a graph that is formed from a single cycle by adding one diagonal, then provided G has no 3-cycles, $M(G)$ has no connected hyperplanes.

Although a matroid need not be hyperplane reconstructible, Brylawski (1981) has shown that its characteristic polynomial, and more generally its Tutte polynomial, can be determined from the multiset of isomorphism classes of its hyperplanes. A matroid M is *minor reconstructible* if $N \cong M$ whenever N is a matroid on $E(M)$ such that $N\backslash e \cong M\backslash e$ and $N/e \cong M/e$ for all e in $E(M)$. Brylawski (1986a) raised the following:

14.4.4 Problem. *Are all matroids minor reconstructible?*

The next problem has become popular recently in view of Robertson and Seymour's (1990) striking result that in any infinite set of graphs there is one that is isomorphic to a minor of another.

14.4.5 Problem. *Is there an infinite set of $GF(q)$-representable matroids none of which is isomorphic to a minor of another?*

For matroids in general the answer to this question is negative: clearly no member of the set $\{PG(2,p) : p \text{ prime}\}$ is a minor of any other. Another interesting such set was noted by Kahn (private communication). For all $r \geq 2$, let N_r be the vector matroid of the matrix $[I_{2r}|J_{2r}-I_{2r}]$ over $GF(2)$. Let M_r be a matroid that is obtained from N_r by relaxing a pair of complementary circuit–hyperplanes. Then no member of $\{M_r : r \geq 2\}$ is a minor of any other. Moreover, no member of this set has a $U_{2,5}$- or a $U_{3,5}$-minor. The question as to whether there is an infinite set of matroids having no $U_{2,4}$-minor such that none is isomorphic to a minor of another is equivalent to the case $q = 2$ of 14.4.5. This is the case of 14.4.5 which seems the most likely to yield a positive answer.

The following problem of Brylawski (1986a) and Robertson (private communication) is closely related to 14.4.5.

14.4.6 Problem. *Let M_1 and M_2 be arbitrary matroids. Is the set of minor-minimal matroids having both M_1- and M_2-minors always finite?*

An equivalent question was posed by Welsh (private communication) when he asked: If $\mathcal{M}_1$ and $\mathcal{M}_2$ are minor-closed classes of matroids each of which is characterized by a finite set of excluded minors, is $\mathcal{M}_1 \cup \mathcal{M}_2$ characterized by a finite set of excluded minors?

Let A be the vertex–edge incidence matrix of a graph G. If A is viewed as a matrix over $GF(2)$, then clearly $M[A]$ is the cycle matroid of G. The *even-cycle* or *factor matroid* of G is the vector matroid of A when A is viewed as a matrix over $\mathbb{R}$. An alternative derivation of such matroids was given in Exercise 13 of Section 12.2. If G is a bipartite graph, then its even-cycle matroid equals its cycle matroid. Whitney's 2-Isomorphism Theorem (5.3.1) specifies precisely when two graphs have the same cycle matroid. Wagner (1988) considered the corresponding problem for even-cycle matroids and solved a special case of it.

14.4.7 Problem. *Determine when two graphs have the same even-cycle matroid.*

We close this section by noting two problems concerning connectivity that are examples of the many such problems prompted by graph-theoretic results. Jackson (1980) proved a conjecture of Hobbs by showing that if G is a simple 2-connected graph with minimum degree exceeding three, then G has a cycle C such that $G\backslash C$ is 2-connected. This suggests the following question for matroids.

14.4.8 Problem. *Let M be a simple connected binary matroid in which every cocircuit has at least four elements. Does M have a circuit C such that $M\backslash C$ is connected?*

If M is not required to be binary, then the answer to the last question can certainly be negative. For instance, $U_{3,6}$ is connected and has every cocircuit of size 4, yet $U_{3,6}\backslash C \cong U_{2,2}$ for all circuits C. Moreover, Jackson (1980) and Robertson (in Jackson 1980) independently gave a graphic example to show that the answer to 14.4.8 can also be negative if M is not required to be simple.

Halin (1969) showed that every minimally n-connected graph has a vertex of degree n and this result was strengthened by Mader (1979) who showed that, in every such graph G, the number of degree-n vertices is at least $[(n-1)|V(G)| + 2n]/(2n-1)$.

14.4.9 Problem. *For $n \geq 2$, if M is a minimally n-connected matroid having at least $2(n-1)$ elements, does M have an n-element cocircuit?*

For $n = 2$ and $n = 3$, Murty (1974) and Wong (1978) answered this question affirmatively; Oxley (1981a, 1984c) and Lemos (1989) extended these results. Some of the matroid results obtained here suggested new conjectures for graphs which were then shown to hold. Thus the links between graphs and matroids have been successfully exploited in *both* directions to give not only new results for matroids but also new results for graphs.

The only progress on Problem 14.4.9 for $n \geq 4$ was made by Wong (1978) who showed that if a 4-connected matroid M with at least eight elements has an element e such that neither $M\backslash e$ nor M/e is 4-connected, then M has a 4-element circuit or a 4-element cocircuit.

Table 14.1. The numbers of non-isomorphic matroids, simple matroids, and binary matroids on an n-element set for $0 \leq n \leq 8$.

n	0	1	2	3	4	5	6	7	8
matroids	1	2	4	8	17	38	98	306	1724
simple matroids	1	1	1	2	4	9	26	101	950
binary matroids	1	2	4	8	16	32	68	148	342

14.5 Enumeration

Blackburn, Crapo, and Higgs (1973) found all non-isomorphic simple matroids on at most eight elements, and Acketa (1984) used their list to determine all non-isomorphic matroids and all non-isomorphic binary matroids on at most eight elements. Table 14.1 summarizes these results.

Let $f(n)$ denote the number of non-isomorphic matroids on an n-element set. Asymptotic bounds on $f(n)$ were noted on page 13. A longstanding conjecture of Welsh (1969a), which Blackburn, Crapo, and Higgs (1973) say is 'surely correct', has still not been proved.

14.5.1 Conjecture. *For all non-negative integers m and n,*

$$f(m+n) \geq f(m)f(n).$$

Let $f_r(n)$ denote the number of rank-r non-isomorphic matroids on an n-element set. Table 14.2 (Acketa 1984) gives the value of $f_r(n)$ for all $n \leq 8$. This table supports the next two conjectures, both of which are due to Welsh (1971a and private communication).

14.5.2 Conjecture. *$(f_r(n) : 1 \leq r \leq n)$ is unimodal.*

Table 14.2. The number of non-isomorphic matroids of rank r on an n-element set.

r \ n	0	1	2	3	4	5	6	7	8
0	1	1	1	1	1	1	1	1	1
1		1	2	3	4	5	6	7	8
2			1	3	7	13	23	37	58
3				1	4	13	38	108	325
4					1	5	23	108	940
5						1	6	37	325
6							1	7	58
7								1	8
8									1

14.5.3 Conjecture. *$f_r(n)$ is a maximum when r is $\lfloor n/2 \rfloor$.*

Let $a(n)$ denote the number of non-isomorphic binary matroids on an n-element set. Noting that

$$a(n) \leq n2^{n^2/2},$$

Welsh (1971b) asked:

14.5.4 Problem. *What is the asymptotic behaviour of $a(n)$?*

In the enumeration of non-isomorphic simple matroids on fewer than 9 elements (Blackburn *et al.* 1973), paving matroids predominate. Welsh (1976) asked whether this holds in general.

14.5.5 Problem. *Is it true, for all non-negative integers n, that more than half of the non-isomorphic simple matroids on an n-element set are paving matroids?*

The final conjecture of this section (9.3.12) compares the numbers $b(M)$ and $c(M)$ of bases and circuits in a simple binary matroid M.

14.5.6 Conjecture. *Let M be a rank-r simple binary matroid having no coloops. Then*

$$b(M) \geq \tfrac{1}{2}(r+1)c(M).$$

14.6 Matroid union

Recski (1985) has given a thorough survey of unsolved problems related to matroid union and progress towards their solutions. This short section presents four of these problems beginning with the oldest (12.3.9), which was originally posed by Welsh (1971b).

14.6.1 Problem. *Characterize all irreducible matroids.*

Cunningham (1977) conjectured that a binary matroid can be uniquely decomposed as a union of irreducible matroids if and only if it is cosimple, and this conjecture was proved by Dawson (1985). For matroids in general, the unique decomposition problem remains open.

14.6.2 Problem. *Among the reducible matroids, characterize those for which the decomposition as a union of irreducible matroids is unique.*

The next two conjectures are due to Recski (1985).

14.6.3 Conjecture. *Let M_1 and M_2 be graphic matroids and suppose that $M_1 \vee M_2$ is non-graphic. Then $M_1 \vee M_2$ is non-binary.*

14.6.4 Conjecture. *Let M and N be given matroids, N being binary, and consider the matroid equation $M \vee X = N$. If there is a matroid X for which this equation is satisfied, then there is a unique freest such matroid.*

Recski (1981) has proved the last conjecture when N is graphic. In addition to the many references cited by Recski (1985), the reader interested in the last conjecture should note the work of Dawson (1984, 1988) on the solution to the above matroid equation.

14.7 Gammoids

In general, gammoids seem rather difficult to handle. Ingleton and Piff (1973) have shown that a matroid of rank at most three is a gammoid if and only if it is a strict gammoid. Moreover, Mason (1972a) has given a criterion for determining when a matroid is a strict gammoid (see also Kung 1978). Thus one can determine when a rank-3 matroid is a gammoid. However, the following problem seems to be open even for rank-4 matroids.

14.7.1 Problem. *Give an algorithm to test whether or not a given matroid is a gammoid.*

Although the excluded minors for the class of quaternary matroids have not been determined, there may be a chance of characterizing some special classes of quaternary matroids. The next two problems seek such results. Recall from Section 13.4 that the set of excluded minors for the class of gammoids, while it has yet to be determined, has been shown to be infinite. However, both the class of binary gammoids and the class of ternary gammoids have been characterized by finite sets of excluded minors.

14.7.2 Problem. *Characterize the class of quaternary gammoids by excluded minors.*

The classes of binary transversal and ternary transversal matroids have been characterized by excluded series minors (de Sousa and Welsh 1972, Oxley 1986b).

14.7.3 Problem. *Characterize the class of quaternary transversal matroids by excluded series minors.*

The last problem of this section was again originally posed by Welsh (1971b).

14.7.4 Problem. *Characterize those transversal matroids whose duals are also transversal.*

Jensen (1978) has shown that a binary transversal matroid is cotransversal if and only if it is principal transversal. However, he also noted that $U_{2,4} \oplus_2 U_{2,4}$ is a (ternary) transversal cotransversal matroid that is not principal transversal.

14.8 A miscellany

We conclude this chapter with a collection of problems that do not seem to belong in any of the earlier sections.

Ingleton (1971a) has noted that every simple rank-3 matroid can be embedded in some projective plane. However, it is not known whether this plane can be chosen to be finite (Welsh 1971b).

14.8.1 Problem. *Let M be a simple rank-3 matroid. Does there exist a finite projective plane of which M is a restriction?*

Some recent work on this problem has been done by Wanner and Ziegler (1991).

Poljak and Turzík (1982) have called a matroid N *sticky* if an amalgam of M_1 and M_2 exists whenever M_1 and M_2 are extensions of N whose ground sets meet in $E(N)$. It follows by Theorem 12.4.10 that every modular matroid is sticky. Poljak and Turzík conjecture that the converse of this also holds:

14.8.2 Conjecture. *Every sticky matroid is modular.*

Partial progress towards this conjecture has been made by Bachem and Kern (1988).

The next problem and conjecture, stated earlier in Exercise 9 of Section 9.1 and Exercise 9 of Section 4.3, relate to circuits and cocircuits.

14.8.3 Conjecture. *If $k \geq 4$ and the matroid M has a k-element set that is the intersection of a circuit and a cocircuit, then M has a $(k-2)$-element set that is a circuit–cocircuit intersection.*

14.8.4 Problem. *Characterize those matroids in which every 4-element set is contained in a circuit or a cocircuit.*

Seymour (1986a) solved the following problem for binary matroids, thereby extending graph results of Chakravarti and Robertson (1980) and Watkins and Mesner (1967). The problem is still open for matroids in general.

14.8.5 Problem. *Let e, f, and g be elements of a 3-connected internally 4-connected matroid M. Characterize when M has no circuit containing $\{e, f, g\}$.*

A 3-element subset of the ground set of a matroid M is in a $U_{2,4}$-minor of M if and only if it is the intersection of a circuit and a cocircuit. Seymour (1981b) conjectured that, for every 3-element set X in a 4-connected non-binary matroid M, there is a $U_{2,4}$-minor of M containing X. Counterexamples to this conjecture were given by Kahn (1985) and Coullard (1986). Kahn's example is obtained from a binary matroid by relaxing a circuit–hyperplane; Coullard's example comes from a binary matroid by relaxing two complementary circuit–hyperplanes.

14.8.6 Problem. *Characterize all 4-connected non-binary matroids that have a 3-element set which is not in any $U_{2,4}$-minor.*

Many matroid problems seek to characterize all those matroids M with a certain property P such that, for all elements e of M, neither $M\backslash e$ nor M/e has P. A variant of such problems tries to determine all such M so that, for all elements e, $M\backslash e$ *or* M/e does not have P. In particular, Oxley (1990b) found all non-binary matroids M such that, for all elements e, $M\backslash e$ or M/e is binary. Moreover, Gubser (1990) determined all graphic matroids M such that, for all elements e, $M\backslash e$ or M/e is the cycle matroid of a planar graph. These results suggest numerous problems, two of which are the following.

14.8.7 Problem. *Find all binary matroids M such that, for all elements e, $M\backslash e$ or M/e is graphic.*

14.8.8 Problem. *Find all non-regular matroids M such that, for all elements e, $M\backslash e$ or M/e is regular.*

Seymour (private communication) suggested the following variant of the last problem. Clearly one could pose analogous problems for many other classes of matroids.

14.8.9 Problem. *Find all 3-connected non-regular matroids M such that M has an element e for which both $M\backslash e$ and M/e are regular.*

The only 3-connected non-binary matroid having an element so that both the deletion and contraction of this element are binary is $U_{2,4}$ (Oxley 1987a).

The next problem was noted earlier in Exercise 13 of Section 11.3.

14.8.10 Problem. *Let M be a non-binary 3-connected matroid having a minor isomorphic to $M(\mathcal{W}_4)$. Does M have a 9-element non-binary 3-connected minor having an $M(\mathcal{W}_4)$-minor?*

White (1980b) discussed a number of open conjectures involving basis exchanges. A special case of one of these is the following.

14.8.11 Conjecture. *Let B_1 and B_2 be bases of a regular matroid M. Then there is an element e of B_1 such that there is a unique element f of B_2 for which both $(B_1 - e) \cup f$ and $(B_2 - f) \cup e$ are bases of M.*

Finally, we refer the reader seeking still more open problems to Kung's (1991) comprehensive survey of extremal matroid theory.

References

Acketa, D. M. (1984). A construction of non-simple matroids on at most 8 elements. *J. Combin. Inform. System Sci.* **9**, 121–132. [14.5]

Acketa, D. M. (1988). On binary paving matroids. *Discrete Math.* **70**, 109–110. [11.2]

Ádám, A. (1957). Über zweipolige elektrische Netze. I. *Magyar Tud. Akad. Mat. Kutató Int. Közl.* **2**, 211–218. [5.4]

Aigner, M. (1987). Whitney numbers. In *Combinatorial geometries* (ed. N. White), pp. 139–160. Cambridge University Press, Cambridge. [14.2]

Aigner, M. and Schoene, J. (1987). On the logarithmic concavity of the Whitney numbers (submitted). [14.2]

Akkari, S. (1988). *On matroid connectivity.* Ph.D. thesis, Louisiana State University. [8.1, 10.2]

Akkari, S. (1991). A minimal 3-connectedness result for matroids. *Discrete Math.* (to appear). [10.2]

Akkari, S. and Oxley, J. (1991). Some extremal connectivity results for matroids. *J. Combin. Theory Ser. B* **52**, 301–320. [10.2]

Appel, K. and Haken, W. (1976). Every planar map is four-colorable. *Bull. Amer. Math. Soc.* **82**, 711–712. [14.3]

Artin, E. (1957). *Geometric algebra.* Interscience, New York. [6.3]

Asano, T., Nishizeki, T., and Seymour, P. D. (1984). A note on non-graphic matroids. *J. Combin. Theory Ser. B* **37**, 290–293. [13.3]

Asche, D. S. (1966). Minimal dependent sets, *J. Austral. Math. Soc.* **6**, 259–262. [1.4]

Bachem, A. and Kern, W. (1988). On sticky matroids. *Discrete Math.* **69**, 11–18. [12.4, 14.8]

Baclawski, K. and White, N. (1979). Higher order independence in matroids. *J. London Math. Soc.* (2) **19**, 193–202. [13.4]

Baer, R. (1952). *Linear algebra and projective geometry.* Academic Press, New York. [6.3]

Basterfield, J. G. and Kelly, L. M. (1968). A characterization of sets of n points which determine n hyperplanes. *Proc. Camb. Phil. Soc.* **64**, 585–588. [6.9]

Birkhoff, G. (1935). Abstract linear dependence in lattices. *Amer. J. Math.* **57**, 800–804. [1.0, 1.7]

Birkhoff, G. (1967). *Lattice theory.* Third edition. Amer. Math. Soc., Providence, Rhode Island. [6.1, 6.9]

Bixby, R. E. (1972). *Composition and decomposition of matroids and related topics.* Ph.D. thesis, Cornell University. [8.3]

Bixby, R. E. (1974). ℓ-matrices and a characterization of non-binary matroids. *Discrete Math.* **8**, 139–145. [11.3]

Bixby, R. E. (1976). A strengthened form of Tutte's characterization of regular matroids. *J. Combin. Theory Ser. B* **20**, 216–221. [9.3, 13.4]

Bixby, R. E. (1977). Kuratowski's and Wagner's theorems for matroids. *J. Combin. Theory Ser. B* **22**, 31–53. [5.4, 11.2, 13.4]

Bixby, R. E. (1979). On Reid's characterization of the ternary matroids. *J. Combin. Theory Ser. B* **26**, 174–204. [6.5, 10.0, 10.3]

Bixby, R. E. (1981). Matroids and operations research. In *Advanced techniques in practice of operations research* (eds. H. J. Greenberg, F. H. Murphy, and S. H. Shaw), pp. 333–458. North-Holland, New York. [1.2, 1.8, 13.2]

Bixby, R. E. (1982). A simple theorem on 3-connectivity. *Linear Algebra Appl.* **45**, 123–126. [8.4]

Bixby, R. E. and Coullard, C. R. (1986). On chains of 3-connected matroids. *Discrete Appl. Math.* **15**, 155–166. [8.4, 11.3]

Bixby, R. E. and Coullard, C. R. (1987). Finding a smallest 3-connected minor maintaining a fixed minor and a fixed element. *Combinatorica* **7**, 231–242. [8.4, 11.3]

Bixby, R. E. and Cunningham, W. H. (1980). Converting linear programs to network problems. *Math. Oper. Res.* **5**, 321–357. [13.2]

Bixby, R. E. and Cunningham, W. H. (1987). Short cocircuits in binary matroids. *European J. Combin.* **8**, 213–225. [13.4]

Bixby, R. E. and Rajan, A. (1989). A short proof of the Truemper–Tseng theorem on max-flow min-cut matroids. *Linear Algebra Appl.* **114/115**, 277–292. [11.3]

Björner, A. and Ziegler, G. M. (1992). Introduction to greedoids. In *Matroid applications* (ed. N. White), pp. 284–357. Cambridge University Press, Cambridge. [14.0]

Björner, A., Las Vergnas, M., Sturmfels, B., White, N., and Ziegler, G. (1992). Oriented matroids. Cambridge University Press, Cambridge. [13.4, 14.0]

Blackburn, J. E., Crapo, H. H., and Higgs, D. A. (1973). A catalogue of combinatorial geometries. *Math. Comp.* **27**, 155–166, with loose microfiche supplement A12–G12. [6.4, 7.2, 14.5]

Bland, R. G. and Las Vergnas, M. (1978). Orientability of matroids. *J. Combin. Theory Ser. B* **24**, 94–123. [13.4]

Bollobás, B. (1978). *Extremal graph theory.* Academic Press, London. [14.3]

Bondy, J. A. (1972). Transversal matroids, base-orderable matroids, and graphs. *Quart. J. Math. Oxford Ser.* (2) **23**, 81–89. [3.2, 13.4]

Bondy, J. A. and Murty, U. S. R. (1976). *Graph theory with applications.* Macmillan, London; North-Holland, New York. [1.8, 2.3, 8.2]

Bondy, J. A. and Welsh, D. J. A. (1971). Some results on transversal matroids and constructions for identically self-dual matroids. *Quart. J. Math. Oxford Ser.* (2) **22**, 435–451. [2.1, 2.4, 9.2]

Borůvka, O. (1926). O jistém problému minimálním. *Práce Mor. Přírodověd Spol. v Brně (Acta Societ. Scient. Natur. Moravicae)* **3**, 37–58. [1.8]

Bose, R. C. (1947). Mathematical theory of the symmetrical factorial design. *Sankyhā* **8**, 107–166. [6.5]

Brown, T. (1971). Deriving closure relations with exchange property. Notes (and an editorial appendix) by H. Crapo and G. Roulet. In *Möbius algebras* (Proc. Conf. Univ. Waterloo, 1971), pp. 51–55. University of Waterloo, Waterloo. [7.3]

Brown, W. G., ed. (1980). *Reviews in graph theory. Vols. 1–4.* Amer. Math. Soc., Providence, Rhode Island. [14.0]

Brualdi, R. A. (1969). Comments on bases in dependence structures, *Bull. Austral. Math. Soc.* **1**, 161–167. [1.2, 12.2]

Brualdi, R. A. (1970). Admissible mappings between dependence spaces. *Proc. London Math. Soc.* (3) **21**, 296–312. [12.3]

Brualdi, R. A. and Scrimger, E. B. (1968). Exchange systems, matchings, and transversals. *J. Combin. Theory* **5**, 244–257. [12.3]

Bruck, R. H. and Ryser, H. J. (1949). The nonexistence of certain finite projective planes. *Canad. J. Math.* **1**, 88–93. [6.1]

Brylawski, T. H. (1971). A combinatorial model for series–parallel networks. *Trans. Amer. Math. Soc.* **154**, 1–22. [7.1, 13.4]

Brylawski, T. H. (1972). A decomposition for combinatorial geometries. *Trans. Amer. Math. Soc.* **171**, 235–282. [4.3, 9.3]

Brylawski, T. H. (1973). Some properties of basic families of subsets. *Discrete Math.* **6**, 333–341. [1.4]

Brylawski, T. H. (1974). Reconstructing combinatorial geometries. In *Graphs and combinatorics* (eds. R. A. Bari and F. Harary). Lecture Notes in Math. Vol. 406, pp. 226–235. Springer-Verlag, Berlin. [6.1, 14.4]

Brylawski, T. H. (1975a). On the nonreconstructibility of combinatorial geometries. *J. Combin. Theory Ser. B* **19**, 72–76. [6.1, 14.4]

Brylawski, T. H. (1975b). Modular constructions for combinatorial geometries. *Trans. Amer. Math. Soc.* **203**, 1–44. [6.9, 12.4]

Brylawski, T. H. (1975c). A note on Tutte's unimodular representation theorem. *Proc. Amer. Math. Soc.* **52**, 499–502. [6.6]

Brylawski, T. H. (1975d). An affine representation for transversal geometries. *Studies in Appl. Math.* **54**, 143–160. [6.5, 12.2, 13.4, 14.3, 14.6]

Brylawski, T. H. (1981). Hyperplane reconstruction of the Tutte polynomial of a geometric lattice. *Discrete Math.* **35**, 25–38. [14.4]

Brylawski, T. H. (1982a). The Tutte polynomial. Part I: General theory. In *Matroid theory and its applications* (ed. A. Barlotti), pp. 125–275. Liguori editore, Naples. [9.3]

Brylawski, T. H. (1982b). Finite prime-field characteristic sets for planar configurations. *Linear Algebra Appl.* **46**, 155–176. [6.8]

Brylawski, T. H. (1986a). Constructions. In *Theory of Matroids* (ed. N. White), pp. 127–223. Cambridge University Press, Cambridge. [1.3, 7.0, 12.4, 14.4]

Brylawski, T. H. (1986b). Appendix of matroid cryptomorphisms. In *Theory of matroids* (ed. N. White), pp. 298–312. Cambridge University Press, Cambridge. [2.1]

Brylawski, T. H. and Kelly, D. (1980). *Matroids and combinatorial geometries.* Department of Mathematics, University of North Carolina, Chapel Hill. [1.5, 6.4, 6.8]

Brylawski, T. H. and Lucas, D. (1976). Uniquely representable combinatorial geometries. In *Teorie Combinatorie* (Proc. 1973 Internat. Colloq.), pp. 83–104. Accademia Nazionale dei Lincei, Rome. [6.3, 6.4, 10.0, 10.1]

Brylawski, T. H. and Oxley, J. (1992). The Tutte polynomial and its applications. In *Matroid applications* (ed. N. White), pp. 123–225. Cambridge University Press, Cambridge. [14.3]

Bush, K. A. (1952). Orthogonal arrays of index unity. *Ann. Math. Statist.* **23**, 426–434. [6.5]

Cameron, P. J. (1980). Extremal results and configuration theorems for Steiner systems. In *Topics on Steiner systems.* Ann. Discrete Math. **7**, pp. 43–63. North-Holland, Amsterdam. [11.2]

Camion, P. (1963). Caractérisation des matrices unimodulaires. *Cahiers Centre Études Rech. Opér.* **5**, 181–190. [13.1]

Casse, L. R. A. (1969). A solution to Beniamino Segre's "Problem $I_{r,q}$" for q even. *Atti Accad. Naz. Lincei Rend. Cl. Sci. Fis. Mat. Natur.* (8) **46**, 13–20. [6.5]

Chakravarti, K. and Robertson, N. (1980). Covering three edges with a bond in a non-separable graph. In *Combinatorics* 79. *Part I* (eds. M. Deza and I. G. Rosenberg) Ann. Discrete Math. **8**, p. 247. North-Holland, Amsterdam [14.8].

Chartrand, G. and Harary, F. (1967). Planar permutation graphs. *Ann. Inst. H. Poincaré Sect. B* (N. S.) **3**, 433–438. [11.2]

Cheung, A. L. C. (1974). *Compatible extensions of a combinatorial geometry.* Ph.D. thesis, University of Waterloo. [7.2]

Coullard, C. R. (1985). *Minors of* 3-*connected matroids and adjoints of binary matroids.* Ph.D. thesis, Northwestern University. [8.4, 11.3]

Coullard, C. R. (1986). Counterexamples to conjectures on 4-connected matroids. *Combinatorica* **6**, 315–320. [11.3]

Coullard, C. R. and Oxley, J. G. (1992). Extension of Tutte's wheels-and-whirls theorem. *J. Combin. Theory Ser. B* (to appear). [11.3]

Coxeter, H. S. M. (1958). Twelve points in PG(5,3) with 95040 self-transformations. *Proc. Roy. Soc. London Ser. A* **247**, 279–293. [11.2]

Crapo, H. H. (1965). Single-element extensions of matroids. *J. Res. Nat. Bur. Standards Sect. B* **69B**, 55–65. [7.2]

Crapo, H. H. (1967). Structure theory for geometric lattices. *Rend. Sem. Mat. Univ. Padova* **38**, 14–22. [7.3]

Crapo, H. H. and Rota, G. -C. (1970). *On the foundations of combinatorial theory: combinatorial geometries* (preliminary edition). M.I.T. Press, Cambridge, Mass. [1.7, 3.3, 12.0, 14.3]

Cunningham, W. H. (1973). *A combinatorial decomposition theory.* Ph.D. thesis, University of Waterloo. [8.3, 10.2, 13.2]

Cunningham, W. H. (1977). Chords and disjoint paths in matroids. *Discrete Math.* **19**, 7–15. [12.3, 14.6]

Cunningham, W. H. (1979). Binary matroid sums. *Quart. J. Math. Oxford Ser.* (2) **30**, 271–281.

Cunningham, W. H. (1981). On matroid connectivity. *J. Combin. Theory Ser. B* **30**, 94–99. [8.2, 13.3]

Cunningham, W. H. and Edmonds, J. (1980). A combinatorial decomposition theory. *Canad. J. Math.* **32**, 734–765. [8.3]

Dawson, J. E. (1980). Optimal matroid bases: an algorithm based on cocircuits. *Quart. J. Math. Oxford* (2) **31**, 65–69. [2.1]

Dawson, J. E. (1981). A simple approach to some basic results in matroid theory. *J. Math. Anal. Appl.* **84**, 555–559. [2.1]

Dawson, J. E. (1983). Balanced sets in an independence structure induced by a submodular function. *J. Math. Anal. Appl.* **95**, 214–222. [12.1]

Dawson, J. E. (1984). Solubility of the matroid sum equation. *Ars Combinatoria* **17A**, 103–116. [14.6]

Dawson, J. E. (1985). Decomposition of binary matroids. *Combinatorica* **5**, 1–9. [14.6]

Dawson, J. E. (1988). Some necessary conditions for sums of matroids. *Quart. J. Math. Oxford Ser.* (2) **39**, 281–283. [14.6]

Dembowski, P. (1968). *Finite geometries.* Springer-Verlag, New York. [6.1]

de Sousa, J. and Welsh, D. J. A. (1972). A characterisation of binary transversal matroids. *J. Math. Anal. Appl.* **40**, 55–59. [13.4, 14.8]

Dilworth, R. P. (1944). Dependence relations in a semimodular lattice. *Duke Math. J.* **11**, 575–587. [1.7]

Dirac, G. A. (1952a). A property of 4-chromatic graphs and some remarks on critical graphs. *J. London Math. Soc.* **27**, 85–92. [5.4]

Dirac, G. A. (1952b). Some theorems on abstract graphs. *Proc. London Math. Soc.* (3) **2**, 69–81. [14.4]

Dirac, G.A. (1960). In abstrakten Graphen vorhande vollständige 4-Graphen und ihre Unterteilungen. *Math. Nach.* **22**, 61–85. [5.4]

Dirac, G. A. (1967). Minimally 2-connected graphs. *J. Reine Angew. Math.* **228**, 204–216. [4.3]

Doob, M. (1973). An interrelation between line graphs, eigenvalues, and matroids. *J. Combin. Theory Ser. B* **15**, 40–50. [12.2]

Dowling, T. A. (1973a). A class of geometric lattices based on finite groups. *J. Combin. Theory Ser. B* **14**, 61–86; erratum **15**, 211. [10.3]

Dowling, T. A. (1973b). A q-analog of the partition lattice. In *A survey of combinatorial theory* (eds. J. N. Srivastava *et al.*), pp. 101–115. North-Holland, Amsterdam, 1973. [10.3]

Dowling, T. A. (1980). On the independent set numbers of a finite matroid. In *Combinatorics* 79. *Part I* (eds. M. Deza and I. G. Rosenberg) Ann. Discrete Math. **8**, pp. 21–28. North-Holland, Amsterdam. [14.2]

Dowling, T. A. and Wilson, R. M. (1974). The slimmest geometric lattices. *Trans. Amer. Math. Soc.* **196**, 203–215. [14.2]

Dowling, T. A. and Wilson, R. M. (1975). Whitney number inequalities for geometric lattices. *Proc. Amer. Math. Soc.* **47**, 504–512. [14.2]

Duffin, R. J. (1965). Topology of series–parallel networks. *J. Math. Anal. Appl.* **10**, 303–318. [5.4]

Duke, R. (1981). *Freedom in matroids.* Ph.D. thesis, The Open University. [12.1]

Duke, R. (1987). On binary reducibility. *European J. Combin.* **9**, 109–111. [12.3]

Edmonds, J. (1965a). Minimum partition of a matroid into independent sets. *J. Res. Nat. Bur. Standards Sect. B* **69B**, 67–72. [12.3]

Edmonds, J. (1965b). Lehman's switching game and a theorem of Tutte and Nash-Williams. *J. Res. Nat. Bur. Standards Sect. B* **69B**, 73–77. [12.3]

Edmonds, J. (1970). Submodular functions, matroids and certain polyhedra. In *Combinatorial structures and their applications* (Proc. Calgary Internat. Conf. 1969), pp. 69–87. Gordon and Breach, New York. [12.1, 12.3]

Edmonds, J. (1979). Matroid intersection. In *Discrete optimization I* (eds. P. L. Hammer, E. L. Johnson, and B. H. Korte). Ann. Discrete Math. **4**, pp. 39–49. North-Holland, Amsterdam. [12.3]

Edmonds, J. and Fulkerson, D. R. (1965). Transversals and matroid partition. *J. Res. Nat. Bur. Standards Sect. B* **69B**, 147–153. [1.6]

Edmonds, J. and Fulkerson, D. R. (1970). Bottleneck extrema. *J. Combin. Theory* **8**, 299–306. [2.1]

Edmonds, J. and Rota, G.-C. (1966). Submodular set functions (Abstract). *Waterloo combinatorics conference.* [1.3, 12.0, 12.1]

Faigle, U. (1987). Matroids in combinatorial optimization. In *Combinatorial geometries* (ed. N. White), pp. 161–210. Cambridge University Press, Cambridge. [12.3, 13.2]

Fenton, N. E. and Vámos, P. (1982). Matroid interpretation of maximal K-arcs in projective spaces. *Rend. Mat.* (7) **2**, 573–580. [6.5, 14.1]

Folkman, J. and Lawrence, J. (1978) Oriented matroids. *J. Combin. Theory Ser. B* **25**, 199–236. [13.4]

Ford, L. R. and Fulkerson, D. R. (1958). Network flow and systems of representatives. *Canad. J. Math.* **10**, 78–84. [12.3]

Fournier, J. -C. (1971). Représentation sur un corps des matroïdes d'ordre ≤ 8. In *Théorie des matroïdes* (ed. C. P. Bruter). Lecture Notes in Math. Vol. 211, pp. 50–61. Springer-Verlag, Berlin. [6.4]

Fournier, J. -C. (1974). Une relation de séparation entre cocircuits d'un matroïde. *J. Combin. Theory Ser. B* **16**, 181–190. [9.1]

Fournier, J. -C. (1981). A characterization of binary geometries by a double elimination axiom. *J. Combin. Theory Ser. B* **31**, 249–250. [9.1]

Fournier, J. -C. (1987). Binary matroids. In *Combinatorial geometries* (ed. N. White), pp. 28–39. Cambridge University Press, Cambridge. [6.8]

Gale, D. (1968). Optimal assignments in an ordered set: an application of matroid theory. *J. Combin. Theory* **4**, 176–180. [1.8]

Gallai, T. (1959). Über reguläre Kettengruppen. *Acta Math. Acad. Sci. Hungar.* **10**, 227–240. [11.3]

Garey, M. R. and Johnson, D. S. (1979). *Computers and intractability. A guide to the theory of NP-completeness.* Freeman, San Francisco. [12.3]

Gerards, A. M. H. (1989). A short proof of Tutte's characterization of totally unimodular matrices. *Linear Algebra Appl.* **114/115**, 207–212. [13.0, 13.1]

Gordon, G. (1984). Matroids over F_p which are rational excluded minors. *Discrete Math.* **52**, 51–65. [6.7]

Gordon, G. (1988). Algebraic characteristic sets of matroids. *J. Combin. Theory Ser. B* **44**, 64–74. [14.1]

Graver, J. E. (1966). *Lectures on the theory of matroids.* University of Alberta. [13.3]

Graver, J. E. and Watkins, M. E. (1977). *Combinatorics with emphasis on the theory of graphs.* Springer-Verlag, Berlin. [8.2]

Greene, C. (1970). A rank inequality for finite geometric lattices. *J. Combin. Theory* **9**, 357–364. [6.9]

Greene, C. (1971). *Lectures on combinatorial geometries* (with notes, footnotes and other comments by Daniel Kennedy). N.S.F. advanced science seminar, Bowdoin College, Brunswick, Maine. [5.3]

Greene, C. (1973). A multiple exchange property for bases. *Proc. Amer. Math. Soc.* **39**, 45–50. [1.4]

Gubser, B. (1990). Some problems for graph minors. Ph.D. thesis, Louisiana State University. [14.8]

Gulati, B. R. and Kounias, E. G. (1970). On bounds useful in the theory of symmetrical factorial designs. *J. Roy. Statist. Soc. Ser. B* **32**, 123–133. [6.5]

Halin, R. (1969). A theorem on n-connected graphs. *J. Combin. Theory* **7**, 150–154. [14.4]

Hall, D. W. (1943). A note on primitive skew curves. *Bull. Amer. Math. Soc.* **49**, 935–937. [11.2]

Hall, M. Jr. (1986). *Combinatorial theory.* Second edition. Wiley, New York. [6.1]

Hall, P. (1935). On representatives of subsets. *J. London Math. Soc.* **10**, 26–30. [12.2]

Hamidoune, Y. O. and Salaün, I. (1989). On the independence numbers of a matroid. *J. Combin. Theory Ser. B* **47**, 146–152. [14.2]

Harary, F. (1969). *Graph theory.* Addison-Wesley, Reading, Mass. [4.1, 9.3]

Harary, F. and Tutte, W. T. (1965). A dual form of Kuratowski's Theorem. *Canad. Math. Bull.* **8**, 17–20, 373. [5.2]

Harary, F. and Welsh, D. J. A. (1969). Matroids versus graphs. In *The many facets of graph theory.* Lecture Notes in Math. Vol. 110, pp. 155–170. Springer-Verlag, Berlin. [12.3]

Hartmanis, J. (1959). Lattice theory of generalized partitions. *Canad. J. Math.* **11**, 97–106. [2.1]

Hausmann, D. and Korte, B. (1978a). Oracle algorithms for fixed point problems — an axiomatic approach. In *Optimization and operations research.* Lecture Notes in Econom. and Math. Systems Vol. 157, pp. 137–156. Springer-Verlag, Berlin. [9.3]

Hausmann, D. and Korte, B. (1978b). Lower bounds on the worst-case complexity of some oracle algorithms. *Discrete Math.* **24**, 261–276. [9.3]

Hausmann, D. and Korte, B. (1981). Algorithmic versus axiomatic definitions of matroids. *Math. Prog. Study* **14**, 98–111. [9.3]

Helgason, T. (1974). Aspects of the theory of hypermatroids. In *Hypergraph seminar* (eds. C. Berge and D. K. Ray-Chaudhuri). Lecture Notes in Math. Vol 411, pp. 191–214. Springer-Verlag, Berlin. [12.1]

Heller, I. (1957). On linear systems with integral valued solutions. *Pacific J. Math.* **7**, 1351–1364. [13.4]

Heron, A. P. (1972a). *Some topics in matroid theory.* D.Phil. thesis, University of Oxford. [9.3]

Heron, A. P. (1972b). Matroid polynomials. In *Combinatorics* (eds. D. J. A. Welsh and D. R. Woodall), pp. 164–202. Institute of Math. and its Applications, Southend-on-Sea. [14.2]

Heron, A. P. (1973). A property of the hyperplanes of a matroid and an extension of Dilworth's theorem. *J. Math. Anal. Appl.* **42**, 119–132. [6.9]

Higgs, D. A. (1966a). A lattice order on the set of all matroids on a set. *Canad. Math. Bull.* **9**, 684–685. [7.3]

Higgs, D. A. (1966b). Maps of geometries. *J. London Math. Soc.* **41**, 612–618. [7.3]

Higgs, D. A. (1968). Strong maps of geometries. *J. Combin. Theory* **5**, 185–191. [7.3]

Hirschfeld, J. W. P. (1983). Maximum sets in finite projective spaces. In *Surveys in combinatorics.* London Math. Soc. Lecture Notes **82**, pp. 55–76. Cambridge University Press, Cambridge. [6.5]

Hoffman, A. J. and Kruskal, J. B. (1956). Integral boundary points of convex polyhedra. In *Linear inequalities and related systems* (eds. H. W. Kuhn and A. W. Tucker) Ann. Math. Studies **38**, pp. 223–246. Princeton University Press, Princeton. [13.0]

Hoffman, A. J. and Kuhn, H. W. (1956). On systems of distinct representatives. In *Linear inequalities and related systems* (eds. H. W. Kuhn and A. W. Tucker) Ann. Math. Studies **38**, pp. 199–206. Princeton University Press, Princeton. [1.6]

Horn, A. (1955). A characterisation of unions of linearly independent sets. *J. London Math. Soc.* **30**, 494–496. [12.3]

Hughes, D. R. and Piper, F. (1973). *Projective planes.* Springer-Verlag, New York. [6.1]

Hungerford, T. W. (1974). *Algebra.* Springer-Verlag, New York. [6.3]

Ingleton, A. W. (1971a). Representation of matroids. In *Combinatorial mathematics and its applications* (ed. D. J. A. Welsh), pp. 149–167. Academic Press, London. [6.1, 6.7, 6.8, 14.8]

Ingleton, A. W. (1971b). A geometrical characterization of transversal independence structures. *Bull. London Math. Soc.* **3**, 47–51. [12.2, 13.4]

Ingleton, A. W. (1977). Transversal matroids and related structures. In *Higher combinatorics* (ed. M. Aigner), pp. 117–131. Reidel, Dordrecht. [13.4]

Ingleton, A. W. and Main, R. A. (1975). Non-algebraic matroids exist. *Bull. London Math. Soc.* **7**, 144–146. [6.7]

Ingleton, A. W. and Piff, M. J. (1973). Gammoids and transversal matroids. *J. Combin. Theory Ser. B* **15**, 51–68. [2.4, 14.7]

Inukai, T. and Weinberg, L. (1978). Theorems on matroid connectivity. *Discrete Math.* **22**, 311–312. [8.1]

Inukai, T. and Weinberg, L. (1981). Whitney connectivity of matroids. *SIAM J. Alg. Disc. Methods* **2**, 108–120. [8.2]

Jackson, B. (1980). Removable cycles in 2-connected graphs of minimum degree at least four. *J. London Math. Soc.* (2) **21**, 385–392. [14.4]

Jacobson, N. (1953). *Lectures in abstract algebra. Volume II. Linear algebra.* Van Nostrand, Princeton. [6.1]

Jaeger, F. (1979). Flows and generalized coloring theorems in graphs. *J. Combin. Theory Ser. B* **26**, 205–216. [13.4]

Jaeger, F., Vertigan, D. L., and Welsh, D. J. A. (1990). On the computational complexity of the Jones and Tutte polynomials. *Math. Proc. Camb. Phil. Soc.* **108**, 35–53. [7.1]

Jensen, P. M. (1978). Binary fundamental matroids. In *Algebraic methods in graph theory* (eds. L. Lovász and V. T. Sós), Colloq. Math. Soc. János Bolyai **25**, pp. 281–296. North-Holland, Amsterdam. [14.8]

Jensen, P. M. and Korte, B. (1982). Complexity of matroid property algorithms. *SIAM J. Comput.* **11**, 184–190. [9.3]

Kahn, J. (1982). Characteristic sets of matroids. *J. London Math. Soc.* (2) **26**, 207–217. [6.8]

Kahn, J. (1984). A geometric approach to forbidden minors for GF(3). *J. Combin. Theory Ser. A* **37**, 1–12. [10.0, 10.3]

Kahn, J. (1985). A problem of P. Seymour on nonbinary matroids. *Combinatorica* **5**, 319–323. [8.4, 11.3, 14.8]

Kahn, J. (1988). On the uniqueness of matroid representations over GF(4). *Bull. London Math. Soc.* **20**, 5–10. [10.1, 14.1]

Kahn, J. and Seymour, P. (1988). On forbidden minors for GF(3). *Proc. Amer. Math. Soc.* **102**, 437–440. [6.5, 10.0, 10.3]

Kantor, W. (1975). Envelopes of geometric lattices. *J. Combin. Theory Ser. A* **18**, 12–26. [10.2]

Kelly, P. J. (1942). On isometric transformations. Ph.D. thesis, University of Wisconsin. [14.4]

Knuth, D. E. (1974). The asymptotic number of geometries. *J. Combin. Theory Ser. A* **17**, 398–401. [1.1]

Korte, B., Lovász, L., and Schrader, R. (1991). *Greedoids.* Springer-Verlag, Berlin. [14.0]

Krogdahl, S. (1977). The dependence graph for bases in matroids. *Discrete Math.* **19**, 47–59. [10.2]

Kruskal, J. B. (1956). On the shortest spanning tree of a graph and the traveling salesman problem. *Proc. Amer. Math. Soc.* **7**, 48–50. [1.8]

Kung, J. P. S. (1977). The core extraction algorithm for combinatorial geometries. *Discrete Math.* **19**, 167–175. [7.3]

Kung, J. P. S. (1978). The alpha function of a matroid – I. Transversal matroids. *Studies in Appl. Math.* **58**, 263–275. [14.7]

Kung, J. P. S. (1986). *A source book in matroid theory.* Birkhäuser, Boston. [1.0, 14.0]

Kung, J. P. S. (1987). Excluding the cycle geometries of the Kuratowski graphs from binary geometries. *Proc. London Math. Soc.* (3) **55**, 209–242. [14.3]

Kung, J. P. S. (1990). Combinatorial geometries representable over GF(3) and GF(q). I. The number of points. *Discrete Comput. Geom.* **5**, 83–95. [14.1]

Kung, J. P. S. (1991). Extremal matroid theory (submitted). [14.8]

Kung, J. P. S. and Oxley, J. G. (1988). Combinatorial geometries representable over GF(3) and GF(q). II. Dowling geometries. *Graphs Combin.* **4**, 323–332. [14.1]

Kuratowski, K. (1930). Sur le problème des courbes gauches en topologie. *Fund. Math.* **15**, 271–283. [2.3, 5.2, 5.4, 6.6]

Lang, S. (1965). *Algebra.* Addison-Wesley, Reading, Massachusetts. [6.7]

Las Vergnas, M. (1970). Sur les systèmes de représentants distincts d'une famille d'ensembles. *C. R. Acad. Sci. Paris Sér. A-B* **270**, A501–A503. [12.2, 13.4]

Las Vergnas, M. (1980). Fundamental circuits and a characterization of binary matroids. *Discrete Math.* **31**, 327. [9.1]

Lawler, E. (1976). *Combinatorial optimization: networks and matroids.* Holt, Rinehart and Winston, New York. [1.8, 12.3, 13.2]

Lazarson, T. (1958). The representation problem for independence functions. *J. London Math. Soc.* **33**, 21–25. [6.5]

Lehman, A. (1964). A solution of the Shannon switching game. *J. Soc. Indust. Appl. Math.* **12**, 687–725. [4.3, 9.1, 9.3]

Lemos, M. (1988). An extension of Lindström's result about characteristic sets of matroids. *Discrete Math.* **68**, 85–101. [14.1]

Lemos, M. (1989). On 3-connected matroids. *Discrete Math.* **73**, 273–283. [8.4, 10.2, 14.4]

Lemos, M. (1990). Matroids having the same connectivity function (submitted). [8.1]

Li, Weixuan. (1983) On matroids of the greatest W-connectivity. *J. Combin. Theory Ser. B* **35**, 20–27. [8.2]

Lindström, B. (1983). The non-Pappus matroid is algebraic. *Ars Combinatoria* **16B**, 95–96. [6.7]

Lindström, B. (1984a). A simple non-algebraic matroid of rank three. *Utilitas Math.* **25**, 95–97. [6.7]

Lindström, B. (1984b). On binary identically self-dual matroids. *European J. Combin.* **5**, 55–58. [13.4]

Lindström, B. (1985a). A desarguesian theorem for algebraic combinatorial geometries. *Combinatorica* **5**, 237–239. [6.7]

Lindström, B. (1985b). *On the algebraic representations of dual matroids.* Department of Math., Univ. of Stockholm, Reports, No. 5. [6.7]

Lindström, B. (1985c). On the algebraic characteristic set for a class of matroids. *Proc. Amer. Math. Soc.* **95**, 147–151. [6.7, 14.1]

Lindström, B. (1985d). *More on algebraic representations of matroids.* Department of Math., Univ. of Stockholm, Reports, No. 10. [6.7]

Lindström, B. (1986a). The non-Pappus matroid is algebraic over any finite field. *Utilitas Math.* **30**, 53–55. [6.7]

Lindström, B. (1986b). A non-linear algebraic matroid with infinite characteristic set. *Discrete Math.* **59**, 319–320. [6.7]

Lindström, B. (1987a). An elementary proof in matroid theory using Tutte's coordinatization theorem. *Utilitas Math.* **31**, 189–190. [6.6]

Lindström, B. (1987b). A reduction of algebraic representations of matroids. *Proc. Amer. Math. Soc.* **100**, 388–389. [6.7]

Lindström, B. (1987c). A class of non-algebraic matroids of rank three. *Geom. Dedicata* **23**, 255–258. [6.7]

Lindström, B. (1988a). A generalization of the Ingleton–Main lemma and a class of non-algebraic matroids. *Combinatorica* **8**, 87–90. [6.7]

Lindström, B. (1988b). Matroids, algebraic and non-algebraic. In *Algebraic, extremal and metric combinatorics* 1986 (eds. M. -M. Deza *et al.*) London Math. Soc. Lecture Notes **131**, pp. 166–174. Cambridge University Press, Cambridge. [6.7, 14.1]

Lindström, B. (1989). Matroids algebraic over $F(t)$ are algebraic over F. *Combinatorica* **9**, 107–109. [6.7]

Löfgren, L. (1959). Irredundant and redundant boolean branch-networks. *IRE Transactions on Circuit Theory*, **CT-6**, Special Supplement, 158–175. [13.3]

Lovász, L. (1977). Matroids and geometric graphs. In *Combinatorial surveys: Proceedings of the sixth British combinatorial conference* (ed. P. J. Cameron), pp. 45–86. Academic Press, London. [12.1]

Lovász, L. and Plummer, M. D. (1986). *Matching theory.* Ann. Discrete Math. **29**. North-Holland, Amsterdam. [12.1]

Lovász, L. and Recski, A. (1973). On the sum of matroids. *Acta Math. Acad. Sci. Hungar.* **24**, 329–333. [12.3]

Lucas, D. (1975). Weak maps of combinatorial geometries. *Trans. Amer. Math. Soc.* **206**, 247–279. [7.3]

Mac Lane, S. (1936). Some interpretations of abstract linear dependence in terms of projective geometry. *Amer. J. Math.* **58**, 236–240. [1.0]

Mac Lane, S. (1938). A lattice formulation for transcendence degrees and p-bases. *Duke Math. J.* **4**, 455–468. [1.0, 6.7]

Mader, W. (1979). Connectivity and edge-connectivity in finite graphs. In *Surveys in combinatorics* (ed. B. Bollobás) London Math. Soc. Lecture Notes **38**, pp. 66–95. Cambridge University Press, Cambridge. [14.4]

Mahoney, C. (1985). On the unimodality of the independent set numbers of a class of matroids. *J. Combin. Theory Ser. B* **39**, 77–85. [14.2]

Mason, J. H. (1971). Geometrical realization of combinatorial geometries. *Proc. Amer. Math. Soc.* **30**, 15–21. [1.5, 7.2]

Mason, J. H. (1972a). On a class of matroids arising from paths in graphs. *Proc. London Math. Soc.* (3) **25**, 55–74. [2.4, 3.2, 14.7]

Mason, J. H. (1972b). Matroids: unimodal conjectures and Motzkin's theorem. In *Combinatorics* (eds. D. J. A. Welsh and D. R. Woodall), pp. 207–220. Institute of Math. and its Applications, Southend-on-Sea. [14.2]

Mason, J. H. (1973). Maximal families of pairwise disjoint maximal proper chains in a geometric lattice. *J. London Math. Soc.* (2) **6**, 539–542. [6.9]

Mason, J. H. (1977). Matroids as the study of geometrical configurations. In *Higher combinatorics* (ed. M. Aigner), pp. 133–176. Reidel, Dordrecht. [7.0]

Mason, J. H. and Oxley, J. G. (1980). A circuit covering result for matroids. *Math. Proc. Camb. Phil. Soc.* **87**, 25–27. [4.3]

Matthews, L. R. (1977). Bicircular matroids. *Quart. J. Math. Oxford Ser.* (2) **28**, 213–228. [12.1, 12.2]

McDiarmid, C. J. H. (1973). Independence structures and submodular functions. *Bull. London Math. Soc.* **5**, 18–20. [12.1]

McDiarmid, C. J. H. (1975a). Extensions of Menger's theorem. *Quart. J. Math. Oxford Ser.* (2) **26**, 141–157. [12.2]

McDiarmid, C. J. H. (1975b). An exchange theorem for independence structures. *Proc. Amer. Math. Soc.* **47**, 513–514. [12.3]

McDiarmid, C. J. H. (1975c). Rado's theorem for polymatroids. *Math. Proc. Camb. Phil. Soc.* **78**, 263–281. [12.1]

Mendelsohn, N. S. and Dulmage, A. L. (1958). Some generalisations of the problem of distinct representatives. *Canad. J. Math.* **10**, 230–242. [12.2]

Menger, K. (1927). Zur allgemeinen Kurventheorie. *Fund. Math.* **10**, 96–115. [8.2, 11.3]

Minty, G. J. (1966). On the axiomatic foundations of the theories of directed linear graphs, electrical networks and network programming. *J. Math. Mech.* **15**, 485–520. [9.1, 11.3, 13.4]

Mirsky, L. (1971). *Transversal theory.* Academic Press, London. [12.2, 12.3]

Motzkin, T. (1951). The lines and planes connecting the points of a finite set. *Trans. Amer. Math. Soc.* **70**, 451–464. [6.9]

Murty, U. S. R. (1969). Sylvester matroids. In *Recent progress in combinatorics* (ed. W. T. Tutte), pp. 283–286. Academic Press, New York. [10.2]

Murty, U. S. R. (1970). Matroids with Sylvester property. *Aequationes Math.* **4**, 44–50. [10.2]

Murty, U. S. R. (1974). Extremal critically connected matroids. *Discrete Math.* **8**, 49–58. [4.3, 10.2, 14.4]

Murty, U. S. R. (1976). Extremal matroids with forbidden restrictions and minors. In *Proc. Seventh Southeastern Conf. on Combinatorics, Graph Theory and Computing.* Congressus Numerantium **17**, pp. 463–468. Utilitas Mathematica, Winnipeg. [13.4]

Nash-Williams, C. St. J. A. (1961). Edge-disjoint spanning trees of finite graphs. *J. London Math. Soc.* **36**, 445–450. [12.3]

Nash-Williams, C. St. J. A. (1966). An application of matroids to graph theory. In *Theory of graphs* (Internat. Sympos., Rome), pp. 263–265. Dunod, Paris. [12.2, 12.3]

Negami, S. (1982). A characterization of 3-connected graphs containing a given graph. *J. Combin. Theory Ser. B* **32**, 69–74. [11.1]

Nešetřil, J., Poljak, S., and Turzík, D. (1981). Amalgamation of matroids and its applications. *J. Combin. Theory Ser. B* **31**, 9–22. [12.4]

Nešetřil, J., Poljak, S., and Turzík, D. (1985). Special amalgams and Ramsey matroids. In *Matroid theory* (eds. L. Lovász and A. Recski) Colloq. Math. Soc. János Bolyai **40**, pp. 267–298. North-Holland, Amsterdam. [12.4]

Ore, O. (1955). Graphs and matching theorems. *Duke Math. J.* **22**, 625–639. [12.2]

Ore, O. (1967). *The four-color problem.* Academic Press, New York. [5.2, 5.3]

Oxley, J. G. (1978a). Colouring, packing and the critical problem. *Quart. J. Math. Oxford Ser.* (2) **29**, 11–22. [6.2]

Oxley, J. G. (1978b). Cocircuit coverings and packings for binary matroids. *Math. Proc. Camb. Phil. Soc.* **83**, 347–351. [9.3]

Oxley, J. G. (1979). On cographic regular matroids. *Discrete Math.* **25**, 89–90. [13.3]

Oxley, J. G. (1981a). On connectivity in matroids and graphs. *Trans. Amer. Math. Soc.* **265**, 47–58. [8.4, 10.1, 10.2, 14.4]

Oxley, J. G. (1981b). On matroid connectivity. *Quart. J. Math. Oxford Ser.* (2) **32**, 193–208. [8.1, 8.4]

Oxley, J. G. (1981c). On 3-connected matroids. *Canad. J. Math.* **33**, 20–27. [8.4]

Oxley, J. G. (1981d). On a matroid generalization of graph connectivity. *Math. Proc. Camb. Phil. Soc.* **90**, 207–214. [8.2]

Oxley, J. G. (1983). On the numbers of bases and circuits in simple binary matroids. *European J. Combin.* **4**, 169–178. [9.3]

Oxley, J. G. (1984a). On the intersections of circuits and cocircuits in matroids. *Combinatorica* **4**, 187–195. [5.4, 9.1, 11.2, 11.3]

Oxley, J. G. (1984b). On singleton 1-rounded sets of matroids. *J. Combin. Theory Ser. B* **37**, 189–197. [7.2, 11.3]

Oxley, J. G. (1984c). On minor-minimally-connected matroids. *Discrete Math.* **51**, 63–72. [10.2, 14.4]

Oxley, J. G. (1986a). On the matroids representable over GF(4). *J. Combin. Theory Ser. B* **41**, 250–252. [6.5]

Oxley, J. G. (1986b). On ternary transversal matroids. *Discrete Math.* **62**, 71–83. [13.4, 14.8]

Oxley, J. G. (1987a). On nonbinary 3-connected matroids. *Trans. Amer. Math. Soc.* **300**, 663–679. [11.2, 11.3]

Oxley, J. G. (1987b). The binary matroids with no 4-wheel minor. *Trans. Amer. Math. Soc.* **301**, 63–75. [11.2]

Oxley, J. G. (1987c). A characterization of the ternary matroids with no $M(K_4)$-minor. *J. Combin. Theory Ser. B* **42**, 212–249. [11.1, 11.2, 13.4]

Oxley, J. G. (1988). On circuit exchange properties for matroids. *European J. Combin.* **9**, 331–336. [9.1, 11.2]

Oxley, J. G. (1989a). The regular matroids with no 5-wheel minor. *J. Combin. Theory Ser. B* **46**, 292–305. [11.1, 11.2]

Oxley, J. G. (1989b). A characterization of certain excluded-minor classes of matroids. *European J. Combin.* **10**, 275–279. [11.2, 11.3]

Oxley, J. G. (1990a). On an excluded-minor class of matroids. *Discrete Math.* **82**, 35–52. [11.1]

Oxley, J. G. (1990b). A characterization of a class of non-binary matroids. *J. Combin. Theory Ser. B* **49**, 181–189. [14.8]

Oxley, J. G. (1991). Ternary paving matroids. *Discrete Math.* **91**, 77–86. [11.2]

Oxley, J. G. (1992). Infinite matroids. In *Matroid applications* (ed. N. White), pp. 73–90. Cambridge University Press, Cambridge. [2.0]

Oxley, J. G. and Reid, T. J. (1990). The smallest rounded sets of binary matroids. *European J. Combin.* **11**, 47–56. [11.3]

Oxley, J. and Row, D. (1989). On fixing elements in matroid minors. *Combinatorica* **9**, 69–74. [11.3]

Oxley, J. and Whittle, G. (1991). A note on the non-spanning circuits of a matroid. *European J. Combin.* **12**, 259–261. [7.3]

Oxley, J., Prendergast, K., and Row, D. (1982). Matroids whose ground sets are domains of functions. *J. Austral. Math. Soc. Ser. A* **32**, 380–387. [12.1]

Parsons, T. D. (1971). On planar graphs. *Amer. Math. Monthly* **78**, 176–178. [5.2]

Perfect, H. (1968). Applications of Menger's graph theorem. *J. Math. Anal. Appl.* **22**, 96–111. [12.3]

Perfect, H. (1969). Independence spaces and combinatorial problems. *Proc. London Math. Soc.* (3) **19**, 17–30. [12.2]

Piff, M. J. (1969). The representability of matroids. Dissertation for Diploma in Advanced Math., University of Oxford. [6.7]

Piff, M. J. (1972). *Some problems in combinatorial theory.* D.Phil. thesis, University of Oxford. [3.2, 6.7, 13.2]

Piff, M. J. (1973). An upper bound for the number of matroids. *J. Combin. Theory Ser. B* **14**, 241–245. [1.1]

Piff, M. J. and Welsh, D. J. A. (1970). On the vector representation of matroids. *J. London Math. Soc.* (2) **2**, 284–288. [6.8, 12.2, 12.3]

Plummer, M. D. (1968). On minimal blocks. *Trans. Amer. Math. Soc.* **134**, 85–94. [4.3]

Poljak, S. and Turzík, D. (1982). A note on sticky matroids. *Discrete Math.* **42**, 119–123. [12.4, 14.8]

Poljak, S. and Turzík, D. (1984). Amalgamation over uniform matroids. *Czech. Math. J.* **34**, 239–246. [12.4]

Purdy, G. (1986). Two results about points, lines and planes. *Discrete Math.* **60**, 215–218. [14.2]

Pym, J. S. and Perfect, H. (1970). Submodular functions and independence structures. *J. Math. Anal. Appl.* **30**, 1–31. [12.3]

Rado, R. (1942). A theorem on independence relations. *Quart. J. Math. Oxford* **13**, 83–89. [12.2]

Rado, R. (1957). Note on independence functions. *Proc. London Math. Soc.* **7**, 300–320. [6.7, 9.1]

Rado, R. (1967). On the number of systems of distinct representatives of sets. *J. London Math. Soc.* **42**, 107–109. [12.2]

Read, R. C. (1968). An introduction to chromatic polynomials. *J. Combin. Theory* **4**, 52–71. [14.2]

Recski, A. (1981). On the sum of matroids, III. *Discrete Math.* **36**, 273–287. [14.6]

Recski, A. (1985). Some open problems of matroid theory, suggested by its applications. In *Matroid theory* (eds. L. Lovász and A. Recski) Colloq. Math. Soc. János Bolyai **40**, pp. 311–325. North-Holland, Amsterdam. [12.3, 14.6]

Recski, A. (1989). *Matroid theory and its applications in electrical network theory and in statics.* Springer-Verlag, Berlin. [13.2, 14.0]

Reid, T. J. (1988). *On roundedness in matroid theory.* Ph.D. thesis, Louisiana State University. [11.3]

Reid, T. J. (1991). Triangles in 3-connected matroids. *Discrete Math.* **90**, 281–296. [11.3]

Richardson, W. R. H. (1973). Decomposition of chain-groups and binary matroids. In *Proc. Fourth Southeastern Conf. on Combinatorics, Graph Theory and Computing*, pp. 463–476. Utilitas Mathematica, Winnipeg. [8.1]

Robertson, N. and Seymour, P. D. (1984). Generalizing Kuratowski's Theorem. *Congressus Numerantium* **45**, 129–138. [7.1, 11.2, 12.4]

Robertson, N. and Seymour, P. D. (1990). Graph minors XV. Wagner's conjecture (submitted). [14.4]

Robinson, G. C. and Welsh, D. J. A. (1980). The computational complexity of matroid properties. *Math. Proc. Camb. Phil. Soc.* **87**, 29–45. [9.3]

Rockafellar, R. T. (1969). The elementary vectors of a subspace of R^N. In *Combinatorial mathematics and its applications* (eds. R. C. Bose and T. A. Dowling), pp. 104–127. University of North Carolina Press, Chapel Hill. [13.4]

Rota, G. -C. (1971). Combinatorial theory, old and new. In *Proc. Internat. Cong. Math.* (Nice, Sept. 1970), pp. 229–233. Gauthier-Villars, Paris. [6.5, 10.3, 14.1, 14.2]

Schrijver, A. (1986). *Theory of linear and integer programming.* Wiley, Chichester. [13.0, 13.2]

Segre, B. (1955). Curve razionali normali e k-archi negli spazi finiti. *Ann. Mat. Pura Appl.* (4) **39**, 357–379. [6.5]

Seymour, P. D. (1975). *Matroids, hypergraphs and the max-flow min-cut theorem.* D.Phil. thesis, University of Oxford. [14.2]

Seymour, P. D. (1976). The forbidden minors of binary clutters. *J. London Math. Soc.* (2) **12**, 356–360. [9.1]

Seymour, P. D. (1977a). A note on the production of matroid minors. *J. Combin. Theory Ser. B* **22**, 289–295. [4.3, 11.3]

Seymour, P. D. (1977b). The matroids with the max-flow min-cut property. *J. Combin. Theory Ser. B* **23**, 189–222. [11.3]

Seymour, P. D. (1978). Some applications of matroid decomposition. In *Algebraic methods in graph theory* (eds. L. Lovász and V. T. Sós), Colloq. Math. Soc. János Bolyai **25**, pp. 713–726. North-Holland, Amsterdam. [11.3]

Seymour, P. D. (1979). Matroid representation over GF(3). *J. Combin. Theory Ser. B* **26**, 159–173. [6.5, 10.0, 10.1, 10.3]

Seymour, P. D. (1980a). Packing and covering with matroid circuits. *J. Combin. Theory Ser. B* **28**, 237–242. [1.4]

Seymour, P.D. (1980b). Decomposition of regular matroids. *J. Combin. Theory Ser. B* **28**, 305–359. [5.2, 6.6, 8.3, 8.4, 11.1, 11.2, 12.4, 13.2, 13.3, 13.4]

Seymour, P. D. (1980c). On Tutte's characterization of graphic matroids. In *Combinatorics* 79. *Part I* (eds. M. Deza and I. G. Rosenberg) Ann. Discrete Math. **8**, pp. 83–90. North-Holland, Amsterdam. [13.2, 13.3]

Seymour, P. D. (1981a). Recognizing graphic matroids. *Combinatorica*, **1**, 75–78. [6.4, 9.3]

Seymour, P. D. (1981b). On minors of non-binary matroids. *Combinatorica* **1**, 387–394. [8.3, 9.1, 11.3]

Seymour, P. D. (1981c). Matroids and multicommodity flows. *European J. Combin.* **2**, 257–290. [11.2, 11.3, 13.2]

Seymour, P. D. (1981d). On Tutte's extension of the four-colour problem. *J. Combin. Theory Ser. B* **31**, 82–94. [13.2, 14.3]

Seymour, P. D. (1982). On the points–lines–planes conjecture. *J. Combin. Theory Ser. B* **33**, 17–26. [14.2]

Seymour, P. D. (1985a). Minors of 3-connected matroids. *European J. Combin.* **6**, 375–382. [11.2, 11.3]

Seymour, P. D. (1985b). Applications of the regular matroid decomposition. In *Matroid theory* (eds. L. Lovász and A. Recski) Colloq. Math. Soc. János Bolyai **40**, pp. 345–357. North-Holland, Amsterdam. [8.4, 13.2]

Seymour, P. D. (1986a). Triples in matroid circuits. *European J. Combin.* **7**, 177–185. [8.4, 9.3, 14.8]

Seymour, P. D. (1986b). Adjacency in binary matroids. *European J. Combin.* **7**, 171–176. [13.2]

Seymour, P. D. (1988). On the connectivity function of a matroid. *J. Combin. Theory Ser. B* **45**, 25–30. [8.1]

Seymour, P. D. (1992). Matroid minors. In *Handbook of combinatorics* (eds. R. Graham, M. Grötschel, L. Lovász). Elsevier (to appear). [11.1, 11.2, 11,3]

Seymour, P. D. and Walton, P. N. (1981). Detecting matroid minors. *J. London Math. Soc.* (2) **23**, 193–203. [9.3]

Shameeva, O. V. (1985). Algebraic representability of matroids (in Russian). *Vestnik Moskov. Univ. Ser. I. Mat. Mekh.* 40 no. 4, 29–32. [6.7]

Stonesifer, J. R. (1975). Logarithmic concavity for edge lattices of graphs. *J. Combin. Theory Ser. A* **18**, 36–46. [14.2]

Tan, J. J.-M. (1981). *Matroid 3-connectivity.* Ph.D. thesis, Carleton University. [11.1, 11.3]

Thas, J. A. (1968). Normal rational curves and k-arcs in Galois spaces. *Rend. Mat.* (6) **1**, 331–334. [6.5]

Truemper, K. (1980). On Whitney's 2-isomorphism theorem for graphs. *J. Graph Theory* **4**, 43–49. [5.3]

Truemper, K. (1982a). Alpha-balanced graphs and matrices and GF(3)-representability of matroids. *J. Combin. Theory Ser. B* **32**, 275–291. [10.0]

Truemper, K. (1982b). On the efficiency of representability tests for matroids. *European J. Combin.* **3**, 275–291. [9.3]

Truemper, K. (1984a). Partial matroid representations. *European J. Combin.* **5**, 377–394. [6.4, 11.3]

Truemper, K. (1984b). Elements of a decomposition theory for matroids. *Progress in graph theory* (eds. J. A. Bondy and U. S. R. Murty), pp. 439–475. Academic Press, Toronto. [13.2]

Truemper, K. (1985a). A decomposition theory for matroids. I. General results. *J. Combin. Theory Ser. B* **39**, 43–76. [7.2, 13.2]

Truemper, K. (1985b). A decomposition theory for matroids. II. Minimal violation matroids. *J. Combin. Theory Ser. B* **39**, 282–297. [13.2]

Truemper, K. (1986). A decomposition theory for matroids. III. Decomposition conditions. *J. Combin. Theory Ser. B* **41**, 275–305. [11.3, 13.2]

Truemper, K. (1988). A decomposition theory for matroids. IV. Graph decomposition. *J. Combin. Theory Ser. B* **45**, 259–292. [13.2]

Truemper, K. (1990). A decomposition theory for matroids. V. Testing of matrix total unimodularity. *J. Combin. Theory Ser. B* **49**, 241–281. [13.2]

Truemper, K. (1992a). A decomposition theory for matroids. VI. Almost regular matroids. *J. Combin. Theory Ser. B* (to appear). [13.2]

Truemper, K. (1992b). A decomposition theory for matroids. VII. Analysis of minimal violation matrices. *J. Combin. Theory Ser. B* (to appear). [13.2]

Truemper, K. (1992c). *Matroid decomposition.* Academic Press, Boston. [14.0]

Tseng, F. T. and Truemper, K. (1986). A decomposition of the matroids with the max-flow min-cut property. *Discrete Appl. Math.* **15**, 329–364. [11.3]

Tutte, W. T. (1954). A contribution to the theory of chromatic polynomials. *Canad. J. Math.* **6**, 80–91. [14.3]

Tutte, W. T. (1958). A homotopy theorem for matroids, I, II. *Trans. Amer. Math. Soc.* **88**, 144–174. [6.5, 6.6, 9.1, 10.3, 11.2, 13.1]

Tutte, W. T. (1959). Matroids and graphs. *Trans. Amer. Math. Soc.* **90**, 527–552. [6.6, 13.2, 13.3]

Tutte, W. T. (1960). An algorithm for determining whether a given binary matroid is graphic. *Proc. Amer. Math. Soc.* **11**, 905–917. [13.2]

Tutte, W. T. (1961a). A theory of 3-connected graphs. *Nederl. Akad. Wetensch. Proc. Ser. A* **64**, 441–455. [8.2, 11.2]

Tutte, W. T. (1961b). On the problem of decomposing a graph into n connected factors. *J. London Math. Soc.* **36**, 221–230. [12.3]

Tutte, W. T. (1965). Lectures on matroids. *J. Res. Nat. Bur. Standards Sect. B* **69B**, 1–47. [4.3, 6.6, 9.1, 9.2, 10.2, 13.3, 13.4]

Tutte, W. T. (1966a). *Connectivity in graphs.* University of Toronto Press, Toronto. [5.3]

Tutte, W. T. (1966b). Connectivity in matroids. *Canad. J. Math.* **18**, 1301–1324. [4.3, 8.1, 8.4, 11.1, 11.2, 13.3]

Tutte, W. T. (1966c). On the algebraic theory of graph colorings. *J. Combin. Theory* **1**, 15–50. [14.3]

Tutte, W. T. (1977). Bridges and hamiltonian circuits in planar graphs. *Aequationes Math.* **15**, 1–33. [8.2]

Tutte, W. T. (1984). *Graph theory.* Cambridge University Press, Cambridge. [11.2]

Vámos, P. (1968). On the representation of independence structures (unpublished manuscript). [2.1, 6.1]

Vámos, P. (1971a). A necessary and sufficient condition for a matroid to be linear. In *Möbius algebras* (Proc. Conf. Univ. Waterloo, 1971), pp. 166–173. University of Waterloo, Waterloo. [6.5]

Vámos, P. (1971b). Linearity of matroids over division rings (notes by G. Roulet). In *Möbius algebras* (Proc. Conf. Univ. Waterloo, 1971), pp. 174–178. University of Waterloo, Waterloo. [6.8]

van der Waerden, B. L. (1937). *Moderne Algebra Vol. 1.* Second edition. Springer-Verlag, Berlin. [1.0, 6.7]

Wagner, D. K. (1985). On theorems of Whitney and Tutte. *Discrete Math.* **57**, 147–154. [5.3, 13.0, 13.3]

Wagner, D. K. (1988). Equivalent factor matroids of graphs. *Combinatorica* **8**, 373–377. [14.4]

Wagner, K. (1937a). Über eine Eigenschaft der ebenen Komplexe. *Math. Ann.* **114**, 570–590. [5.2]

Wagner, K. (1937b). Über eine Erweiterung eines Satzes von Kuratowski. *Deut. Math.* **2**, 280–285. [5.2]

Wagner, K. (1960). Bemerkungen zu Hadwigers Vermutung. *Math. Ann.* **141**, 433–451. [11.1]

Walton, P. (1981). *Some topics in combinatorial theory.* D.Phil. thesis, University of Oxford. [11.2, 13.4]

Walton, P. N. and Welsh, D. J. A. (1980). On the chromatic number of binary matroids. *Mathematika*, **27**, 1–9. [11.2, 14.3]

Wanner, T. and Ziegler, G. M. (1991). Supersolvable and modularly complemented matroid extensions. *European J. Combin.* **12**, 341–360. [14.8]

Watkins, M. E. and Mesner, D. M. (1967). Cycles and connectivity in graphs. *Canad. J. Math.* **19**, 1319–1328. [14.8]

Welsh, D. J. A. (1969a). A bound for the number of matroids. *J. Combin. Theory* **6**, 313–316. [1.6, 14.5]

Welsh, D. J. A. (1969b). On the hyperplanes of a matroid. *Proc. Camb. Phil. Soc.* **65**, 11–18. [13.3]

Welsh, D. J. A. (1969c). Euler and bipartite matroids. *J. Combin. Theory* **6**, 313–316. [9.3]

Welsh, D. J. A. (1971a). Generalized versions of Hall's theorem. *J. Combin. Theory Ser. B* **10**, 95–101. [12.2]

Welsh, D. J. A. (1971b). Combinatorial problems in matroid theory. In *Combinatorial mathematics and its applications* (ed. D. J. A. Welsh), pp. 291–306. Academic Press, London. [12.3, 14.2, 14.4, 14.5, 14.6, 14.7, 14.8]

Welsh, D. J. A. (1976). *Matroid theory.* Academic Press, London. [1.3, 2.0, 2.1, 5.2, 6.7, 9.3, 12.0, 12.1, 12.2, 12.3, 14.0, 14.2, 14.5]

Welsh, D. J. A. (1982). Matroids and combinatorial optimisation. In *Matroid theory and its applications* (ed. A. Barlotti), pp. 323–416. Liguori editore, Naples. [13.2]

White, N. (1971). *The bracket ring and combinatorial geometry.* Ph.D. thesis, Harvard University. [6.9, 9.1]

White, N. (1980a). The transcendence degree of a coordinatization of a combinatorial geometry. *J. Combin. Theory Ser. B* **29**, 168–175. [6.5]

White, N. (1980b). A unique exchange property for bases. *Linear Algebra Appl.* **31**, 81–91. [14.8]

White, N., ed. (1986). *Theory of matroids.* Cambridge University Press, Cambridge. [2.2, 7.3]

White, N., ed. (1987a). *Combinatorial geometries.* Cambridge University Press, Cambridge. [1.2, 14.0]

White, N. (1987b). Coordinatizations. In *Combinatorial geometries* (ed. N. White), pp. 1–27. Cambridge University Press, Cambridge. [6.5, 6.6, 6.8, 14.0]

White, N. (1987c). Unimodular matroids. In *Combinatorial geometries* (ed. N. White), pp. 40–52. Cambridge University Press, Cambridge. [13.4]

White, N., ed. (1992). *Matroid applications.* Cambridge University Press, Cambridge. [14.0]

Whitney, H. (1932a). Non-separable and planar graphs. *Trans. Amer. Math. Soc.* **34**, 339–362. [5.2]

Whitney, H. (1932b). Congruent graphs and the connectivity of graphs. *Amer. J. Math.* **54**, 150–168. [5.3, 8.2]

Whitney, H. (1933). 2-isomorphic graphs. *Amer. J. Math.* **55**, 245–254. [5.3, 13.3]

Whitney, H. (1935). On the abstract properties of linear dependence. *Amer. J. Math.* **57**, 509–533. [1.0, 1.1, 1.8, 6.0, 6.6, 9.1]

Wilson, R. J. (1973). An introduction to matroid theory. *Amer. Math. Monthly* **80**, 500–525. [Preface]

Witt, E. (1940). Über Steinersche Systeme. *Abh. Math. Sem. Univ. Hamburg* **12**, 265–275. [11.2]

Wong, P. -K. (1978). On certain n-connected matroids. *J. Reine Angew. Math.* **299/300**, 1–6. [8.4, 14.4]

Woodall, D. R. (1974). An exchange theorem for bases of matroids. *J. Combin. Theory Ser. B* **16**, 227–229. [1.4]

Zaslavsky, T. (1987a). The Möbius function and the characteristic polynomial. In *Combinatorial geometries* (ed. N. White), pp. 114–138. Cambridge University Press, Cambridge. [14.2]

Zaslavsky, T. (1987b). The biased graphs whose matroids are binary. *J. Combin. Theory Ser. B* **42**, 337–347. [12.2]

Zaslavsky, T. (1989). Biased graphs. I. Bias, balance, and gains. *J. Combin. Theory Ser. B* **47**, 32–52. [12.2]

Appendix
Some interesting matroids

This appendix is intended to serve as a quick reference for some of the matroids that appear throughout this book. It contains numerous redundancies to facilitate its use. All the matroids listed here are 3-connected except where otherwise noted. Most of the matroids are specified via geometric representations. The list begins with one special 4-element matroid and follows this with certain 5-, 6-, $\cdots$, 13-element matroids. It concludes with several infinite families of matroids.

Recall that a matroid is regular if and only if it is representable over every field. Evidently a matroid and its dual have the same automorphism group. The automorphism group G of a matroid having ground set E is *doubly transitive* if G is transitive on ordered pairs of distinct elements of E. Similarly, G is *triply transitive* if it is transitive on ordered triples of distinct elements of E.

$U_{2,4}$

- The 4-point line; isomorphic to $\mathcal{W}^2$, the rank-2 whirl.
- The unique excluded minor for the class of binary matroids (6.5.4, 9.1.5).
- F-representable if and only if $|F| \geq 3$ (6.5.2).
- An excluded minor for the classes of graphic (6.6.5, 13.3.1), cographic (6.6.5), and regular matroids (6.6.4, 13.1.1).
- Transversal, a strict gammoid, a gammoid.
- Identically self-dual.
- Automorphism group is the symmetric group.
- The unique 3-connected non-binary matroid M having an element e such that both $M\backslash e$ and M/e are binary (Oxley 1987a).
- The unique matroid M with at least four elements such that $\{M\}$ is 2-rounded (Oxley 1984b).
- Algebraic over all fields.

$U_{2,5}$, $U_{3,5}$

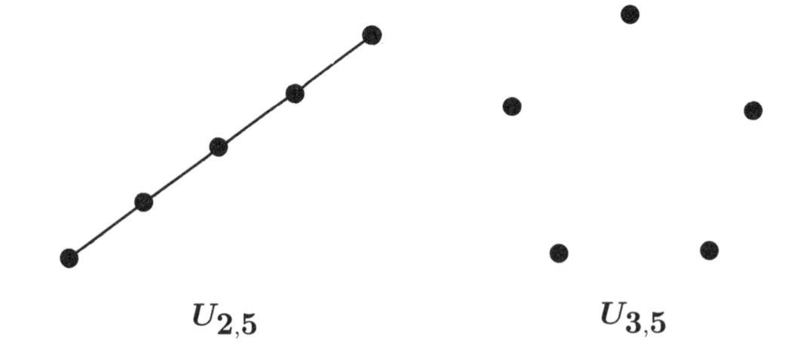

- $U_{2,5}$ is the 5-point line; $U_{3,5}$ is five points freely placed in the plane.
- Each is F-representable if and only if $|F| \geq 4$ (6.5.1, 6.5.2).
- The excluded minors for the class of ternary matroids are $U_{2,5}$, $U_{3,5}$, F_7, and F_7^* (6.5.7, 10.3.1).
- Each is not graphic, not cographic, and not regular.
- Both are members of the classes of transversal matroids, strict gammoids, and gammoids.
- A pair of dual matroids.
- Each has the symmetric group as its automorphism group.
- Each is algebraic over all fields.

$M(K_4)$

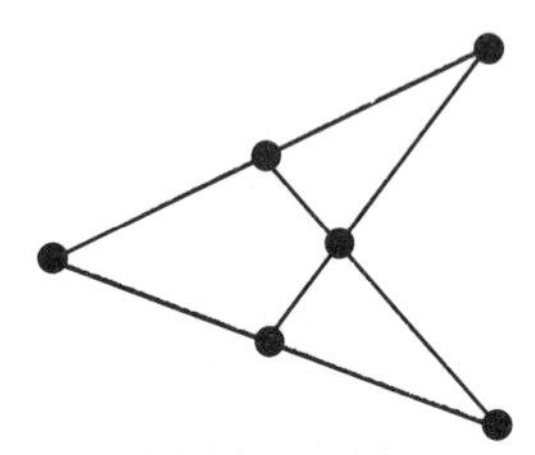

- Isomorphic to $M(\mathcal{W}_3)$, the rank-3 wheel.
- Regular.
- Graphic, cographic.
- A minor-minimal non-transversal matroid (Section 12.2).
- A minor-minimal matroid that is not a strict gammoid.
- A minor-minimal non-gammoid.
- Self-dual but not identically self-dual.
- Transitive automorphism group.
- A minor of every 3-connected binary matroid with at least four elements (11.2.14).
- A connected non-empty binary matroid is a series–parallel network if and only if it has no $M(K_4)$-minor (11.2.15).
- For all elements e, neither $M(K_4)\backslash e$ nor $M(K_4)/e$ is 3-connected (8.4.5, 11.2.2).
- Algebraic over all fields.

$\mathcal{W}^3$

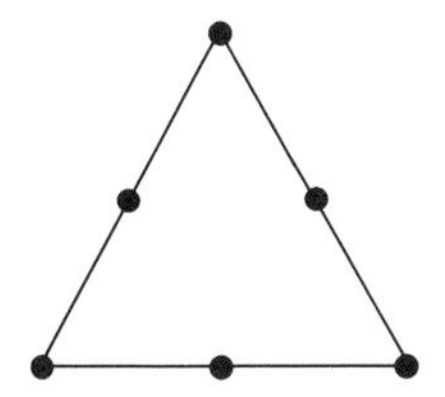

- The rank-3 whirl.
- F-representable if and only if $|F| \geq 3$.
- Not graphic, not cographic, not regular.
- Transversal, a strict gammoid, a gammoid.
- Self-dual but not identically self-dual.
- Non-transitive automorphism group.
- For all elements e, neither $\mathcal{W}^3 \backslash e$ nor $\mathcal{W}^3/e$ is 3-connected (8.4.5, 11.2.2).
- The unique relaxation of $M(K_4)$.
- Algebraic over all fields.

Q_6

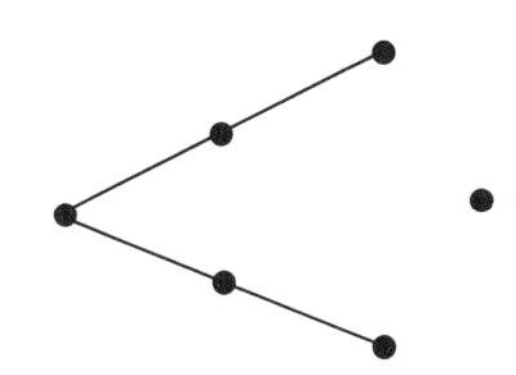

- F-representable if and only if $|F| \geq 4$.
- Not graphic, not cographic, not regular.
- Transversal, a strict gammoid, a gammoid.
- Self-dual but not identically self-dual.
- Non-transitive automorphism group.
- The unique relaxation of $\mathcal{W}^3$.
- If N is a matroid having at least four elements, then $\{N\}$ is 1-rounded if and only if $N \cong U_{2,4}$, $M(\mathcal{W}_2)$, or Q_6 (Oxley 1984b).
- Algebraic over all fields.

P_6

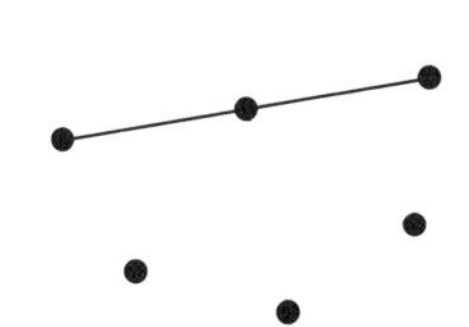

- An excluded minor for $GF(4)$-representability (Section 6.5).
- F-representable if and only if $|F| \geq 5$ (6.5.8).
- Not graphic, not cographic, not regular.
- Transversal, a strict gammoid, a gammoid.

- Self-dual but not identically self-dual.
- Non-transitive automorphism group.
- The unique relaxation of Q_6.
- Algebraic over all fields.

$U_{3,6}$

- Six points freely placed in the plane.
- F-representable if and only if $|F| \geq 4$ (Section 6.5).
- Not graphic, not cographic, not regular.
- Transversal, a strict gammoid, a gammoid.
- Identically self-dual.
- Automorphism group is the symmetric group.
- The unique relaxation of P_6.
- Every 3-connected matroid of rank and corank exceeding two has a minor isomorphic to one of $M(K_4)$, $\mathcal{W}^3$, Q_6, P_6, and $U_{3,6}$ (11.2.20).
- Every 3-connected non-binary matroid of rank and corank exceeding two has a minor isomorphic to one of $\mathcal{W}^3$, P_6, Q_6, and $U_{3,6}$ (11.2.19).
- Algebraic over all fields.

R_6

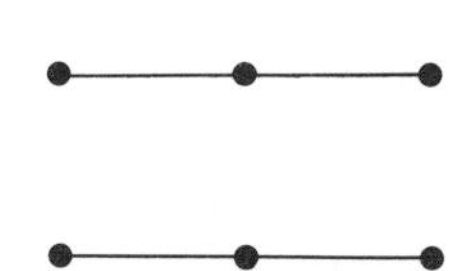

- Isomorphic to $U_{2,4} \oplus_2 U_{2,4}$.
- Connected but not 3-connected.
- F-representable if and only if $|F| \geq 3$
- Not graphic, not cographic, not regular.
- Transversal, a strict gammoid, a gammoid.
- Identically self-dual.
- Transitive automorphism group.
- Has P_6 as a relaxation.
- A connected $GF(4)$-representable matroid is uniquely $GF(4)$-representable if and only if it has no R_6-minor (10.1.11).
- Algebraic over all fields.

$\boldsymbol{F_7}$, $\boldsymbol{F_7^*}$

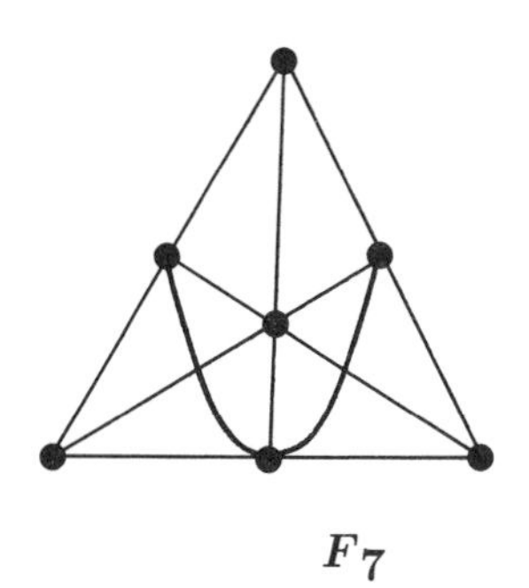

F_7

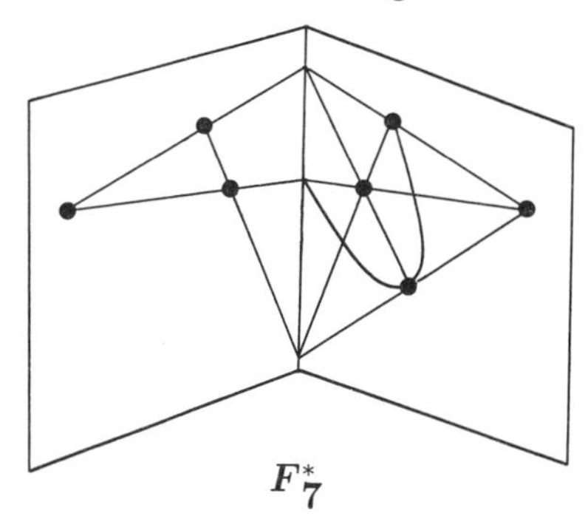

F_7^*

- The Fano matroid and its dual.
- F_7 is the smallest projective plane, $PG(2,2)$ (Section 6.1) and is isomorphic to the unique $S(2,3,7)$ (Section 11.2).
- Each is F-representable if and only if F has characteristic two (6.4.8).
- Each is an excluded minor for F-representability if and only if F has characteristic other than two (6.5.5).
- Both are excluded minors for the classes of graphic matroids (6.6.5, 13.3.1), cographic matroids (6.6.5), and regular matroids (6.6.4, 13.1.1).
- Each is not transversal, not a strict gammoid, not a gammoid.
- Their common automorphism group is doubly transitive.
- Every single-element deletion of F_7 and every single-element contraction of F_7^* is isomorphic to $M(K_4)$ (Section 1.5).
- F_7 is a splitter for the class of binary matroids with no F_7^*-minor, and F_7^* is a splitter for the class of binary matroids with no F_7-minor (11.2.3).
- For M in $\{F_7, F_7^*\}$, M is algebraic over F if and only if F has characteristic two (Lindström 1985b,c).

$\boldsymbol{F_7^-}$, $(\boldsymbol{F_7^-})^*$

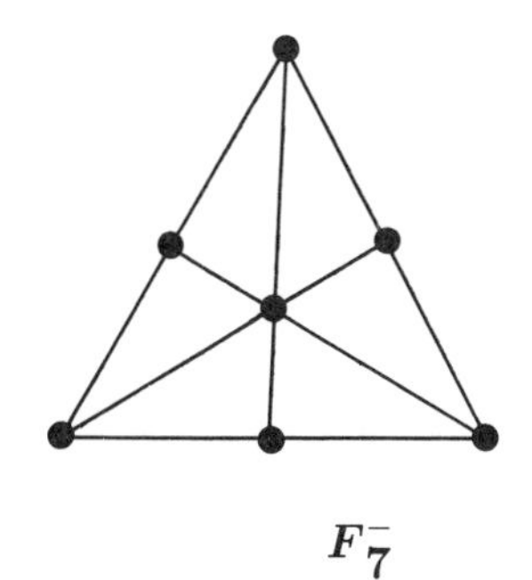

F_7^-

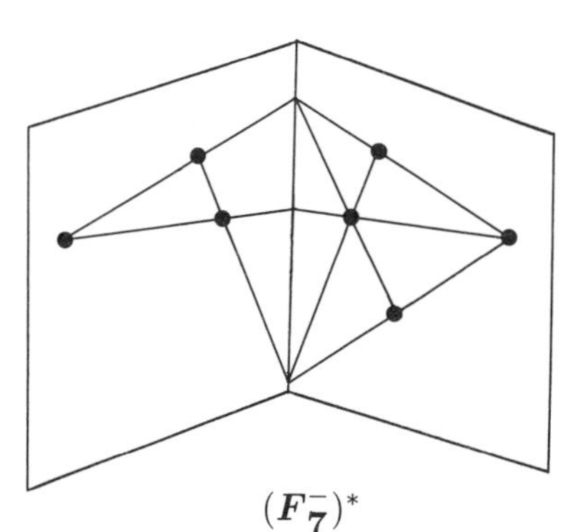

$(F_7^-)^*$

- The non-Fano matroid (1.5.12) and its dual.
- Each is F-representable if and only if F has characteristic other than two (6.4.8).

• Each is an excluded minor for F-representability if and only if F has characteristic two and $F \neq GF(2)$ (6.5.6).
• Each is not graphic, not cographic, and not regular.
• Each is not transversal, not a strict gammoid, not a gammoid.
• Their common automorphism group is non-transitive.
• Every single-element deletion of F_7^- and every single-element contraction of $(F_7^-)^*$ is isomorphic to $M(K_4)$ or $\mathcal{W}^3$.
• F_7^- is the unique relaxation of F_7, and $(F_7^-)^*$ is the unique relaxation of F_7^*.
• Each is algebraic over all fields.

P_7

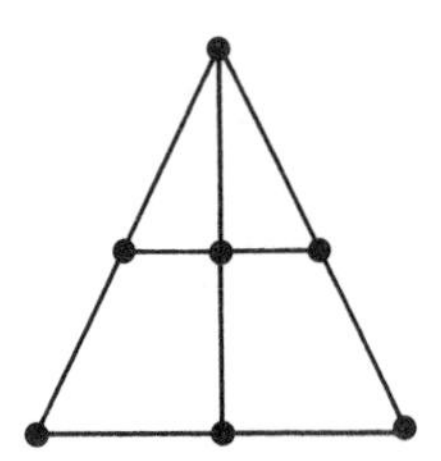

• F-representable if and only if $|F| \geq 3$ (6.4.13).
• Not graphic, not cographic, not regular.
• Not transversal, not a strict gammoid, not a gammoid.
• Has no $M(K_4)$-minor.
• Base-orderable, strongly base-orderable.
• The excluded minors for the class of ternary gammoids are $U_{2,5}$, $U_{3,5}$, $M(K_4)$, P_7, and P_7^* (Oxley 1987c).
• Algebraic over all fields.

$\mathcal{W}_+^3, (\mathcal{W}_+^3)^*$

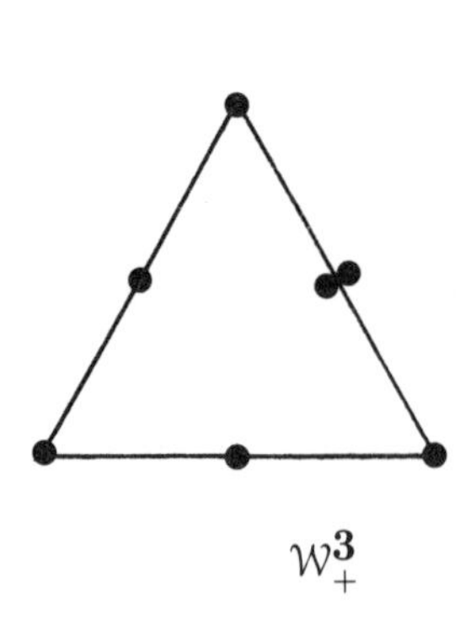

$\mathcal{W}_+^3$

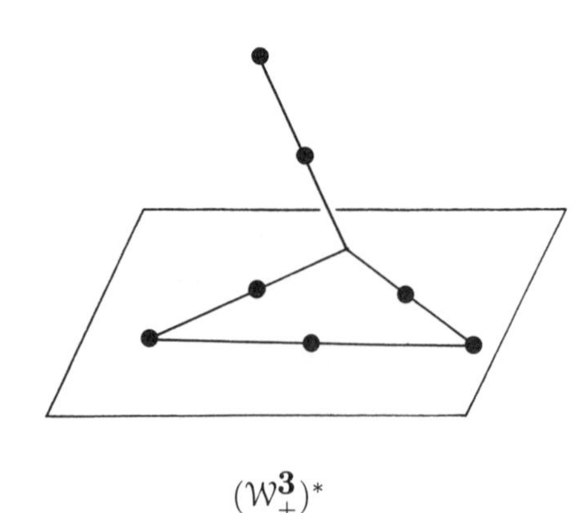

$(\mathcal{W}_+^3)^*$

• Each is connected but not 3-connected.
• Each is F-representable if and only if $|F| \geq 3$.

- Each is not graphic, not cographic, and not regular.
- $\mathcal{W}^3_+$ is non-transversal but all its proper series minors are transversal (Oxley 1986b).
- $\mathcal{W}^3_+$ is a strict gammoid and a gammoid.
- $(\mathcal{W}^3_+)^*$ is not a strict gammoid but all its proper parallel minors are strict gammoids.
- $(\mathcal{W}^3_+)^*$ is transversal and is a gammoid.
- Their common automorphism group is non-transitive.
- Each is algebraic over all fields.

$\boldsymbol{AG(3,2)}$

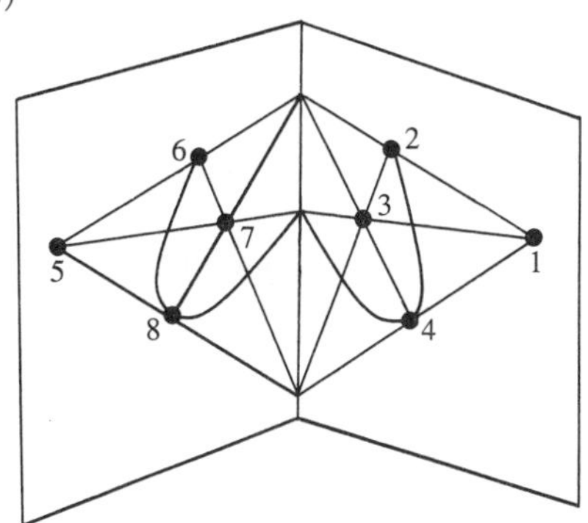

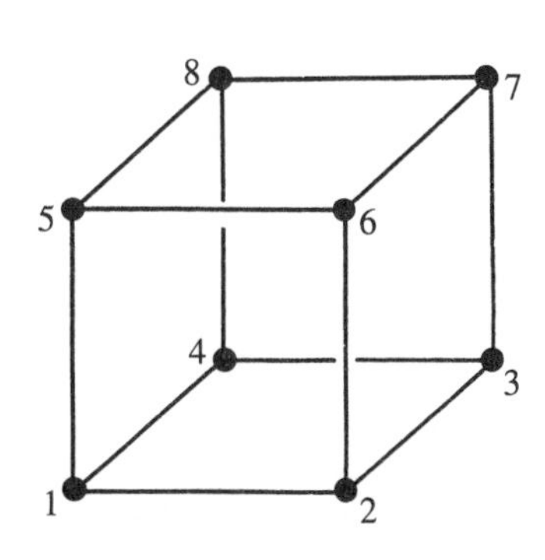

In the figure on the right, the 4-point planes are the six faces of the cube, the six diagonal planes such as $\{1,2,7,8\}$, and two twisted planes: $\{1,8,3,6\}$ and $\{2,7,4,5\}$.

- The binary affine cube; isomorphic to the unique $S(3,4,8)$.
- F-representable if and only if F has characteristic two.
- Not graphic, not cographic, not regular.
- Not transversal, not a strict gammoid, not a gammoid.
- Identically self-dual.
- Automorphism group is triply transitive.
- Every single-element deletion is isomorphic to F_7^* and every single-element contraction is isomorphic to F_7.
- This matroid and S_8 are the only 8-element, 3-connected, binary non-regular matroids (11.2.4).
- The unique splitter for the class of binary paving matroids (Acketa 1988).
- Represented over $GF(2)$ by $[I_4|J_4 - I_4]$.
- The complement of F_7 in $PG(3,2)$.
- Algebraic over F if and only if F has characteristic two.

$AG(3,2)'$

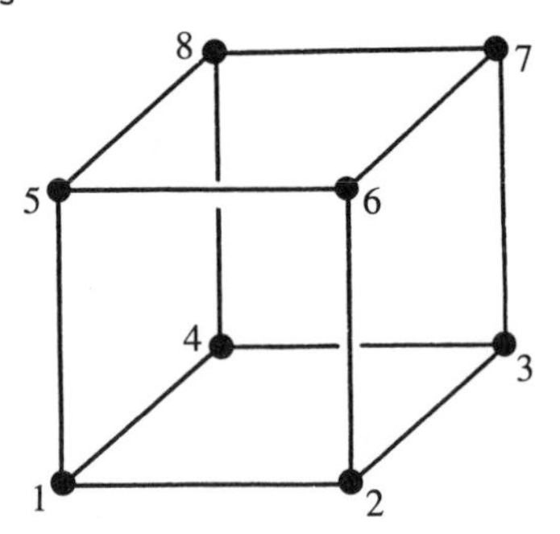

4-point planes are the six faces of the cube, the six diagonal planes such as $\{1,2,7,8\}$, and one twisted plane, $\{1,8,3,6\}$.

- The unique relaxation of $AG(3,2)$.
- A smallest non-representable matroid.
- Not a (minimal) excluded minor for F-representability for any field F.
- Not graphic, not cographic, not regular.
- Not transversal, not a strict gammoid, not a gammoid.
- Self-dual but not identically self-dual.
- Non-transitive automorphism group.
- Every single-element deletion is isomorphic to F_7^* or $(F_7^-)^*$ and every single-element contraction is isomorphic to F_7 or F_7^-.
- Non-algebraic.

R_8

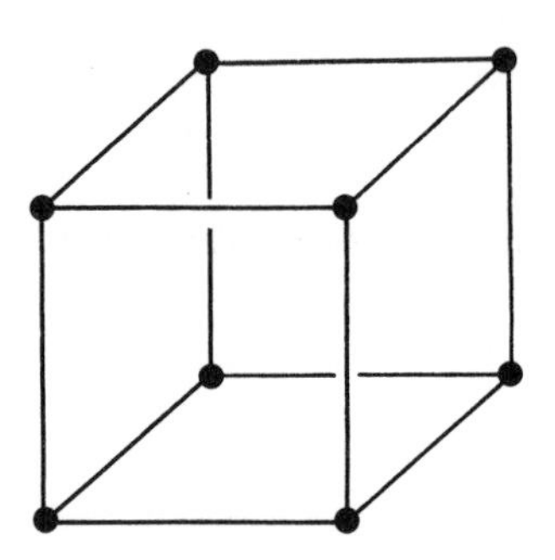

4-point planes are the six faces of the cube and the six diagonal planes.

- The real affine cube (Section 2.2, Exercise 8); it is represented over all fields of characteristic other than two by $[I_4|J_4 - 2I_4]$.
- F-representable if and only if the characteristic of F is not two.
- Not graphic, not cographic, not regular.
- Not transversal, not a strict gammoid, not a gammoid.
- Identically self-dual.
- Transitive automorphism group.
- Every single-element deletion is isomorphic to $(F_7^-)^*$ and every single-element contraction is isomorphic to F_7^-.

• A relaxation of $AG(3,2)'$.
• The unique matroid that can be obtained from $AG(3,2)$ by relaxing two disjoint circuit–hyperplanes.
• Algebraic over all fields.

F_8

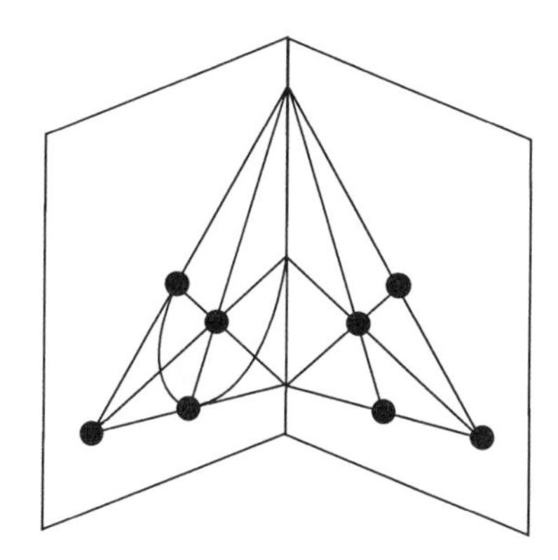

• $P_N(F_7, F_7^-)\backslash E(N)$ where $N \cong U_{2,3}$.
• A smallest non-representable matroid.
• Not a (minimal) excluded minor for F-representability for any field F.
• Not graphic, not cographic, not regular.
• Not transversal, not a strict gammoid, not a gammoid.
• Self-dual but not identically self-dual.
• Non-transitive automorphism group.
• The only two relaxations of $AG(3,2)'$ are F_8 and R_8.
• Has exactly three non-isomorphic single-element deletions including F_7^* and $(F_7^-)^*$, and has exactly three non-isomorphic single-element contractions including F_7 and F_7^-.
• Non-algebraic.

Q_8

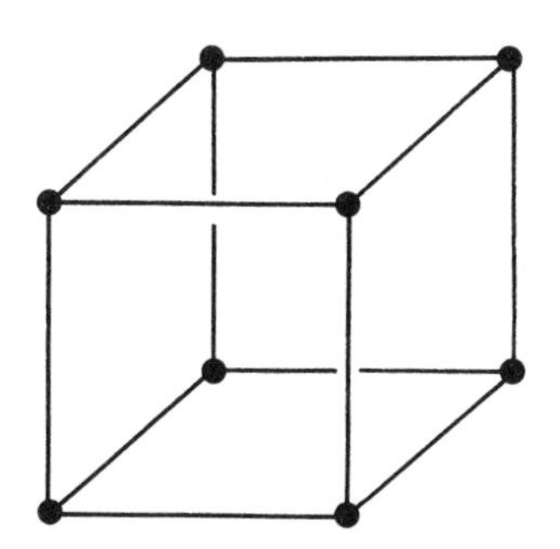

4-point planes are the six faces of the cube and exactly five of the six diagonal planes.

• The unique relaxation of R_8 (6.5.10).
• A smallest non-representable matroid.
• An excluded minor for F-representability if and only if $|F| \geq 5$ and F

does not have characteristic two.

- Not graphic, not cographic, not regular.
- Not transversal, not a strict gammoid, not a gammoid.
- Self-dual but not identically self-dual.
- Non-transitive automorphism group.
- Non-algebraic.

L_8

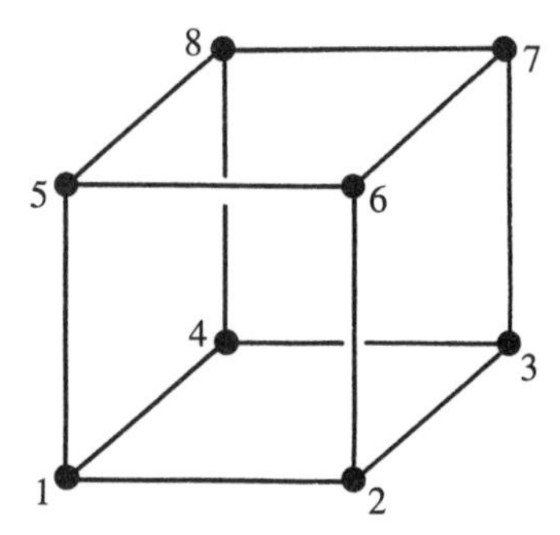

4-point planes are the six faces of the cube plus the two twisted planes, $\{1, 8, 3, 6\}$ and $\{2, 7, 4, 5\}$.

- F-representable if and only if $|F| \geq 5$.
- Not a (minimal) excluded minor for F-representability for any F.
- Not graphic, not cographic, not regular.
- Not transversal, not a strict gammoid, not a gammoid.
- Identically self-dual.
- Transitive automorphism group.
- A splitter for $EX(\mathcal{W}^3, P_6)$ (Oxley 1990a).
- Algebraic over all fields.

S_8

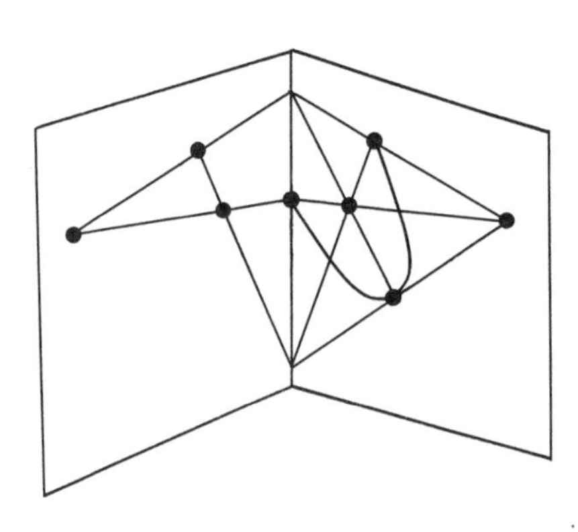

$$\left[\begin{array}{c|cccc} & 0 & 1 & 1 & 1 \\ I_4 & 1 & 0 & 1 & 1 \\ & 1 & 1 & 0 & 1 \\ & 1 & 1 & 1 & 1 \end{array}\right]$$

This matrix represents S_8 over $GF(2)$.

- F-representable if and only if F has characteristic two.
- Not graphic, not cographic, not regular.
- Not transversal, not a strict gammoid, not a gammoid.

- Self-dual but not identically self-dual.
- Non-transitive automorphism group.
- Has a unique element x such that $S_8\backslash x \cong F_7^*$ and a unique element y such that $S_8/y \cong F_7$. Moreover, $y \neq x$.
- S_8 and $AG(3,2)$ are the only 3-connected binary single-element coextensions of F_7.
- $\{U_{2,4}, F_7, F_7^*, S_8\}$ is 2-rounded (Section 11.3).
- The complement of $M(K_4) \oplus U_{1,1}$ in $PG(3,2)$.
- Algebraic over F if and only if F has characteristic two.

V_8

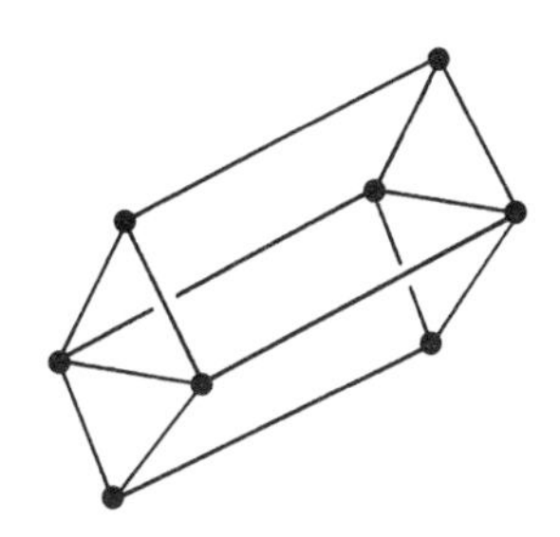

- The Vámos matroid (or Vámos cube) (2.1.22).
- A smallest non-representable matroid (6.1.10, 6.4.10).
- An excluded minor for F-representability if and only if $|F| \geq 5$.
- Not graphic, not cographic, not regular.
- Not transversal, not a strict gammoid, not a gammoid.
- Self-dual but not identically self-dual.
- Non-transitive automorphism group.
- Non-algebraic (Section 6.7).

T_8

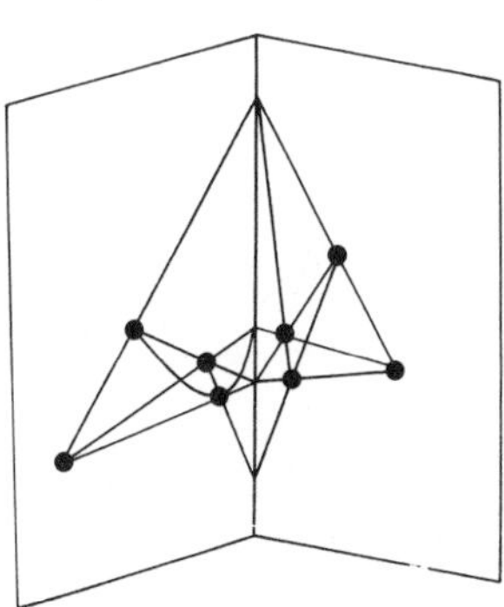

- Represented over $GF(3)$ by $[I_4|J_4 - I_4]$ (Section 2.2, Exercise 8).
- F-representable if and only if F has characteristic three.
- An excluded minor for F-representability for all F whose characteristic is not two or three.
- Not graphic, not cographic, not regular.
- Not transversal, not a strict gammoid, not a gammoid.

- Self-dual but not identically self-dual.
- Non-transitive automorphism group.
- T_8, R_8, $PG(2,3)$, and $S(5,6,12)$ are the only splitters for the class of ternary paving matroids (Oxley 1991).
- Algebraic over F if and only if F has characteristic three (Lindström 1985b,c).

J

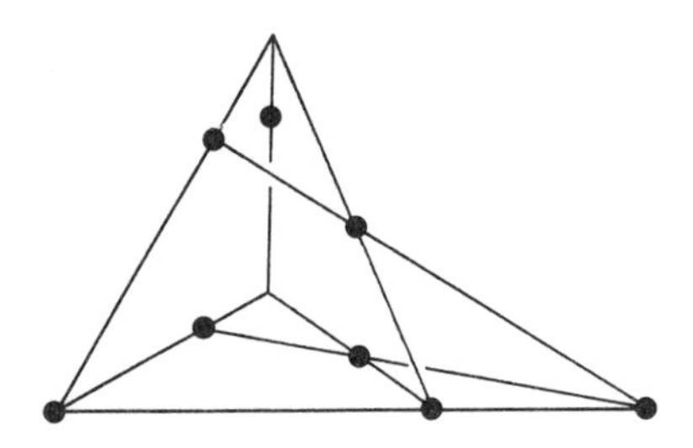

- F-representable if and only if $|F| \geq 3$.
- Not graphic, not cographic, not regular.
- Not transversal, not a strict gammoid, not a gammoid.
- Not self-dual.
- Non-transitive automorphism group.
- J and $S(5,6,12)$ are the only 3-connected splitters for the class of ternary matroids with no $M(K_4)$-minor (11.2.22).
- Not base-orderable or strongly base-orderable.
- The excluded minors for the class of ternary base-orderable matroids are $U_{2,5}$, $U_{3,5}$, $M(K_4)$, and J (Oxley 1987c).
- Algebraic over all fields.

P_8

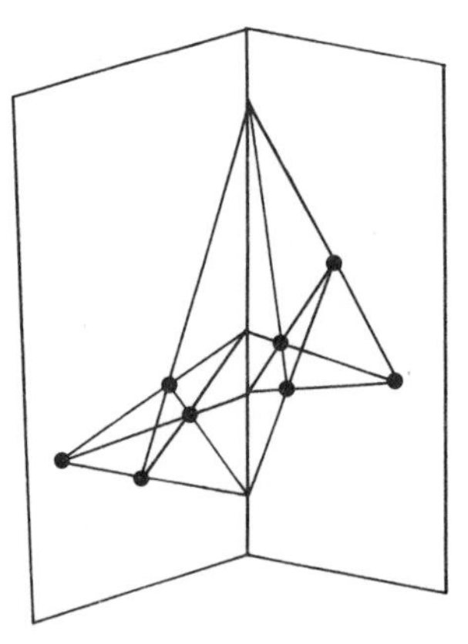

$$\left[\begin{array}{c|cccc} & 0 & 1 & 1 & -1 \\ I_4 & 1 & 0 & 1 & 1 \\ & 1 & 1 & 0 & 1 \\ & -1 & 1 & 1 & 1 \end{array}\right]$$

This matrix represents P_8 over $GF(3)$ (2.2.25).

- F-representable if and only if the characteristic of F is not two (6.4.14).
- An excluded minor for F-representability if and only if F has characteristic two and $F \neq GF(2)$ (6.5.6).

- Not graphic, not cographic, not regular.
- Not transversal, not a strict gammoid, not a gammoid.
- Self-dual but not identically self-dual.
- Automorphism group is transitive on the points but not on the hyperplanes.
- Every single-element contraction is isomorphic to P_7 and every single-element deletion is isomorphic to P_7^*.
- Base-orderable but not strongly base-orderable.
- The excluded minors for the class of ternary strongly base-orderable matroids are $U_{2,5}$, $U_{3,5}$, $M(K_4)$, P_8, and J (Oxley 1987c).
- Algebraic over all fields.

$M(\mathcal{W}_4)$

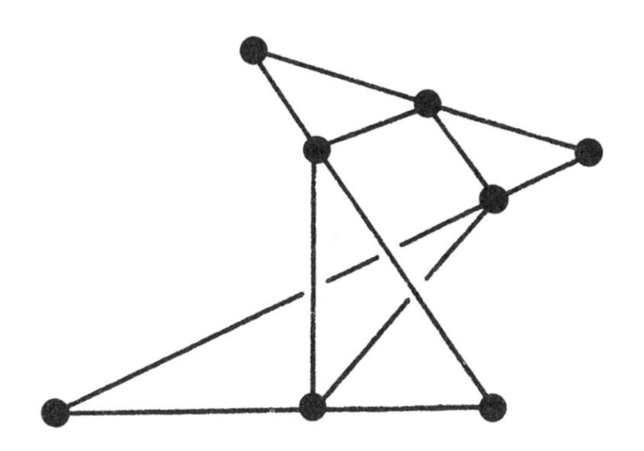

- The rank-4 wheel.
- Regular.
- Graphic, cographic.
- Not transversal, not a strict gammoid, not a gammoid.
- Self-dual but not identically self-dual.
- Non-transitive automorphism group.
- Every 3-connected regular matroid with at least seven elements has an $M(\mathcal{W}_4)$-minor (11.2.21).
- A 3-connected binary matroid M has no $M(\mathcal{W}_4)$-minor if and only if, for some $r \geq 3$, M is isomorphic to a minor of the vector matroid of the matrix $[I_r|J_r - I_r]$ over $GF(2)$ (11.2.21).
- For M binary having at least four elements, $\{U_{2,4}, M\}$ is 2-rounded if and only if $M \cong M(\mathcal{W}_3)$ or $M(\mathcal{W}_4)$ (Oxley and Reid 1990).
- Algebraic over all fields.

$\mathcal{W}^4$

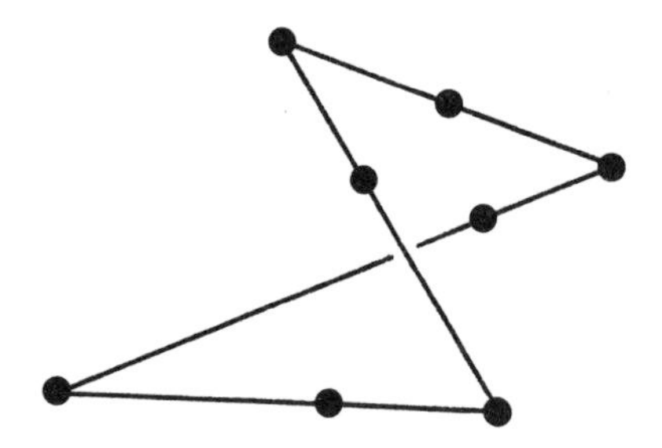

- The rank-4 whirl.
- F-representable if and only if $|F| \geq 3$.
- Not graphic, not cographic, not regular.
- Transversal, a strict gammoid, a gammoid.
- Self-dual but not identically self-dual.
- Non-transitive automorphism group.
- The unique relaxation of $M(\mathcal{W}_4)$.
- Algebraic over all fields.

$M^*(K_{3,3})$, $M(K_{3,3})$

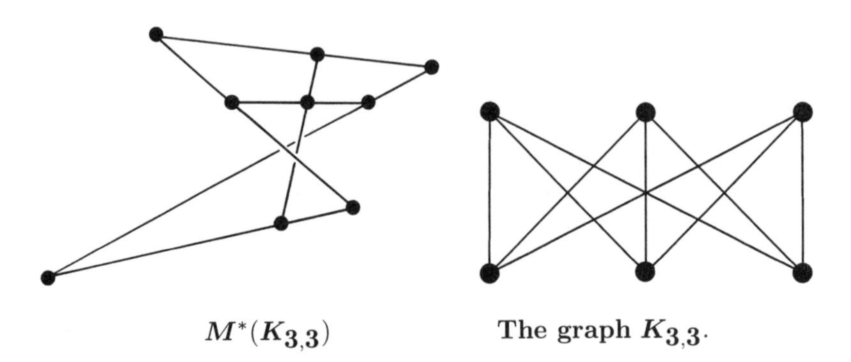

$M^*(K_{3,3})$ **The graph $K_{3,3}$.**

- Both are regular.
- $M^*(K_{3,3})$ is an excluded minor for the class of graphic matroids (6.6.5, 13.1.1) and $M(K_{3,3})$ is an excluded minor for the class of cographic matroids (6.6.5).
- Each is not transversal, not a strict gammoid, and not a gammoid.
- Their common automorphism group is transitive.
- $M^*(K_{3,3})$ has no 3-connected regular single-element extensions.
- A matroid is isomorphic to the cycle matroid of a planar graph if and only if it has no minor isomorphic to $U_{2,4}$, F_7, F_7^*, $M(K_{3,3})$, $M^*(K_{3,3})$, $M(K_5)$, or $M^*(K_5)$ (6.6.6).
- Each is algebraic over all fields.

***AG*(2, 3)**

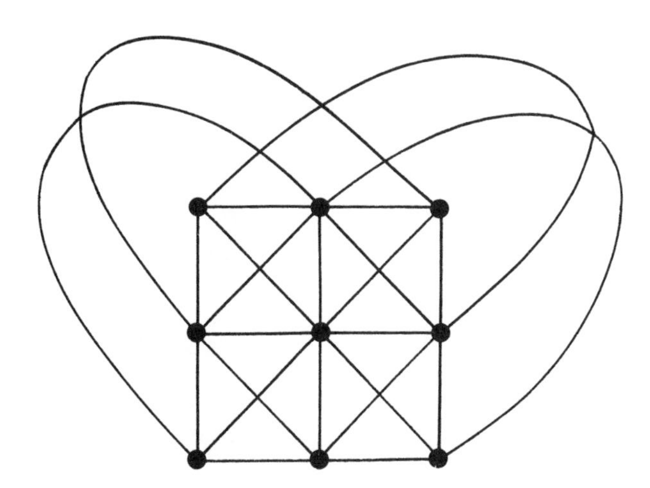

- The ternary affine plane; isomorphic to the unique $S(2, 3, 9)$.
- F-representable if and only if F contains a root of the equation $x^2 - x + 1 = 0$. In particular, $GF(q)$-representable if and only if q is a power of 3, q is a square, or $q \equiv 1 \pmod 3$.
- Not graphic, not cographic, not regular.
- Not transversal, not a strict gammoid, not a gammoid.
- Doubly transitive automorphism group.
- Base-orderable and strongly base-orderable.
- The unique maximal simple 3-connected rank-3 ternary matroid having no $M(K_4)$-minor.
- The complement of $U_{2,4}$ in $PG(2, 3)$.
- Algebraic over all fields (Gordon 1988).

$Q_3(GF(3)^*)$

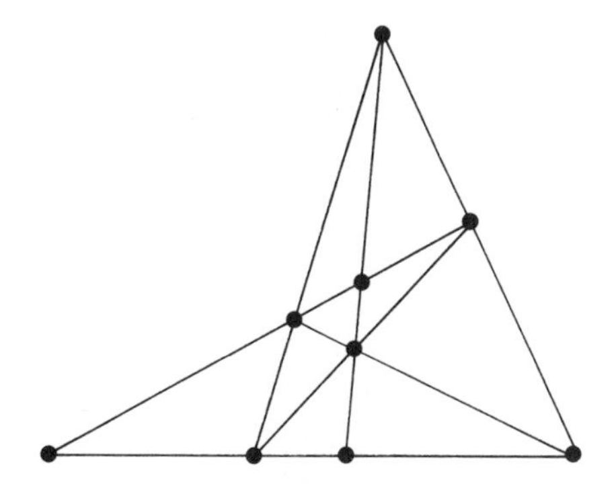

- The rank-3 ternary Dowling geometry.
- F-representable if and only if the characteristic of F is not two.
- Not graphic, not cographic, not regular.

- Not transversal, not a strict gammoid, not a gammoid.
- Non-transitive automorphism group.
- The complement of $U_{3,4}$ in $PG(2,3)$.
- This matroid and $AG(2,3)$ are the only ternary rank-3 simple matroids with more than eight elements that are representable over some field of characteristic other than three (Kung and Oxley 1988).
- Algebraic over all fields.

R_9

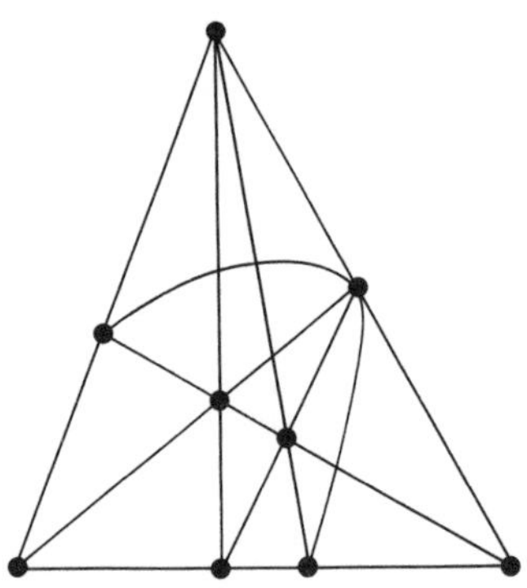

- The ternary Reid geometry.
- F-representable if and only if F has characteristic three.
- Not graphic, not cographic, not regular.
- Not transversal, not a strict gammoid, not a gammoid.
- Non-transitive automorphism group.
- The complement of $U_{2,3} \oplus U_{1,1}$ in $PG(2,3)$.
- Algebraic over F if and only if F has characteristic three (Gordon 1988).

Pappus

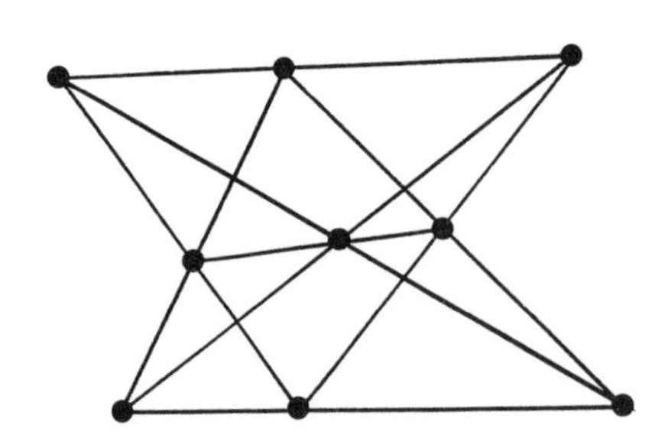

- The Pappus matroid (1.5.14).
- F-representable if and only if $|F| \geq 4$.
- Not graphic, not cographic, not regular.
- Not transversal, not a strict gammoid, not a gammoid.
- Transitive automorphism group.
- Algebraic over all fields.

non-Pappus

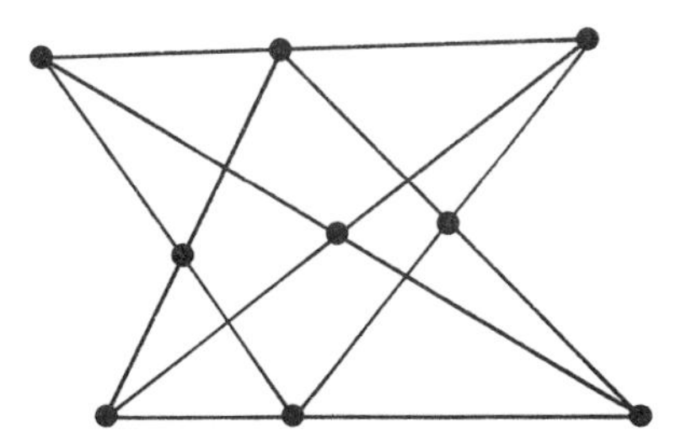

- The non-Pappus matroid (1.5.14).
- Non-representable (6.1.11).
- An excluded minor for F-representability if and only if $|F| \geq 5$.
- Not graphic, not cographic, not regular.
- Not transversal, not a strict gammoid, not a gammoid.
- Non-transitive automorphism group.
- The unique relaxation of the Pappus matroid.
- Algebraic over a field F if and only if F has non-zero characteristic (Ingleton 1971a, Lindström 1986a).

$\boldsymbol{M(K_5)}$, $\boldsymbol{M^*(K_5)}$

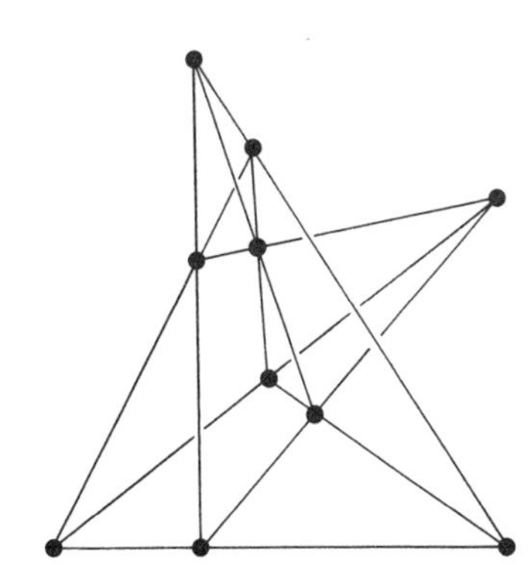

- $M(K_5)$ is the 3-dimensional Desargues configuration shown above.
- Both are regular.
- $M(K_5)$ is an excluded minor for the class of graphic matroids (6.6.5, 13.1.1), and $M^*(K_5)$ is an excluded minor for the class of cographic matroids (6.6.5).
- Each is not transversal, not a strict gammoid, and not a gammoid.
- Their common automorphism group is transitive.
- $M(K_5)$ is the unique rank-4 simple regular matroid with the largest number of elements among such matroids (Heller 1957).
- $M(K_5)$ is a tangential 2-block over $GF(2)$ (Section 14.3).
- $M(K_5)$ is a splitter for the class of regular matroids with no minor isomorphic to $M(K_{3,3})$ (Seymour 1980b).
- Each is algebraic over all fields.

R_{10}

$$\left[\begin{array}{c|ccccc} & -1 & 1 & 0 & 0 & 1 \\ & 1 & -1 & 1 & 0 & 0 \\ I_5 & 0 & 1 & -1 & 1 & 0 \\ & 0 & 0 & 1 & -1 & 1 \\ & 1 & 0 & 0 & 1 & -1 \end{array}\right]$$

This matrix represents R_{10} over all fields.
(In characteristic two, $-1 = 1$.)

- The unique 10-element regular matroid that is neither graphic nor cographic (13.2.5).
- The unique splitter for the class of regular matroids (Section 13.2).
- The unique simple regular matroid that is neither graphic nor cographic and has rank at most five (13.2.5).
- Not transversal, not a strict gammoid, not a gammoid.
- Self-dual but not identically self-dual.
- Doubly transitive automorphism group (Seymour 1980b).
- Every single-element deletion is isomorphic to $M(K_{3,3})$, and every single-element contraction is isomorphic to $M^*(K_{3,3})$ (11.2.8).
- Represented over $GF(2)$ by the ten 5-tuples that have exactly three ones each.
- Algebraic over all fields.

non-Desargues

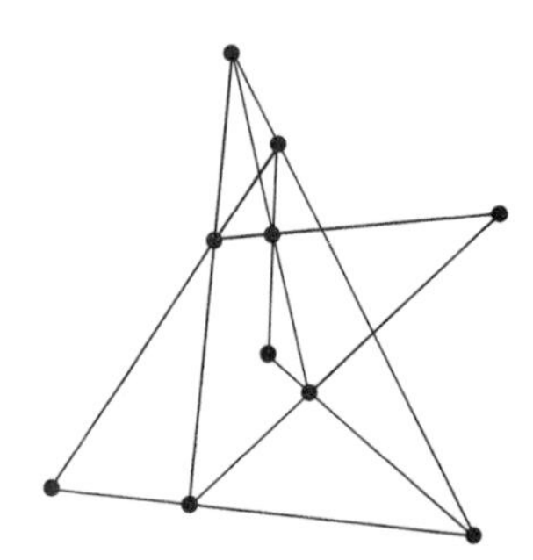

- The non-Desargues matroid; it has rank three.
- Not representable over any division ring (Section 6.1).
- An excluded minor for F-representability if and only if $|F| \geq 7$.
- Not graphic, not cographic, not regular.
- Not transversal, not a strict gammoid, not a gammoid.
- Non-transitive automorphism group.
- Non-algebraic (Lindström 1985a).

R_{12}

$$\left[\begin{array}{c|cccccc} & 1 & 1 & 1 & 0 & 0 & 0 \\ & 1 & 1 & 0 & 1 & 0 & 0 \\ & 1 & 0 & 0 & 0 & 1 & 0 \\ I_6 & 0 & 1 & 0 & 0 & 0 & 1 \\ & 0 & 0 & 1 & 0 & 1 & 1 \\ & 0 & 0 & 0 & 1 & 1 & 1 \end{array}\right]$$

This matrix represents R_{12} over $GF(2)$.

- Regular.
- Not graphic, not cographic.
- Not transversal, not a strict gammoid, not a gammoid.
- Self-dual but not identically self-dual.
- Non-transitive automorphism group.
- A 3-connected regular matroid is either graphic or cographic, or has a minor isomorphic to R_{10} or R_{12} (13.2.2).
- Algebraic over all fields.

$S(5, 6, 12)$

$$\left[\begin{array}{c|cccccc} & 0 & 1 & 1 & 1 & 1 & 1 \\ & 1 & 0 & 1 & -1 & -1 & 1 \\ & 1 & 1 & 0 & 1 & -1 & -1 \\ I_6 & 1 & -1 & 1 & 0 & 1 & -1 \\ & 1 & -1 & -1 & 1 & 0 & 1 \\ & 1 & 1 & -1 & -1 & 1 & 0 \end{array}\right]$$

This matrix represents $S(5, 6, 12)$ over $GF(3)$.

- The unique Steiner system with these parameters (Section 11.2).
- F-representable if and only if F has characteristic three.
- Not graphic, not cographic, not regular.
- Not transversal, not a strict gammoid, not a gammoid.
- Identically self-dual.
- Automorphism group is the 5-transitive Mathieu group M_{12}.
- A splitter for the classes $EX(U_{2,5}, U_{3,5}, M(K_4))$ and $EX(M(K_4), P_6, Q_6)$ (Section 11.2).
- Every contraction of three elements is isomorphic to $AG(2, 3)$.
- Base-orderable but not strongly base-orderable.

$PG(2,3)$

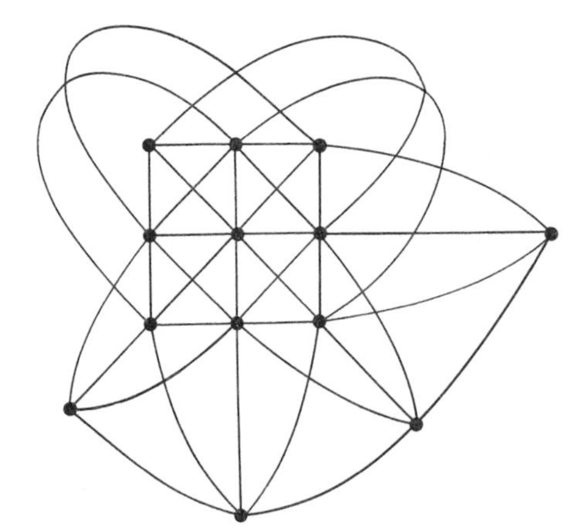

• The second smallest projective plane; this matroid is isomorphic to the unique $S(2,3,13)$.
• F-representable if and only if F has characteristic three.
• Not graphic, not cographic, not regular.
• Not transversal, not a strict gammoid, not a gammoid.
• Doubly transitive automorphism group.
• Every simple ternary matroid of rank at most three is a restriction of this matroid.
• Algebraic over F if and only if F has characteristic three (Gordon 1988).

$M(\mathcal{W}_r), \mathcal{W}^r$

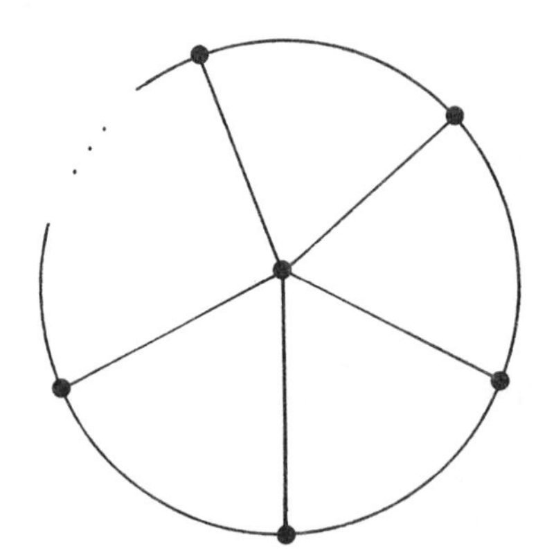

The graph $\mathcal{W}_r$, the r-spoked wheel.

• The rank-r wheel and rank-r whirl ($r \geq 2$).
• The whirl is the unique relaxation of the wheel.
• $M(\mathcal{W}_r)$ is graphic, cographic, and regular.
• $\mathcal{W}^r$ is not graphic, not cographic, and not regular; it is F-representable if and only if $|F| \geq 3$.
• For $r \geq 3$, $M(\mathcal{W}_r)$ is not transversal, not a strict gammoid, and not a gammoid.

• $\mathcal{W}^r$ is transversal, a strict gammoid, and a gammoid.
• Both are self-dual but not identically self-dual.
• Except for $\mathcal{W}^2$, which is isomorphic to $U_{2,4}$, these matroids have non-transitive automorphism groups.
• The only 3-connected matroids for which no single-element deletion and no single-element contraction is 3-connected are the matroids $M(\mathcal{W}_r)$ and $\mathcal{W}^r$ for $r \geq 3$. $\mathcal{W}^2$ is 3-connected, but $M(\mathcal{W}_2)$ is not.
• Algebraic over all fields.

$U_{r,n}$

$$\boldsymbol{E} = \{\mathbf{1}, \mathbf{2}, \ldots, \boldsymbol{n}\}$$
$$\mathcal{B}\,(\boldsymbol{U_{r,n}}) = \{\boldsymbol{X} \subseteq \boldsymbol{E} : |\boldsymbol{X}| = \boldsymbol{r}\}$$
$$\mathcal{C}\,(\boldsymbol{U_{r,n}}) = \{\boldsymbol{X} \subseteq \boldsymbol{E} : |\boldsymbol{X}| = \boldsymbol{r}{+}\mathbf{1}\}$$

• The rank-r uniform matroid on an n-element set.
• $U_{2,q+2}$ and $U_{q,q+2}$ are excluded minors for $GF(q)$-representability. For general r and n, F-representability has not been determined (Section 6.5).
• Not graphic, cographic, or regular unless r or $n-r$ is 0 or 1, in which case $U_{r,n}$ is graphic, cographic, and regular.
• Transversal, a strict gammoid, a gammoid.
• Identically self-dual if and only if $n = 2r$; otherwise not self-dual, the dual being $U_{n-r,n}$.
• Automorphism group is the symmetric group.
• Every minor is also uniform.
• 3-connected unless $n > 0$ and r or $n-r$ is 0, or $n > 3$ and r or $n-r$ is 1.
• Algebraic over all fields.

$PG(r-1, q)$

• The rank-r projective geometry over $GF(q)$ (Section 6.1).
• The simple matroid associated with $V(r,q)$, the r-dimensional vector space over $GF(q)$.
• For $r = 2$, isomorphic to $U_{2,q+1}$.
• For $r \geq 3$, F-representable if and only if F has $GF(q)$ as a subfield.
• For $r \geq 3$, not graphic, not cographic, not regular.
• For $r \geq 3$, not transversal, not a strict gammoid, not a gammoid.
• For $r \geq 3$, never self-dual.
• For $r \leq 2$, automorphism group is the symmetric group. For $r \geq 3$, automorphism group is doubly transitive.
• Every simple $GF(q)$-representable matroid of rank at most r is a restriction of $PG(r-1,q)$.
• For $r \geq 3$, algebraic over a field F if and only if F has the same characteristic as $GF(q)$.

$AG(r-1,q)$

- The rank-r affine geometry over $GF(q)$ (Section 6.2).
- Obtained by deleting a hyperplane from $PG(r-1,q)$.
- For $r=2$, isomorphic to $U_{2,q}$.
- $AG(2,2) \cong U_{3,4}$; for $AG(2,3)$ see above. For $r=3$ and $q \geq 4$, and for $r \geq 4$, $AG(r-1,q)$ is F-representable if and only if F has $GF(q)$ as a subfield.
- For $r,q \geq 3$, not graphic, not cographic, not regular.
- For $r,q \geq 3$, not transversal, not a gammoid, not a strict gammoid.
- For $r \geq 3$, identically self-dual if and only if $(r-1,q)=(3,2)$; otherwise never self-dual.
- For $r \leq 2$ or $(r-1,q)=(2,2)$, automorphism group is the symmetric group. In all other cases, automorphism group is doubly transitive.
- The simple matroid associated with every proper contraction is isomorphic to $PG(k-1,q)$ for some $k<r$.
- 3-connected unless $(r-1,q)$ is $(1,2)$ or $(2,2)$.
- For $r=3$ and $q \geq 5$, and for $r \geq 4$, $AG(r-1,q)$ is algebraic over F if and only if F has the same characteristic as $GF(q)$.

Notation

Operations

Index